Modern Developments *in* Animal Breeding

If then Socrates we find ourselves . . . unable to make our discourse in every way wholly consistent and exact you must not be surprised. Nay, we must be well content if we can provide an account not less likely than another; we must remember that I who speak and you who are my audience are but men and should be satisfied to ask for no more than the likely story.

Modern Developments *in* Animal Breeding

By

I. MICHAEL LERNER and H. P. DONALD

Department of Genetics
University of California
Berkeley, California, U.S.A.

Agricultural Research Council
Edinburgh, Scotland

ACADEMIC PRESS
London and New York · 1966

ACADEMIC PRESS INC. (LONDON) LTD
BERKELEY SQUARE HOUSE
BERKELEY SQUARE
LONDON, W.1.

U.S. Edition published by
ACADEMIC PRESS INC.
111 FIFTH AVENUE
NEW YORK, NEW YORK 10003

Library of Congress Catalog Card Number: 66–16699

PRINTED IN GREAT BRITAIN AT
THE UNIVERSITY PRESS
ABERDEEN

FOREWORD

The scientific revolution, viewed by Bernal some twenty-five years ago as a twentieth century parallel to the industrial revolution, has since the war assumed proportions unprecedented in human history. The rate of change and magnitude of its effects are altering the daily life and attitudes of most of mankind. In turn, the social changes brought about, and other effects such as population pressure, influence the workings of science. For one thing, they affect scientific goals and for another, they modify the ways in which scientific information is spread and the political institutions through which it is exploited. As far as biology, and more specifically genetics, is concerned, there has been within the last few decades such progress, such growing power over nature, as to warrant a determined effort by all who have an interest in the future pattern of society to consider where this technical progress is taking us. The startling achievements of the physical and chemical sciences should not be allowed to obscure the fact that biology also brings its gifts which if properly used can improve the lot of man. Given the prospect of directed control by man over the evolution of his own species and of his agricultural crops and plants, it is surely wise to consider who is exercising this power.

Agriculture, in common with other industries, is experiencing an intensification which shows itself in the rising output of a worker and the falling number of individual enterprises. It is also to be seen in the trend towards integration. Livestock production is not immune to these changes and its positive response to the economic advantages of large businesses has created a need for a new and vigorous technology. While the structure of animal production is being modified, its genetic efficiency is a matter of deep concern and, at least in some circumstances, a factor of sufficient importance to alter the outcome of the process. It is our hope that this book will help to ensure that the urgent need for more animal protein will be communicated to those who are influential to any degree in increasing the speed at which new and better livestock can be developed. To this end a survey of the issues faced in the organisation and carrying out of animal improvement by state as well as by private and quasi-private agencies is opportune, for surprisingly enough this task has not hitherto been undertaken. Despite the abundance of publications on animal production, there is a marked lack of discussion about the interplay between the technology and the economics of breeding. Although the repercussions of such developments as artificial insemination and performance testing on the structure of agricultural industries have been common knowledge for many years, recent textbooks on animal production and breeding have little or no mention of them. Students, to the extent that they have to depend on literature, cannot readily consider what may be learned from the poultry industry that is valid for larger animals, nor, indeed, are they given a hint that the logic of performance testing, to say nothing of the pressure on food supplies, may require modifications in the current systems of breeding improved livestock.

There are several restraints circumscribing both the intent and the execution of this volume which are due to the limitations of our training and experience. Economics, political theory, sociology, and psychology are all woven into the fabric of the issues to be examined. Since we cannot claim even amateur status in any of these disciplines, there have to be gaps in our treatment of the subject. Hence the work must be viewed as a contribution by geneticists to a topic of broader significance than genetics or animal breeding can alone encompass. It is presented in the belief that similar approaches by social scientists will contribute to a fuller analysis and so to better prediction of the consequences of different policies and perhaps to identifying the best courses to follow, should "best" prove to be capable of an acceptable definition. Just as the advances in human biology create social and moral problems exceeding the competence of specialists in biology, so do the advances in the production of livestock pose problems in politics, economics and sociology that cannot be solved by research workers acting independently.

It is, therefore, open to doubt whether population genetics as currently taught is as effective for guiding the practice of animal breeding as it could be. Among modern economists there is a similar dissatisfaction with classical economics which is revealing itself as an attempt to develop a science of political economics. Classical genetics and classical economics both have a model form that is logical and beautiful but often takes no account of the realities of organised capitalism or socialism. Both are concerned with making predictions yet make them badly because of inadequate theory and information. Starting from the desired ends, political economics considers what policies and what instruments of policy would secure them. It is the science of a participant rather than that of an observer. In genetics a corresponding development would have the purpose and ambition to apply itself to the attainment of the goals of animal breeding. Like any other discipline, it would seek to establish a code of principles from which would flow a logically unified body of knowledge: by critical and relevant research, it would provide the theory and stimulate the practice of co-ordinating the necessary political and administrative actions with the biological facts.

In surveying the opportunities and problems of livestock genetics, our aim has been to integrate, not to present new data. Controversial topics have not been avoided but treated openly so that the discussions may serve as starting points for constructive thinking. Since we have no revelations to offer, and only limited truths to go by, a trial and error process of examining issues is the only way to approach a fuller understanding. Although our familiarity with them is largely based on experience in the United Kingdom and the United States, we have often supplemented it with reports from other countries.

Only a most inadequate acknowledgment of the multitude of research workers on whom we have relied can be made. With some reluctance, a policy of referring mainly to recently published papers has been followed in order to direct readers to fuller sources of information should they want them. We would have enjoyed developing each topic as it came, savouring each report and watching the unfolding of man-controlled evolution, but it has not been possible to turn over half a library in the making of this book. Like many others before us, we have not found a concise way of communicating the necessary technicalities of genetics and animal breeding to readers unfamiliar with these fields. If, however, they are prepared to take some parts on trust, they will find that the rest will offer no difficulty. In a treatise that covers the interactions of the science and practice of animal production, there are bound to be errors of fact and of judgment, but we hope they will not be so serious as to vitiate our purpose. In brief, it is to examine some of the problems raised by the recent developments in science in the light of our belief that animal breeding has a significant role to play in the welfare of mankind. The primary question to be explored is how this role could be most adequately fulfilled in the best interests of society as a whole. Animal breeding does not exist for the sake of supporting breeders, geneticist breed societies, recording or supervising organisations, but all of them serve society through animal breeding.

We wish to express our appreciation to the Agricultural Research Council for travel grants and contributions towards subsistence, and to the staff of the Natural History Museum, London, for their hospitality during preliminary work on the manuscript.

January 1966

I. M. Lerner
H. P. Donald

CONTENTS

Part I: The Objectives of Animal Breeding

CHAPTER 1

ANIMAL BREEDING IN ITS SOCIAL CONTEXT

Forecasts of the rate of expansion of the human population of the world are generally familiar. Stimulating and thorough discussions of population pressure and food supply may be found in many sources. In particular, the report of the Stanford Research Institute (1959), the symposium of the Royal Statistical Society (1963) and the volume edited by Mudd (1964) on which much of the material in this chapter is based, may be recommended.

TABLE 1.1

Past and future estimates of world population (from Stanford Research Institute, 1959, and Mudd, 1964)

Year	Population in millions
1000	275
1100	306
1200	348
1300	384
1400	373
1500	446
1600	486
1650	545
1750	694
1800	906
1850	1171
1900	1550
1925	1907
1950	2497
1975	3828
2000	6267

Table 1.1 provides dramatic evidence of the acceleration of population growth rate in this century. Even if drastic measures to control the pressure of the expected numbers were soon adopted, it is too late to achieve much reduction in the projected figure for the year 2000. Urban alienation of land now used for food growing makes matters worse, and it is difficult to escape the conclusion of Slater (1963), that by the end of our century "the farmers of the world will be faced with producing on three-quarters of an acre of cultivated land and on one acre of permanent grassland what they now produce on an acre of cultivated land and two acres of permanent grassland." No one can foresee now whether an increase of one-third of the current production on cultivated acreage and a doubling of that on grassland will be attained in thirty-five years, or whether other developments, such as the production of acceptable synthetic foods, will reduce the menace of starvation. Meantime, finding ways and means of increasing the world's food supply, and even more important, the efficiency of food production, presents one of the most challenging problems mankind has yet faced. It is, perhaps, unseemly for those who are well-fed to object that appetites grow by what they feed on, and that there is, therefore, no hope of catching up on the shortage of food. Although there may be room for argument concerning the definition of a shortage, there will surely be a greater shortage than there is now unless food production is raised to match the increasing number of mouths. Some authorities fear a losing battle against want but dissenting voices should also be noted. Thus, Clark (in Wolstenholme, 1963) estimates that 45,000 million people could be readily supported on earth, and, if survival only at the minimum caloric level is desired, ten times as many. Similarly, Mayer (1964) thinks that arable farming could be so extended and fertilisers so much more liberally applied that the problem might turn into the disposal of surpluses. It is assumed that technological advances such as chemical synthesis of food, taking salt out of sea water, and heating lakes by atomic energy to encourage cloud formation will ensure food for everyone. Furthermore, since space research must include the nutrition of astronauts there is the possibility of a scientific fall-out to benefit man on earth. But even if these minority views are accepted, the problem of significantly and speedily augmenting current food supplies remains. The times demand a strategy for the war against want on the agricultural front.

I. World Food Needs

It has been estimated that about two-thirds of the present population in the world suffer from malnutrition. To correct this deficiency and to provide adequate food for the expected increase in numbers of people calls for a steady annual increase of 2·25% in the total food production in the world. In recent years the average increase has been estimated at not more than 1%. Although some areas have exceeded this rate (North America produces about 120% of its caloric requirement), others, such as parts of Asia (with the Far East producing less than 90% of its needs) actually show a decrease in food production per head when 1957–58 is compared with 1937–38.

TABLE 1.2

World protein consumption

Region	Total protein consumption g/caput per day	% of protein of animal origin	% of total diet of animal origin (milk, meat, eggs, fish)
Far East	56	14	5
Near East	76	18	9
Africa	61	18	11
Latin America	67	37	17
Europe	88	41	21
North America	93	71	40
Oceania	94	68	?
World	68	29	?

The problem is acute in respect of protein, and particularly animal protein. As Table 1.2 shows, regions of the world vary considerably in the proportion of protein in the total diet and of protein supplied by animals. North America enjoys a proportion of animal protein in the average diet eight times higher than that of a Far Eastern diet, while Europeans exist on half the North American amounts.

If the quantities of animal products in food are too small, protein malnutrition results. It has been shown that the mortality rates of children tend to rise when their protein consumption, as estimated in the last column of Table 1.2 goes down. There are, of course, many reasons other than protein deficiency for high infant mortality. Deaths from this cause for children between 1 and 4 years of age, occur more frequently than do earlier deaths partly because of the transition from breast feeding to adult diets. It is, therefore, perturbing to find that the ratio of death rates in South India to those of England and Wales is six for the age group of 0–1 years, but rises to over 25 for the age group of 1–4 years.

That the problem of food shortage is not only one of distribution but also of production is clearly demonstrated by Table 1.3. The so-called food surpluses in Europe, North America and Australasia shrink into insignificance when viewed in the context of the world deficits of food.

TABLE 1.3

Estimated world food deficit (from Aylward in Ovington, 1963)

	Deficit	
Type of food	in millions of metric tons	as % of U.S. agricultural production
Animal protein as non-fat milk solids	1·8	35
Pulse protein as dry beans and peas	0·4	40
Other protein as wheat	35·6	120 (Other protein and remaining calorie deficit combined)
Remaining calorie deficit as wheat	8·6	

II. Shortening the Food Chain

There are many means, some obvious and some obscure, by which the amount and the efficiency of food production could be increased. More than a quarter of a century ago, Kunkel (1938), among others, pointed out the extreme inefficiency of producing food through a lengthy chain, that is, the series of steps in the transformation of materials by living organisms which are necessary to convert energy and nutrients into edible forms. If the 1 million calories per year that a man requires could be obtained directly from mineral and synthetic sources, there would be no food chain. Should humans need in addition a source of protein which they could get directly from plants, there would be one or more links to the chain. Animals interpose links which considerably lengthen the food chain in many instances. Kunkel, who conservatively assumes that efficiency of conversion at each link is about 20% (and 10% may be a closer figure for animals), calculates that an Eskimo must consume 5 lb of seal to gain a pound in weight, each pound of seal being derived from 5 lb of fish, each pound of fish from 5 lb of shrimp or other invertebrates, each of which in turn is produced by 5 lb of algae. In short, it takes 625 lb of algae to make one pound of Eskimo, with a loss of at least 99·84% of the originally available energy.

It seems that one of the readiest ways of increasing efficiency of food production is to shorten the food chain by placing greater reliance on plant products supplemented by synthetic additives to correct for deficiencies of amino acids and other nutritional essentials.

Indeed, the prospects of increasing food supplies from plants are in many ways brighter than those from domesticated animals. For instance, the efficiency in terms of man-hours required to produce a given amount of beef has increased by 13% between 1948 and 1960 (Byerly, 1962). Milk has done better with an increase of 42% and poultry better still, with 117%. Both, however, are inferior to feed grains which have shown an increase of 168% attained through improvements in varieties, crop husbandry and agricultural machinery.

Plants yield more calories per acre than do animals. Figure 1.1 from Mangelsdorf (1961) shows this in a striking fashion. Plant products are more easily transported, require less labour and are consequently cheaper. Against these formidable advantages, however, can be placed a tendency to be bulky, less digestible, and deficient in certain amino

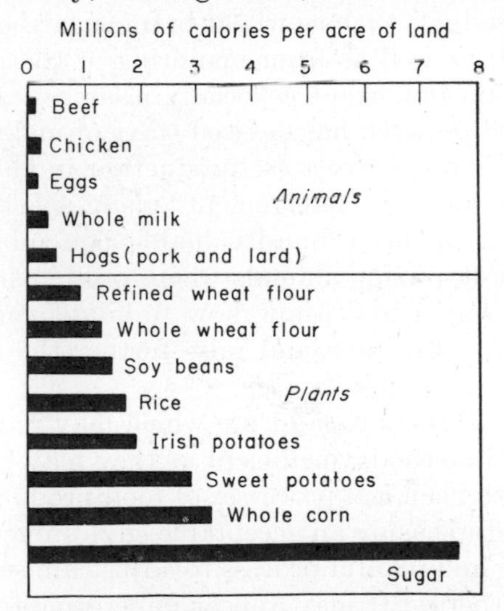

Fig. 1.1. Caloric yield of plant and animal products per acre (Reproduced with permission from Mangelsdorf, 1961).

acids, although recently a mutation remedying this fault in part has been found in maize (Mertz *et al.*, 1965). There is nothing insuperable about these objections, but so far no way has been found of evading the fact that people like meat, milk, fish and eggs.

According to some writers, the increasing world demand for food will lead to such an expansion and perfecting of industrial methods of protein production, that they will quantitatively overshadow traditional farming. Whilst it might be rash to deny this possibility, there seems to be no reason for expecting any sudden responsiveness to existing needs that would make animal production superfluous. As yet margarine has not displaced butter, nor synthetic fibres, wool. Such has been the recent increase in production on each acre of land in Europe and North America that the power of livestock to compete with chemical industry may in fact have the effect of postponing the development of substitute methods. This would be regrettable if it meant less progress towards enough protein for all.

While, therefore, on a pessimistic view, the future may lie with the vegetarians, there is good reason now for the omnivorous to fight a strenuous rear-guard action by improving the production of animal protein from fish and livestock. Furthermore, the maximum output of protein and calories may be achieved by mixing animal and crop husbandry. To some extent this follows from the advantages of using grass in rotations, and to some extent from the use by animals of surpluses or of by-products such as wheat bran, sugar-beet tops, fish meal, and poultry offal, as well as from manuring of the ground. Livestock are likely to remain important especially in areas unsuitable for cropping, at least until research has devised ways of making them suitable. Meantime research helps livestock production in these areas through discoveries about grazing management, plant species, and nutrition. Urea, for instance, has been found valuable as a substitute for part of the protein needs of grazing animals where protein supplements would be expensive. Research may show how to adapt rumen flora to cope with larger quantities of urea and raise further the nutritive value of rough pastures.

There are social aspects to land use which may require the maintenance of traditional methods, inefficient as they may eventually become in comparison with chemical processes of food production. Soil fertility must be conserved to ensure an acceptable environment for man and he may find it a long and painful process to adjust himself to the vicarious new world of synthetics. It may appear imperative to the next generation, if not to this one, to protect the countryside from pollution and the by-products of industry, from depopulation, from erosion and misuse (Udall, 1963). Ecological stability in some areas may be best achieved by farming methods involving grazing animals, and this stability may have to be maintained by a tax on urban activities. Involved in this is the widespread social problem posed by the declining status and income of the small farmer unable to raise the output and efficiency of his farming (Hoogschagen, 1958; Organisation for Economic Cooperation and Development, 1964).

The post-war expansion in science and technology is raising the general standard of living in which those who reside in the country wish to share. Rural industry, like urban industry, is moving towards the use of more capital, less labour, and higher outputs. Within the context of small family farms, these changes are difficult to accomplish but there are a number of adjustments facilitating them (Ahlgren, 1962). Farmers may increase the size of their farms or, alternatively, seek part-time work elsewhere. They may embark on contract farming so that risk capital can be provided by larger enterprises. They may place increased reliance on purchasing and marketing co-operatives. By some means augmented productivity and higher real income must be achieved for the diminishing country populations. This is not primarily a genetic problem but there is a genetic element in it. Even if the bulk of research effort is devoted to the development of industrial food production or of stock especially adapted to intensive enterprises, traditional kinds of livestock on the farm could also be subjected to improvement as a matter of national agricultural policy.

III. Genetics and Breeding

In the course of a long history, animal breeding has experienced many changes in its theories and its practices. But until recently it has never been obliged to adapt itself precipitately to new conditions. All at once, however, events seem to conspire to destroy the old order. Genetics is becoming indispensable in livestock breeding. Economic pressures to intensify agricultural production, rising standards of living in the country as well as in the towns, and technical aids such as artificial insemination, computers and reproduction of data have combined with the maturing of the science of heredity to put the system of pedigree breeding under severe strain. This system has been an integral and essential part of the industry that produces milk and meat. Consequently, everyone in a world already short of animal protein has an interest in its future. At this juncture in the affairs of countries where some version of the pedigree system has been used, it might then be useful to consider how well it fits in with the needs of modern animal production, and whether changes are desirable from the standpoint of general welfare.

As farming enterprises grow larger, their managements have to equip themselves with information and resort to technologists to help them reach decisions and plan for more distant goals. Industrial developments of this kind widen the range of farming activities, since the old style farmer, sensitive to local markets and operating on hunches, remains as a contrast to those for whom farming is rapidly becoming more of a programme than a way of life. There is now manifest a spectrum of personal attitudes varying from frankly revolutionary to backward-looking. Many livestock breeders still do not know Mendel's laws announced 100 years ago, so that in the struggle to feed the world's growing population one of the main industries producing protein is found to be largely based in practice on breeding methods that owe little to modern genetics.

In Western countries the completely remodelled poultry industry exists along with a sheep industry notably reluctant to modernise itself. In many of them, including some with advanced industrial technologies, the administrative machinery of governments and agricultural organisations relative to animal breeding appears to be roughly coeval with sheep breeding methods. Yet all parts of all industries must thrust and strive, or lapse into stagnation. To adapt the outmoded ways of thinking that are enshrined in official attitudes and policies is therefore often as urgent a task as to improve the efficiency of livestock. Agriculture has long been regarded as conservative in its outlook and indeed in the less intensively farmed areas the adjective is probably justified. However, in the more intensively farmed areas, there have been many developments which stand comparison with those in the most progressive of other industries.

Statistics and genetics have no more destroyed animal breeding than chemistry has destroyed biology (Commoner, 1961). They are powerful allies and should be treated as allies by animal breeders. If geneticists advise that a trait has negligible heritability, then it is a waste of time selecting for or against it until some way is found of increasing the available genetic variation. If statistics show that twenty offspring are needed for a meaningful progeny test, then it is no good pretending that two or three are enough. Without aims, however, genetics and statistics are futilities for they cannot generate them. They are incomplete in themselves and are not an industry.

A reasonable scepticism regarding the assertions of scientists is proper. The operative word however is "reasonable". Scepticism is often a cloak for ignorance and as such is unprofitable. A well-informed scepticism that reveals and advertises weaknesses in theory or in factual support can provoke further research that justifies or demolishes it. It is too valuable a weapon to lose by getting it confused with obscurantism, ignorance or unwillingness to face facts.

Among scientists there are individuals of varied abilities, and diverse training. Not only are they far from equal in skill at expressing themselves but they differ in degree of self-criticism and in native intelligence. There is no need to assume that they are individually or collectively always right or always wrong. But it is operationally wise to recognise when they are expressing well established and generally held scientific beliefs which are as close as one can get to real if limited truths.

A geneticist can be as much a prisoner of his theories as a breeder. To shrug off conflict between the biologist and the practising farmer by saying that there is something to be said for both sides is not enough. Not one but many conflicts occur and their best resolutions are not always compromises. Breeders are shackled by tradition, by doubts about genetics, by financial worries, and by personal objectives. On their side, geneticists may err by oversimplifying, by relying on inadequate data, by their personal scientific ambitions, and by the fact that many of them carry no burden of financial or administrative responsibility. For the creative scientist, as for the creative breeder, reason is the handmaiden not the master. Both are artists and they should understand each other well. Every man in fact is in part an artist shaping his own life, moving and touching other lives. Since human affairs cannot always be intelligible unless seen as a whole, it would probably be better if genetic theories were more fully tested and exposed by a well-informed opposition. With such a paradoxical impossibility, it is incumbent upon geneticists to keep the edge of self-criticism sharp. As Price (1964) points out, there can be no acceptable non-scientific critics of science. The mystique of science is such that opinions about it are sought automatically from eminent scientists. The generals of science must find from their own ranks the strategists, administrators, historians and economists although their qualifications for making *ex cathedra* statements in these capacities are dubious. Their public image notwithstanding, scientists are not cold and objective men in white coats proceeding with infallible scientific method and impersonal conclusions. With so much ambient mythology, there is a risk that when misplaced enthusiasms or erroneous claims made by individual scientists are discovered, they will be attributed to science itself.

An example from the past is easy to find. Not only did many of the early Mendelians, in the excitement of the dawn of the new era in the study of heredity, insist on having the key to animal improvement, but it also took the next generation of geneticists, now working on the population level, some time to arrive at reasonably sober estimates of the power of their genetic tools (compare, for instance, Lerner, 1950, with Lerner, 1958). The occasional yielding to the temptation that scientists felt to meet adverse propaganda with half-truths and scepticism with unjustifiable assurance has undermined the research worker's special claim to be heard. Animal breeding is not the exclusive preserve of either pedigree breeders or of scientists. It is appropriate to recall in this context Oppenheimer's (1963) gentle advocacy of a proper humility in scientists: ". . . it would perhaps be good if in talking . . . (to our friends in other walks of life) we could count on a greater recognition of the quality of our certitudes where we are dealing with scientific knowledge that really exists, and the corresponding quality of hesitancy and doubt when we are assessing the probable course of events, the way in which men will choose and act. . . ."

IV. Social Implications

The word revolution is not too strong for the consequences of artificial insemination in dairy cattle breeding or of big scale operations in the poultry industry (including the technologist's mode of thinking about breeding), but it is hardly applicable to large animal production. In spite of all the spectacular advances over the whole range of agricultural problems, world productivity rises relatively slowly. To integrate and exploit an advance, even a simple one, within the enormous complexity of agriculture takes time. It could happen and probably does happen that technical progress on occasion outpaces human adaptability in some directions, but to damp down research and development on this score may well lead to difficulties in re-animating it when the need arises.

The results of applying science to the breeding industry cannot be properly assessed in isolation from the other elements, such as education, the availability of computers, the changing character of consumption and the economic policies of governments. One instance is the expansion of scientific endeavour which has coincided with an increase in the size of individual breeding establishments in the post-war years. Not only do these phenomena represent a trend characteristic of the times but more important, they interact with each other.

The growth of scientific information places a premium on the operational size of a breeding enterprise, since only the bigger establishments can obtain and apply much of it. Reciprocally, the larger and the more integrated an economic unit dealing with animal improvement is, the more it can contribute to the pool of new knowledge.

Interaction of this sort is a universal phenomenon. Modern biologists (exemplified by Dobzhansky, 1962) look upon human evolution as a continuing process involving much more than either biology or a history of cultures. A mutual feedback between biological and social factors is what moulds the history of the human species. With reference to domestic animals, this is a particularly fitting concept, since they have always been regarded by man in an anthropomorphic way, as if sharing his attitudes and feelings. Class distinctions, the prohibited degrees of relationship between mates, emigrations and immigrations occur both in man and the animals under his control with varying frequency or intensity depending on what has gone on before. The future of animal breeding, therefore, will not be the outcome of theorizing nor the result of wishful thinking but a response produced by the interaction of biological realities of livestock production with changing human culture. Neither the exact challenge nor the actual responses are predictable, but there is certainly no dearth of possibilities.

In general, the social implications of science have attracted increasing attention since Bernal (1939) as a natural scientist emphasised them a quarter of a century ago (see Goldsmith and Mackay, 1964). Although much has happened to bring nearer his vision of a "socialized integrated scientific world organisation", there are obviously many economists as well as humanists who deplore such a future. But whatever view may be taken of socialism on the one hand, or the gaps between the two (or more) cultures on the other, there is agreement that the scientific method should be used to limit and preferably eliminate such a palpable evil as starvation, and to improve man's lot on earth. Bernal himself considered that no culture could stand apart from the dominating practical ideas of an age without becoming pedantic and futile. He was not considering any specific patterns of behaviour, let alone that which governs activities in the field of animal breeding, but what he says of the whole applies equally forcefully to the part.

Culture has been viewed by Bose (1961) as an adaptation aiding the survival of communities which is based on a constellation of ideas and emotions. When new adaptations are required in the name of biological fitness, cultures must change and become more complex although not in any predetermined way. Over the centuries there has been a sequence of ideas and emotions on the subject, each giving way in its turn to another more in harmony with its time. Perhaps it has always been the fate of animal breeding to suffer some degree of disharmony with the trend of events. Geared as they are to the slow rhythm of the seasons and compelled to endure long gestations, domestic livestock engender a certain tempo of thought and action in those who breed them. This is unfortunate in Europe and North America but doubly so in Africa and Asia. Old ideas and institutions tend to outlive their usefulness and become over-valued. They lose initiative but continue to be supported out of habit or pride. Yet there is an urgent need to adapt them to changed conditions and at times to have the courage to abandon some of them completely. There is no aspect of animal breeding that research has left untouched; there is no breed that is perfect; and there is no system of breeding that could not be improved. Adaptations, of course, will come sooner or later, but sooner and more smoothly if consciously sought by flexible minds.

V. Decision Making

One of the most pressing and yet neglected problems in animal breeding concerns decision making. It is easy to look back and see how technical achievements lead to the concentration of power in few hands. What has happened in other industries is also happening in agriculture. The growth of large businesses based on vertical or horizontal integration, the breeding of a million cattle by a single artificial breeding organisation, and the control of prices by government action, are three well-known examples that leave no small-scale breeder untouched. In each case, a few people make decisions affecting many. This is a social problem with considerable effects on animal breeding because of its susceptibility to being organised. Breeding has been so vague in its objectives, so inefficient in its methods, and so rudimentary in organisation that it offers great scope for the modern decision-maker. In many situations this may turn out to be the scientist, unconscious perhaps of his influence, but a decision-maker none the less. Churchman (1961), whose studies of this subject make enthralling reading, goes so far as to say that science *is* decision-making. In his sense, the sense that scientists determine what information they will collect and how they will interpret it, this conclusion is inescapable. Scientists, however, are not the only decision-makers in the new forms of animal breeding although they obviously influence the nature and the quantity of data provided for those who do make decisions.

Those who are at home on the frontiers of knowledge can become very powerful, but they are few and often do not communicate easily with other people. There are sometimes problems of communication between management and technologists; how much wider, therefore, is the gulf between the latter and the ordinary farmer. Important though it is to those concerned, this question is but one facet of the vastly greater one of informing the voter in a democracy about the political issues involving science.

The organisation of animal breeding has heretofore been a haphazard matter. Neither the responsibilities nor the prerogatives of scientists and managers, breeders and bankers, private concerns and the state, are sufficiently clearly defined anywhere in the world to allow a simple analysis of the genetic consequences of the various developments taking place. Yet the combination of big science and big business results in the emergence of a whole set of new problems of vital import to the future of mankind under any form of government. The genetic management of our resources of useful or potentially useful animals is one of such problems, and discussion of it is the main purpose of the present work.

CHAPTER 2

THE WAR ON HUNGER

The purpose of this chapter is to describe briefly the agricultural background against which the potentialities of animal breeding are to be considered. Into the intricate fabric of agriculture are woven many minor patterns which, being all functional parts of a whole, cannot be comprehended separately. If animal breeding is to meet the calls on it, those who are responsible must ensure that its objectives are generally acceptable to society and that its methods are consonant with its purpose. There can be no hope of bringing in a few painless adjustments intended to stave off radical changes. As a reminder of the exciting developments in many agricultural practices which will influence and be influenced by types of livestock, a few pages have been culled from the works of recent historians, research workers and commentators. Inadequate though they are to give more than hints of the ferment of ideas, they may serve to show the need for a continuing adaptation of all the sciences and practices to each other. Only a little imagination is required to foresee changes when a strong demand, a flourishing technology and a relatively underdeveloped industry are in conjunction.

I. Improvement in a Developing Agriculture

During the eighteenth century the invention of machinery and the exploitation of coal in Britain led to the growth in size of urban populations and to fundamental changes in the methods of agriculture including the use of enclosures, the provision of winter feed and pastures of clover and artificial grasses. Later, the activities of Robert Bakewell and his immediate successors laid the foundations for the structure of animal breeding as it is known today. These breeders owed their opportunity to the increased demand of city markets and to improvements in road, rail and water transport both within the country and for export of live animals from Britain. The techniques of breeding thus established then began spreading to other countries. In many instances, livestock production depended on locally developed methods, but, by and large, where it became organised, its structure was based mainly on the British model.

TABLE 2.1

Trends in yields of crops, milk and eggs in the United Kingdom

Item	Pre-war average	3-year average 1960–63	% increase
Wheat	17·8 cwt/acre	31·3	76
Sugar beet	8·2 tons/acre	13·0	63
Potatoes	6·7 tons/acre	8·9	33
Milk	560 gall/cow	777	39
Eggs	149 eggs/hen	191	28

The nineteenth century brought a steady increase in productivity through the spread of improved husbandry. At the same time considerable changes were effected in the conformation of animals before as well as after the compiling of herdbooks began. How much genetic change in milk yield has taken place is problematical, but greater fleece weights (in Merinos and Romneys), smaller fleece weights (in Wiltshire Horns), more eggs (in ducks and hens), and faster growth (in pigs) have certainly been achieved in the comparatively recent past. All these adaptations took place in the context of developing markets and changing husbandry. In his book Ernle (1961) has traced the main lines of development in agriculture in the United Kingdom which in many ways are also typical of other countries. The one word which comprehends all the changes, regardless of time or continent is intensification. During the eventful period 1910–50, the tempo of change quickened. Under the impact of scientific discoveries coming ever faster and being applied promptly, the old rigidity of both plant and animal production, expressed in fixed rotations and other rules of good husbandry, had to give way to a new adaptability all over the world (Russell, 1954). Advances are still coming rapidly. In the year ending May 1964, net output in Britain was nearly a quarter more than it was ten years earlier. In this same period, the number of agricultural workers fell by over a quarter due to the mechanising of traditional farm processes and the saving of labour by modern methods. As with output per man, production from both crops and animals is rising, but more slowly for the latter. Table 2.1 shows how the output of several products has grown in the United Kingdom since 1939, and similar figures are available for other countries.

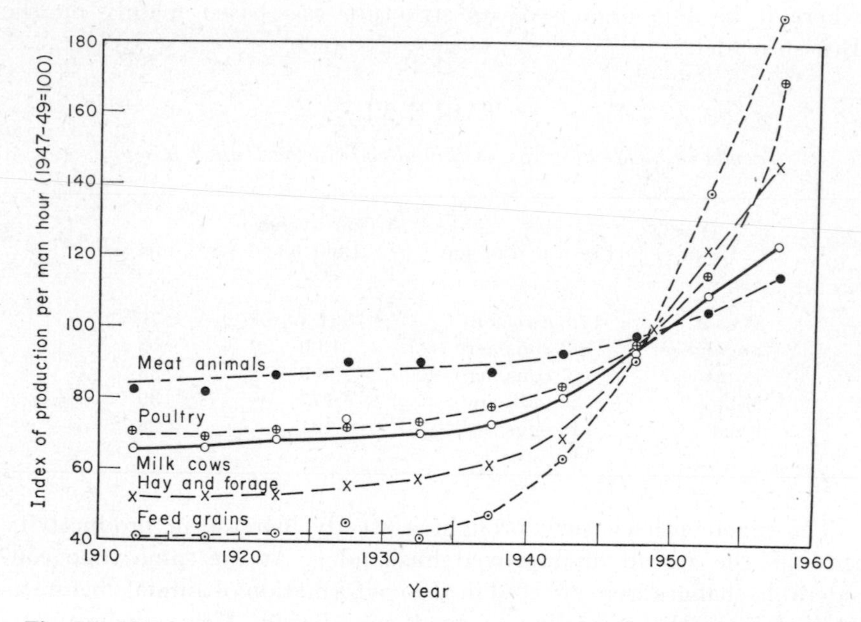

Fig. 2.1. Man-hour output in livestock and crop production (from Byerly, 1964; reproduced with permission of the author and publisher).

For instance, a very similar process has been going on in the United States. Figure 2.1 (from Byerly, 1964) shows how the output of five major products per man-hour took a marked upward trend between 1940 and 1959. Between 1925 and 1940, the annual number of eggs per laying hen rose from 112 to 134, an increase of 1·5 eggs per year. By 1960 the number of eggs rose to 209, the rate of increase having speeded up to 3·75 eggs per year.

Again, in the countries with advanced dairy industries there has been a general upward trend in yield of milk (or total fat) and in herd size, since 1945 (Table 2.2). Since this trend is so fast, so widespread, and independent of breed, it seems likely that much of the increase is attributable to higher levels of husbandry, such as more liberal corn feeding of cows in the United States.

Although net financial outcome of farming operations is the major factor governing them, it is not the only one. Livestock farmers have often a loyalty or personal attachment to their stock which makes them

TABLE 2.2

Trends in dairy cow production

Country	Breed	Period (years)	Increase per year kg butterfat	gall milk
Denmark	Red Dane	15	2·7	14·8†
	Jersey	15	2·7	14·8†
New Zealand	Jersey	16	1·3	7·2†
Holland	Friesian	12		5·3
England	Friesian	15		10·0
	Ayrshire	10		9·2
Scotland	Friesian	7		7·4
	Ayrshire	7		8·1
U.S.A.	Holstein	4		16·2
	Ayrshire	4		18·5

Comparisons between countries are not recommended owing to the varying methods of estimating average yields and to the unequal periods for which data are available.

† Based on 4% butterfat equivalent.

reluctant to abandon either their profession or their animals contrary to what economic pressures tell them to do. Others hope vaguely (and, usually, vainly) that things will improve if they can hold on long enough. There has been, consequently, a time lag in the multiplication of economically superior animals and in the demise of inferior ones. As farming becomes more of a business, this time lag is likely to become shorter. With the growth of performance testing and the intelligent analysis of the results, the merits and demerits of particular breeds and crosses of sheep and cattle could become widely and quickly known as has happened with pigs and poultry. It may be some time yet before new varieties of the larger livestock are bred to meet specific conditions of production as poultry are now, but they are in clear prospect.

Poultry are now bred largely by careful combinations of lines to have a uniform rate of growth and mature body size, to produce a large number of eggs of specified size and colour, to have high food efficiency, and to present an attractive carcass. All these characteristics have been notably improved while the cost of eggs and of poultry meat has been reduced in the last two decades. The eye judgment of breeders played no part in this process. In the task of improvement of such traits as disease resistance, appetite, ability to utilise various diets (for instance, soya bean meal which is deficient in methionine), laboratory techniques applied purposefully on an adequate scale are necessary.

Many of those who produce other types of stock still have to learn the lessons from poultry: costs of production at all stages must be decreased as far as possible; market requirements met accurately; growth rate and production increased; food efficiency improved; the number of separate products or qualities reduced to a minimum; husbandry mechanised; fertility raised and mortality lowered; overheads spread over a large production and all unprofitable activities eliminated.

Since many livestock farmers do not yet see the need for records of performance of their stock and thus have only a vague idea of the strong and weak points of their businesses, it might appear that they, in general, have a long way to go before they can take advantage of recent technical developments. This was true also of the numerous small poultry farmers who no longer exist as independent producers. What happened in their case was that they first provided a market for improved varieties and then were forced by economic pressures to quit the industry or become absorbed by more efficient enterprises.

Books on medieval agriculture show that stockmen flourished long before specialist breeders. They had the task of drawing a usable surplus from their animals under very primitive conditions, and in skill and devotion to their charges concede nothing to their successors. With the advent of shows and then pedigree breeding further opportunities were created for them to develop their expertise and authority to the point where they reigned practically supreme in their departments of the farms. They could no doubt extract the best out of their herds and flocks by a fine sensitivity to the needs of individual animals in a farming world which could be lavish with cheap labour. It is to be feared that, like many other kinds of craftsmen, these artists are losing status in industrial farming. Where dairymen are judged by the number of cows milked in an hour, there is no place for the slow milking cow or the man who will patiently milk her out. There is no place for the time-consuming hurdle flock of sheep, for the small flock of chickens maintained under extensive conditions, or for the sow that must be watched while she farrows. By degrees all classes of stock are being subjected to selection which favours animals that need a minimum of individual attention. The contrast between the heterogeneity of colour and form of pre-purebred flocks and herds, which made visual identification of individual animals easy, with the monotonously uniform appearance of strain crosses of chickens is a reflection of this process.

Just as new varieties of crop plants have to be suitable for the new machines to sow and harvest them, and, more recently, for the animals that will eat them, so the test of a better animal must be conducted under advanced conditions of livestock production. There seems little doubt that stockmen and breeders will also have to prove themselves under the same conditions.

The basic difficulties about raising livestock production in either developed or under-developed countries are sociological rather than economic or scientific. They include religion, systems of land tenure, and a failure by "economic man" to live up to his name. The extent to which farmers are unresponsive to the profit motive would make an interesting study. Whatever the reasons, however, many are strongly resistant to infringements of their chosen occupation. When the result of this is a failure to increase or cheapen production, the world is the poorer and, where there is no surplus, so is the taxpayer or consumer.

II. Recent Advances

A. Husbandry

Much of the increased productivity of the land has come from drainage, provision of winter feed for stock, improved cultivations, especially from the speed and power of tractors, fencing and fertilisers. Although these techniques are hardly recent in conception, they are still being refined and are capable of increased utilisation (Cuthbertson, 1963). In particular, the use of fertilisers can be extended as higher yielding crop plants and grasses become available and as heavier stocking rates compatible with good health become possible.

While seeing some scope for reclamation of both high-lying and low-lying ground, Russell (1954) thinks that the greatest potential for increasing production lies in more effective use of the land already occupied. No doubt this would require the transformation of nomadic and subsistence farming with all the social consequences of such change. In order to place farming on the spiral of increasing productivity, systems of land tenure and of providing capital would have to be adjusted to permit the full exploitation of the soil. Russell supposes that this transformation has already taken place in Northwest Europe and North America but the process has only begun.

Advances in the sciences of plant breeding and animal nutrition and animal health are of the greatest significance to animal breeders. In fact, for a well-balanced assessment of future prospects in animal production they can hardly be considered separately. As might be expected, this is most evident where growth is fastest (whether it be growth of animals or growth in human skill) and this is at the most intensive end of the scale of agricultural activities. Pig and poultry producers have accordingly appreciated, as have dairy farmers, the newer knowledge of nutrition that permits them to use well-balanced rations that are based on the mineral, amino-acid, and vitamin needs of high-producing stock. Sheep and cattle farmers are gradually taking advantage of the better yielding and more palatable strains of crops, including grass. All kinds of stockmen become increasingly dependent on the ability of the veterinarians to steer them clear of disease.

Machinery for automatic feeding, watering, ventilation and cleaning of animals is already in common use for poultry and for beef cattle in feedlots. There seems to be no insuperable obstacle in the way of extending such methods to other classes of livestock which repay intensive rearing. Another well-known development is the use of antibiotics and other drugs for augmenting production and for hygiene. Just what the future holds in this respect is not obvious, but it is certain that as new methods of husbandry arise there will be a need for coping with diseases that may not have been important formerly but have become important under new intensive conditions. Minimum-disease and specific-pathogen-free pigs have an obvious bearing on this problem. This potential seems to be high. Disease takes a heavy toll of intensively kept animals and it seems likely that future housing will be associated with means of obtaining and maintaining animals which are free of certain diseases. Among the features of these buildings will be automatic feeding and temperature control. The latter is important because it affects efficiency of food utilisation and, through it, rate of maturing. Temperature control may also be of significance in the kinds and levels of incidence of disease which stock experience.

B. Artificial insemination

Since 1945 artificial insemination (A.I.) of dairy cattle has provided new and great opportunity for advances in livestock production. It soon justified itself on economic grounds. But contrary to original expectation, finding suitable sires for artificial insemination centres proved very difficult. As a result, the recording of performance had to become highly organised. In the long run, the repercussions of artificial insemination of dairy cattle on other kinds of livestock, through the complementary development of operational research, deep freezing of semen, and progeny and performance testing, and the attitudes of farmers to them, may be more important than artificial insemination itself.

Naturally enough, the main benefit of using this technique accrues to those who control the intellectual and physical machinery of its operation and to their employers. Some breeders of pedigree dairy cattle, however, suffer a loss of trade for young bulls, and also a loss of prestige, since their breeding methods seem to be inadequate when put to the test of performance. With the spread of testing to pigs and sheep, there will be the same disagreeable revelations about the way these species are being bred. The effects of A.I. in cattle are more fully discussed in Chapter 6.

C. Genetic theory

During the last thirty years, the theory of population genetics has become a solid basis for industrial planning. Generations of students have been taught its precepts and speak its language. As a result of their activities, a large store of genetic information about livestock has been collected and made available to all. Statistical methods have been devised for use in the specific problems of animal breeding and widely applied in many countries. With the coming of machinery for the handling of data, culminating in the computer, the geneticist became possessed of very powerful tools for manipulating livestock populations. Examples of his use of these tools to good effect is to be seen in the dairy, pig and poultry industries. In addition, publicity in technical and popular journals for scientifically gathered and interpreted data has created a new social atmosphere in which breeders must now work. The mystique of stock sense has become an uncertain refuge, as has the green thumb in mechanised plantations.

A comprehensive theory has been provided for understanding the biological evolution of livestock. As it becomes more refined, its strength will increase, but it is already too firmly established to be wrecked or even shaken by personal distaste for its conclusions. Although there may well be on occasion some sentimental or political objections to their logical consequences (for instance, financial embarrassment to pedigree breeders), these conclusions will not on that account be less valid. Chapter 3 is devoted to a fuller discussion of modern genetic theory.

D. Marketing

Off the farm, marketing has been growing in complexity with the aid of cold storage, grading, rapid transport and packaging. Farmers and the breeders who supply them with breeding animals must reckon with the results of consumer research and advertising. They must expect that elimination of hard physical labour, propaganda of a medical character, super-marketing, deep freezing and freeze-drying, and a growing interest in quality and hygiene will have repercussions on farms such as the practice of contracting for specific products. For evidence of this they need only recall the increasing preference for meat that is free of fat and is tender. Consumers have also obliged sheep farmers to share their wool markets with manufacturers of artificial fibres. Furthermore, rapid traffic in both live animals and carcass meat which is now possible between countries makes a producer sensitive to changes in supplies occurring far from his own parish. Yet it also provides opportunities of selling in wider markets such as were taken by the United States broiler and the Danish bacon producers. Political and economic considerations may determine whether markets are protected and prices buffered, but biological facts are also determinants of uniformity of products, efficiency of production, and the long-term outlook for the development of superior strains.

These are, perhaps, the principal changes of immediate interest to animal breeders but they will be followed by others in the near future. Evolution is not mere change. It is the replacement of simpler systems by more intricate and more complex ones. There is no prospect, therefore, that any reversion to former and less exacting concepts will occur.

It is as much a law of livestock breeding as it is of natural evolution and of the history of human institutions that the most rapid development occurs when new ecological or social opportunities arise. In livestock breeding, the opportunities are greater than ever, since the authority of formalism in breeding has been broken at the moment when scope for exercising the new liberty is expanding. Animals must now be bred not for their fair appearance but to fit the machinery of mass production.

III. What Is Improvement?

Possibly one of the greatest difficulties facing breeders, agricultural organisations, and geneticists is how to reach agreement on what changes in performance constitute improvement. This is hardly surprising. Three elements, each complex, are involved:

1. pedigree breeders' incomes, unlike those of users of unregistered stock, include the sale of pedigree breeding stock;
2. performance has many components, some of which may be incompatible with each other;

3. economic and environmental conditions of production are inconstant, prices and husbandry varying with time and locality.

Pedigree breeders have a readily understandable point of view. As long as breeding stock is bought on appearance, they are there to supply the market. They must foster the popular type by whatever breeding, husbandry or advertising methods they have at hand. They are fully entitled to give it as their opinion that a certain balance of characteristics, that is, type, is desirable and that they are providing it. Others are equally entitled to different opinions.

The owners of commercial cattle or pigs or sheep are aware that their net income is affected by many items in a profit and loss account other than milk yield per cow, the number of pigs in a litter, or the weight of a weaned lamb. They have not, however, been very well informed by research workers about the importance of other components of net profits, such as outputs per man and per acre, mortality and veterinary expenses, the salvage value of worn-out breeding females, and their rate of depreciation. Although familiar, these items are rarely studied by geneticists. There is consequently not much information about the amount of genetic variation in them; nor is there likely to be much until suitable data are collected. On the assumption that they are of low heritability and unlikely to deteriorate as an unfortunate side effect of selection for weight of meat or milk, it is thought by some to be reasonable to ignore them in breeding.

Many advocates of positive eugenics, who have a vision of breeding a superior race of men, find it hard to persuade others where superiority and breeding value lie. Improvers of livestock face the same difficulty. Since there is room for disagreement, the pessimists can be expected to opt for attention in breeding to numerous criteria and a preference for no change. They have the advantage of being able to point out how uncertain markets are and how unreliable the estimates of costs of production. Changes in public taste can alter quality differentials in milk or meat very quickly. Technical developments in prospect could alter prices and costs substantially as, for instance, in such items as the housing of sheep, the timing of lambing, or the preparation and marketing of convenience foods instead of carcass meat.

The various authorities influencing animal breeding (poultry now excepted) have not seen fit to secure much help from those trained in the methods and approach of economics—a remarkable omission in a large industry. How does it come about that agricultural economics has not penetrated at least into research on animal breeding? Could it be that a combination of disdain by scientists and lack of imagination in economists has produced this result?

A negative attitude to change, no matter how well-founded, is still unpalatable to optimists and to most scientists who live by it. They can point out that over the centuries during which agriculture has reached its present stage of productivity, the trend has consistently been towards high performance and risk-taking. Occasional disastrous droughts have not prevented Australians from increasing wool production from Merinos, even at the cost of some merit in meat, milk and fertility. Claims for great hardiness have never helped a breed such as Soay sheep or Longhorn cattle that lacked potential for a saleable commodity. Unfavourable genetic correlations, making it impossible to combine all the components of perfection, no doubt exist but they may not be serious. Some of them may be counteracted by husbandry, others overcome by selection, while still others may require that some losses are put up with in order to enjoy greater gains. Although some attempts to raise livestock production by breeding may fail, those which succeed will lead to the replacing of breeds that remain static.

In the last resort, discussion about objectives in improvement resolves itself frequently into disagreements about land use. Good growth rate and food efficiency may be desirable everywhere, but the conditions under which these traits are to be manifested vary widely. Beef production, for instance, may take place in feed lots with high energy diets permitting slaughter at 1–1½ years of age, or on extensive grazings with low energy diets, periods of shortage, and a suitable weight for slaughter at 2–3 years. In southwest Scotland, for example, Galloway cattle are bred for adaptation to wet cold winters with silage feeding. To fit in with the farming system in use, the cattle must be of the right size and fatness to kill when 2–2½ years old. Too much emphasis on growth rate might disqualify them for this purpose. Since, however, there is nothing immutable in farming systems, those who control the fortunes of this breed are faced with the risk that, if husbandry intensifies further, a faster growing breed might prove more popular.

Efficiency is also wanted in straight-bred beef cattle or in various crossbreds from dual-purpose and dairy cows. Selection for growth rate, therefore, might be imposed on all breeds, but the attention paid to other things would vary in kind and intensity depending on the use to be made of land. For the British Friesian breed internal tensions are created in this way, since it is the source of much British home-produced beef. Some users require steers for a production system aimed at rapid turnover and early maturity for slaughter, others require steers for pasture and silage feeding, while still others apply intermediate methods. Type varies with rate of growth and age, as well as with beef and dairy character. General agreement on type, therefore, for an animal that is primarily a dairy cow, but also an important beef producer, is evidently impossible. As history shows, there are several ways of escaping from this conflict of purposes: splitting the breed into sub-breeds, organised crossbreeding, and changing to another breed.

Faced with the growing world shortage of animal protein aggravated by the increasing demand for beef as standards of living rise, the administrators and sponsors of long term genetic research with livestock have a problem in deciding what resources to devote to the question of raising beef supplies from dairy cattle. As Plowman (1964) and many others have pointed out, more beef can be had from a fixed population of dairy cattle by keeping cows longer, raising all surplus male and female calves to greater live weights, using bulls of the largest breeds, and reducing losses from disease and infertility. Except for crossbreeding, these methods are all well-known and applied during the growth phase of any breed when prices are high. More beef, however, means less of something else, and beef is the most expensive form of animal protein to produce. An indirect expense that may be incurred is the loss of efficiency in improving the production of milk protein, especially if the criteria for selecting dairy cows are modified to allow for more pressure on meat qualities. This difficulty does not arise if selection for meat is carried out on a beef breed used for crossing with dairy cattle.

Similar problems of defining priorities face the breeders of sheep and pigs. Breeders, producers, manufacturers and consumers all have different approaches to the question of what is wanted and, if agreement by all to every change was required, it might well be impossible to reach in respect of any numerically strong breed. There is no group of people with special insight who can be relied on to have the right answer, although in retrospect it is easy to see that in the past some were nearer to it than others. During the past 200 years alone, many hundreds of local varieties must have failed and, at least, in some cases it was probably not bad luck but bad judgment that brought misfortune. Breeders of Percheron horses, Middle White pigs and Merinos could hardly be blamed for becoming redundant in the United States, in the United Kingdom, and in the hills and plains of New Zealand, where economics demanded tractors, less fat, and export lamb; but the Clydesdale horse with feathered feet, the Vermont Merino with overwrinkled skin, the rainbow type hog or the long resistance to polled varieties of cattle, were surely errors of judgment.

Such minor tragedies, nevertheless, had, and have, a major advantage. Most of them occur within breeds, but some, like those just mentioned, affect whole breeds because official breed type happened to be wrongly conceived. It is the price of progress that many attempts to find it should fail. To be limited to making one attempt either because of authority at the centre of a breed association or of another type of organisation, or because numbers of animals in a breed are small, means taking a grave risk that errors of judgment will not readily be detected or corrected.

IV. Short-range Objectives

Making the best of the material and institutions now available involves a critical appraisal of animals, enterprises and industries. All three have attracted recent scrutiny from farmers, biologists of various kinds, economists, and business men. As a result, livestock production is in the process of abandoning many of its old principles and of trying to discover new ones in an intensive search for efficiency in the use of land, labour and capital.

A. Improving animals

To say that the immediate task of breeders is to improve efficiency is not very helpful, since numerous changes might be claimed to lead to higher efficiency. To select for certain traits involves risks of misreading future markets but risks must be taken by someone. Waiting for full knowledge, or striving to maintain or "improve" animals in all respects simultaneously may be the correct policy for a fraction of a breed. Should other more positive policies succeed, this fraction can then be discarded.

Until the appearance of large breeding enterprises animal breeding was a form of self-expression for many of those attracted to it, of whom some could be relied upon to be individualistic in their aims and methods. Only for short periods during the formative stages of a breed would unity of effort be achieved. This is one cause of inefficiency of the current pedigree system, for it is organised to suppress all but minor deviations from official type. Therefore, while securing the advantages of diversity is a long-range problem, the present need is to encourage the breeders who will concentrate on improving performance at some cost, if need be, to type as officially defined. This kind of breeder is more likely to be found among the young and among those whose herds and flocks have currently a limited value. If they succeed in creating better adapted varieties, it is probable that type will be redefined to suit the stock they produce.

These remarks do not apply equally to all classes of livestock. In poultry breeding there has already been both a ruthless sifting of breeds and varieties for suitability to markets and husbandry, and remarkable improvements in both broilers and egg-layers. Further improvements are becoming increasingly difficult to obtain. As a result, the question of securing genetic diversity looms much larger in the poultry industry than it does with cattle, sheep or pigs, which are in a more rudimentary stage of organisation.

TABLE 2.3

Changes in rate of gain and efficiency of feed utilisation in pigs (from Ellis, 1964)

Year	Average daily gain, lb	Feed per pound gain, lb
1910	1·1	4·5
1920	1·3	4·6
1930	1·2	4·5
1940	1·4	4·1
1950	1·5	3·6
1960	1·6	3·5

Given the need for protein and the revulsion against fat, there is good reason for breeders to try to develop the rapid production of lean meat and milk. In pigs, beef cattle and poultry, many studies have shown that a strong positive correlation between rate of growth and economy of food utilisation exists and that both are of fairly high heritability (see Smith, King and Gilbert, 1962, for a genetic study). Substantial increases in efficiency in the course of the last twenty-five years have been claimed for pigs (Table 2.3). Gains of this magnitude can probably be attributed to improvement in nutrition, housing, and disease control as well as to breeding. If for any reason such as successful selection or the use of feed with higher energy content, pigs can be induced to grow faster, they will use less feed for maintenance. Furthermore, constant selection pressure for less fat has been effective and has, therefore, probably helped to lower food costs.

Direct selection for efficiency of food use, as Taylor and Young (personal communication) have pointed out, is likely to prove a complicated process. Estimates of genetic variance and correlations are highly dependent on the feeding regime adopted and purposeful direct selection for food efficiency will eventually require some constancy of the feeding system and of the conditions of production such as rearing, fattening and age or weight at slaughter. One of the points to keep in mind is that growing bacon or pork pigs are not the only members of a herd that must be fed: pigs that die of disease and accident, pigs that suffer chronic infections, pigs among the breeding stock that fail to mate or conceive promptly and completely, all waste food. Another point to remember is that an efficient pig eating only 3·2 lb of dry feed for each pound of gain in live weight consists very largely of bone, gut, skin, water and fat and only 10% of it comprises edible protein.

As would be expected, the efficiency of food use by pigs, chickens, wool sheep or dairy cows is better for the higher producers than it is for the lower producers which have higher maintenance costs per unit of production. Most of the investigations upon which this statement is founded have not, however, taken account of the food costs of maintaining breeding stocks. Notwithstanding this lacuna in knowledge, a positive correlation between efficiency of food use and output is one of the chief arguments for intensifying livestock production. Apart, however, from the consideration that net profit is not the only thing to be desired (in a hungry world output per acre would certainly be more popular) the intensification of farming processes is the only open-ended policy, that is, a policy of unlimited future. Those who would seek net profit under conditions of inefficient food use can certainly not hope to increase net profits very far by accepting inefficiency. They cut themselves off from the possibilities of exploiting brains, machinery and land.

Another aspect of the efficiency of food use arises from the sex dimorphism in the size of animals. It is well known that uncastrated mature males of cattle, sheep and pigs are approximately one and a half times as big as the females of the species. The rate of growth of males is faster than that of females, and their efficiency of food use should also be greater. For this reason, to say nothing of reducing pain and encouraging lean meat production, castration may be increasingly regarded as anti-social.

Because of their need for roughage, ruminants are less amenable than pigs to the recording of food consumed. That there will be differences in efficiency among them, however, is not to be doubted. Whether these will be found with respect to proneness to lay on fat, growth rate, milk production, rumen flora or appetite level, is a question for the future to answer. It is also a moot point whether efficiency in milk production, and, particularly, in the protein fraction, has anything in common with efficiency in meat production. Not surprisingly in the circumstances, much of the variation among animals in efficiency is unwanted and has unknown sources. Even dairy cows still present many problems concerning both the aims of breeding and the methods of reaching them (Lush, 1960). The present organisation and methods of animal breeding are not designed (except in the poultry industry) specifically or effectively to increase animal protein production. There is a regrettable lack of information about the most efficient ways of producing protein. The inter-relationships of breed, body size, feed levels and food composition are likely to be complex but will have to be determined for intelligent production.

The dairy cow is a case in point. Her breeding is often more highly organised than that of any other form of livestock except poultry, and she is a very important source of proteins as milk and meat. Unfortunately, there are no quick and cheap methods of measuring her breeding value from this point of view, and she continues to be selected for type, milk yield, and butterfat percentage. For discussion of the complexities of the genetics of milk protein production and its relation to butterfat production, Robertson, Waite, and White (1956), and Berge (1963) should be consulted.

TABLE 2.4

Apparent digestibility of organic matter of foods by different species in relation to fibre content of food (from Blaxter, 1960)

Species	% fibre		
	0	15	30
Ox	86	75	63
Sheep	89	76	63
Pig	94	70	46
Hen	86	57	27

If the battle between food production and human reproduction (Luck, 1957) proceeds as predicted, the proportions of the different kinds of animals raised may have to change. Species differ in ability to digest food. Table 2.4 shows how increasing amounts of fibre reduce the apparent digestibility of organic matter far more in pigs and poultry than in cattle and sheep.

Pigs and poultry which have simple stomachs and prefer concentrated food similar to that of man, may eventually suffer in competition for food supplies with him. Ruminants would appear to have good long-term prospects of maintaining or increasing their numbers because they eat fibrous foods and can make protein with the aid of their microflora out of foodstuffs impossible for man to digest. Thus in spite of the fact that in the ruminant, only 50–70% of the metabolisable energy of food is available, whereas in monogastric species the comparable figure is 80–90% (Mitchell, 1964), they are in no danger of being eventually set aside in favour of pigs and poultry. If the horizon, for policy purposes, is brought closer, the prospects for applying the financing, breeding, husbandry, and sales methods of the poultry industry for the improvement of larger animals are probably better for pigs than for sheep or cattle. Pigs have not only a much higher reproductive rate and therefore higher potential selection differentials, but also have already gone part of the way to achieve the necessary testing facilities, intensified husbandry and simplified objectives.

As Table 2.5 shows, the ox, pig and hen differ from each other in efficiency of gain on any one kind of food, but they do not maintain the same relative efficiencies from food to food. From this particular point of view the best use to be made of food of high energy content and low fibre is to feed it to pigs and hens. However, as demonstrated by the prices paid for intensively reared beef, the laws of supply and demand are quite capable of overriding this biological consideration.

TABLE 2.5

Comparison of species with respect to net energy values of different feeds (from Blaxter, 1960)

Feeds	Absolute values kcal energy gained per gram dry matter			Relative values		
	Ox	Pig	Hen	Ox	Pig	Hen
Maize grain	2·04	3·01	3·05	100	147	150
Earthnut cake	1·77	2·61	2·28	100	147	129
Wheat bran	1·50	1·71	1·06	100	114	71
Wheat chaff	0·63	0·28	—	100	44	—

Considerable changes might be possible with sheep. A ewe normally milks only four months, and the milk is usually converted into lamb instead of directly consumed by man. While she is milking, a ewe is also producing wool that could, if necessary, be reduced and replaced by synthetic fibre. Under such a change in objective, the Wiltshire breed, that produces little more than enough wool to protect itself, may become important.

Like other animals, a lamb is most efficiently growing when growing fast. Since the average growth rate of about 0·75 lb/day is well below that attainable by some lambs, there would appear to be room for improving both mothers and offspring. The cost of production of protein (which amounts to about 5 lb in a 40 lb carcass) in lambs includes the maintenance of a ewe. Hence, it is natural to wonder whether there is opportunity for improvement by (a) reducing the size of ewe, (b) increasing her fertility, (c) improving pre- and post-natal environment to give lambs a quick start, and (d) reducing mortality. Stratified cross-breeding already does much in these directions and could probably do more by emphasising the reproductive qualities on the female side of a cross and growth rate, and carcass quality, especially meatiness, on the male side (Rae, 1964).

TABLE 2.6

Annual feed requirements of ewes and lambs for production (from Wallace, 1955)

		Feed consumption (lb T.D.N.)		
	Weight of lamb in lb	Per lb carcass weight of lamb	Per lb protein in lamb	Per 1000 calories in lamb
Ewe with 1 lamb	77	18·5	149	14·8
	112	14·9	134	9·4
	141	13·8	137	7·5
Ewe with 2 lambs	130	12·8	99	11·6
	192	11·0	95	7·7
	284	10·5	104	5·7

Interesting estimates of the feed cost of producing lamb meat have been made by Wallace (1955). The figures in Table 2.6 have been extracted from his data. These figures show the advantage of twins in lowering costs of production, but also demonstrate how expensive animal protein is in terms of total digestible nutrients (T.D.N.). Table 2.7 shows that lamb, pork and beef compare unfavourably with milk and eggs in this respect.

Demand for more productive sheep to match the increased output from pastures and rising cost of labour and for types specially adapted for intensive production in sheep houses is focusing attention on breeding for higher fertility, year-round lambing, good appetite, and appropriate carcass quality. All these objectives seem achievable in the near future in terms of crosses between two or three specialised breeds. Lengthening the breeding life of ewes, however, although it would by itself save feed, reduce costs of depreciation and release more young females for immediate consumption, is less promising as a genetic proposition, because of the low heritability of this trait (Bayley *et al.*, 1961) and its dependence on fertility.

TABLE 2.7

Milk, meat and eggs produced from 100 lb feed and ratio of protein produced to protein in the feed (adapted from Ellis, 1964)

Product	Total product	lb protein in product / lb protein in the feed
Milk	105·0	·25
Eggs	32·7	·25
Broiler	44·4	·22
Turkey	28·7	·16
Pork	27·7	·14
Beef	12·5	·10
Lamb	11·5	·08

Total product expressed as live weight or whole egg or milk and protein expressed as edible.

In Italy and France the improvement of milk production from ewes is an object of testing and recording schemes and it could happen that, although the breeds concerned are maintained primarily for milk production, what is learned from them may influence the thinking of breeders concerned with lamb production. For the future, ewes may be needed which have higher milk yields and shorter lactations so as to obtain fast growth in lambs during the first month of life. For a review of the quantitative aspects of milk production in sheep, Boyazoglu (1963) can be consulted.

Finally, it may be noted that cattle and sheep, as well as pigs, still leave much to be desired with respect to their fertility and viability. Although apparently not highly responsive to genetic selection, these traits are economically important. Death rates up to 30% occur between birth and age at slaughter and good sense dictates that ways be found of keeping these losses down. Something more can be done to reduce these losses, including the use of appropriate breeding systems.

B. Improving breeding methods

Although the principles of population genetics are well explored, the genetic details have not necessarily been filled in for the larger domestic animals. For obvious reasons, the study of inheritance has been limited to traits which are commonly recorded and, on a smaller scale, to those observed in specially selected herds. The result is that information about pigs and chickens is more extensive than about beef cattle and sheep; that among dairy cattle, more is known about inheritance of milk yield than about the genetics of its composition; and that in all countries there are breeds about which nothing specific is known. Serious attempts to improve them require an increase in selection differentials, and this, in turn calls for groups of animals larger than normal-sized herds. Furthermore, reduction of intervals between generations and application of performance and progeny testing when appropriate are also required.

Breeders have been often urged to reduce as far as they can the number of traits on which selection is based. Very uneven attention has been paid to these suggestions so far, but the trend is towards their wider acceptance. Oddly enough, it is the so-called practical breeder who shows reluctance to increase the directness and efficiency of his methods. The state of affairs in sheep breeding in the United States has been described by Terrill (1958) in a review of fifty years of progress in sheep breeding research. The growth of scientific knowledge is documented by him and due reference made to the creation of new breeds, but no evidence of genetic improvement in old ones is or could have been given. Instead, Terrill mildly observes that "Commercial producers of wool and lambs sometimes seem to be more aware of the need to apply selection pressure on production traits than are ram breeders and purebred breeders." The same could be said in most if not all other countries.

Although one cause for inefficiency of present-day selection procedures stems from the fact that unimportant selection criteria are often employed, another lies in the use of indirect methods. The time may be soon coming when it will be necessary to pay attention to some traits that are at present ignored merely because of the difficulty of measuring them. Temperament has much to do with the performance of domestic animals. Yet it is very difficult to reduce variation in this character to some scale of values which can be handled statistically, although many studies on genetic determination of temperament in laboratory animals are in progress (Fuller and Thompson, 1960). Another most important characteristic that is as yet difficult to measure in larger animals is disease resistance. The bases of disease resistance are not well understood but when they are, adequate methods of measuring them will, no doubt, be developed.

For the time being it is true that efficiency of food utilisation in meat animals can best be studied in terms of growth rate, but this stage is probably not permanent. In the future it should be possible to distinguish the various causes of inefficiency. To be able to detect differences quickly and easily would end the present necessity for long periods of recording of consumption which limits the number of animals tested. It is to be hoped that sooner rather than later some connection will be found between observable or readily measurable biochemical traits and metabolic processes that have a demonstrable connection with performance. Because of the comparative ease of controlling their environment and studying very large numbers, it is likely that these developments will take place first with chickens and only much later with such large animals as cows. One example is provided by the study of Wilcox, Van Vleck and Shaffner (1962) of the apparent relation between the level of serum alkaline phosphatase and egg production (see Chapter 9).

Carcass quality, including flavour, has so far proved very resistant to description in quantitative terms. Slight variations important only to comparatively few educated palates should not be allowed to divert research effort from the important problems of learning how to recognise animals of superior efficiency as protein producers and how to manage them properly. This includes testing preconceived ideas about correct conformation to see whether they help or hinder these objectives.

The process of adapting animals to their intended purpose has always been steadily changing and is likely to continue so. Development of techniques for pre-digesting or tenderising meat with papain or other enzymes could have a great influence on the quality that is required in carcass meat. Changes in consumer preference for the amount of fat in meat and their consequences are well known. At the present time, the tendency is away from fat, and, since protein and not fat is in short supply, it seems likely that for the foreseeable future the trend will be in the direction of more protein and less fat in meat. From the same argument it could well be inferred that, provided adequate cheap methods for measuring protein in milk are evolved, the process of changing the composition of milk in the direction of higher protein content would receive impetus. The slaughter of young animals for meat, to have them tender and free of fat, could be reversed by advances in meat technology if it proves possible to treat meat to conform to public taste without killing animals at very young ages. This would require alteration in selection procedures, so that animals would be bred to reach their maximum economic protein production at greater ages, in order to reduce the overhead costs of maintaining breeding stock by producing a greater amount of meat from each of their progeny.

Technical developments create new problems but they also make some old ones irrelevant. For instance, drugs for disease control have altered the importance of resistance to worms, footrot and blow fly in sheep. Better understood mineral requirements have made possible the prevention of some metabolic diseases. New ideas in housing and in husbandry, however, are likely to bring new breeding problems in their train, because of the changes they involve in lighting, in the introduction of heavy feeding, of damp atmospheres, and of closer social relationships among animals which were free to avoid each other under the old systems. The ambitions of those who would breed better animals and of those concerned with their husbandry are thus mutually interdependent, and call for advances in methods of controlling both nature and nurture.

By intensifying animal production on land that will permit it, or in buildings, the need for locally adapted races might be reduced. Much will depend on the reasons for having local races. Requirements for adaptation to local markets or local diseases may or may not be changed much, but ability to stand a harsh climate or inadequate food might no longer be an essential. The dense housing of poultry, for instance, has brought in its wake a new set of health problems and the development of appropriate counter-measures. Performance testing of pigs on feeding systems capable of maintaining very high growth rates has also brought problems such as watery meat and foot weakness. With this experience as a guide, it would be wise to prepare to deal with nutritional and pathological issues in housed sheep as well as with those in swine and cattle. Much attention is being given to what have been called the emerging diseases (Food and Agriculture Organisation, 1963) that are spreading to new territories or increasing in frequency as stock rearing becomes more intense. There is, however, a long way to go before the collaboration of veterinary, nutritional, genetic and other specialists accords with the need. Such machinery as exists for collecting appropriate research data on these several aspects of Johne's disease, enterotoxaemia in sheep, and blue tongue is pathetically inadequate, and these are only three of the more important and worsening diseases.

As man's control of animal environments grows more efficacious, differences in husbandry practices will shrink. In tropical countries anxious to enjoy the benefit of livestock of high productive capacity from temperate regions, there is every reason to hope that the control of environment, which now permits modern poultry keeping, can be extended to other classes of stock, especially dairy cows.

Most discussion of livestock improvement centres on the aims and methods of selection processes applied within breeds. As such it is only tactical in scope. On the strategic level, a hypothetical high command would concern itself with selection between breeds, and the relations between dairy cattle, beef cattle, pigs, sheep, chickens and turkeys. What producers spend on their own development is their business, but the amount of taxpayers' money to be spent on advisory and research work aimed at increasing productivity is apparently not governed by any known set of principles. These six sub-industries are competitive in this respect as well as in food markets. Those prepared to help themselves, especially poultry and dairy cattle, have shown advances in efficiency and production to repay efforts put into research and development. No doubt sheep and beef cattle could do likewise. But it is perhaps debatable whether scarce resources of advisory and research personnel should be diverted to improving them if the breeders do not wish it. However, should improvement be in the public interest, breeders cannot be allowed to impede it while demanding protection from competitors.

Although it might be possible to compute a regression function relating improvement to expenditure on development as a useful guide for those who have to decide on the amounts to spend on conflicting claims for assistance, such a function cannot be devised for basic research. This means that many research projects must be supported in order to make sure of supporting the few which turn out really valuable in improving current breeding methods. A further discussion of this problem is contained in Part IV.

C. Improving enterprises

Raising the efficiency of livestock-rearing enterprises has been the purpose of advisory and extension workers for a long time. They are helped now by a more general awareness of the need for increased efficiency if producers are to maintain their financial status. They are also helped by new techniques, such as gross margins, production functions and linear programming.

There would appear to be much more that could be done forthwith, even in the relatively narrow field of animal breeding. On the assumption that there is a case for maintaining traditional small private breeding herds, it is desirable to know more about the cost of producing breeding stock and of recording, testing, and culling it. As yet, neither breeders nor geneticists have developed any close collaboration with economists with a view to finding out what are the greatest economic weaknesses of the various breeds available. Selection indexes need realistic economic values not now available. The economics of breeding, including the cost of improvements and their value, ought to be known, no matter whether carried out by breeders or large organisations as a hobby, for a living, or as a public duty.

The making of new breeds is a sign of virility in agriculture and is to be welcomed. Those who embark upon such a venture, however, should remember the lessons of history and make sure that their stock is not left to speak for itself. Advertising is not a substitute for performance but in our society it is an essential complement. Promotion nowadays usually begins when, or even before, a new breed is started. This sometimes applies also to a proposal to import a foreign variety. By the time there is a sufficient surplus to permit sales, much has been done to create the impression that there is a need for new stock of high performance. Since at least some importations have much merit in their home lands, the claims made by promoters of new breeds are quite likely to be justifiable in some sector of the livestock industry. No objective trials of new breeds are routinely encouraged anywhere, and therefore the testing of these claims is usually a prolonged and chancy empirical process that could easily be replaced by a formal and adequate procedure sponsored by a progressive industry.

There is no genetic difficulty in making a new breed. Such problems as there are arise from promoting it. Sometimes it is an advantage to have a recognisable type as, for instance, in polled versions of horned breeds or crossbreeds, especially if recognition by Government inspectors is required, as it is in Britain. The official commending of colour-marked cattle helped the Hereford and Angus breeds. A new breed with a dominant trait would have a corresponding advantage by advertising its use in crossbreds as well as purebreds. Where, however, it is desirable from the promoter's point of view to limit multiplication of his breed to his own stock, a recessive colouration is preferable since it would disappear in at least some crossbreds. These elementary considerations are no longer of interest in the highly integrated poultry industry but they are still important in cattle and sheep breeding.

New breeds have been justified on at least two grounds. Firstly, there may be economic and environmental niches to which existing breeds or crossbreeds are not well adapted, or could not easily and quickly be adapted. Numerous creations of this kind have been successful, including the Lacombe pig, Corriedale sheep, and Santa Gertrudis cattle. Although extreme examples might be convincing, objective assessment of the case for a new breed as an alternative to improving an existing one is usually impossible.

Secondly, some useful genes may, for practical purposes, be absent from one breed and become available only by starting a new breed on a crossbred foundation, as has been done several times in order to produce polled varieties of cattle. Sheep suitable for year-round breeding might be best designed with one of the recognised breeds of extended breeding season as a component. No one knows, however, how much genetic variation there would be in standard breeds if they were given the chance to show it, nor, therefore, how easily they would be modified by selection. How much more or less effective selection within an existing breed would be than selection in a new breed of crossbred origin is usually a rather academic question. The real reason for the making of new breeds is that it is difficult to attain effective selection within an existing breed. There are several explanations for this. In some countries breed structure may be too ill-defined to serve as a basis for a concerted effort. In others, the restrictive practices of breed associations which are thought to be essential to survival preclude too much enterprise among members. One type is made official and any other is frowned upon. What perhaps should cause anxiety is that breed associations might learn the risks attached to forcing progressives to set up rival associations, and may smother them with kindness within the family. In genetic terms, the rate of improvement may be greater if sought in inter-breed rivalry than in intra-breed orthodoxy.

Although many experts in the past have been willing to announce beforehand whether or not a new breed or sub-breed will be economically desirable, necessary, or potentially useful, there is, in reality, no way of knowing. Since impartial and comprehensive evaluations of whole breeds are not available, none of the existing, let alone new ones, can be classed as necessary or unnecessary. In brief, the value of a breed lies in the minds and ambitions of the owners. No conditions for success or failure can be laid down but several with a bearing on the outcome of an importation or a new breed can be culled from past experience:

1. the number and significance of people interested in the results;
2. the variety of conditions under which trials are made and experience gained;
3. the susceptibility of immigrants to local diseases, undernutrition or malnutrition;
4. the undesirable traits that accompany the desirable;
5. the size and genetic variance of the samples tested.

Animal breeding has been looked upon in the present discussion as a form of evolution, man-controlled evolution certainly, but a process which still requires genetic diversity and selection for consequent change. Animals, whether wild or domestic, which have ceased to evolve have as a rule begun the downward path to extinction. Darwin came to this conclusion a century ago and there seems to be no reason at present to come to any other for domestic animals.

V. Unorthodox Food Sources

Pirie (in Royal Statistical Society, 1963) suggests that some species of plants and animals not now utilised to a great extent in human nutrition could be exploited for that purpose. Such species must be acceptable as foodstuffs and must not compete for energy sources with man or with current links in man's food chain. More extensive use of horse and dog meat could be encouraged in the West, but because of the kind of food these animals require they cannot make much of a contribution to the solution of the problem. There might be some judicious harvesting of wild animals (Ovington, 1963). Granted that venison, grouse, pronghorns and many other species are now being cropped, there is still a great number of relatively unexplored possibilities. For instance in an economic study carried out on an African ranch (Dasmann, 1964), it is claimed that on a given area, presumably capable of producing an annual yield of 94,500 lb of beef were it devoted to this purpose, 118,300 lb of game meat might be obtained by providing a little protection from predators and some careful culling. Other suggestions for harvesting wild animals are shown in Table 2.8.

Game ranching seems to have something in common with sea fishing. But as yet it has neither the capital investment nor the skilled labour. It has also to learn how to live with migrations, epizootics, droughts, fires and predators. As Dasmann (*loc. cit.*) puts it, the argument for game ranching in Africa, as in the United States, is that it is a means towards conserving the fauna. Well established wild populations, like domestic ones, can tolerate and even benefit by periodic reduction of numbers and if, in the process of conservation, some financial return through shooting licences or sales of meat is to be had, so much the better.

In addition to cropping wild animals, domestication of some African species may be possible for augmenting supplies of animal protein, should any of them be found to be effective users of the available feed. Bigalke (1964) notes that 15 of the 22 domesticated species of animals are of the order Artiodactyla. There are some 90 other species belonging to this group which have not been properly utilised and which might be profitably domesticated (some of them are actually being tested in

Texas for raising on marginal land unsuitable for cattle grazing). It is probably not far from the truth to say that the ratio of the number of species of animals used as a source of food to the number of known species is not more than one to three or four thousand, and this ignores the relatively unused potential of sea farming.

TABLE 2.8

Yields of wild and domestic animals (from Pirie in Royal Statistical Society, 1963)

Type of land	Measure of yield	Domestic animals	Wild animals
Range	Kg live weight/animal/day	0·14	0·19–0·24
Savannah	Kg live weight maintained/hectare	20–28	157
Bushland	Kg live weight maintained/hectare	3·7–13·5	52·5
Chernozem	Animals/mile^2	34	134
Depleted land	Kg live weight/hectare	15·5 (further depletion)	31 (recovery)

As Reed (1959) and Zeuner (1963) explain, the origins of domesticated animals are remote and exceedingly vague. The behaviour patterns of man having changed in the meantime, the study of domestic animals today is a most uncertain approach to the circumstances in which domestication was brought about. Some arts, now little cultivated, but based on imprinting or other facets of animal psychology may have been practised widely in the distant past. Domestication is not likely to have been of a once-for-all nature and modern ethology may help to mark the end of the second millennium A.D. as a notable era in this process.

In addition to harvesting what may be called sea-crops in the form of protista, crustaceans, fish and mammals (extensively documented in Ovington, 1963), intensive artificial culture of fish has recently come into prominence in Europe, Asia and Africa (Hickling, 1962). There seem to be exceedingly fruitful potentialities in replacing haphazard fishing methods by stock-raising in ponds of a great variety of species of fish, either directly edible by man or usable as a protein source for conversion into beef or poultry meat. The food chain in the latter case, although relatively long, would not involve competition between man and livestock. Indeed, intensive attempts to establish efficient selection schemes for improvement in yield and quality are already under way (see Wohlfarth *et al.*, 1965).

Although the food reserves in the salt seas seem comforting, it is well to remember the potential difficulties in using them. Apart from technical problems, including management of hatcheries, use of fertilizers, and control of predators and competitors, that are as yet unsolved, there is the prospect of international rivalries of the kind that make the future of whales look bleak.

With land animals, it may be a question as to how appetising zebras, steenbocks, kudus, impalas or other similar species would be to urban populations. Indeed, the human palate, at least in short-range terms, may prove to be a severe limiting factor in utilising other untapped animal resources. People tend to choose foods rather than nutrients and their choice is conditioned by racial and other prejudices, rearing, religion, and magic as well as by availability and attractiveness (Campbell and Cuthbertson in Cuthbertson, 1963). It is not impossible to influence the diets of those who habitually suffer qualitative deficiencies but it will often be a long task. Pirie (1962) has emphasised this point after presenting a long catalogue of such potential contributors to the dining table as the capybara, the kangaroo, the hippopotamus and other aquatic mammals, the guinea fowl, various reptiles and amphibia, fresh water and marine fish not currently used on a sufficient scale, and invertebrates including snails, locusts and larvae of various insects. It should also be noted that these and many sources of vegetable protein such as Pirie's (1958) leaf extract, that are for the time being unacceptable to man, could still find utilisation in livestock feeding.

Welcome as new sources of food from land animals would be, there is no apparent prospect that any of them could relieve much of the world's needs of animal protein. It is necessary therefore to consider what can be done to raise production from existing livestock industries. Much effort is properly being applied to questions of health, nutrition, and food preservation. In this book, however, discussion is limited to matters arising primarily from the genetic aspects of animal production.

Part II: The Theory of Animal Breeding

CHAPTER 3

GENETIC THEORY

The foundations of modern genetic theory were laid by Mendel in 1865. He was successful because he chose for experimental breeding, probably on the basis of *a priori* deduction, contrasting characters that were simple, easily recognised, and, in modern terms, very highly heritable. He suggested that his pairs of alternative characters were essentially controlled by corresponding pairs of "factors" in the parents, and that in the production of a fertilised egg one of each pair of factors was contributed by each parent. Subsequently, his factors became known as genes, and their physical nature and mode of transmission of genetic information from ancestors to descendants established. Particulate theories of heredity were not new, but this one, involving pairs of alternative factors, explained the regularities of inheritance which Mendel observed. These regularities are now enshrined in Mendel's laws. Their existence has provided a mathematical basis for predicting results of matings and for testing the laws in a great variety of animals and plants under all manner of conditions. During the past century no multicellular animal or plant has been found to ignore them entirely. Abundant exceptions, deviations and complex combinations of them have, however, been found.

After investigating what at first appeared to be failures of the laws, geneticists have modified and extended them. A body of genetic theory is therefore now available which, although by no means explaining everything, does remove much of the mystery from animal breeding and provides a far better foundation for action than was formerly available.

Population genetics is a logical development from these basic principles of inheritance and variation. It attempts to describe in algebraic terms the results of transmission of genes from generation to generation and to predict the behaviour of future generations in enormously more intricate situations than those considered by Mendel.

Whereas Mendel worked nearly entirely with a few genes with clear-cut effects, the population geneticist often extends his studies to genes the individual effects of which are often too small to be discerned. With the aid of computers and appropriate algebra, he can work out the consequences of his hypotheses and test them on large populations whether real or simulated. Fortunately, much of the research that produced the theory behind population genetics originated from livestock data and, consequently, there is no dearth of evidence that the theory goes far to account for the variation in populations of animals.

During its history, genetic theory has had to accommodate new facts as they were revealed. This process has not ended. In most characters of livestock that are of high economic importance there is always a large part of the variation which is not attributable to any specific genetic or environmental cause and this part is vaguely called environmental variance. When the heritability of milk yield is stated to be 25%, it is implied that variance of non-genetic origin (or genetically non-additive) is 75% of the total. Some of this, as twin research shows, is attributable to season of calving, inaccuracy of observation and other specific factors, but much is still unaccounted for. It is eminently possible, therefore, that this fraction could be found to contain variability among cows due to the interaction between heredity and environment, that is, due to a varying response to varying challenges from pathogens, mineral deficiencies and other unrecorded items of the environment. It is possible (but not yet demonstrated) that breeders of cattle might occasionally recognise some such interaction and, by acting accordingly, attain better results in a particular herd than would the approximations of genetic theory. Better results will also be due sometimes to chance.

Under most practical circumstances there is no incompatibility between efforts to exploit genetic variation by selective breeding and efforts to remove some of the environmental variation. It is sometimes urged that because the latter is for many important characters relatively large, attempts to eliminate it would offer better prospects of improvement than selective breeding. Certainly, if performance can be economically improved by altering some aspects of management, it would be intelligent to do so, although not necessarily intelligent to ignore heredity at the same time. Unfortunately, the causes of environmental variation are often unknown and therefore not easily removed. Some are known (for example, age or season) but cannot be eliminated. In the best-regulated herds of dairy cows, environment accounts for about 75% of the variation in lactation yield. At present, there is remarkably little that practical breeders can do about it. This state of affairs, no doubt, will gradually yield to research and, as it does, the heritability of milk yield will rise and with it the efficacy of selective breeding.

Twin research with cattle has already shown how this may come about. The extraordinary similarity of identical twins, not only in colour and form but in growth rate and milk production, turns out to be due not only to their genetic identity but also in part to the reduced range of environments that most twin experiments afford (Donald, 1959). Because twins are housed together and fed alike, all the physical and other external variables on a farm which a given pair experiences, provide only a limited sample of possible environments. Not all characters are equally affected by this uniformity of experience. As might be expected, live weights at 18 months of age, which sum up the preceding nutritional experiences of a pair of twins, are more influenced by season of birth, plane of nutrition and juvenile diseases than are the numbers of services for successful conceptions just at that age. Experimental control thus achieves more for the former than for the latter.

I. The Genetic Basis of Animal Breeding

It may be appropriate at this point to examine the major assertions and postulates of animal breeding theory upon which the genetic approach to animal improvement is based. Fundamentally, they are rooted, as has been stated, in Mendelism, as extended (with a few exceptions of straight-forward traits controlled by single genes) to polygenically determined characters. The following account is semi-technical and is not essential for many of the subsequent chapters. Readers who wish more information on the subject are referred to several books available in English including those by Lush (1945), Mather (1949), Li (1955), Lerner (1958) and Falconer (1960). Those whose interest in the technical aspects of genetics is limited, and those intimidated by the specialised vocabulary should omit the remainder of this chapter.

Perhaps the first few of the basic formulations to be presented are by now commonplace, and will be taken for granted by most readers acquainted with genetics. Nevertheless, it may be well to devote some space to them, since subsequent statements, which may not be as generally accepted, are extensions of them.

To start with, it must be said that non-Mendelian inheritance, such as that based on the transmission of cytoplasmic particles, has so far not been shown to have much significance in animal improvement. Although evidence on possible cytoplasmic effects have been reported for domestic animals (e.g. by Allen, 1962, for chickens), it appears to be of a tenuous kind. This is not to say that differences between reciprocal crosses due to maternal effects are of no importance, nor is it to deny the role non-genic (cf. Sonneborn, 1964) or non-chromosomal material (Sager, 1964) may play in genetic processes.

A rejection of the claims for what has been called by the Lysenko school vegetative hybridisation of animals seems to be needed. A voluminous literature has accumulated which attempts to present a case for the possibility of transforming hereditary endowment of animals by such techniques as blood transfusions or by exposure to foreign genic material (DNA). But the results of the best controlled experiments on the subject (Buschinelli, 1961; Burger, Shoffner and Roberts, 1961; Kosin and Kato, 1963; Billett, Hamilton and Newth, 1964) were completely negative. Yoon (1964) attributes the failure of the experiments in which DNA was used for the attempted transformation to the low rate of its incorporation into recipient cells. Since in his tests on mice, the injections were not only of homologous material, but also intragonadal rather than intra-peritoneal, as in some allegedly successful transformations, the interpretation of the latter has to be made even more cautiously. The most that can be said of the positive evidence presented in publications emerging from other than Eastern Europe (those by Benoit, LeRoy, Vendrely, 1958; LeRoy, 1962; and Stroun *et al.*, 1963), is that indications suggesting the utility of incorporating these techniques are not ready for use in improving livestock.

Before proceeding with a more formal exposition of the accepted principles of animal inheritance, a reference to the recent developments in genetics, deriving largely from work with lower organisms, is called for. The units of various kinds of genetic function, such as cistrons, mutons, recons, or the more recently suggested polarons, as well as the concepts of modulators, of operator, regulator and structural genes and controlling elements in general, the relationship between hormones and genes, the phenomena of position effect, paramutation and gene conversion, which are being built as a superstructure on simple Mendelian theory (see e.g. Stent, 1963, Herskowitz, 1965, and other modern textbooks of genetics) are of undoubted great significance in understanding the working of the hereditary apparatus. It is true that analogies between these mechanisms in bacteria and in higher organisms (e.g. control of haemoglobin structure in man) have been proposed, but these aspects of the newer genetics have not yet made an impact on the breeding of domestic animals. There is a need to undertake and expand genetic studies of chromosome aberrations, so actively pursued in man, and of cell activities on a molecular level in economic animals; for the present, however, the postulates underlying genetic improvement of stock do not include the models and ideas developed from such studies.

Present genetic notions of animal breeding rest on a number of propositions which can be formulated as follows:

1. *Inheritance in domestic animals is particulate in nature.* Metric or continuously distributed traits usually depend on a large number of such particulate units or genes located in chromosomes.

2. *The genotypes of domestic animals are diploid, containing two sets of chromosomes each derived from one of the two parents.* Exceptions to this general rule include parthenogenetic development (for example, in turkeys: Olsen, 1960 and 1965) in which both members of the diploid set derive from the same parent. Animals of this sort are still only of experimental interest.

3. *Gametes may be viewed as random samples of chromosomal complements of the parents and they unite at random to produce zygotes.* This proposition is not quite firm because non-randomness in the production of gametes has definitely been established in some animals (Sandler and Novitski, 1957; Lewontin and Dunn, 1960). In useful economic traits no concrete evidence for departure from a random assortment of chromosomes or of gametes is available. This, however, does not mean that it does not exist. But a systematic search for it would be extremely difficult. For instance, inequalities in the proportions of the two sexes at conception may trace to such a phenomenon. Furthermore, the assumption that an adult population is a random sample of diploids from the parental gene pool is very unlikely to be sound. The elimination of individuals between fertilisation and birth or attainment of breeding age is very often genotype-dependent. Hence the proposition as a whole must be viewed only as an idealised condition, and departures from it should be checked wherever possible and whenever they may have important consequences in breeding decisions. A very simple example of this arises when non-pedigree mass mating of selected individuals is being used to propagate the next generation and the assumption is made in evaluating the results of selection that each of the selected parents is equally represented by offspring.

4. *The organisation of genes into chromosomes results in linkage affecting the genetic composition of the population.* This may be of particular importance in relation to selection (Bodmer and Parsons, 1962) and heterosis (Lewontin, 1964). It cannot be said that the full practical consequences of linkage have been worked out and the postulate, though valid, needs further experimental examination before its full significance can be known. Theoretical studies

indicate that linkage disequilibrium can have significant effects on selection results.

5. *All phenotypic differences between individuals are determined by genetic factors resident in the chromosomes and by environmental factors resident outside of the chromosomes themselves, as well as by interactions between the two types of determinants.* Under this definition cytoplasmic and maternal effects would be classified in the environmental portion of variability. The interactions may arise between genotype and location, and between genotype and year or generation. Individuals or inbred lines tested for performance at one station may rank differently at another or at the same location at different times. The switching may be due to variation in climate or other external environment or to variations in the internal environment of the animals themselves (for instance, immunological reactions to subclinical diseases and infections).

6. *Genotypic variance is divisible into additive and non-additive portions.* The first of these consists of variability contributed by additive effects of genes. This is the portion which will contribute directly to selection progress. The second type of variability is produced by interaction between genes. Interaction between alleles, that is, dominance effects, may fall into one or the other category depending on gene frequency. Effects of over-dominance and interactions between genes at different loci which may be grouped under the general heading of epistasis are non-additive. This postulate is of considerable significance in selection theory, since prediction of advances under selection depends on Mendelian algebra addressed to the additive variability.

7. *Environmental variance is divisible into a portion common to the whole population under observation and a portion which may be common to some members but not to all members of the population.* A simple example is the maternal effect which offspring of the same dam share but which does not contribute to the similarity between half-sibs by the same sire. Another example is a byre or herd effect, present when the offspring of a given sire are housed or maintained together and thus become phenotypically more similar to each other than they would be from strictly genetic considerations.

8. *The mean additively genetic value of a population depends on gene frequency.* This is simply to say that if differences between genes at a given locus are translated into phenotypic effects, what may be called the expected average merit of the population is merely the average of the sums of such individual effects of all genes over all individuals.

9. *The non-additive genetic value of the population depends (in addition to gene frequency) on the breeding structure of the population.* The diploid combinations of various genes determine the genetic endowment of the individual but not necessarily the genetic endowment of its offspring. Under inbreeding it is possible that more diploids of one particular genotype are present in a population than would be expected to occur in random mating. Similarly, under out-crossing certain other diploid combinations may be found at a non-random frequency. The distinction between the additive and non-additive genotypic value of a population is of fundamental importance in projecting the effects of different kinds of selection. Indeed, it is this distinction that makes those systems of improvement based on heterosis and those based on ordinary additive gene action so different from each other.

10. *A randomly mated large population will in the absence of disturbing pressures maintain constant gene frequencies.* This is the so-called Hardy-Weinberg equilibrium discovered in 1908 or earlier. The process of improvement calls for increasing the frequencies of genes deemed desirable, and, hence, for applying pressures disturbing the equilibrium.

11. *The significant changes in gene frequencies of domestic populations depend on three types of processes: systematic, random and unique.* Since change in gene frequency is the most important contributing factor to the improvement of populations at least at present, this postulate needs somewhat more extended discussion than the others. A comprehensive analysis of the forces entering these processes is given by Wright (1955). Here, factors of significance in domestic rather than natural populations are emphasised. Thus, although in the latter, recurrent mutation is of considerable import, the relatively short-term aspect of domestic breeding populations makes this systematic process only of retrospective significance. Detrimental or deleterious genes may be present in any given population because of recurrent mutation in the past, but in the course of existence of a flock and herd the newly appearing recurrent ones can play only a small role. In other words, the significance of recurrent mutation in animal improvement relates to the elimination of undesirable mutant genes that have previously arisen and to propagation of advantageous ones. But this, of course, is the process of selection. Hence, one may say that among the systematic processes selection is the most important force operating on gene frequency and hence in the improvement of additively genetic traits.

Recurrent immigration which may involve cross-breeding is another systematic process relevant to the improvement of domestic animals. The introduction of new genes and increases in frequencies of genes already present can be obtained by this method.

Classified under random processes are various fluctuations in gene frequency and, in particular, those due to accidents of sampling. Since changes in gene frequency depend on selection or introduction of diploid individuals (even if it is done in the form of importing sperm or by using artificial insemination) there are sampling variations in both selection and migration. Their magnitude may be practicable from knowledge of the population size but their direction cannot be predicted for any particular case. It is clear that the importance of such random processes increases as population size decreases. In one-sire herds, for instance, the effect of chance would be much greater than in a population utilising a hundred or more sires.

The extreme form of this effect may be, in a sense, a unique event. Accidental fixation of an allele may possibly occur if the breeding population is small. In the formation of the early breeds the initial gene pool was a restricted one, so that the genetic variability was severely circumscribed. This situation is very much akin to the evolutionary *founder principle* of Mayr (1957) who describes it in the following words (for "colony" read "breed"): "The founders of a new colony of a species contain inevitably only a small fraction of the total variation of the parental species. . . . All subsequent evolution will proceed from this original endowment. How important this restriction is, is evident from recent selection experiments in which several lines were exposed to the same selection pressure. Almost invariably the end results were different in the different lines." Of other possible unique events, the origin of a mutation novel to the breed, such as double muscle, or the chance selection of an exceedingly good sire at one end of the frequency distribution of genotypes or, yet again, an epizootic which would suddenly reduce the breeding size of the population, may be mentioned.

One more category of modes of change of gene frequency has been considered by Wright. It includes changes in the system of coefficients determining the magnitude of the other processes, changes which occur because the gene pool of the population has been undergoing a transformation or because the environment is no longer the same. In an artificial selection programme it is unlikely that these modes need to be considered independently of the factors already discussed. However, one point must be stressed here and that is that natural selection continues to produce effects even in domestic environments where populations are subjected to artificial and directed forms of selection. In some instances, natural and artificial selection may work hand in hand, as perhaps, is the case with viability. In others, they may be antagonistic because of genetic homoeostasis (see postulate 16) as in an experiment on selection for long shanks in chickens (Lerner and Dempster, 1951). In general, it must be realised that, while natural selection can occur without artificial selection in the wild, the laboratory or the farm, the reverse is not true: probably all artificial selection (that is, selection imposed on a population by man for whatever purposes) is accompanied by natural selection, which is purposeless and is merely a process involving differential reproductive ability of animals with unlike genotypes (see discussion by Lerner, 1958, 1959).

12. *Heterosis may result when the chromosome complements contributed by each parent derive from populations of different origin.* It does not matter whether heterosis is due to single locus overdominance, to combined dominance at many loci, or to other reasons. From the standpoint of the application of hybrid vigour in animal improvement, the details may differ depending on which one of the explanations for heterosis is correct, but the general principle of producing heterosis by combining gametes of unlike origin is still valid.

13. *Inbreeding degeneration depends on an increased proportion of homozygous individuals in the population, which is the result of severe reduction in effective population size.* Neither the particular basis for heterosis nor *a priori* specification of the characters which will display inbreeding degeneration is included in this formulation. However, speculations on the relations between reproductive traits and characters which are apt to show the greatest effects of inbreeding have been made on an empirical basis among others by Robertson (1955) and Lerner (1958). Further reference to this subject is to be found in the last section of this chapter.

14. *Phenotypic correlations between different traits are the result of combination of genetic and environmental correlations.* Genetic correlations may arise from pleiotropic action of genes, from linkage, or because of common introduction of the genes involved into the population. Environmental correlations are a result of the fact that two or more traits develop in the same animal and are thus exposed to the same environment. This postulate or proposition is

of utmost importance because the theory of index selection, one of the highly significant contributions of population genetics to animal improvement (discussed in Chapter 4), is founded on it.

15. *As a result of selection for one character, correlated responses in others occur.* Empirically, this fact was well known to Darwin but its genetic basis arises from the theory of genetic correlation subsumed in the previous paragraph. Of a special importance in breeding improvement are the correlated responses in reproductive characters which often seem to occur after prolonged selection. Their basis has been discussed, among others, by Mather (1943) and Lerner (1958), and will also be considered in the next chapter.

As a corollary to this and the previous postulate, it follows that secular changes in the values of genetic correlations will occur under selection. In other words, traits which may not have been originally correlated may become correlated (particularly in a negative way) as a result of selection for one or both of them. The practical consequence of this situation is that the parameters which have to be used in constructing a selection index are subject to change whenever genetic correlations appear or change in value (see the section on selection indexes in Chapter 4). In general, whatever deficiencies in practice result from the application of propositions 14 and 15 must be laid at the door of deficiencies in knowledge rather than in their erroneous nature. It has been repeatedly stressed by proponents of population genetics that more precision in determining the various parameters is required. This precision clearly depends not on matters of principle but on obtaining additional information.

16. *Individual genotypes resulting from prolonged selection show a considerable degree of genetic balance.* Further selection in some specific direction, not for total fitness but for traits which may be useful only to man rather than to the animal itself, may lead to deterioration in reproductive performance with a consequent antagonistic action of natural against artificial selection. Suspension of selection under these circumstances may be expected to result in a full or partial return of the selected character to an unimproved level and in restoration of reproductive fitness, provided the inbreeding has not been too intensive. Lerner (1954 and 1958), and Robertson (1956) have discussed the various aspects of this phenomenon of genetic homoeostasis as well as conditions under which it may occur. It should be noted that the empirical facts regarding the regression of characters displaying genetic

homoeostasis are independent of any particular theory or hypothesis (such as heterozygous superiority suggested by Lerner, 1954). Furthermore, it cannot be predicted on *a priori* grounds which characters will exhibit this behaviour unless their relation to reproductive fitness is known. There are examples of highly additive characters which as expected do not show it, but there are also experiments on record in such traits as egg number which, by the time selection becomes ineffective, is very likely based on largely non-additive gene action and regresses relatively little (see the section on selection in poultry).

These sixteen propositions can be put in a still more general form. They can be condensed into four broad contributions of genetics to the techniques of animal improvement:

1. the mathematical model of Mendelian inheritance,
2. prediction of selection gains,
3. heterosis,
4. balance and co-adaptation.

II. Genetics and Improvement

The mathematical model of Mendelian inheritance extended to polygenic characters, and the partitioning of total phenotypic variability into its various genetic and environmental fractions are still the bases of selection in closed or in partially isolated populations. This is true, even though some specific problems have not been completely solved. Optimal or minimal effective population size, for instance, or optimum selection intensity which may depend on whether a short term or long term outlook is considered, need more study.

The theory of prediction of selection gains and hence the possibility of *a priori* comparison of different breeding systems has been worked out. In many ways this still has considerable weaknesses, largely because of the uncertainty about many parameters and the assumptions that have to be made. Indeed, perhaps the biggest problem that technologists of animal breeding have to face is that of finding signals which would indicate that a selection method has reached the point of diminishing returns or at which it is no longer as efficient as some other method. Because of the fact that a good many traits much subject to non-additive gene effects have also high environmental variability, progress in any given population is not easy to check and the approach of a trait to a plateau is not immediately detectable.

There is now an appreciation of the economic possibilities of heterosis and the associated idea of selecting for combining ability, that is to say, for a high performance of crosses between lines, breeds, or strains. One of the most spectacular transformations which has occurred recently in animal breeding has been the spread of this system of improvement in poultry breeding. But it is predictable that no one system will last indefinitely, for the balance sheet of genetic variability dictates that gains are achieved at the cost of reducing variance. This means that whatever system is used (and the more efficient the system may be, the quicker this will happen), sooner or later genetic variability which responds to this particular system will approach exhaustion, and either new variants or alternation between the known systems of selection and mating will have to be resorted to. Indeed, this is a direct consequence of the fourth and final contribution of population genetics and that is the development of the concept of balance and co-adaptation in Mendelian populations. Essentially, this idea refers to the resistance of an adapted population to change and, in some instances, to the regression of gains under suspension of selection. One of the forms in which this phenomenon expresses itself is the reduction in reproductive capacity following intensive selection for some useful traits. For example, the great increase in the body size of turkeys has been accompanied by a serious drop in fertility.

At the moment, the immediate future developments arising from the principles that have been discussed call for research along several lines. As part of the operational research to be carried out by industry, more precise information on various parameters entering decision functions in animal breeding is required. As part of the operational research by academic institutions, removal of various restrictions on the genetic model must be studied. What is called the Monte Carlo type of investigations of the consequences of selection and mating systems (methods which involve computer-based theoretical investigations of the kinds of outcome to be expected from any system in terms of probabilities) may be suggested. In the same area of research, clarification and validation or discard of the various assumptions that have been discussed might also be listed. Finally, in terms of purely academic or basic research, all of these assumptions, as well as any new ones that may be proposed, should be investigated not only in terms of statistics but also in the light of newer knowledge in molecular biology and biochemical genetics. It is not at all improbable that the newer concept of the gene as a stretch of DNA coded for a protein may call for a drastic revision of mathematical genetics based on the so-called beanbag model of fifty years ago.

There is no need to elaborate the points of difference between the foregoing approach to breeding and that which was common before population genetics developed. Such notions as identity of genotype and phenotype, or, at a later stage, the notion that an identical pedigree implies an identical hereditary endowment, the confounding of genetic and environmental variation, the goal of producing single superior individuals rather than improving the average level of a population, the refusal to use as a parent an individual of unsatisfactory performance but with high combining ability with another parent of equally unsatisfactory performance, are still prevalent among large animal breeders, and all are incompatible with the current concepts of population genetics. The question must, however, be asked whether or not the more modern ideas are an improvement on the old ones. A number of experiments have addressed themselves to this problem. Most of these have been carried out on laboratory animals in pilot experiments, largely because of the length of time and the high cost which experiments on economic traits would require. But some indications are available with respect, firstly, to the validity of the general ideas of polygenic inheritance based on the Mendelian scheme and, secondly, regarding more specific agreement or lack of it with prediction on the basis of the propositions that have been listed earlier. The literature on the subject is exceedingly large and it would serve little purpose to attempt a full discussion of it here.

With regard to laboratory animals the reader may be referred to the reviews by Chapman (in Hodgson, 1961) and Kojima and Kelleher (in Hanson and Robinson, 1963), who have considered *Drosophila*, *Tribolium*, mice and other laboratory animals that have been used for experiments on selection and the general principles underlying such studies. In general, what emerges from the results is that, whereas in many cases prediction of exact gains from a theoretical consideration of the methods followed departs considerably from the actual results, the selection methods whether they be individual, family or for combining ability, are appropriate where theory so indicates. But, generally speaking, for short terms, that is to say, several generations, the realised results of given mating and selection practices do not seem to vary greatly from what is theoretically expected. Furthermore, a good many of the possible reasons for discrepancy between expectation and actual results have become evident as a result of such experiments and are being further checked. Opinions dissenting from this generalisation (e.g. Sheldon, 1963) should, nevertheless, be noted.

III. Verification of Selection Results

In larger animals this problem is more complex. The issue whether or not there is adequate evidence that genetic principles work in livestock or poultry has been often raised. It has proved difficult to obtain precise and useful information on this point. Indeed, it is rarely an easy matter to discriminate between the great variety of explanations that may be advanced for changes in performance observed over a period of time.

Perhaps the best that can be done is to consider each individual case after generalisations as to the effectiveness of alternative breeding systems have been reached. Thus, if it can be established that with respect to, say, litter size in mice, a particular system of selection is entirely inadequate, it might be concluded that for pigs the same system is not likely to have been responsible for whatever results occurred in the course of its use. Such an attitude may not satisfy everybody. It does point strongly to the necessity of continued experimentation and of further accumulation of data on this matter in a variety of species. The difficulty is that experiments which might lead to conclusive results are exceedingly elaborate, expensive and require a long time with all but very short lived animals.

A number of considerations relevant to tests on genetic improvement can be set out without even mentioning staff, facilities, time and cost. In part, the discussion of this matter has been adapted from the review by Chapman (in Hodgson, 1961). However, his ideas have been expanded and somewhat modified in what is to follow.

Firstly, any experiment aiming at conclusive results must have adequate population size and a number of replications of whatever procedures are being tried out. In the past this has been a crucial defect in most work dealing with larger animals. Indeed, it may be questioned whether the expense and effort involved in meeting these requirements with such animals as beef cattle would be worthwhile in terms of the knowledge gained. Secondly, the previous history of the base population, or rather its genetic biography, is essential for the eventual interpretation of the data. Such knowledge is obviously very difficult to obtain. In an ideal experiment, a population would be established and kept for a good many generations before selection was attempted on it in an experimental way, but in practice, base populations have to be taken from whatever material is available. Even when purebred populations are chosen, that is to say, populations which have a documented history in herdbooks, it may be sometimes rather difficult to discover just what the genetic peculiarities of the biography were. Thirdly most types of tests need to have a close control of environment. This desideratum is very often shattered when an unbidden infection or epizootic appears in spite of all precautions. For a proper evaluation of field results not only a constant environment but also a variable environment comparable to that which the animals would eventually encounter in the field might have to be provided. There are so many intangibles even in a "constant" environment for larger animals (for instance, uncontrollable variation in quality of food, day by day or year by year) that to meet this requirement fully is probably out of the question. Finally, the keeping of accurate individual pedigrees throughout the experiment is clearly necessary.

Various kinds of information must be collected in order to have a chance of interpreting the results. At least ten may be mentioned irrespective of the character under selection. It must also be kept in mind that in most situations of interest to the animal breeder more than one character is used in selection. This fact introduces tremendous complications that make the recording of the necessary data very cumbersome.

1. *The kind of selection used.* That means determining the exact emphasis placed on the individual, the pedigree and family, be they sibs or progeny, and also on the various aspects of performance.
2. *The mating system.* The results will depend on whether the mating of selected animals is at random, whether like are frequently mated with like, whether inbreeding, out-crossing or back-crossing to a superior individual is practised. There are some complications here with many large animals in quantifying precisely what has been done. In pilot experiments with mice or fruit flies it is often possible to set up a rigid system of mating, such as mating only double first cousins. In such large animals as cattle, however, this becomes very difficult because the correct mates may not be always available when they are needed.
3. *The extent of inbreeding.* This is normally possible to compute, at least in theory, from the pedigrees and often from the nature of the mating system. The actual increase in homozygosity which is the really interesting information is usually not available. In *Drosophila* it is possible to approach the answer if suitable "trick" stocks are used, but there is no immediate hope of utilising such methods in the larger domestic animals (see the section on trick breeding in Chapter 4).
4. *The amount of selection attempted and realised.* It is of vital importance to know how much selection pressure is being applied, that is to say, how superior the animals used for breeding are to the other animals of their own generation. There are different ways in which this superiority may be expressed. It could be shown on a phenotypic scale as the performance of the individual animals selected relative to the average of their generation. With

sex-limited characters this cannot be done in one of the sexes and a scale based on progeny test or family record has to be used instead. Sometimes a selection based on pedigree has to take into account the superiority of the immediate ancestors. In addition to the specification of the amount of selection applied, it is necessary to know whether in fact the attempt has been successful. It may happen that the artificial selection pressure is interfered with by a counter-pressure of natural selection so that the breeder overestimates the actual amount of selection applied. He might choose the top 25% of his population for producing the next generation, but find that the very best animals turn out to be sterile. Under other circumstances he may prove to have been too pessimistic. Wherever possible, it is important to find out the reasons for any discrepancy.

5. *The degree of heritability and the changes in it during selection.* This implies some prior knowledge of the genetics of the trait involved. Indirect ways of estimating or making first approximations for heritability are usually possible without elaborate experimentation. If no such previous information is available, the kind of data discussed here should be able to provide it.

6. *The extent of non-additive genetic variation.* In addition to additive genetic variation, measured by the degree of heritability in its narrow sense, which is also an index of how effective mass selection can be, there is in most characters genetically determined variation which is non-additive and does not respond to ordinary mass selection. Estimation of this type of variability is not a simple matter since it involves a number of theoretical assumptions and requires large numbers of individuals. However, first approximations can undoubtedly be made in some instances.

7. *Performance.* Knowledge not only of the average performance and variability of each generation or cohort is required, but also of the performance of contemporary controls. Sometimes the control takes the shape of selection in an opposite direction but for a number of reasons unselected controls are to be preferred (see section on control lines in Chapter 8). The extent of operation of natural selection on it should be reported.

8. *Degree and kind of correlated responses.* As already noted, selection for almost any character carries in its wake changes in other traits. Linkage or pleiotropy are perhaps the most common causes for such correlated effects. Clearly, the success and efficiency of breeding improvement in one character must be evaluated in terms not only of changes in that trait itself but also in terms of

possible changes in other traits. Since in many tests of selection, correlated responses adversely affecting reproductive characters have been observed, it is particularly important to provide records of the various components of reproduction in both selected and control populations (see section on correlated responses in Chapter 4).

9. *Effects of suspending or reversing selection.* Test populations initiated from the main population at different points in time after selection has been started, in order to determine whether the results of selection are permanent or whether the population tends to regress to its original level if selection is suspended, are necessary to obtain this information. It will be readily seen how much more complicated any experiment is made in terms of design, space and number of test populations by this requirement.

10. *Results from crosses.* Although this is, perhaps, of subsidiary interest, information on what happens when the various selected, suspended or reversed lines are crossed could be very helpful in interpreting the outcome of the rather comprehensive experiment to which all these desiderata refer. Crosses between replicate populations selected in a given direction may provide clues to the selection limits to be expected.

It seems exceedingly unlikely that any experiment with larger animals that has been undertaken to date or that would be practicable in the immediate future could possibly meet all of these requirements. Since experiments so meticulously and thoroughly carried out are not feasible, the interpretation of those that have been made is beset by some difficulties and uncertainties. It becomes virtually impossible to make an absolutely conclusive claim regarding the efficiency of one or another selective procedure applied to useful properties of livestock and poultry. The best that can be done is to make educated guesses and compare roughly the various systems that have been tried; for instance, mass selection for egg number in chickens can be with reasonable assurance shown to be less effective than selection on basis of family testing. Similarly, selection for increased butterfat production within dairy cattle breeds can be asserted to have more effect when it is carried out on the basis of yield and not of coat colour, but the finer shades of discrimination among various selection methods are unfortunately exceedingly difficult to make.

There are still further complications in interpreting results of even the idealised experiment which would meet the criteria so far discussed. The situation with respect to linkage of genes involved in determining the character under selection needs to be known. Is there linkage disequilibrium? How many effective segregating factors are there? Non-Mendelian effects may be important in determining a given character. It is true that maternal influence can be often estimated and allowed for in interpreting results, but this does not necessarily apply to non-pathogenic infections that may be present, nor to incompatibilities between parent and offspring unless information on this point is available from other sources. Further difficulties may arise out of non-additive genetic variation, the analysis of which itself poses problems. In particular, the scale of measurement which is used to describe the trait under selection may be misleading. All in all, it should be clear that unequivocal demonstrations of the superiority of one or another method of genetic improvement of large animals are unlikely to be obtained, especially if the systems compared differ only a little in their expected results. A system which appears to be the most efficient at one stage may not be the most efficient at a later stage. Consider selection for annual egg record which is based only on early performance. A population in which such selection has not been previously practised will, in the light of several experiments, respond to this method. But a population in which additive genetic variance of the early egg production has been exhausted as a result of selection and, perhaps, inbreeding, will not respond. Indeed, there may also be negative genetic correlations generated between early and later performance so that improvement in the first trait will lead to a decrease in the second, with the result that the total annual egg record will show no change at all.

IV. Selection in Poultry

Among the useful domestic animals there is no doubt that poultry have been the subject of the greatest number of experimental investigations relating to selection. Probably no economic aspect of the chicken's life cycle has been neglected. But, unfortunately, especially in the early experiments, adequate statistical analysis of the procedures used and the results obtained has been lacking. Some selection experiments lasting dozens of generations are on record, yet it is not possible to say from the data published (or for that matter from the data available) whether or not the results completely conform with expectations on the basis of the postulates of population genetics. This is true even of many simply devised experiments in which metric characters of high heritability were utilised.

Perhaps the two experimental flocks that have been subjected to most thorough statistical analysis have been the White Leghorns selected for egg production at the University of California and at

Purdue University. Full details regarding the first of these and a variety of analyses carried out have been reviewed by Lerner (1958) in a schematic form. Should more detailed data be desired, the original papers cited by him should be consulted, as well as the article by Abplanalp (1962). The Purdue data were analysed by Yamada, Bohren, and Crittenden (1958) and in some other aspects by Bohren and McKean (1964). In the briefest possible form the results of both experiments may be summarised in the following fashion.

Selection for increased egg production on the basis of family records appears to follow predictions based on selection differentials used in the early generations of a selection programme. After that, either a decrease or complete cessation of gains is observed in spite of the fact that selection differentials still remain high. The explanation for the change appears to lie in the exhaustion of the kind of genetic variability which is drawn upon by the particular methods of selection used. Relaxation or suspension of selection for egg number does not seem to result in a regression to the unimproved level (although, as is noted later, some drop occurs), but it does seem to be accompanied by correlated responses (even though some selection for the characters thus responding is also involved) in such other traits of reproduction as number of female offspring produced per dam. It is also possible that selection based on part-performance eventually results in a decrease in the remaining part as has been observed in a long-term experiment by Morris (1963). Comparisons of breeding systems, differing for instance, in the use of pullets instead of hens, or in the application of such techniques as multiple shifts of cockerels, indicate that results are in general accordance with those predicted from the general theory. This conclusion is founded partly on the data from the flocks mentioned here and partly results obtained in other populations (see, for instance, Dempster and Lerner, 1957).

Another experiment with poultry also described by Lerner (1958) dealt with mass selection for longer shanks. The original heritability of this trait appears to have been considerably higher than that of egg production and selection was extremely effective until a plateau was reached. Here, however, the plateau seemed to be due less to the exhaustion of genetic variability than to the opposition of natural selection to artificial selection. Apparently the balance of the individual genotypes was destroyed by moving the shank lengths too far from the optimal level for this particular population. As a result, suspension of selection did indeed lead to an immediate regression in the selected trait while reproductive performance improved rapidly in spite of the fact that inbreeding was continued at the same rate as in the selected line.

A more refined test of the adequacy of selection theory in improved flocks has been described by Dickerson (1962). His analysis suggests that when selection has been directed towards multiple objectives for a long time, gains that are expected on the basis of population genetics theory fail to be realised. In every generation of selection offspring from selected parents performed better than unselected controls, but no accumulation of egg production gains was observed.

Dickerson attributes the failure of improvement to continue to negative correlations between components of performance and to genotype-environment interaction which produce at least an operational overdominance. In his opinion modification of the genetic theory for populations that have already been improved is called for.

Basically, the results described in all of these reports are in agreement with the previous conclusions derivable from *Drosophila* experiments (Clayton, Morris and Robertson, 1957; Clayton and Robertson, 1957), that while short-term selection gains are predictable from the simplified theory, long-term ones are not. Furthermore, replicate populations started from a small number of individuals may produce substantially different results as has been noted in the discussion of the founder principle. Still, it should be realised that even short-term prognoses (e.g. twelve generations in the *Drosophila* experiments of Sen and Robertson, 1964) may refer to a period ranging from less than a decade to forty to fifty years depending on the species of domestic animal in question.

To sum up, then, it may be said that the postulates described are, generally speaking, valid for unimproved populations of chickens, and, after the populations reach a certain improved level, they still may hold if the parameters involved have not been changed in the course of selection. The big problem in breeding practice is to discover indications of changes in the rate of selection progress early enough to modify the breeding system.

What has been said so far applies to selection in closed populations. The evidence with respect to other breeding systems in poultry is much less complete. Although a large number of breeders have put into effect selection for combining ability, which, if operational criteria are a guide, must have been successful, the published experimental evidence is rather scanty. It is thus not yet possible to check the exact assumptions which are made in the theory of breeding for combining ability.

The last point that need be given brief mention in this section is the effect of relaxation or suspension of selection. The theoretical aspects of the problem were considered by Robertson (1956) who introduced the concept of homoeostatic strength, a parameter measuring the degree to which the mean of a selected population returns towards the original equilibrium. Experimental evidence in chickens and a review of related literature is given by Nordskog and Giesbrecht (1964). Apparently, for rate of egg production, a decline of the order of 0·75–2·25%, exclusive of inbreeding effects, may be expected per generation after suspension of selection. This corresponds to a loss of three to nine eggs per bird per year. With respect to egg weight, which is normally maintained by artificial selection at a level higher than that of the point of maximum reproductive fitness (Lerner, 1951), the decline may be even more severe. Indeed, one of the important problems in poultry breeding practice is how to avoid having to sacrifice some selection pressure for egg number in order to maintain egg size. The moral is that selection for many traits, even after a plateau in performance has been reached, must be continued lest regression occurs. The thought from *Alice Through the Looking Glass* of running fast in order to stay in the same place involuntarily comes to mind.

V. Evidence from Larger Animals

For obvious reasons, the extensive material available to laboratory geneticists, to plant breeders and even to workers with poultry, has not been available to geneticists interested in the larger animals. The number of controlled selection experiments on livestock conducted on a sufficiently large scale to permit accurate comparison of the results obtained and the results expected is very small. As a rule, other types of data must then be relied upon to demonstrate that there are no special forms of inheritance peculiar to livestock and that the main tenets of genetic theory derived from other organisms are equally applicable to cattle, pigs and sheep.

Most selection experiments with these animals suffer from a variety of complications (lack of controls, concurrent close inbreeding, small numbers, environmental changes and others). It is difficult to find convincing examples of selection results for dairy cattle, in spite of the long-continued programmes of research on the effect of selection of superior sires, of progeny testing, and of artificial insemination.

For beef cattle, Brinks, Clark and Kieffer (1965) have recently reported on a study of selection in a closed herd of Herefords in which thirty-three sires were used over a period of twenty-five years to sire over 2000 weaned offspring. Three of their conclusions having a bearing on the present discussion may be quoted:

1. "Estimates of genetic changes, calculated by subtracting the environmental trend from the phenotypic trend, indicate that substantial genetic progress was obtained for birth weight, gain from birth to weaning, weaning weight, and weaning score. Estimates of total genetic response from 1934 through 1959 were 9·7 lb in birth weight, 21 lb in gain from birth to weaning, 30 lb in weaning weight and 6·5% in weaning score."

2. "Comparisons of expected and estimated genetic response from 1943 through 1959 indicate that slightly more response than expected was obtained in birth weight (106%) and weaning score (116%), and considerably more than expected was obtained in gain from birth to weaning (139%) and weaning weight (130%). Adjustment for detrimental effects of inbreeding were not included in these values. If adjustments for the effects of inbreeding had been made, estimates of considerably more progress than expected would have been obtained in these traits."

3. "Comparisons of expected genetic response and actual phenotypic response for postweaning traits from 1943 through 1959 indicate that, in general, the actual response was as great as or greater than expected. Although environment is a factor in these trends, there apparently has been some genetic response, and the data do not suggest that this response is less than expected."

For swine, Dettmers, Rempel and Comstock (1965) have reported a successful eleven-year programme of developing in Minnesota a strain of miniature pigs for medical research. Body weight at 140 days was reduced by at least 29%, the latest average (for 99 pigs farrowed in 1961) being 38·6 lb. Here again, there were neither controls nor selection in opposite directions. Another experiment carried out in Illinois by Craig, Norton and Terrill (1956) did include two-way selection, for large (ten generations) and for small (eight generations) size. The two lines showed increasing divergence which rose to 50 lb at 180 days of age in the last comparison made. The results were consistent with a heritability of 0·16–0·17 for the trait, falling (contrary to the Minnesota experiment) somewhat short of the usual heritability estimates for body weight.

For sheep, a successful outcome of two-way selection (plus a control line) for multiple births in New Zealand Romneys was reported by Wallace (1958). Although there are statistical complications in interpreting his results, Table 3.1 shows that the lambing percentages of the high line, the control and the low line showed increasing divergences. Earlier heritability values for this character were so low as to suggest that selection in it would be futile. They seem to have been underestimates for this population, as well as for Australian Merinos in which

Turner *et al.* (1962) were also able to increase the percentage of multiple births.

Finally, interim results of selection in Scottish Blackface sheep (A. F. Purser, personal communication) may be given to illustrate a test of genetic theory on the basis of a more precisely designed experiment than is usually possible with larger animals. Four selection and two control lines of about 250 ewes each were started from the same base stock. Two were devoted to changing cannon bone length (as measured at 6 weeks of age) in opposite directions. In two other lines, plus and minus selection pressures for the amount of medullation

TABLE 3.1

Lambing % difference in a two-way selection experiment (from Wallace, 1958)

	Lambs born/ewes lambing			
	High line — control		High line — low line	
Age	First 5 years of experiment	Last 5 years of experiment	First 5 years of experiment	Last 5 years of experiment
2	6	9	7	10
3	7	11	8	12
4	8	21	13	23
5	14	18	18	29
6	31	42	39	42

in samples of mid-side wool at 6 weeks of age were applied. Fortunately, both characters showed, as previous work had forecast, sufficiently high heritability to permit effective selection on each individual's own performance. Heritability for medullation was 0·55 and for cannon-bone length 0·59. With this information and a knowledge of the measurements of the animals chosen for breeding, it is possible to estimate the progress that should be made if no attention is paid to any other character (as is the case here). The comparison of observed and expected gains is shown in Fig. 3.1, which shows the divergence between the pairs of complementary lines. There is no reason to suppose that the results obtained deviate markedly from those expected. There may be a hint that realised progress is tending to fall somewhat behind expectation, but it is too soon to be sure whether this is, indeed, so. If it is, there will be no shortage of possible reasons.

Other types of evidence of the correspondence between genetic theory and the actual working of inheritance in livestock come from pedigree analysis, from twin data, and from the results of inbreeding.

The first of these deals with Mendelian traits. These are superficially simple characters such as blood groups, which can be clearly followed from one generation to another. Their presence is easily recognised and they are demonstrably transmitted according to the rules which Mendel discovered. Although various qualifications are usually attached, the rules account for most of the observed variation and permit sound predictions to be made. Because these are now generally acceptable,

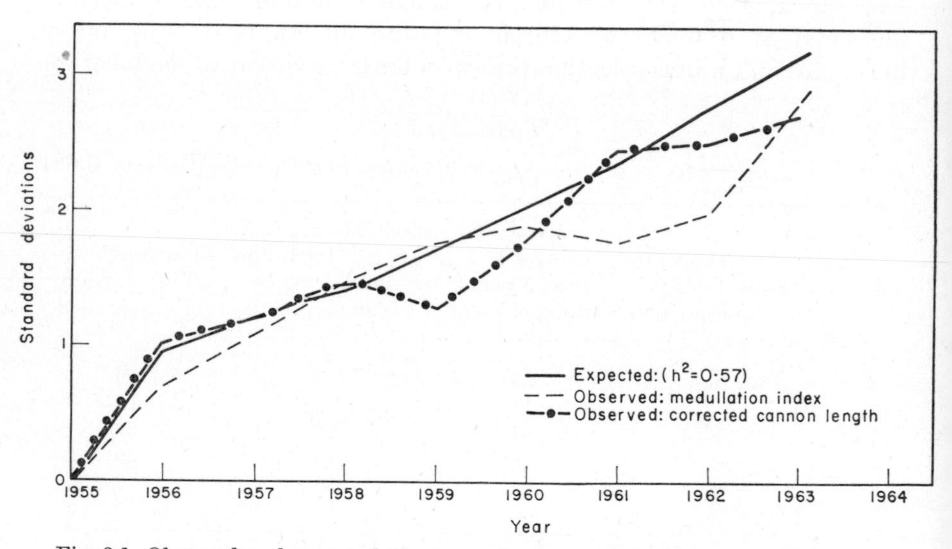

Fig. 3.1. Observed and expected divergence in sheep selected in opposite directions for cannon length and amount of medullation (from A. F. Purser, unpublished).

the use of blood groups in clearing up ambiguities of parentage has become possible. Of recent years there has been a notable growth in the number of biochemical variants known to behave in Mendelian fashion so that the prospect of being able to identify linkage groups is drawing closer. Apart from the numerous blood groups found in cattle, pigs and sheep, there are recognisable varieties of protein in blood serum and milk (discussed in Chapter 9), the familiar colour patterns, and many other simply inherited characters (Hutt, 1964). Besides such textbook examples, there are numerous lethals, sublethals and deleterious genes, most of which, unlike those responsible for dwarfism in beef cattle, are not being fully investigated. Many of them would no doubt be found to behave in a simple Mendelian fashion.

The use of twins in genetic research is based on the fact that the closer the genetic relation between individuals, the greater is the phenotypic resemblance between them. There are two kinds of twins in cattle:

identical (monozygotic) and fraternal (dizygotic). As in man, the members of an identical pair are exactly alike in all traits, such as blood groups, that are wholly determined by heredity. By contrast, fraternal twins, with some exceptions caused by common foetal circulation, are no more alike in such traits than full brothers or sisters. Comparisons of resemblance within pairs of the two kinds of twins can be extended by investigating pairs of less related individuals, such as half-sibs or second cousins, or completely unrelated animals. If all kinds of pairs are born at the same time and studied on the same farm,

TABLE 3.2

Intra-pair correlations in Ayrshire cows milked for at least 100 days
(H. P. Donald, unpublished)

Kind of pair	Monozygotic twins		Dizygotic twins		Half-sisters		Unrelated pairs	
Lactation	1	2	1	2	1	2	1	2
Days in milk	0·57	0·40	0·48	0·46	0·33	0·63	0·20	−0·38
305-day milk yield	0·85	0·65	0·59	0·35	0·34	0·53	0·25	−0·13
70-day milk yield	0·73	0·49	0·48	0·29	0·11	0·49	0·15	0·13
Butterfat %	0·96	0·91	0·57	0·69	0·16	0·13	0·18	0·07
S.N.F. %	0·90	0·81	0·66	0·51	0·37	0·56	0·14	0·22
Casein %	0·91	0·84	0·63	0·48	0·21	0·35	0·11	0·34
Non-casein protein %	0·87	0·57	0·52	0·55	0·57	0·24	0·07	0·24
Lactose %	0·79	0·51	0·82	0·50	0·26	0·60	0·28	0·25
Body weight at 18 months	0·90	0·90	0·53	0·54	0·55	0·52	0·33	0·32
Number of pairs	26	21	28	22	28	21	36	26
Value of correlation at 5% level of significance	0·38	0·43	0·37	0·42	0·37	0·43	0·33	0·38
Average correlation	0·83	0·68	0·59	0·49	0·32	0·45	0·19	0·12

they can all be expected to show that an animal resembles its own specific mate to some degree simply because of the common environment that has been imposed. Beyond this, however, it would be expected on theoretical grounds that the more closely related the two members of a pair, the more alike they will be. This in fact is just what has been found, as Table 3.2 shows.

In it are entered the intra-pair correlations (as a measure of resemblance of pair mates) for four categories of pairs ranging from identical through fraternal and half-sister to unrelated pairs. All were reared in one herd on the same farm from the age of about seven days. With such limited numbers of pairs the correlations have large sampling errors, but as the average values at the bottom of the table show, the diminishing similarity in performance as the relationship weakens is quite obvious.

Since there is some resemblance in unrelated pairs, they may be providing a measure of the extent to which the two environments of a pair are correlated because they have been reared and milked together and thus shared, at least in part, the same ups and downs of life.

How steeply the gradient of similarity rises towards the identical pairs depends on how heritable the characters studied are, on the importance of prenatal environment (which is the same for twins but not for non-twins), and on whether or not environmental correlations are the same for all kinds of pairs. About this last point not much is known, but it is an absorbing question. Each animal is an important part of the environment of the other member of the pair and yet its nature is largely determined by heredity. Therefore, to some extent, the environment of an animal is determined by its own genotype. The same sort of complexity could arise with infections or induced behaviour patterns. For this and other reasons, it is not to be supposed that "on the same farm" means that all animals have the same environments. Those born at same season, and of similar mothers, and penned together will have much more highly correlated environments than those which share no more than the soil, climate, vegetation and general management. For more details on cattle twins, Donald (1959) and Johansson (1959) may be consulted.

Resemblance between cattle twins is only a special case of resemblance between litter-mates. The mass of information about resemblance between relatives and most heritability estimates come, however, from the comparison of groups of half-sisters by the same sire. When heritabilities are very low, half-sibs resemble each other hardly more than unrelated animals, and when they are high, half-sibs show a much closer similarity than unrelated animals (Lush, 1945; Lerner, 1958; Falconer, 1960).

The remaining fruitful source of information on the mechanism of inheritance in farm animals comes from inbreeding investigations. Although the unfortunate consequences that follow the mating of close relatives were discovered a long time ago, the practice of inbreeding during the formation of some breeds, especially the Shorthorn, gave rise to the notion that in the hands of a constructive breeder, inbreeding had at least a therapeutic value due to cleansing the stock of bad characters while at best it concentrated good characters. This view persists widely to this day. With the development of the theory of inbreeding by Wright (1921), however, a quantitative approach to inbreeding data became possible.

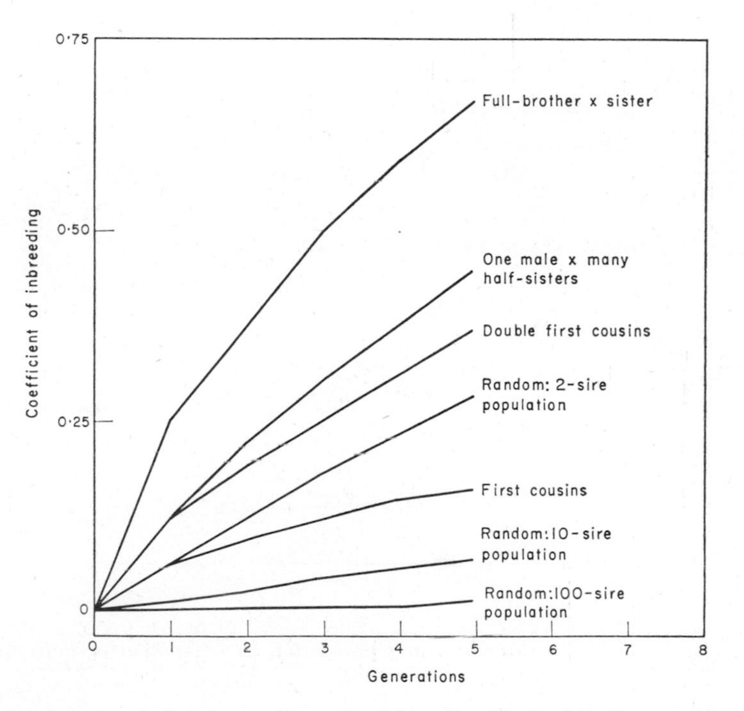

Fig. 3.2. Increase in homozygosity under inbreeding (derived by Lerner, 1958, from Wright's classical study, 1921, and reproduced with permission of the publisher).

Figure 3.2 summarises the expected genetic results of different degrees of inbreeding. They are measured in terms of a coefficient, F, lying between 0 and 1. Technically, F expresses the probability that the two alleles at a locus which enter a zygote are derived from an ancestral allele common to both; or, alternatively, it can be thought of as the proportionate decline in the average number of heterozygous loci carried by a randomly chosen individual within a given population. To use an extreme but common example, a one-sire herd that is self-contained must in due course use a son of that sire on female descendants of that sire. Each mating then traces back to one instead of two or more sires. Such a mating is shown in Fig. 3.2 as one male × many half-sisters. This is not the strongest type of inbreeding but it brings about a loss of almost half the available genetic variation if kept up for six generations. When breeders speak of inbreeding or line breeding they rarely mean a mating scheme as intense or as long-continued as this. Because heterozygotes are more vigorous, natural selection may operate in their favour so that the computed degree of homozygosity is probably often an overestimate of the attained level.

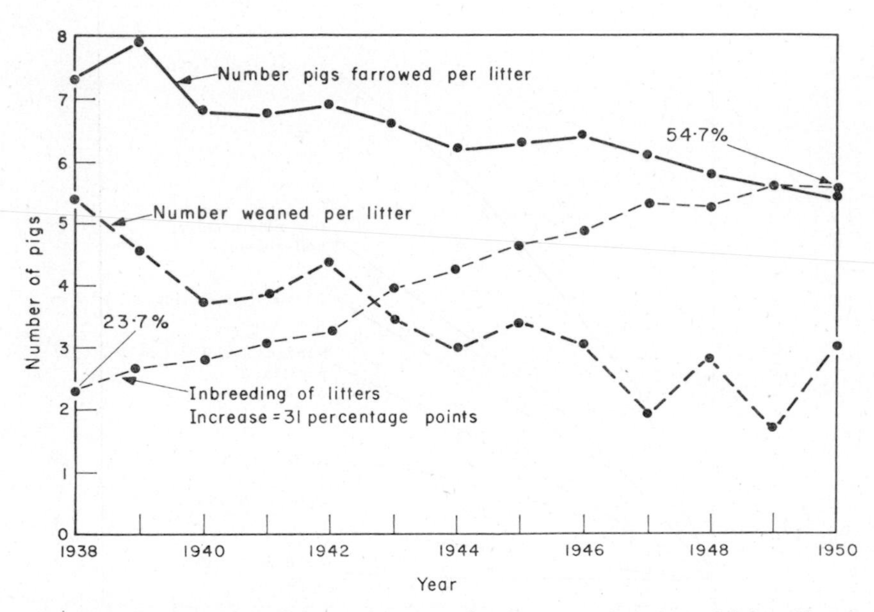

Fig. 3.3. Decline of performance in reproduction as a function of inbreeding for Poland-China one-sire herds (from Craft, 1953).

What Fig. 3.2 does not show is that increasing the proportion of homozygous loci is accompanied by a loss not only of genetic variability (which may be desired) but also of vigour, viability, fertility and milk production (which is usually not desired). Many experiments (for example, Brinks, Clark and Kieffer, 1965, for beef cattle) have confirmed that what is called inbreeding degeneration in livestock is roughly proportional to the amount of inbreeding. It appears not to matter much which of the lines in Fig. 3.2 is followed. What matters is the size of the coefficient. Any breeder therefore who inbreeds, can be sure (save for accidents of sampling) of some deterioration. In well-investigated subjects, such as litter size in pigs, he can be advised of the likely loss he will incur by his inbreeding. An example of decline under inbreeding is shown in Fig. 3.3. Other instances and a technical discussion of inbreeding degeneration are to be found in Lerner (1954).

A breeding programme based on a very few highly selected males clearly leads very quickly to inbreeding. It then becomes a question of compromising somewhere between the advantages of rigorous selection and the disadvantages of inbreeding. Looked at in a slightly different way, inbreeding and crossbreeding are deviations on either side of outbreeding which are likely to develop as selection tightens.

There is no serious alternative to genetic theory as a basis for practical animal breeding. Livestock have been the stimulus and the material for the development of the large part of it known as population genetics. In the preceding paragraphs, an attempt has been made to give a rough idea of the sort of evidence available but it is not to be supposed that the various categories discussed have any basic genetic independence of each other, or that they hint at more than a small fraction of the known facts, many of which are drawn from animals and plants of limited agricultural importance. The risk of labouring this point is taken because there is sometimes a temptation for practical breeders and officials to suppose that what they do not know about breeding livestock is not knowledge, or that because the particular breeds they are interested in have not been subject to trial and experiment, nothing is known about their behaviour.

In general, it is easy enough to fault current theory and it is often done. But the theory will improve with use. Although no amount of algebra or genetics will remove the influence of chance, forecast economic changes, or insure against political acts, there is no going back to the mystique of pedigree breeding which had these same deficiencies.

Since there is no immediate prospect of reducing generation intervals substantially, fears of disastrous errors due to applying breeding theory to practical situations have little foundation. Rates of change in the genetic constitution of animal populations are so slow that it is difficult to establish whether or not a change is taking place in any direction. The only reason for trying to apply genetic principles is to gain a better measure of control and thereby a better response to selective breeding. None of the principles, however, will secure progress if they are nullified by lack of consistent objectives or by over-complex specifications. Knowledge of genetic correlations is admittedly fragmentary and it is often open to anyone to assume favourable or unfavourable consequences of a selection policy as he wishes. Whether or not to risk an unfavourable outcome is largely a matter of temperament. But most progress is made by the risk-takers.

On the evidence, there can be no reasonable doubt that the main assertions of population genetics are correct in essentials and are a sound basis for industrial development. Although there is much to be done by geneticists in strengthening and developing the structure, its grand design is unlikely to change much.

Where there is controversy between breeders and geneticists, it is mostly about aims, less about methods and not at all about theory. Aims are for breeders, large or small, to fix but if they insist on including in their aims not only characters of low economic value, but also methods such as eye judgment and small breeding groups of low efficiency, they must expect to have to defend themselves against criticism from other interested groups. They must expect also to have to meet competition from livestock bred with different aims and methods.

CHAPTER 4

IMPROVEMENT METHODS

A full discussion of breeding methods now in use falls outside the scope of this book. There are many texts which cover this subject. Descriptions written by different authorities and dealing with beef cattle, dairy cattle, swine, sheep and goats, horses, and poultry may be found in the volume edited by Cole (1962). The chapter on livestock breeding in Byerly (1964) also gives a summary of the generally accepted practices in the United States.

There is a great deal of variation both among the methods employed for the different species in a given country and among countries for the same species. In part, this variation is traceable to the dissimilarities in the biological properties of each kind of animal, and, in part, to the differences in the level of technology achieved. Not surprisingly, the most complex and thorough procedures are found in poultry, and the least advanced methods are used (with some exceptions in countries where they are of great economic significance) in sheep. In between, lie swine, dairy and beef cattle, the improvement of which depends on techniques, sometimes as organised as the elaborate Danish performance testing of pigs (Lush, 1936) or progeny testing of bulls used in artificial insemination (Johansson, 1961), and sometimes harking back a century or more to the show-ring. The upper section of Table 5.2, which gives a broad survey of the situation in the different species of livestock, may be consulted for a comparison of the usual breeding methods applied to them.

This chapter deals mainly with certain features of the techniques and problems encountered in animal improvement which are significant in the context of this book. In essence, breeding procedures must rest on four points:

1. there should be material on which the breeder can exercise his powers of selection, i.e. genetic variability is a prerequisite to genetic improvement;
2. the methods available to the breeder (primarily selection and control of the mating plan) must be based on adequate information, including performance testing and evaluation;
3. decisions as to whether the breeder is concerned with short or long-range results must be taken;
4. the breeding scheme must be chosen on the basis of minimum costs per unit gain.

The following sections deal with some of the issues arising from these principles.

I. Performance Testing

Probably the most important single events in the history of performance testing were the invention of the trapnest for recording egg production, the introduction of the Babcock test for assaying the percentage of butterfat in milk, and the development of probes and ultrasonic devices for measuring the amount of backfat in pigs. All of these techniques are applicable not only in testing stations but are also suitable for on-the-farm tests. The merits of these two forms of recording depend on the purpose of testing and on the animal concerned. In poultry, each breeder usually depends on his own recording for selection, sometimes entirely on his own farm and sometimes (especially for testing hybrid or cross-bred combinations) on private egg-production farms in different locations. For public comparisons with other breeders (which basically implies advertising purposes), central egg-laying tests are used. In dairy cattle, with the increasing use of progeny testing of bulls for artificial insemination, on-the-farm recording is widely employed. Indeed, the American Dairy Herd Improvement Association considers farm testing as reliable as central testing (which, of course, can be carried out in a standardised environment). Furthermore, the costs of centralised testing of larger animals in sufficient numbers (examined for European conditions by Johansson, 1960) seem to be prohibitive.

The Danish progeny testing scheme for swine, however, is based on central testing stations, as are feed-lot tests of cattle in several parts of the United States. Tests on sheep have been conducted both on the individual farms (e.g. in Wisconsin, Ohio and England), and in central stations (e.g. in Texas and Utah).

The superiority of one or the other method hinges mainly on three issues: cost, heritability and genotype-environment interaction. The relevance of the first of these items is self-evident. As for the second, traits with high heritability, such as growth rate on a given diet, fleece weight and egg size are usually not much affected by location, and, therefore, can be accurately evaluated on the breeders' own premises. Characters of lower heritability, such as egg production, may be more informatively tested if there is a variety of conditions. This can apply only to inter-line or inter-strain selection and not to selection within populations except for progeny tests carried out on the scale made possible by artificial insemination.

With respect to genotype-environment interaction, that is to say, a differential response of a genotype in different environments, objections have been raised to relying only on testing stations on the grounds that the genetic value of locally adapted types is underestimated when all recording is done under one system of management and one set of environmental conditions. Evidence that some breeds or strains are superior under some conditions and inferior under others is not difficult to find, especially, if the conditions and the genotypes are, respectively, widely disparate. Statistical evidence of interactions from comparisons with narrower limits can also be found (e.g. Dickerson, 1962b), but the amount of variation attributable to these interactions is relatively small. Furthermore, to identify the sources of weakness and strength of particular strains is proving very difficult and quite impossible to predict in advance. It seems likely that genotype-environment interactions will be found to be numerous, sometimes exaggerating, sometimes counteracting each other, and influencing mainly those characters which are most subject to inbreeding depression. These are traits such as fertility and mortality and other components of fitness which have very low heritabilities and show very large amounts of apparently non-heritable variation. It might be found, in due course, that most of the genotype-environment interactions occur at the individual rather than at the strain or breed level. It also could happen that genotype-environment interactions are self-cancelling in the sense that sometimes they are favourable and sometimes unfavourable during the course of the history of a single individual. Young animals may be reacting to infections of one sort or another with a partially or unequally developed apparatus of immunological tolerance or protection. Later on, they may be choosing their own environments in such a way as to run more risks than others of a similar kind and with similar opportunities. Accident proneness could be a significant characteristic of animals.

An allied problem in testing arises from interactions between individuals within a population. Every animal in a flock, herd or testing station is part of the environment of other individuals (see the general reviews of this issue as it occurs in natural populations in Andrewartha and Birch, 1954, and Wynne-Edwards, 1962, and for domestic animals, Hafez and Lindsay, 1965). This is self-evident, but nowhere more so than in the behaviour of identical twins, both human and bovine. Moreover, some of the environment of one such twin which the other provides is to a degree genetically determined. Consequently, a part of the environment of a twin is conditioned by its own genotype. The same will apply to any pair of animals but with lessening force as relationship becomes more remote.

Other well known examples come from the behaviour pattern of many animals from man downwards known as the peck order. Fitness in a particular environment must include the correct and successful responses to social and actual stimuli offered by other members of a group. As a result, performance is connected with the position of an animal in the intra-group social order (see McBride, 1964, for chickens, and McBride, James, and Wyeth, 1965, for pigs). This problem also exists on the strain level. The question may be asked whether different lines of chickens under test should be raised separately, thus introducing differences between their pen environments, or together, thus injecting social rank-performance genotype interactions, which would not be present under the actual conditions of production.

In general, whenever changes in the environment of domestic animals occur, it could happen that temperament and behaviour would be highly important factors in finding adapted strains or breeds. Individually fed pigs have a different order of merit from group fed pigs; indoor sheep will differ from outdoor sheep; intensively kept poultry from those maintained under extensive conditions. Genotype-environment interactions of this sort may be commoner than is realised. Domestication itself provides the outstanding evidence of the way they can be turned to good use.

The main issue that faces the breeder is to find a compromise between accuracy, on the one hand, and costs of testing, on the other. It is evident that a single lactation record will not evaluate a cow's genotype for milk yield exactly, but at the same time, a full 305-day record for each lactation is not necessary. It is equally clear that trapnesting a chicken seven days a week is a wasteful procedure when doing it only three days a week provides sufficient accuracy in selection. The particular compromises adapted for the different traits of the various species are, of course, a technical matter. They are considered in the general references to breeding methods already cited (see also Chapter 5).

II. Progeny and Sib Tests

The most common method of choosing the parents of the next generation from a population is mass or individual selection. Aids for evaluating the genotype, and, hence, the breeding worth of individuals are provided by progeny and sib tests. When a character under selection has very high heritability, information on the performance of relatives, in addition to that on the individual considered as a potential parent, will augment the accuracy of the estimate of its genotype only a little. Important exceptions are sex-limited traits, such as milk or egg production, and characters (e.g. carcass quality or performance as a steer in a feed lot) which cannot be measured on a breeding animal. In contrast, when heritability is low, the phenotype of an individual gives only a very limited clue to its genotypic merit, and data on its relatives (usually progeny or sibs) must be relied upon in selection.

The question of whether progeny and sib tests are to be employed and, if so, to what degree, is, therefore, an operational one. It was first thoroughly explored by Dickerson and Hazel (1944), who viewed the problem in the light of gains to be expected from different selection methods, including those relying on individual records, pedigrees, sib tests and progeny testing. In addition to heritability, the interval between generations (i.e. the average age of the parents when the offspring are produced) and the reproductive rate need be known for a valid comparison of the different selection methods. The use of such comparisons for poultry has been described by Dempster and Lerner (1947). But neither they nor Dickerson and Hazel have considered another feature of operational significance, namely, the financial aspects of carrying out the different forms of testing. Yet another factor is the decrease in the potential limits of improvements which is caused by the use of family records (see the section on costs of change in this chapter).

In their original study, Dickerson and Hazel concluded that regular progeny testing is unlikely to increase, and may actually reduce, rate of genetic gain, "unless (1) the progeny-test information becomes available early in the test animal's lifetime, (2) the reproductive rate is low, and (3) the basis for making early selections is relatively inaccurate." However, they also pointed out that extension of artificial insemination to large populations would increase the advantages of progeny testing for sheep and cattle. This is what happened in the selection of dairy cattle sires, wherever organisation of artificial breeding has become extensive, as in Britain where at the time that the Dickerson and Hazel article was published, the number of artificially inseminated cows stood at about 4000 a year, whereas twenty years later, the comparable figure was over 2 million.

Selection under these circumstances must then rely more and more on evaluation of genetic merit of sires on the basis of progeny tests. Some questions of crucial operational significance immediately present themselves. Thus, Henderson (1964) addressed himself to the method of selecting bulls for testing and concluded that only a partial solution to the problem is possible. The obvious place to find promising young sires is among the sons of sires already proven to be of outstanding merit, out of dams who are themselves by outstanding sires and of high performance. Yet, even this method of selection offers no certainty that the young sires will be good, since the practical upper limit to the correlation between the estimates of genetic merit on basis of ancestry and the actual genotypic worth of young sires is only about 0·67. This means that, at most, $(0{\cdot}67)^2$, or 45% of the genetic variation among untested sons can be accounted for by the fullest information on the parents.

How to determine the number of daughters needed from sires under test was examined in a comprehensive theoretical fashion by Skjervold (1963) and Skjervold and Langholz (1964, 1964a), who obtained general expressions for optimal breeding structure of populations reproduced by artificial insemination, taking into account testing capacity, selection intensity, degree of heritability, maternal effects, and inbreeding depression. More specifically, Van Vleck (1964), who made a study of bull-testing schemes for the artificial insemination centre at Cornell, found that the maximum genetic improvement will be obtained when a large number of young sires are judged by twenty to fifty daughters each. It also turned out that the schemes producing the highest genetic gains were the most profitable ones.

It will be readily appreciated that in a changing situation specific answers to questions of this type can have no finality. A running re-evaluation, not only of the parameters to be used in solving the mathematical equations, but also of the formulae themselves is desirable. It seems indisputable that applied research of this type will be needed to ensure that the money allocated for improvement is spent to the best advantage. When dealing with progeny testing on the scale now developing in cattle, playing by ear will no longer do. This is a lesson that future decision makers for other kinds of breeding enterprises should remember.

III. Selection Indexes

When several criteria are employed in the estimation of genetic merit, they may be combined into a single figure known as the selection index.

The use of indexes of various kinds is not a new practice. Simultaneous selection for egg number and viability in chickens has been based on family average egg production per original bird housed, essentially, an index. However, the genetic theory which made it possible to assign proper weights to each criterion so as to maximise genetic gains was worked out considerably later (Smith, 1936; Hazel, 1943). Various aspects of index selection, including methods of construction, are discussed by Hazel and Lush (1942), Lerner (1950), Dickerson (1955), and Kempthorne (1957). More recently, the relative efficiencies of index selection and of other methods were examined by Young (1961) and by Finney (1962).

TABLE 4.1

Values of the parameters needed in computing a selection index for bacon pigs
(adapted from compilation of Smith and Ross, 1965)

Traits	1	2	3	4	5	6	7	8	9
. Daily gain	*·42*	·73	—·17	·07	—·07	·13	—·13	·10	—·03
. Feed efficiency	·76	*·48*	·05	·04	—·19	·17	·16	·17	·16
. Dressing-out %	—·19	—·01	*·32*	—·19	·19	—·06	·29	·22	·15
. Carcass length	·14	·08	—·40	*·62*	—·22	·37	—·14	—·19	—·05
. Backfat thickness	—·15	—·21	·28	—·30	*·54*	—·34	·13	·12	—·13
. Carcass conformation	·28	·30	0	·46	—·52	*·28*	—·17	·36	·24
. Belly thickness	—·03	—·10	·25	—·17	·22	—·13	*·38*	·07	—·07
. Ham score	·14	·24	·34	—·23	—·26	·37	·19	*·36*	·27
. Eye muscle area	—·11	·34	·36	—·08	—·28	·35	—·16	·44	*·42*

Ieritabilities are shown on the diagonal; phenotypic correlations appear above diagonal; genetic orrelations below. Note that feed efficiency here refers to weight gained per pound of food conumed.

The biological information necessary to construct an index includes the heritability of the relevant traits (Spector, 1956, has a compilation of estimates for domestic animals), and the genetic and phenotypic correlations among them (Tables 4.1 and 4.2 give examples for pigs and cattle, respectively). In addition, the economic value of each character has to be known. Usually, it is impracticable, except in very large establishments, for an individual breeder to obtain the appropriate heritability and correlation figures. For many unimproved populations, however, they appear to fall within similar ranges of magnitude and, therefore, published values may be used. In highly improved populations, these figures may no longer be applicable: successful selection may have reduced heritabilities and greatly changed correlations between traits (see the section on correlated responses). Hence, a breeder using a selection index needs to verify periodically whether the figures used in its construction are still correct.

TABLE 4.2

Estimates of correlations between milk production and other characters of dairy cows (based on compilations by Farthing and Legates, 1957, and Johansson, 1961a)

Character		Correlation Phenotypic	Genetic
Body size		0·31	0·26
Growth rate		+	?
Muscle development		—	—
Services per conception		0	?
Incidence of oviduct cysts		+	?
% butterfat:	Jersey	−0·36	−0·50
	Guernsey	−0·32	−0·57
	Guernsey	−0·34	−0·77
	Holstein	−0·22	−0·38
	Holstein	−0·25	−0·21
	Holstein	−0·10	−0·52
	Ayrshire	−0·14	−0·20

The problem of weighting for economic worth is even more difficult. For some characters economic values must be mere guesses, while for others they may fluctuate widely from year to year. The results of selection using a given index may not be available for years and by then, the economic values may be quite different from those used in its construction. Thus, before the first World War, an index used in the selection of Yorkshire pigs would have carried a positive economic weight for thickness of backfat; after the war, the sign of the economic value for this character would have become negative.

An even more complex problem is how to decide on the economic value whenever the objectives include both traits of importance to the purchaser of stock and to the breeder. As an example, the grower of broiler chicks is interested in growth rate and carcass quality. He has no concern with fertility and hatchability. Yet the breeder's financial success depends as much on the reproductive capacity of his stock as on its average genotype for broiler qualities. In one case, the criterion is based on the value the grower will obtain if he purchases stock from the breeder; in the other, the value depends on the money the breeder can make by selling his stock to the growers. For poultry at least, the argument that good fertility is important to growers because it reduces their costs is not very strong, since the initial outlay per chick forms a rather small part of the total cost of raising broilers.

Despite the various complications, indexes have been constructed for all classes of livestock, and, within each, for a variety of purposes. In poultry, selection indexes have been extensively used in commercial practice (Dickerson, 1962). With many traits included in an index (and some characters might appear in an index several times in the shape of individual records and different kinds of family averages) and populations containing thousands of individuals, for each one of which an index value must be calculated, index selection is an operation for a computer, and a costly one. Hence, it is important that the gains in increased efficiency expected from index selection be compared with the costs of using this method before it is adopted.

There are four kinds of purposes to which selection indexes can be put:

1. In selection for a single trait, an index incorporating information on the individual and on its various relatives, ancestors, collaterals or descendants, increases the accuracy of estimation of the animal's genetic merit, especially for traits of low heritability. Because improvement in only one character is sought, the task of estimating correlations between traits and of assigning relative economic merit is avoided. However, it is unrealistic to suppose that selection of this type can have much practical significance. Reproductive capacity, health, efficiency of feed utilisation and adaptation to environment, all must enter in one way or another into selection for any economically important objective. In fact, every economic property is no doubt a compound of many traits. Basing selection on some combination of these does not make it a "single character".

2. Selection may be directed primarily to one trait, but the index may incorporate information on other traits as an aid in identifying genetic merit. If the trait selected for has low heritability but has genetic correlations of appreciable degree with other characters, the latter may be appropriately included in the index, with zero economic values assigned to them.

3. A similar situation prevails when selection directed towards one trait produces undesirable shifts in other traits genetically correlated with it. Indexes in which zero economic weights are

assigned to them may be designed so as to leave their means relatively unchanged, while the character selected for is improved.

4. The most important use for selection indexes is in breeding populations where multiple objectives are pursued. Whether they are expressed in terms of individual characters such as growth rate, conformation or carcass quality, or of a compound such as monetary value of a litter, the index will have several items in it and the same character may appear in more than one of them. Occasionally, superficial paradoxes in construction of indexes of this kind are encountered. For example, in Hazel's (1943) original study of index selection in pigs, he found that carcass score contributed to the total monetary value. But the phenotypic correlation of this character with the economically more important 180-day body weight was higher than the genetic correlation. In these circumstances, because of certain biometrical properties, an index in which market score is given a negative weighting is found to be more efficient than one in which it is ignored.

When selection for several traits is practised, three methods of procedure are available: (a) each trait can be selected for separately and simultaneously (selection by independent culling levels), (b) one trait at a time could be selected in succession (tandem selection), and (c) selection on all traits could be carried out simultaneously by using a selection index. It has been shown by Hazel and Lush (1942), and by Young (1961) that the third method is theoretically never inferior to selection by independent culling levels, which in turn is as at least as efficient as tandem selection. There is no reason to believe that the theory is faulty, though only limited experimental support for it (e.g. Abplanalp, Asmundson, and Lerner, 1960) is available. It must, however, be realised that there are some problems in selection that no index can solve. Thus, if a high negative genetic correlation exists between two traits, because the same genes enter into their determination, it is impossible to improve one without the other suffering deterioration. Selection for short shanks and high body weight of chickens would probably be self-defeating as a result of such a situation (Lerner, Asmundson, and Cruden, 1947). In a similar case, Merritt and Slen (1963) appear to have broken up genetic correlations between body weights at different ages (suggesting that some of the correlation was due to linkage) by selection. All in all, though there are difficulties in striving for multiple goals, the invention of the selection index has done much to minimise them. It is true that in the past breeders have made changes in their flocks and herds without the help of this device. But it is also true that improvement was neither fast nor certain. If there is to be a speeding-up of the rate of progress, more use of statistical aids such as the selection index will be needed.

IV. Heterosis and Crossbreeding

Perhaps the greatest change in animal breeding procedures of recent decades has been the increase in systematic crossbreeding and crossing between lines and strains as an alternative to purebreeding. There are many variants of these procedures. Some may involve selection for combining ability such as the reciprocal recurrent selection illustrated

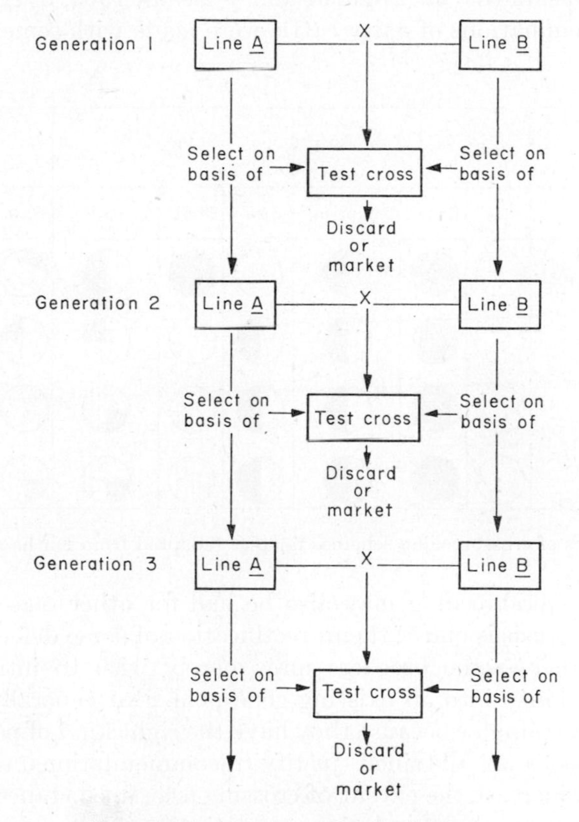

Fig. 4.1. Reciprocal recurrent selection.

in Fig. 4.1. This method may be utilised within a breed as well as between breeds. Other techniques are based on systematic crossing of two or more breeds. Several schemes used in pigs are shown in Fig. 4.2. Still other variants are described by Lerner (1958).

Both the extent and the purposes of crossbreeding vary between species and countries, but livestock production has become heavily dependent on it (see section on stratification in Chapter 6). In the United States, nearly all broilers are of crossbred origin; 80% of the swine come from breed or line crosses; crossbreeding similarly dominates sheep production. The main exception lies in range-bred beef cattle, though some crossing is beginning to take place also in this class. Dairy cattle form a special category: crossbreeding here is founded on the use of beef bulls in order to improve the meat quality of offspring not intended for milking. From the figures published by the Milk Marketing Board, it appears that in England and Wales in 1963, over 40% of all artificial inseminations of dairy cattle were made with semen from beef bulls.

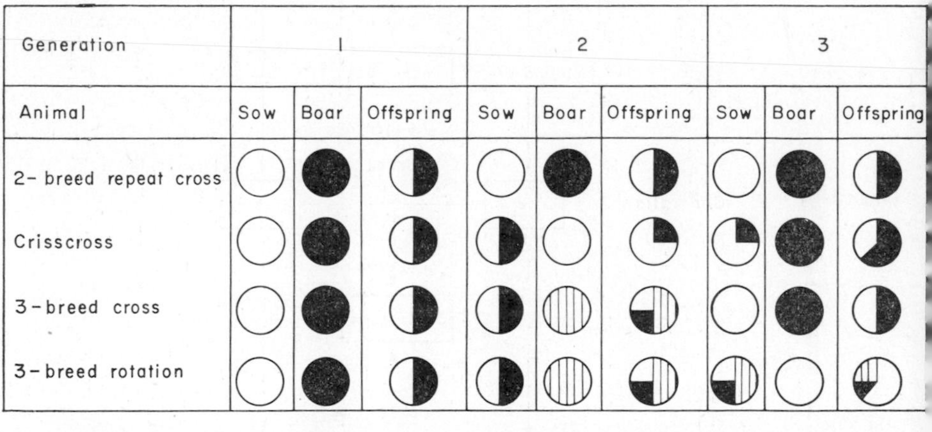

Generation	1			2			3		
Animal	Sow	Boar	Offspring	Sow	Boar	Offspring	Sow	Boar	Offspring
2- breed repeat cross									
Crisscross									
3-breed cross									
3-breed rotation									

Fig. 4.2. Examples of crossbreeding schemes for pigs (adapted from Durham in Cole, 1962).

Resort to crossbreeding may also be had for other reasons. Hybrid vigour or heterosis is one of them; rectification of some defect of a purebred is another. Some breeders may merely wish to introduce new genetic variability into an existing gene pool. But generally speaking, crossbreds are popular because they have the right kind of performance. The advantages so obtained justify recommendations to livestock producers to increase the extent of crossing (see, for instance, Skårman, 1963 and 1965, on sheep and pigs, respectively).

Refinement of crossbreeding techniques has been carried farthest for chickens. Selection may be practised within lines or breeds for general combining ability with any other strain or breed or for combining ability with another specific parent (Merritt and Gowe, 1960, discuss this issue with regard to broilers). Mass creation of inbred lines for eventual crossing (often within a breed) is a widespread tool used by breeders of laying stock.

In larger animals, crossbreeding sometimes replaces selection rather than supplements it. As might be expected, pigs, which have a relatively rapid rate of reproduction and exploit a narrow range of nutritional conditions, lend themselves to poultry methods more easily than cattle. Much effort has been put into developing inbred lines and into performance testing of crossbreds, but so far without any startling success. The principle of emphasising selection for the maternal characteristics of fertility and milk production in one line or breed, and the growth and carcass characteristics had been applied in practice for a long time. This does not mean that it is universally applicable. Smith (1964a) examined the theory of selecting specialised sire (for growth rate and carcass quality) and dam (for reproductive ability) lines in pig breeding, but found that selecting for overall performance in a single line produces almost as good results.

The amount of heterosis obtained varies. In some instances, such as beef×dairy crosses, this is but a minor consideration. For milk yield, estimates of 5 to 10% increase from crossing have been made. For mortality and fertility, the increase may be as much as 15%. But the economic value of crossbreds should not be measured only on one character at a time, for many small advantages combine to make a much larger one. How much superior crossbred performance must be to reach worthwhile magnitude is a question with no simple answer, since it is possible to specify husbandry practices and prices for surplus calves, cull cows, milk and milk quality, which when combined would show any breed or cross to economic advantage. Were it possible to rank the several breeds and crosses by efficiency of food use relative to maintenance and caloric or protein output, a sounder basis for choice of a breed or cross might be made.

V. Lethals and Undesirable Recessives

One of the consequences of the spread of artificial insemination has been an increased fear of proliferating lethal genes. "Undesirable recessive" is now a popular term for pathologically abnormal as well as normal but unwanted varieties of new-born animals, such as red calves in black breeds. Many of them, however, have not been sufficiently investigated to be confidently attributed to recessive genes.

Textbooks on animal breeding often contain detailed lists of lethal or detrimental genes. Although undoubtedly useful for some purposes, these catalogues are of limited practical value. It had been assumed by geneticists, before the mathematical consequences of selection were thoroughly understood, that the presence of lethals in a population presents a growing danger and requires vigorous efforts on the part of breeders to combat it. This, however, is not so. Recessive lethals undoubtedly occur in every species of domestic animal, and deleterious genes of all sorts are abundant. When they are carried in large populations, their manifestation is usually sufficiently rare or they are so difficult to detect as to warrant no action at all on the part of breeders. In small populations, or where inbreeding is practised for the production of lines to be used in crossing, such lethals would become evident very rapidly.

The fear that artificial insemination will spread undesirable genes is not logical. At worst, the bulls used at artificial insemination centres will not transmit more of these genes than the numerous sires they supplant. At best, they may spread many fewer, because bulls known to be potential or actual transmitters are not used. Where a sire has been widely used and later found to be heterozygous for a rare undesirable gene, care can be taken to reduce the risk of using another heterozygote of the same kind in the same area.

It is not to be expected that private breeders will confess publicly that their stock are disseminating a lethal gene, but artificial insemination gives the incidental opportunity to assess sires for such genes, and in large populations even those of low frequency (Johansson, 1961). Commercial dairymen are not slow to complain about deformed calves, and it seems unlikely that artificial insemination will be permitted to spread common deleterious genes very widely. A greater danger might actually be that good bulls are too readily culled on such grounds.

Mason (1964), in a common-sense discussion of breeding policy in relation to recessive lethals, points out that the idea of eliminating them is unrealistic. Whether measures should be taken to reduce the frequency of lethals depends on several considerations: the nature of the defect and the risk to the mother's health, how common it is, and whether any heterozygote advantage is involved. In his survey of artificial insemination practice in nine countries, Mason learned that recessive lethals were a serious problem in none. Sporadic breeding tests are made on suspected carrier bulls but, as a routine procedure, progeny testing specifically for lethals is not operationally sound. It is expensive to organise and can only be used for checking the most feared of all possible lethals. The usual attitude of artificial insemination operators is that any common lethals will advertise themselves among the first few hundred of calves that are bred from each sire in normal use.

An exception to this policy may have to be made if heterozygotes are found to have some advantage over homozygotes, such as might have been the case with a form of dwarfism in beef cattle (Marlowe, 1964). It would then be sensible to measure the extent of the advantage, to compute the equilibrium frequencies of the undesirable allele, and then decide on operational grounds whether or not steps to discriminate against the recessive are necessary, and, if so, what they should be.

With the evolution of more advanced breeding methods, including the formation of small elite stocks of performance-tested animals, it will eventually be insufficient to apply the present crude custom of cutting the throat of any animal judged guilty of being heterozygous for an unpopular gene. In the artificial insemination of cattle it could happen, and, indeed, is certain to happen, that a suitable progeny test will show that a bull capable of raising production transmits some undesirable trait as well. For a balanced judgment, economic values will have to be placed on his ability to raise the average production of all daughters, and on other characters such as milk composition and conformation as well as on the undesirable trait. To this must be added a clear understanding of its mode of inheritance. Where a specific gene has been implicated, its frequency in the population will be an important factor in estimating the damage a heterozygous bull might do. All animals are known to have some undesirable recessives to pass on, in the sense that they have alleles, which when homozygous reduce fertility, milk production, and vitality (the so-called mutational genetic load). To regard them as a well-defined class of genes and to let their discovery induce an automatic negative mental reaction, is to overlook both biological and financial realities.

VI. Trick Breeding

Generally speaking, genetic methods of improvement of populations have a universality in that they are applicable to all species. Yet there are certain properties of lower organisms and of plants which can be utilised in selective breeding but which are not practical to use in larger animals. Most mutagenic techniques fall into this category. Antibiotic-producing fungi can and have been subjected to irradiation, and mutants for increased production have been selected to establish many commercially valuable strains. Induced mutations have also been successfully incorporated in plant breeding programmes. An early very thorough review of the subject may be found in Gustafsson (1947), while later results have been described by Gregory (1961) and by Gaul (1965). Allied to these methods are techniques of producing polyploidy which are also finding their way into agriculture.

In large animals the situation is not as promising. A limited attempt at inducing useful variation in domestic animals was not particularly encouraging (see Chapter 9). Perhaps the scale of a successful experiment of this sort might have to be prohibitively large. Again, methods for inducing polyploidy in larger animals are apparently not in the immediate offing. Nevertheless, it is encouraging that there is one economically important animal, the silkworm, where some of the genetic tricks, so successfully used for experimental purposes by *Drosophila* geneticists are being adopted for commercial purposes.

Investigations on the genetics of the silkworm have been pursued with great intensity by the Japanese (reviewed by Tazima, 1965). Not only have some 260 genes been studied but of the 28 linkage groups possible, 19 have already been established. Contrary to what has been said about the larger mammals, triploids and tetraploids may be readily produced in this species, and even the problem of the control of sex appears to have been solved (see references in Chapter 9). But the most impressive achievement in the field of trick genetics lies in the development of methods of sex determination of zygotes before the eggs hatch.

In silkworms, males give a higher yield of silk per unit weight of mulberry leaves than females. It is, therefore, of considerable economic advantage to be able to determine the sex before hatching and then eliminate the females. By irradiation, Tazima produced a translocation of a small piece of chromosome 10 to the Y chromosome (in silkworms, females are heterogametic, and hence have an XY constitution, while males are XX). This region carries a normal allele of gene *w-2*, a recessive that causes the pigment of the outer membrane of the egg, which normally darkens after several days, to stay white. The translocation permitted the creation of a strain in which males carry the *w-2* gene on both chromosomes, while females are heterozygous for it. Consequently, in it, dark eggs give rise to females, while light eggs produce only males. An automatic sexing machine has been developed which rejects the female eggs, and thus ensures that only male cocoons are formed.

The same principle has been applied to the problem of recognising the sex of larvae, also of important commercial consequence: since silkworm breeding is based on exploitation of heterosis and, because moths mate very soon after eclosion, it is essential to have an early separation of the males and females.

Techniques of a related kind have been used in poultry. Some breeds, sexually dimorphic at hatch, were developed in the mid-twenties, but autosexing was rendered obsolete by the discovery of a method of determining sex by examining the cloaca of newly-hatched chicks. Genetic techniques of the type depend on the exploration of the effects of single genes, or chromosome mapping, and on making use of chromosomal abnormalities. For obvious reasons, such studies on large animals are prohibitively time-consuming and expensive. Whether or not this will always be the case, remains to be seen.

VII. Correlated Responses

The idea of correlated responses has always loomed large in the minds of both evolutionists and breeders. This is the phenomenon which Darwin (1872) described under the term "correlated variability" by which he meant "that the whole organization is so tied together during its growth and development, that when slight variations in any one part occur, and are accumulated through natural selection, other parts become modified."

Schmalhausen (1949), among modern students of evolution, has stressed the evolutionary importance of correlation and correlated response and has discussed in detail various mechanisms which could account for it. As noted by Simpson (1953), palaeontologists long ago observed phenotypic associations of characters. But their principal sources, at least so far as artificial selection is concerned, became known only after such genetic phenomena as pleiotropy and linkage were elucidated, even though by no means all correlated responses arise exclusively from them.

One form of correlated response, important in long range evolutionary situations, occurs when an integrated system of organs and functions is disrupted by a novel or increased selection pressure applied to some single trait. Correlated components of fitness then become subject to secondary natural selection pressures working to restore the integrity of the function or organisation of the animal. These components are then brought back into harmony with the shifted property or dimension which was the primary focus of natural selection. As a result of this process, clusters of correlations develop among various characters (Berg, 1961).

The significance of both genetic and phenotypic correlations between traits (for a technical, but reasonably simply stated, model of the relationship between the different kinds of correlation, see Searle, 1961) lies in the fact that usually the expressed desiderata of artificial selection are minimal, that is to say, they involve only a few traits rather than a whole complex including fitness itself. This means that successful selection directed to these desiderata must inevitably produce in its wake undesirable changes in at least those characters which were originally at their optimum. The long-standing problem of quality of milk has its origin here. Because of negative genetic correlation between milk yield and percentage fat content in cattle (Table 4.2), intensive and successful selection for one of these characters would lead to a decrease in the other. Since the correlation is about $-0 \cdot 5$, it may be theoretically possible to avoid the consequences of the correlation by concentrating selection on those genes which are concerned only with one of the two characters. Since the available genetic variation is then much less, progress from selection would be slowed down.

The complications that genetic correlations can bring about in artificial selection programmes have been discussed extensively for chickens by Dickerson (1955). Because of the intricate network of positive and negative correlations between the variety of traits entering the production of a commercially successful laying chicken, selection directed toward all of these traits simultaneously may be self-defeating. This can happen when, for instance, the selection pressure applied towards trait A is counteracted by the movement of this very trait in the opposite direction as a result of a correlated response to a selection for trait B. Difficulties of this type are magnified by the fact that it is possible for positively correlated characters to become negatively correlated after a period of selection. This process can occur in a variety of ways, including exhaustion of linkage-free additive variation, as has been noted by Lerner (1958).

In the same publication, he has distinguished two types of correlated response: facultative and obligate. The first depends on a purely fortuitous genetic correlation due to pleiotropy or close linkage (the two are virtually impossible to distinguish in higher organisms) between the selected and unselected characters in the initial population. The direction in which the correlated response would occur is here unpredictable. Indeed, in the example of this type of correlated response cited by Lerner (Prevosti, 1955, working with *Drosophila*) correlated responses to the same kind of selection occurred in opposite directions in different populations.

Obligate responses take place in a specifiable direction. They depend on the fact that the traits affected by selection are physiologically or morphologcally in an optimum combination at the initiation of selection. Under such circumstance, it may be expected that reproductive fitness (and hence breeding efficiency in the population) will be lowered as a result of selection, not necessarily immediately, but sooner or later. There are exceptions to this rule, at least over a short range of time: reproductive fitness does not always suffer as a result of selection for some specific trait. One example is provided by the study of Dempster, Lerner and Lowry (1953), who in selecting for egg number in chickens, did not lower the reproductive fitness of their flock. More recently Rahnefeld *et al.* (1963) selected mice for seventeen generations for post-weaning growth but still found this trait to be genetically positively correlated with litter size.

An interesting case of correlated response was described by Belajev and Trut (1963) in farm-bred silver foxes. In selecting for calm tempera-ment over a period of ten years, they observed a correlated response in the form of prolongation of the oestrus period, which is a highly desirable reproductive trait from the breeder's standpoint. It is not improbable that phenomena of this kind played an important role in the domestication of some of the economically useful animals.

Correlated responses which reduce fitness to the point of near extinc-tion of the population, or, even only to the point of inefficient repro-duction, are of serious consequence to a breeder. They are often compounded by the effects of inbreeding, as might have happened in the classical case of the Duchess strain of Shorthorns (Wright, 1923). They can also be independent of consanguinity, as appears to be the case in a selection experiment on shank length of chickens, in which suspension of selection permitted considerable restoration of fitness despite con-tinued inbreeding (Lerner, 1958).

Enough has been said to show that correlated responses have deep practical and theoretical interest. Their bearing on selection limits and performance ceilings is considered in the next section.

VIII. The Costs and Limits of Change

All breeding improvement is founded on the changing of gene fre-quencies. The aim of the breeder is to increase the proportion of desirable alleles in the gene pool under his control at the expense of the undesirable ones. In theory, the limits of selection on the hypothesis of additive gene action are reached when the desirable alleles are fixed and occur in all members of a population in a diploid state. Beyond this point, all selection can do (unless the objectives are modified) is to continue eliminating undesirable alleles arising from recurrent mutation.

In practice, such a situation rarely, if ever, happens. Some desirable alleles that are originally present at a low frequency may be completely lost by chance from a gene pool. Furthermore, often an allele is desirable or not depending on other genes in a given genotype. Still other genes may display a heterozygous advantage and hence will have equilibrium frequencies somewhere between zero and one and therefore be un-likely to become fixed. However, even with these qualifications, live-stock improvement must be viewed in genetic terms as a process intended to lead to the production of populations with an optimum array of gene frequencies at each locus.

It should be apparent that this goal can be attained only at a price, both biologically and financially. There are two types of biological costs involved. One is based on the fact that, in the replacement of an allele by a more desirable one, many individuals will have to be culled or prematurely removed in choosing animals for reproducing the population. This matter is of considerable importance in natural selection, where it has other implications as well. It was first raised by Haldane and recently reviewed by Van Valen (1963). The second type of biological cost consists of the loss of adaptation through drift in small populations and through undesirable correlated responses. Both types of biological costs result in financial costs. But these have, in addition, at least two other components. As an extension of the biological costs of the first kind, there are charges against the improvement programme in the form of loss of value of stock which has become obsolete or inferior in comparison with the improved model. Secondly, in changing gene frequencies, there are the direct expenses of observation, performance testing and data processing.

The immediate question arises then as to who shoulders the burden of money spent on livestock improvement. Fanciers may accept a return in the form of relaxation and pleasure; pedigree breeders in the form of sales of breeding stock or the rewards of distinction at shows; large scale poultry enterprises in the form of dividends. The only accounting of costs in such establishments is to the shareholders. But when stock improvement becomes a question of investment of public funds, a taxpayer might legitimately ask just what amount or kind of change is to be brought about, what for, what will it cost and what will it be worth. At the present time, he would receive no answers. It is not to be supposed that improvement must be obtained at any price, or even at a stiff price. Whether or not it would be a good thing to improve the speed of greyhounds is probably a matter of opinion, but few could be found to say that the cost of doing so should be borne by the public, or even by all dog owners. A more penetrating but difficult problem is posed by the beef cattle industry. Who should pay for improvements in it? The answer seems likely to vary with political rather than with biological circumstances, and will not be attempted here.

The optimal rate of change in relation to the limits of the modifications possible also involves the question of costs. Early population geneticists assumed that the choice of a breeding plan should be based on a maximisation of the annual rate of genetic change. Therein lay the reason for often preferring heavy dependence on sib testing to reliance on individual performance. It is now becoming clear that improvement rates can be too rapid if a high rate of change per year is achieved at the expense of a reduction in the ultimate ceiling of improvement. On one theory of selection limits (Robertson, 1960), selection based on family data or any other information auxiliary to individual measurements will reduce the long-term possibilities of achievement, the degree of reduction being inversely related to population size. Indeed, on this theory, optimal selection intensity (the proportion of individuals in a population which are allowed to become parents of the next generation) is not the maximum compatible with maintaining population numbers, but is half way between this and no selection at all. Selection that is too intense leads to a rapid decrease of effective population size and consequently to chance losses from the gene pool of desirable genes present in low frequency. This argument, however, cannot be turned to advantage by those who oppose operational principles based on genetics. If a brake on the rate of genetic change has to be applied, its justification must be based on scientific consideration and not on vested interest or prejudice.

The notion of a physiological limit to performance is often aired. In genetic terms, such a limit is simply the highest expression of a character which it is possible to attain from a given gene pool in a given environment. This is merely saying that genes which are needed to transcend the maximum level of performance in that environment are not present in the gene pool or cannot be combined in all members of the population in every generation. For instance, if the maximum expression depends on heterozygosity at some loci, the average of a closed flock or herd can never be equal to the performance of the best individuals.

It must be granted that there may be thermodynamic or even morphological considerations that could determine limits of performance. A cow that could give milk without being fed, or a chicken consisting entirely of breast muscles are unlikely creatures, though the possibility of producing the latter in tissue culture is not too extravagant an idea. But fundamentally, these are limits not of physiology of higher animals but of properties of all living systems. Within them, the problem resolves itself into making available enough genetic variation (Lerner, 1958, and Robertson, 1963a, may be consulted for a further discussion of this question). This explains why no geneticist has succeeded, despite several efforts, in producing rat-size mice, while dog fanciers, not in the least versed in genetics, have produced a fifty-fold range of variability, stretching from the 4-lb Chihuahua to St. Bernards weighing 200 lb (Hubbard, 1954). No detailed answer comes to this question, as we know nothing about the various biochemical, physiological or endocrinological determinants of size and shape in the original canine gene pool. The remoteness of origin, the great supply of variability from hybridisation (Zeuner, 1963, lists five ancestral dog species, not counting the wolves), the sacrifice of reproductive capacity for the sake of fixation of type by inbreeding, the number of generations of trial and error that went into the creation of extant breeds, and similar factors come to mind as partial answers. The question can by no means be considered resolved.

The brief account of the genetic background to livestock improvement that this and the preceding chapters afford will not satisfy those in search of a full exposition of the theory. For them the references will provide guidance in search of further information. The day has long passed, however, when everyone with an interest in some aspect of animal breeding can hope to be familiar with the whole corpus of scientific knowledge pertaining to it. Some things must be left to the experts. On this assumption, the purely genetic treatment of the subject has been curtailed in order to allow more complete discussion of the ways in which science permeates the animal breeding industry.

Part III: The Practice of Animal Breeding

CHAPTER 5

ORGANISATION AND METHODS

Each advance in techniques of measuring passes gradually from being novel to being necessary (Cowan, 1963). This has happened to finger-printing and to blood-typing within the memory of many readers. It is happening to chemical estimates of milk quality and to the measurements of the amount of fat in bacon carcasses. By no great effort of the imagination a student can foresee the day when the elite breeding stocks of sheep will have to be described in terms of their enzymes so that those who breed them for meat and wool can favour the ones with the most desirable properties. For human beings there is already a great mass of information about biochemical individuality (e.g. Williams, 1956) and about its genetic basis (Harris, 1959).

All methods of assessment of performance, including eye judgment of conformation and type, are primarily intended to make a breeding policy effective. However, they quickly tend to generate the very aims of that policy. In fact, if performance testing does not provide or arise from the aims of breeding, it becomes irrelevant as a means. This is an old as well as a very modern problem. In its present form it is created by the setting-up of facilities for the quantitative observation which results in so-called paper records that may in fact be ignored by breeders in the selection of animals. Performance testing under these conditions is make-believe; and yet in a much older form the problem arose as soon as milk recording or egg-laying trials were started. It took a long time before the information produced began to have a significant effect on the ways in which dairy cows and poultry were selected in breeding. Even eye judgment, to the extent that it deals with characters which are of very low heritability or even no heritability at all, such as condition, provided information which was quite meaningless as far as breeding was concerned, though, of course, a judge did not think so at the time.

A breeder who records identities and performance has data that are of potential value not only to himself but also to others. The balance of advantage, however, may lie very much to one side or the other. At one extreme there are records that are erroneous, illegible, or not comparable with others. Although conceivably of value to the individual collecting them, they are worse than useless to anyone else. At the other extreme, there are records (as of routine blood grouping), the direct benefit of which to a herd-owner may be negligible.

Since most records are of an intermediate kind on this scale, problems arise in collecting and using them due to conflicting interests. The history of herd testing and egg-laying trials shows how the balance has by degrees shifted away from the individual towards the group to which he belongs. One by one the devices by which poor results could be hidden are giving way to precise methods aimed at obtaining correct information for general use.

I. The Purposes of Recording

The purely organisational aspects of data handling are often confounded by the need for good public relations. Although this fact can be and often is used to delay or prevent refinements, public relations cannot be neglected unless those responsible for animal breeding do not need the cooperation of farmers for identifying and recording animals. The cost of data on the performance of A.I. bulls to A.I. organisations would rise if fewer farmers felt disposed to test their herds for other reasons. If sheep farmers continue to be less than enthusiastic about collecting data on performance, those who would improve sheep will have to find some way of engaging their interest or else build up and pay for breeding stocks themselves as the poultry breeders do. Long established activities such as herd testing which have in the past depended on breeder cooperation, are likely to be the most intractable to modernise. If adequate farmer cooperation is not forthcoming, new forms of organisation for obtaining data will have to be invented. Eventually, information will have to be paid for by those who want it and as it becomes more expensive and complicated to handle, the question as to who really wants it will have to be answered.

To the individual breeder, the value of records may be quite distinct from their value to the organisers of testing schemes, or to research workers. Farmers unaccustomed to interpreting and using records at the times when they are of significance in husbandry or in breeding may, if they take records, obtain only the indirect benefits that are available to everyone, such as the use of progeny-tested A.I. sires or information from publication of research results. Some may in addition find that recording has a sales value even if the records are otherwise ignored. Thus, any student of milk records can find many herds in which a bull has been allowed to go on siring daughters long after a progeny test has shown that he is depressing yields by 50 gallons or more. With the intensification and higher capitalisation of farming, the proportion of such farmers might diminish as clearer distinctions come to be drawn between records useful in breeding and those with a bearing on efficient management. Breeders who keep no performance records of their stock are, however, still very numerous even in dairy cattle breeding where they tend to be found in the lower strata of the hierarchy. Provided that there is a sufficient price offered by buyers for registered stock without performance records, suppliers will normally present themselves. They will come from among those content with moderate to low levels of management and lacking a wish to have their stock tested or to use anything more than eye judgment in breeding. There may not be a very promising future for these suppliers of stock; still none but themselves can be held responsible for misjudging the coming trends. Yet under the pedigree system the number of breeders in the lower strata far outnumber those who are genetically significant. There is, therefore, within breed societies a great weight of opinion and influence expressed by members who have little to gain personally from progressive policies. In the past, this mattered little, since the tempo of change was very slow, but such unbalance hampers the adoption of forward-looking policies.

For breeders, the question whether to test or not is rather complex. The proportion of cows tested for milk yield varies from about 60% in Denmark and Holland to 25% in the United Kingdom, and 15% in the United States. Many factors contribute to this variation, including the warmth of official encouragement. Where performance records are becoming or have become essential for top-class breeding stock, a decision not to test is a decision not to try to produce that class. This has clear implications for the capital value and depreciation of a herd. Testing, however, may show that a herd is of moderate to poor performance and such a discovery would have the same implications. It would not be surprising if, after a little unfortunate experience of testing, some breeders of beef cattle and pigs denigrated it, and proclaimed the merits of eye judgment.

There is no case, however, for trying to persuade all breeders in a numerically large breed to take records for breeding purposes. Only enough females must be observed to supply and test adequately the young males from which the sires of the elite stocks are to be drawn. For this the whole population may not be necessary, possibly only a half or less. If an adequate test of a young sire calls for 300 inseminations of herd-tested cows, and ten of them are to be tested each year, some 3000 tested cows are required for this purpose alone. The owners of 6000 more may well feel entitled to the use of already proven sires.

II. Accuracy

Most breeders of livestock have acquired a strong belief in the virtue of accuracy. From their point of view, it is one of the unfortunate aspects of population genetics that it compels an objective approach to the subject.

Reconciliation of the breeders' notions of accuracy of recording performance with those of the statistician and geneticist is often needed. The latter are usually concerned with, and think in terms of, groups of populations. Breeders, like lawyers, tend to consider each animal as an individual with its own special circumstances. The biological apparatus of breeding, however, is designed to deal with populations. Even in poultry, it was not long ago that selection was directed towards producing individuals with record performances. Horse racing certainly justifies this approach. The economics of animal production, however, call for consideration of groups or averages and not of individuals. It comes about, therefore, that scientists try to spend the available effort of observation over the most effective numbers of animals often at the price of reducing the frequency or number of observations on each one (Robertson, 1957).

Milk yields of dairy cows offer a good example of the dichotomy in thinking. Enthusiasts among breeders will record milk yields daily in order to measure them exactly; yet with one-thirtieth of the effort they could have estimates so close to the actual yield as to make no important difference for breeding purposes (Johansson, 1961). Greater consciousness of the cost of data, and of the importance of time in sire testing, are leading to the use of milk yields estimated from the first 70 or 100 days of heifer lactations. The heritability of this yield is about the same as that of the full 305-day lactation yield and the genetic correlation between part and full yields is high. Several studies (e.g. Van Vleck and Henderson, 1963) have shown that the ranking of A.I. sires on daughters' first lactations is practically as accurate as ranking them on the first two and much quicker. To judge a cow expected to milk for four or more lactations by performance in 70 days of her first one is, in the opinion of breeders, altogether too rough and ready a procedure. But the technical answer to such an assertion is simple. There are three points to it: (a) 70-day yields are quicker and cheaper to obtain; (b) it is better to make progress at the rate of, say, 2% each year than to achieve 5% every four years; and (c) lactation yields or lifetime yields are themselves only rough estimates of a cow's breeding value. Even identical twin cows calved, fed and milked side by side, differ in yield by about 100 gallons and the consequences of varying the season of calving and age at calving would add to this difference. There is no way in fact to find out what a cow's characteristic yield is within narrow limits, though some day, when egg transfer is perfected, her genotype may be assessed through her daughters as that of a sire may be now. The arguments for limiting and adapting the observations on milk yield to the purpose of breeding for milk yield fall to the ground if the data are not used for this purpose or are needed for some other purpose, such as management. The question of paying for them then takes on another aspect.

Poultry breeders have made a close study of the amount of egg recording they should do and in recent years have effected great economies in collecting and processing testing data from birds on test without sacrificing precision of genotypic assessment. Thus not only part-time recording of egg production but also minimal sampling of egg quality (size, shell strength, freedom from bloodspots, etc.) is used. However, heavy dependence on computers by all leading breeding establishments permits them to make the most of the data collected (Dickerson, 1962).

Milk yields and pig performance are approached in a less refined manner and sheep recording is even more primitive. In due course, however, the principles established in the breeding of poultry will have to be applied to the larger animals.

How much expenditure on data collection is justifiable is a question that cannot be given a general or final answer. Methods tend to evolve from the crude to the refined, from the cheap to the costly, as more and more comes to depend on the information. A first trial in a new field of recording carried out on a temporary grant should perhaps aim primarily at becoming accepted. Later on, more or less detail may be found advisable according to the uses made of the information. It is not at all inconceivable that more expense on recording may be required for publicity purposes than for breeding. Touchberry, Rottensten and Andersen (1960), Johansson (1961), and others discuss these matters in connection with the merits of the special Danish testing stations for the daughters of dairy sires. In the stations the daughter groups can be observed more closely and accurately (at a price) than on their home farms. They can be given known amounts of feed and are readily available as groups for judgments on conformation and temperament. Field data by contrast are less expensive and less accurate in some respects but more plentiful. Furthermore there are difficulties in interpreting the results. Differences between sires are apparently larger and more heritable at the stations than in the field, and the correspondence between station and field rankings of sires is far from perfect. In spite of the fact therefore that the stations exist to assist the selection of A.I. sires, the comparative merits of station data and field data are not to be settled by the heritability of sire differences alone (see section on performance testing in Chapter 4). The vital question is the cost of a unit of progress, and into it will come the amount of selection practised and the information on which it is based. An optimum solution might require a combination of testing methods as well as their refinement to coincide exactly with the realities of bull selection. Where selection is in fact based on numerous criteria, the recording of performance may be mainly of value for farm management, for investigational work, or for watching the performance of an industry, and only to a lesser extent for breeding, since simultaneous selection in many directions is designedly weak in any one of them.

III. Pedigrees

Pedigrees present another source of disagreement about the amount of useful detail. In poultry very little attention is paid to pedigrees and what use is made of them is largely aimed at avoiding close inbreeding. In larger animals there is much variation in the amount of detail shown in herd and flock books: from none at all about females in some sheep flock books to extended pedigrees with performance data on both sides. A common pattern for an individual entry consists basically of an ear number, a herdbook number, and a name of at least two parts (the herd prefix and another referring to the individual and sometimes its male or female ancestors). This is essentially a system of identifying individual animals. Each animal has a number stamped on it or attached to it and reference to this number in a book listing all numbers will give access to supplementary information. Some breed societies no longer publish a herdbook. A filing system and photo-copying provide a substitute. Others face the cost of printing extended pedigrees and long names. What information is wanted depends on who wishes to identify animals and for what purpose. A policeman looking for stolen cattle, a blood-grouper checking pedigrees, a breeder and a buyer will not think alike. Since breeders, however, design and pay for the herdbooks they naturally use systems which they hope will best serve their needs. Anyone else who uses their systems has scant right to criticise long pedigrees, herd prefixes or lack of legitimate performance data.

Ever since Coates began compiling the Shorthorn herdbook, conscientious efforts have been made by all breed associations to ensure that pedigrees are accurate. As the difference between registered and non-registered animals shrinks, the stratification of the breeding population makes it quite immaterial whether the animals in the lower strata have correct pedigrees, or, indeed, whether they have recorded pedigrees at all. For the business of sire testing no more is needed than a symbol to show what breeding population they belong to, and an identity number listed under their sire and dam. Even in the upper strata where identities and parentages may matter, the usual names have merely the purpose of aiding memory, advertising the breeder, and perhaps indicating membership of some saleable but transient concept of family. Pedigrees, however, still have a commercial value and as long as this remains sufficient, animals will be supplied with them, and money expended on recording, checking, and printing. Expenses of this kind seem to have no direct relevance to livestock improvement. Nevertheless, they may be acceptable as a charge since their existence is implied by the structure of the pedigree system. A social or political decision to support the system and help it to compete in the production of better livestock will entail meeting its costs somehow.

With the advent of blood typing, it has become feasible to estimate rather closely the likelihood that the parents of a particular individual are as stated, provided all three animals are available for testing (Stormont, 1959). Since the tests are expensive to carry out, they will not be requested unless there are compelling reasons. Some improvement in accuracy of recording parents may follow the mere threat of blood typing.

If organised crossbreeding and performance testing expand, those responsible for data processing may find it necessary to devise new systems of identifying animals suited to the uses to which they are put. As distant ancestors, farm and breeder fade into insignificance, corresponding changes in the identification system are likely. A greater use of code numbers to indicate performance, sire, blood groups and locality, and to simplify classification and analysis can be foreseen, since much of the need for identifying animals will come from the recording of performance.

Pedigrees as published have been disparaged many times in the last 100 years or more, but they will continue so long as commercial breeders are still willing to pay for them. Pedigree breeders can be expected to protest that the knowledge of pedigrees conveys much more to owners than the plain recital of names. For them it sums up a breeding programme, and recalls a mental picture of developing animals, their relatives and their circumstances. This may well be so, but the argument weakens if pedigree breeding itself lacks conviction. The test of merit is performance, not pedigree, and it is immaterial what a breeder's methods are if his stock are only average performers.

IV. Publishing

The publishing of data is not undertaken solely because of the use to be made of them but also for other reasons, such as (1) to show that the collecting agency is working in an active and enterprising fashion, (2) to promote research, and (3) to provide an information service for breeders (e.g. pedigrees and associated information). In most situations, it is not easy to segregate these reasons, and they also vary in importance. Developmental stages of a new type of recording, for instance, will be followed by routine where research is no longer necessary. Some records, like wool tests, litter records, and milk production data, could be of managerial value to all farmers; others, such as blood groupings, are not. Trimming publication costs is likely to require an assessment of the reasons for publishing when selected data on punched cards or tape are readily available to enquirers interested enough to ask for them.

Although outlays on publishing data are perhaps not serious for publicly financed institutions, they absorb an important fraction of the funds of a breed association. One way of escaping from an ironical situation in which a breed association pays heavily for pretending that all members are important and constructive breeders who must have access to the data, is to encourage the idea that it is proper for the industry as a whole to finance this activity through some separate organisation. This is already practised to varying degrees in many dairy industries (including that of the United States) and in the British pig industry. Ultimately, however, even publicly financed institutions may wonder whether it is necessary to publish a great volume of records, including pedigrees, when only a few breeders are able to put it to good use. Before selective publication is adopted, it will be necessary to identify these few breeders and their younger replacements. This achieved, it will not be long before the science of breeding becomes as arcane as the art of breeding was before it.

V. Computable Models

Just as growth rate in swine is a character that is changing from being a means to being an end of improvement, so the mechanising of data handling has become a factor in deciding which data to collect and, consequently, which objectives to select for. The essential feature of a computer is speed, and it is therefore illogical to use one unless the result is a worthwhile saving of time, or what is the same thing, money. In due course, the computer may be expected to work out its own diet. Already it is well understood that observations on animals have to be transformed and coded to suit it. The further step of adjusting the frequency, accuracy, and variety of observations to maximise the rate of progress in breeding (Dickerson and Hazel, 1944; Lush, 1945; King, 1955; Robertson, 1957; Lerner, 1958; Smith, 1964; Skjervold and Langholz, 1964) is only a short one for the computer itself to take.

The necessity for giving computers their instructions may provoke more clear thinking about the the objectives of large animal breeding than there has yet been. In industries more advanced in computer use than agriculture, the concept of a "computable model" or mathematical description of a complicated process is commonplace. By means of such a model, the computer can simulate all the components, and if so instructed, actually control the activity at all stages. In genetic research, models have been set up for working out the results of breeding methods and to test complex hypotheses (e.g. by Fraser, 1962).

Production equations and linear programming are by no means novel ideas in agriculture (see Heady and Dillon, 1961) but their practical application is hindered by the lack of operational data. A forward looking industry will need to set up machinery to collect and analyse information on the effects of changing the amounts, for example, of food, or fertilisers, or labour, or breed, on output before it can use computers to the limit in planning.

The highest level of model making yet attempted is the description of the workings of an entire economy (Anon., 1964). Although such a model cannot take cognisance of future decisions of a political character, it can work out the consequences of alternative decisions, provide information on the workings of the system, and expose interdependence and feedback among its elements. Such a development may be a distant prospect for agriculture and for animal breeding, but it may not be long before those with a horror of social planning will nevertheless be glad to see some coordination of the unconnected initiatives now characteristic of livestock breeding.

What has been said about the influence of machines on breeding applies also to statistical methods. That the problems of animal breeding led Wright, Lush, Robertson and others to contribute notably to the mathematical formulation of the theory of heredity is now a matter of history. For a long time statistical techniques were designed to fit existing practices in animal breeding but the process is now being reversed. Few breeders, for instance, could now expound the theory behind contemporary comparisons, let alone describe varietal types among them (Searle, 1964), but as a result of their use there is a growing tendency to charge the expenses of analysis, and storage of milk yield data, in excess of what is needed for comparing heifer groups, against management instead of against breeding.

VI. Evolution of Testing and Recording Methods

Records serve many purposes which can be condensed into four categories: (a) selective breeding; (b) management; (c) research; (d) publicity.

Some records can be put to all four purposes to some extent but the proportions vary with time and circumstance. What is worthwhile when a recording system is set up may become redundant as the system provides the basic data for its evolution. What is valuable for management (for example, data on fertility or body condition) may be almost useless for breeding. It seems sensible therefore to have recording practices under constant review so that no superfluous data are collected, and no desirable data omitted. Those who do the reviewing should include scientific persons familiar with advances in data handling in population genetics, and in methods of measuring performance. In the past, recording techniques have at times been allowed to become unnecessarily expensive habits rather than sharp tools. To some extent, no doubt, this has been due to neglect, to compromising, and to distrust of scientists. Where recording is subsidised by the general taxpayer there is also a political aspect depending on governmental attitudes to the financial support of farming and especially to the idea that the farming industry is a desirable agglomeration of small units not able to organise recording systems unaided.

This political aspect may influence the initiating of new forms of recording. Whereas the keeping of pedigrees and organising milk recording were usually the result of farmers' initiative, more recent projects such as blood-typing or progeny testing stations do not flourish as purely local enthusiasms. Given the perpetual urgency of raising food production, it is probably fair to say that national interests cannot await the evolution of thought to the point where enough farmers would be willing to finance a new kind of testing about which they knew little. Moreover, developments, such as performance testing of sheep and beef cattle, cannot be expected to begin where herd testing of dairy cows began half a century or more ago. Those responsible will wish to employ the knowledge that has been gained from dairy herd testing and modern data handling and they will want to see the data put promptly to work. Somehow, therefore, they must acquire access to facilities and finance that cannot be provided by a few enthusiastic breeders. Funds will have to be provided by a dynamic livestock industry if the large sums put into research and development by successful industrial companies and the surviving poultry concerns are any guide to economic fitness. It will fall to large scale organisations, no matter whether private or public or hybrid, to accept the risks inherent in new activities.

TABLE 5.1

Types of records

Dairy cattle	Pigs
Census	Census
Pedigree	Pedigree
Colour	Colour
Milk yield	Litter size and weight
Butterfat %	
Protein %	
Type classification	
A.I. conception rates	
Progeny test	Progeny test
(a) on farms	Performance test
(b) in testing stations	growth rate
	food efficiency
	carcass quality
Blood grouping	Blood grouping

No attempt is to be made here to discuss the numerous kinds of data that are collected. A list of some of the commoner categories of records collected from herds of dairy cattle and pigs will serve to illustrate not only the variety of records but also the kinds of techniques and technicians. Table 5.1 has been compiled in descending order in such a way as to show roughly how data range from mere enumeration to the intricacies of blood grouping; how the number of stock owners directly interested grows smaller; and how the observations require increasing skill and large scale organisation. Since the list is also approximately in order of practical application, it indicates how the methods used by pig and dairy cattle breeders have been modified over the last fifty years.

VII. Evolution of Breeding Methods

The evolutionary trend of breeding procedures in livestock seems to start with local populations of animals which develop a cohesion and later on a breed society or its equivalent. These breeds were often and possibly always built up on a crossbred foundation just as were the new breeds of more recent origin.

The second stage of the process is intra-breed selection which comes when there is agreement on the objectives to be sought in breeding. At the same time as this intra-breed selection is going on, there is a between breed selection which reduces to virtual insignificance all but a few prominent breeds.

The third stage is the appearance within the dominant breeds of strains, sub-populations or isolates which lead to intra-breed between-strain selection, as a result of which, most of the strains are gradually eliminated.

The fourth stage of the process is the inbreeding of these strains accompanied by a further stage of sub-division so as to make a number of inbred lines. The inbred lines in their turn are subject to the same selection, either pure or in crosses. The surviving inbred lines are then exploited in the form of line crosses involving two, three or more of these lines.

In the United States and the United Kingdom sheep and cattle are predominantly at stage two, pigs are moving towards stage three, and poultry have reached stage four. Organisation has to develop to permit moving from one stage to another.

A rough comparison of the major classes of farm livestock is given in Table 5.2. It is intended to apply to both the United States and the United Kingdom but fits neither exactly. How extensively a technique must be applied to rank as "widely used" is of necessity quite arbitrary. The main differences, however, in both purebred and crossbred populations are perhaps clear enough. Poultry have a strong lead in organised crossing of selected strains. Because they are individually of small value, are readily kept in large numbers, and multiply rapidly, they are relatively easy to mass produce. Given mass production, better breeding systems can be designed. It is not a question of a special genetic situation; for there is nothing about their heredity to distinguish poultry from other animals. In all of them size and growth rate are moderately highly inherited as is the fat content of milk or meat. Fertility and viability are generally of low heritability within breeds or lines as Table 5.2 indicates. The heritabilities shown there, it should be pointed out, are merely round figures representative of published values.

TABLE 5.2

A comparison of the main classes of farm livestock
X = *widely and effectively used;* ⊗ = *minor use as yet*

		Dairy cattle	Beef cattle	Pigs	Sheep	Poultry for eggs	Poultry for meat
Closed population	Eye judgment	X	X	X	X		X
	Breed Society	X	X	X	X		
	Performance test	X	⊗	X	⊗	X	X
	Artificial insemination	X	⊗	⊗		⊗	X
	Family testing	X		⊗		X	X
Organised crossing	Eye judgment	X	X	X	X		X
	Performance test			⊗		X	X
	Large scale breeding			⊗		X	X
	Inbred lines					X	⊗
	Control population					X	
Approximate heritability	Amount of product	30	40	40	35	20	40
	Quality of product	45	30	50	40	50	30
	Reproduction and viability	5	5	10	5	10	10
Ranking	Value of individual animals	1	1	3	4	5	5
	Reproductive rate	5	6	3	4	1	1

VIII. The Handling of Records

The methods followed in different parts of the world to integrate the handling of records show much diversity. Some idea of it can be acquired from a small selection of countries with dairy industries. In Scotland, the functions of collecting milk yields, analysing them, using them for choosing and culling bulls for artificial insemination, and keeping pedigrees, are carried out by four independent bodies which collaborate as and when they feel disposed. In England and New Zealand, all these functions, except pedigree registration, fall to the Milk Marketing Board, and the Dairy Board. Details of organisations in fifteen European countries can be found in O'Connor (1962). Breed improvement in the U.S.S.R. is implemented by the State Department of Inspection for Breed Improvement and Herdbooks. Its functions

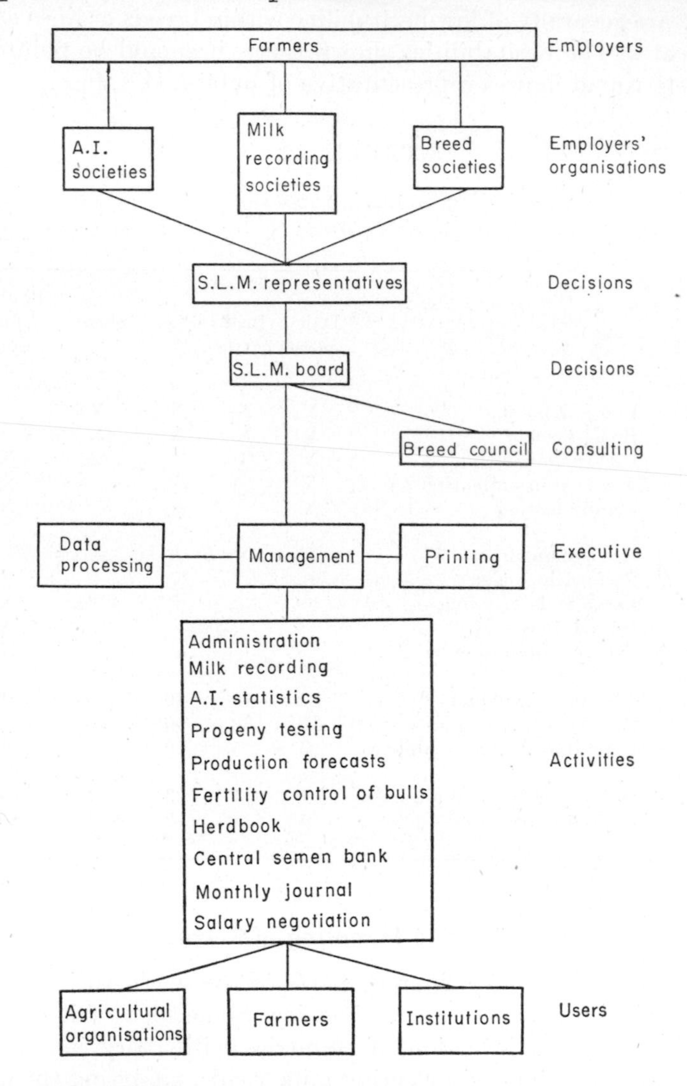

Fig. 5.1. The organisation of the Swedish Livestock Management, a consortium of 26 A.I. societies, 26 milk recording societies and 5 breed societies (from a promotional pamphlet of the Swedish Livestock Management).

include instructions for breed improvement, registration and publishing of State herdbooks. The herdbooks are selective and include only animals of high production whether purebred or not. They also contain a summary of breed performance. Scientific institutions are charged with raising productivity, improving existing breeds, and making new breeds, and it is from the broad ties between agricultural science and practice that the successes of animal breeding in U.S.S.R. are believed to have come.

Integration has been achieved in Sweden in another way. Figure 5.1, adapted from publicity material of Swedish Livestock Management, shows how the several farmers' organisations interested in milk records and breed registrations have combined to form a board which carries out on their behalf a wide assortment of activities ranging from herd

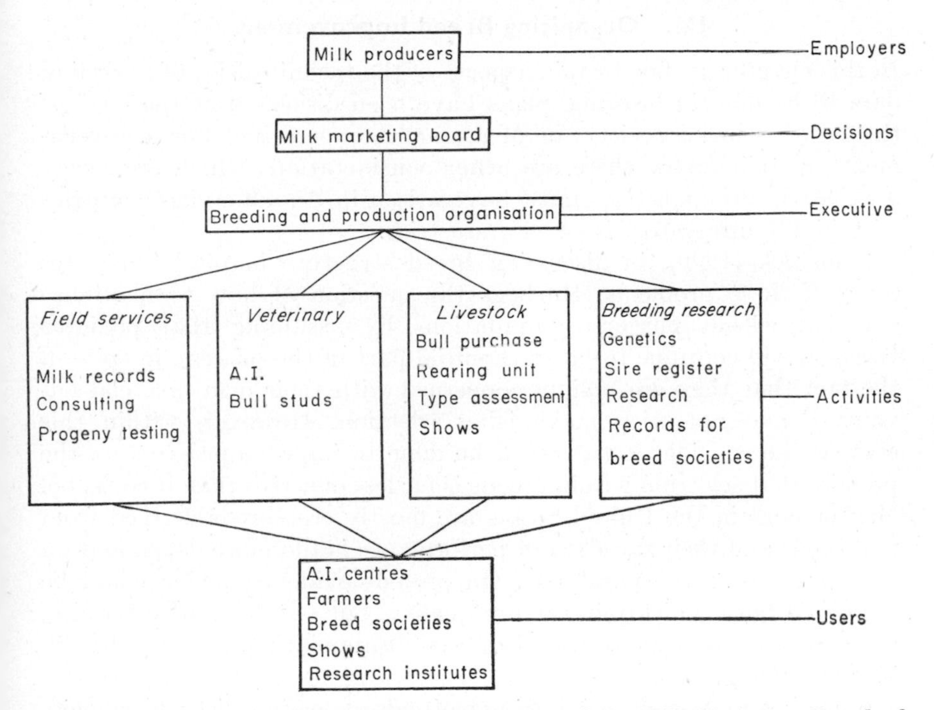

Fig. 5.2. Breeding and production organisation of the Milk Marketing Board of England and Wales.

testing to publishing herdbooks. By so doing, it relieves the farmer organisations of technical supervision of staff, computers, printing and other specialised processes. They have, of course, to finance it, but by combining they achieve more at less cost than they could by independent action. Apart from pedigree registration and publication, the English arrangements have much in common but they are achieved through a milk producers' board (Fig. 5.2). Trends in all countries are away from the original simple herd records that were left in a raw state for the farmer to interpret as best he could. There are several reasons for this. The needs of the artificial insemination industry for data in bull performance, and the exigencies of research on the characteristics of cattle, individually and in groups have brought about the use of expensive machines to cope quickly with incoming data. Almost as a by-product, the herd-owner benefits by a faster service, more detail, and statistics which enable him to compare his herd with those of other owners.

IX. Organising Breed Improvement

In this chapter, so far, certain aspects of the technicalities of recording data to be used in breeding plans have been discussed at the level of the farm or breed society, or of a small co-operative. For a government or an industry there are other considerations which are rarely ventilated, although they must have arisen in Scandinavian countries with highly integrated livestock industries.

Comstock (1960), in discussing breed structure in pigs, illustrates some of these problems. He begs the question of how to maximise selection within purebred populations by assuming that pedigree breeders will continue to be an essential part of the process, in spite of the fact that they are mainly concerned with their own financial advantage and not with maximising selection. However, within this context, he considers what could be done to improve progress on the premise that selection within breeds is far less effective than it could be. Market hogs in the United States are mostly crossbreds derived from purebreds and their standard of performance is therefore dependent on the standard of the purebreds. On present evidence, swine could be improved fairly quickly in carcass quality, but selection for litter size, growth rate and feed efficiency has been disappointing, and the outlook is unpromising.

Comstock proposes the following methods of dealing with the problem of improvement:

1. a more extensive study of crossbreeding and a determined search for breeds that cross well;
2. more intensive selection within purebreds;
3. more research on the improvement of the parents of the crossbred, known as selection for combining ability;
4. new breeds.

All these proposals have led to some action in the past but not enough. Initiative is still needed on a large scale and in places where it matters.

A vast comparison of breeds and crosses is conceivable if the facilities and personnel could be drafted to carry it out. More intensive recording of herds and larger selection differentials are possible but unlikely unless

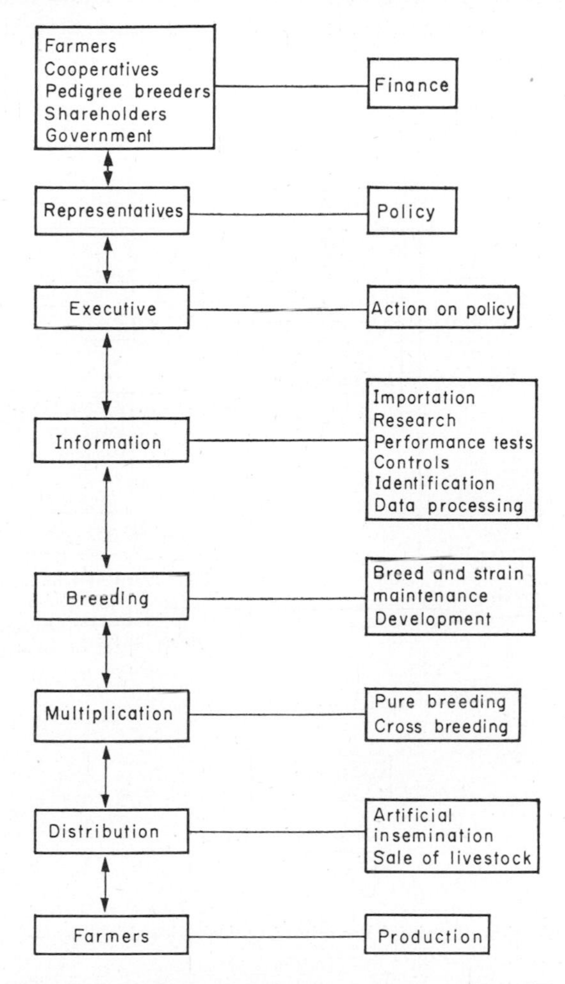

Fig. 5.3. Essentials of industrial organisation of breeding.

enough breeders can see what they get out of it. The same thing might be said of research on combining ability, a laborious task with no guarantee that a favourable outcome would be applied in pedigree breeding. Likewise, new breeds may be desirable in theory, but this is not enough to establish them. It seems therefore that all the proposals require a body with power and finance to carry out a policy on an industrial scale. Obviously, a breed association of the traditional kind will not suffice. Something new is needed. It could develop out of existing associations, or from new ones such as the Performance Registry International which encourages performance testing of beef cattle in the United States. Dairy boards and artificial breeding

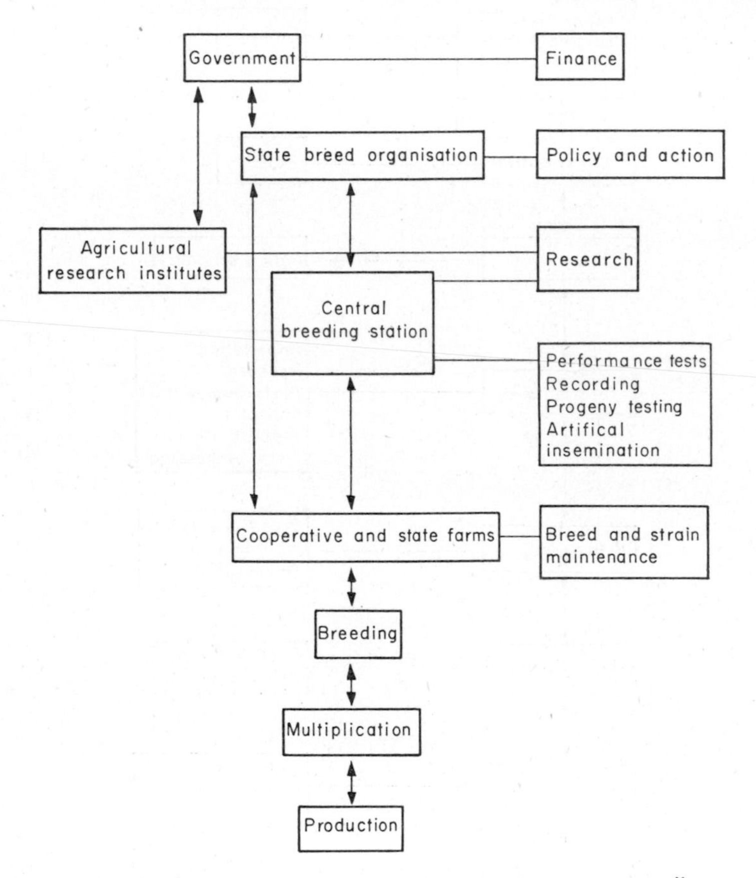

Fig. 5.4. Organisation of animal breeding in Czechoslovakia (based on Šiler, personal communication).

organisations might establish their own testing, pedigree registration and breeding departments. The existence of such a body or bodies is no assurance that a policy will be laid down and the finance provided; but its absence means that there will be neither. Whatever the precise origins and characteristics of the system adopted, it has to provide the essentials of the industrial organisation of breeding as shown in Fig. 5.3, more fully discussed in the section on integration of services (a comparable diagram for a Central European socialist state is shown in Fig. 5.4).

In the United Kingdom a Pig Industry Development Authority was established in 1957 and financed by a levy on carcass meat. The Authority, now to be absorbed by a Meat and Livestock Commission, consists of seventeen representatives (two for pedigree breeders, four for commercial producers, one for agricultural employees, two for bacon producers, one for bacon distributors, two for distributors of fresh pig meat, one for manufacturers of pig products other than bacon, one for employees of processors, marketers, and distributors of pig products, and three independents) and their function is to promote breeding, production, grading and marketing. Of Comstock's four proposals, which have as much validity in the United Kingdom as in the United States, only one has so far received substantial attention from the Authority. Efforts have been made to encourage progeny and performance tests; and the next step is to persuade breeders to apply the data to selection (Smith, 1965). As logic requires, there is a plan to build up a new stratification of the breeding population based on performance testing to replace the old one based on official breed type. But it is a slow process. Enough breeders and the necessary buyers of stock have to be weaned off the old system to make the new one viable.

Although individual members of such an Authority may be enthusiastic about their duties, if they represent diverse interests within the industry, their net effectiveness may be small. Sceptics would wait to see whether, having willed some of the means of swine improvement, the Authority also willed the ends.

X. Local Varieties and Importations

As the speed of transport has arisen, the idea of a locally adapted breed has tended to change. For the purpose of animal breeding, the size of a locality has been growing until whole countries have to be regarded as mere localities. Wise as it may be to look askance at claims of special adaptation to local circumstances (for instance, breeds of beef cattle, such as Welsh Blacks and Highland cattle are characteristic of certain regions, probably more because of tradition than of any biological peculiarities), it should be remembered that at least some forms of local adaptation are bound to have a genetic basis. Evidence of gene-controlled aberrations of metabolism, and of hereditary variation in susceptibility to deficiencies of diets and to infections (Hutt, 1958) is abundant. Progress in nutrition and veterinary science will find ways of adjusting environments to avoid exposing genetic weaknesses, but there remain local hazards of heat and cold, food supplies, and diseases (note postulate 5, in Chapter 3) that will require locally adapted stock, especially when extensively kept.

Breeds of domestic livestock, and man himself, have flourished under many environmental conditions. The Americas, Australia, and New Zealand have all founded their livestock industries on livestock from the Old World. Even in Britain, there may be no significant breed to-day that is not wholly or partially of immigrant origin within the last 200 years. In recent years, the introduction of Landrace pigs and Friesian cattle has been of benefit both directly from their performance and indirectly by the stimulus to pig and cattle breeding. There are good reasons therefore to exploit and rationalise the exchange and testing of breeds and crossbreds between countries as suggested in Chapter 8 (see also articles by Lush and by Phillips, in Hodgson, 1961). In the not very distant future there may well be international centres or schemes for comparative trials.

Migration works best when stock moves towards better conditions as in stratified sheep industries. In the reverse direction, adaptation takes longer. Although the prospects of obtaining new and desirable genes from really exotic varieties may glitter, they may be a long way off. History gives many warnings against too great an optimism in this respect. The failure of British breeds to establish themselves in India and other tropical countries has not been for want of trying, nor has the failure of the fifty-year-long attempt to develop a hybrid of cattle with buffalo that would show improved hardiness and productivity in Northern Canada.

For the winners of the battle of intensively reared breeds, a world market is the prize. Livestock industries, if they enter this battle, need to be organised accordingly. It will not be won by advertising unless backed by convincing performance data. The other side of this coin shows that, unless protected by regulation, local or national breeds will be subject to competition from abroad. This should encourage animal breeders in one country to match improvements in other countries, but, in spite of the historical evidence of the value of importations, their beneficial influence has been allowed to work only sporadically in the past. Until Charollais bulls were brought to England in 1961 and given an extensive trial by the Milk Marketing Board with the aid of artificial insemination, no adequate test of an imported breed had ever been conducted in Western countries. Even this test was conducted only on crossbreds. Still, as the successful emigrations of many breeds of livestock have already shown (for examples, Jersey cattle, Merino sheep, thoroughbred horses and United States-bred poultry), breeds in possession of a territory cannot rely on their adaptation to local conditions to protect them from invasion from abroad. In addition to being adaptable, the invaders may be supported by good records of performance and the power of big business.

XI. Integration of Services

A fully integrated scheme for the breeding of livestock would, on present knowledge, have the basic structure shown in Fig. 5.3 which puts in diagrammatic form the stage which has already been reached in advanced industries. It also exposes the points at which the less advanced industries have fallen behind. How quickly these can catch up is a matter of political and social attitudes to change and they are not the same in all countries or for each type of livestock. The arrangements for obtaining finance and determining policy are crucial since all the rest flows from them. In the figure, five separate possibilities for ensuring the necessary financial basis are shown. These are not intended to be mutually exclusive or to be exhaustive. If, however, the finance and policy are provided by a single corporation (for example a dairy board, an artificial insemination organisation or a government), all stages of the process of breeding and distribution of livestock would be influenced, if not wholly controlled by that corporation. All its costs would ultimately have to be paid for by those who bought commercial stock as semen or as animals, or by the general taxpayer.

It will be observed that this scheme is not dependent on the improvement achieved in the course of breeding of livestock. In this respect it does not differ in any essential from the pedigree system (which is included in the figure). The main deficiency of the pedigree system is that it is inadequate with regard to finance and policy. In consequence, the activities described as information and breeding fall short of national needs in objectives and in scale.

As mere providers of services, breed associations seem to have a rather uninspiring future. If they can make the transition to organised and centrally directed breeding, they will be users of services (and scientists) and will be competitive with other large organisations. This seems one way to exploit the skills and interest of the individual breeder as well as to provide insurance against the possibility that the large breeding organisations will become as conservative as breed associations are now. Whether or not breeders collectively accept this challenge, the stratification of animals by performance is bringing out a new class of elite breeder and, thus, of breed society administrator and councillor. Funds will be increasingly diverted towards performance trials and the incidental costs, and away from the lower ranks that contribute no genes to subsequent generations. No doubt such a metamorphosis of breed societies would offend some of their members, who might prefer to stifle competition in their old sphere of influence in animal breeding, exercise an authority they do not possess, and enjoy a status they have inherited. But the younger members will sometimes risk this disapproval.

If applied research and development become a charge on the industry which profits by them, they should include organised comparisons of breeds and crossbreds. Breeding institutions could use this kind of service in the making of the new elite and multipliers' herds based on performance, and possibly inbred lines, or lines developed by recurrent selection. Should control of genotype-environment interaction also yield to research, another need may arise for specialised lines and the services required to keep and exploit them.

On a longer view, genetical engineering and euphenics (see Chapter 9) will also fall into this category. Meantime, however, the manipulations of the physical basis of heredity and the biochemical adaptation of environment to extract the best out of existing genotypes have still to be achieved by research workers. Meantime, also, the industry still awaits the application of computer technique to comparatively small scale enterprises, as, for instance, to a single breed association running a cooperative breeding venture, and using a miniature computer for drawing instantaneously on a central fund of knowledge of markets and their effect on breeding aims, as well as of performance tests. It may not have to wait much longer.

CHAPTER 6

EVOLUTION OF LARGE SCALE BREEDING

The development of breeding methods did not proceed synchronously in all classes of animals. However, until recently, the histories of breeding of the various kinds of livestock were similar enough to allow a rather general survey to be made for all of them.

I. Historical

From Aristotle onwards, animal husbandmen have never been at a loss for theories on breeding. Berge (1961) has traced them with a touch of humour and a fine detachment. At all times there have been great authorities and from them or their followers it is possible to learn how the doctrines of breeding have changed since the Middle Ages. It is convenient to begin with the coming to Britain of the Arab horse in the seventeenth century. About this time, breeders, after having crossed the imported horses with those locally available, adopted a belief in purity of pedigree. Then there were Buffon (1708–88) who advocated systematic crossbreeding, and Candolle (1778–1841) who thought that some parts of the body were governed by the dam and others by the sire.

Bakewell (1725–95) and others of his day do not seem to have been much influenced by earlier authorities but they laboured with clear objectives and strong selection in a population with plenty of genetic variation. Bakewell was a pragmatic sort of man, a man of methods rather than of principles. Authors who reported on his methods were obliged to infer his principles from his practice since he made no statement of them himself. It has been said (as quoted by Youatt, 1834, p. 191) that he based his work on choice of breed, beauty and utility of form, texture of flesh and propensity to fatness, and gave effect to it by mating like to like, even if related. Until recently, many a breeder of meat animals would have subscribed to this philosophy. It hardly amounts to a revelation of breeding secrets, and this is small wonder, for Bakewell could have had none. His friends provided him with some Emperor's clothes that have worn well. One far-sighted writer, however (see Youatt, *loc. cit.* p. 192), commenting on Bakewell's passion for early maturity, small bones and thick layers of fat observed that "Having painfully, and at much cost, raised a variety of cattle the chief merit of which is to make fat, he has apparently laid his disciples and successors under the necessity of substituting another that will make lean." It is as true now as in the eighteenth century that those who breed the elite and significant animals are but products of their day and generation. Their aims are bound to be superseded but no one knows by which others for there is no predestination in evolution.

In 1794 George Culley reflected "Could any of these people be prevailed upon to make an experiment, they would most probably find that excellence does not depend on the situation or size of horns, or on the colour of faces and legs, but on other more essential properties." He would find that although this is now known, it is also known to be an expensive and possibly unrewarding business for the unaided breeder to try to improve the essential properties.

After the turn of the eighteenth century, there was no intellectual background to the efforts of animal breeders in Europe trying to produce meat animals for the markets created by industrialisation, and to fill the gap there was a resurgence of the theory of pure breeding which came to mean pedigree breeding. To improve credibility, it was asserted that the longer and purer the descent, the more reliable the inheritance. No doubt the convenience of this assumption and the vested interest it created enabled it to survive the palpable objections. This is something to marvel at because it was recognised clearly enough that breed characters were not in fact invariable even among full sisters of the bluest blood. It came nearest to being true for colours, horns, and to a lesser extent conformation, which helps to explain the pre-occupation of breeders to this day with these classes of characters. Getting rid of this concept to clear the field for new ideas has been a long struggle.

During the nineteenth century a "doctrine of the indigenous breed" was popular in Norway. According to this doctrine, locally adapted types of animal were the best to use. It has a very modern sound. Unfortunately, then as now, the difficulty was to define locality or measure the superiority.

Next came the notion of prepotency which had the effect of diverting some attention from the breed or race to the individual animal. It is still being used (Briggs, 1958). As with the preceding notions, this one had elements of truth in it, which are referred to now as homozygosity, dominance or epistasis according to the kind of gene action involved. Erected upon this rather small foundation, however, were some theories about the prepotency (and thus the desirability) of particular animals, or breeds, which seem likely to have transcended the reality or else to have misinterpreted sampling errors.

In 1859, Darwin published "The Origin of Species", and in 1865, Mendel the results of his studies of inheritance in garden peas, but they had no immediate impact on animal breeding. Galton (1822–1911) was also ignored, although his quantitative approach to variation among relatives could have been valuable. That had to wait for Fisher, Wright, Haldane and Lush in the period 1918 to 1939.

While the founders of modern population genetics were being disregarded, breed societies and herdbooks were growing in strength. Studbooks began with the first volume of the General Stud Book for horses in 1808, and was followed by the Coates Herd book for Shorthorns in 1822. During the next hundred years many more were started. Mason (1957) names 24 cattle breeds, 16 pig breeds, 21 horse breeds and 33 sheep breeds in Great Britain (Colburn, 1963, lists 43 breeds of sheep). In the United States of America there are almost as many. Rice *et al.* (1957) give the addresses of many Livestock Registry Associations: 25 for cattle, 16 for swine, 22 for horses and 25 for sheep. Most of these societies or associations still subscribe to the principle that pedigree allied to the use of eye judgment for securing adherence to formalised breed type is the basis of successful breeding. This formalistic approach is intellectually related to the typological view of systematics. For the typologist the type is real and the variation an illusion, whilst for the populationist the type (average) is an abstraction and only the variation is real. Replacing the static idea of type by the dynamic notion of an interbreeding population moving erratically and at changeable speeds towards an unknown future is ranked by Mayr (1963) as perhaps the greatest conceptual revolution in biology.

The beginnings of systematic research on the genetic aspects of animal breeding are of recent origin. Important as they were in moulding the thinking of breeders, the Pennycuik experiments of Cossar Ewart, carried out in the period 1896–1904 do not really fall into the stream of genetic research on methods of animal improvement. In fact, considerable strides towards attaining today's standards of excellence in many kinds of animals were made in pre-Mendelian days, and it may be said that there was genetic research before genetics itself was born. The roots of research in animal breeding are to be found in empirical observations and field trials long before theoretical foundations for them had been laid.

During the late nineteenth century, attempts to measure the success of mass selection in raising egg production in poultry and the experiments on the inbreeding of guinea-pigs, started shortly thereafter by the United States Department of Agriculture (very likely the first use of laboratory animals as pilot subjects for studies specifically directed towards animal breeding), were initiated outside the framework of Mendelism. Nevertheless, they can probably claim to be the earliest systematic efforts to investigate the genetic basis of animal improvement. The rediscovery of Mendel's laws in 1900 had a stimulating effect. The relative ease with which Mendelian inheritance could be investigated for truly monogenic situations and the unbridled opportunities for invoking various types of gene interactions in studies on polygenic characters allowed a great deal of experimental work to be done. Some of it was useful.

It was, however, the coming of population genetics which gave the greatest impetus to the development of animal breeding investigations. Though the foundations of the theory and analytical methods of population genetics are usually attributed to the trinity of Fisher, Haldane and Wright, it was the last who played the most important role in bringing quantitative genetics into livestock improvement. Not only was he himself active in animal breeding (having served as Animal Husbandman in the U.S. Department of Agriculture before becoming a Professor of Zoology), but it was his formulations upon which Lush, the greatest exponent of this approach, built.

The influence of Lush on the development of animal breeding research can hardly be exaggerated. Since the first appearance of "Animal Breeding Plans", hundreds of investigators trained directly under him or under his students, have followed his lines of research. Probably no research institution in the Western world which deals with the genetics of domestic animals lacks one such worker. Contributions from the school of Lush have been many and significant both from the theoretical standpoint and in application to breeding of all classes of livestock, poultry, and more recently, fish.

It was not only because it elaborated the inheritance of continuously distributed traits that population genetics became important. At least two other factors played a highly significant part in the spread of this school of thought. One was the possibility of generalising the theory of quantitative inheritance and so justifying the wider use of rapidly breeding animals such as mice, rats, *Drosophila, Tribolium* or quail for reaching conclusions relevant to larger animals. The other factor was the development of statistical analysis which could be applied to existing accumulations of data. For some purposes it was therefore unnecessary to wait for several generations of breeding.

An attempt at summarising recent evolutionary history of swine breeding has been made by Comstock (1960). His choice of important developments during the last century will serve for all classes of stock.

1. Formation of standard pure breeds and registry associations.
2. Development of Mendelism and population and quantitative genetics theory.
3. Recognition of the value of cross-breeding.
4. Renewed breed formation and the founding of the Inbred Livestock Registry Association.
5. The shift of emphasis in selection to economic traits.
6. The analysis of selection effects.
7. Measurements of backfat on live hogs.

Other compilers might make a different and shorter list. The next two items on his list could be the general adoption of performance testing for breeding stock, and the building up of large breeding populations under unified control.

II. Functional Stratification

Large scale breeding, like any other significant evolutionary or cultural change, can only take place when the circumstances are right for it. In a complex industry, such as livestock production, a change of this order causes repercussions in many directions and calls forth strong reactions. For understanding recent events and for assessing their influence, it is as well to consider briefly the existing organisational structure and the modifications of which it is capable.

There are now three kinds of functional stratification in animal breeding that are easily recognised. Firstly, there is stratification on the basis of geography and land use (Nichols, 1957; Epstein, 1965). Secondly, there is stratification by repute, namely breeders of elite herds, multipliers and users of non-registered stock (see Chapter 7). Although originally this kind of stratification was based, or was intended to be based, on the performance of animals, it has gradually come to apply to the performance of the breeders themselves. Performance stratification is now being re-established as progeny testing and selective breeding from proven sires lead to the formation of highly selected strains and derivatives of varying degrees of relationship to them. There may well be a fourth type of stratification in the near future if minimal disease or specific pathogen-free stocks of animals prove to be economically justifiable.

The importance of geography and land use depends on the kind of stock. Owing to the bulky fibrous food they graze, ruminants are more affected than non-ruminants. Technical control of environment, however, influences even ruminants under extensive conditions. Housing and feeding, the supply of minerals and vermifuges, and intensification generally make the relationship between land and climate and between land and breed less important. Both the need for and the speed of adapting stock to new intensive forms of production are greater than for extensively kept or range stock. Altering the genotype of range sheep or cattle may be a risky proceeding if the physiological nature of adaptation is not known. This is also true of animals for intensive farming, but it may be easier to adjust the environment if the stock prove deficient in some respects. Long continued selection under range conditions has produced genotypes that in some ways are well suited to intensive conditions, for example, in fertility, temperament, or fleece type as the use of range or hill ewes for lowland crossing shows. In other ways, however, they may have become unsuitable, notably in growth rate or conformation.

Dairy cattle, pigs and poultry are reared under a wide variety of conditions, a variety that is probably greater than is realised, since so little is known of the minor challenges to health faced by livestock. Sheep are still markedly regional in distribution (see Portal and Quittet, 1950, about France, and Lall, 1956, about India). Tradition is important here, but adaptation by physiological or psychological means such as temperature regulations or behaviour in bad weather respectively, may sometimes be critical. In contrast to the merino, which apparently has a compulsion to go on producing wool protein even when underfed, other breeds, at home in regions where a regular winter shortage of food coincides with pregnancy, may be found to reduce wool production and to exploit protein catabolism to help them survive.

To many people the most striking effect of geography and land use is to be seen where intensive cropping excludes all livestock. But the classical form of stratification is shown where sheep migrate from less to more productive land and change from purebreeding to crossbreeding in the transition (Fig. 6.1). The eastward movement of Western ewes for crossbred lamb production in the United States, and the downward movement of ewes in Britain from poor hill grazings to land capable of producing fat lambs are well-known. An extension of this system has been forecast for the supply of suitable ewes for indoor production of lamb on the principle that the breeding ewes should be reared on land that is cheaper than the land they are in turn to rear market lambs on.

To unite the advantages of cheap breeding stock with the higher performance that their offspring can show when food supplies are more liberal, it is necessary to use suitable rams. They must combine with the ewes in such a way as to produce a crossbred lamb appropriate to the market. Consequently, they need to be complementary to the ewes, rather than excellent in themselves. To meet varying requirements there is a range of breeds specialising in the production of these sires of which the Hampshire and the Suffolk are examples. Extremes of size are to be had in the Southdown and Oxford, and an extreme in wool

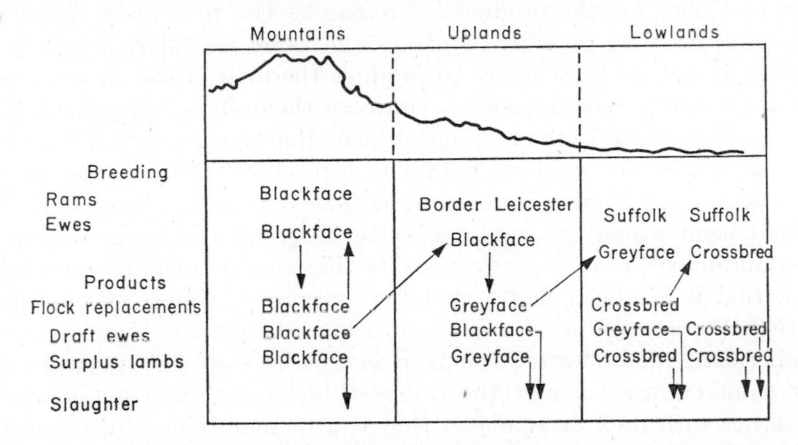

Fig. 6.1. Diagrammatic representation of breed stratification in the sheep industry of Scotland, based on the Scottish Blackface and its cross progeny, the Greyface.

character in the fast growing but fleeceless Wiltshire Horn. In all these breeds an ability to confer carcass quality on the progeny is usually sought. Little seems to have been done as yet, however, to use progeny testing in discovering and multiplying the rams which are meritorious. In future there will be a demand for rams that can be relied upon to produce a carcass of the right weight and quality for contract supplies. For intensive production completely new breeds may be necessary to match with high fertility ewes.

Crossbreeding in sheep seems to depend only to a relatively small extent (0–15% of the parental mean) on heterosis (Sidwell, Everson and Terrill, 1962; Donald, Read and Russell, 1963), but it has not yet been properly studied in conditions where vigour is most valuable. Practical breeders have probably profited somewhat by heterosis in the crossbred ewe. This type of ewe is common in Britain because the numbers of draft ewes from the hill land are well balanced with the demand from low land and have at no time far to travel. These draft ewes become dams of crossbred ewes sired mostly by Border Leicester or Suffolk rams as is shown diagrammatically in Fig. 6.1. Apart from heterosis, the object of the crossing is to produce a ewe that has the size, fleece, fertility and milk production needed in the mother of a fat lamb. It might be hard to find a more complicated task than improving the sire of such a ewe. Yet it might not be wise to be intimidated by it since changes in just one trait might be useful. The real problems are to define the aims and to obtain the necessary data.

Two forms of stratification by performance have appeared since the war, and both are the products of research. The first arose from the organised crossing of inbred lines of the same or different breeds of poultry in one or more stages to produce the final broiler or egg-layer. Although comparatively poor performers themselves, the inbred lines merge successfully in the making of birds that accurately meet specifications, which may include failure to reproduce. Dairy cattle are far from emulating poultry but are nonetheless near to establishing the second form which is a ranking on the basis of their relationship to outstanding sires found by progeny testing and artificial insemination. Now that it has been shown that the most likely source of good bulls is the offspring of proven sires mated to daughters of proven sires (Robertson, 1960a, Carter, 1962), it is a short step to segregating (on paper and on farms as well) the animals which meet these qualifications. Together with their descendants they can be maintained by a continuation of the same process. With several such groups, independent regional policies can be followed. If these policies result in the creation of a set of sub-breeds or lines within a breed, then organised crossing between lines becomes a possiblity as well as the classical inter-group selection process.

III. Large Scale Operations in the Poultry Industry

In the United States the poultry industry has developed into a small group of large-scale integrated enterprises with mechanisation, factory-farming, technical specialist control, and a monopolistic trend in breeding. Sykes (1963) has confidently predicted that this is bound to happen with other classes of livestock throughout the world with, perhaps, a few minor exceptions. Even in some of the smaller new nations of Africa, this may be simply a matter of time.

The details of the causes which have brought about this metamorphosis of the poultry industry in a remarkably short period of time may be different from those which will be important in larger animals. Artificial insemination, for instance, a powerful influence in this direction in cattle, played no significant role in the poultry revolution. But the basic causalities are the same and spring from the necessity for greater efficiency in food production, as outlined in Chapter 2.

As late as the beginning of this century poultry breeding as a business was oriented primarily on poultry shows in which merit was judged by conformity to the breed standard. Commercial qualities of poultry were not being measured precisely and contributed very little to the standing of a breeder as a source of stock. There was a great number of different breeds and the numerous breeders propagating them had their reputations established almost entirely on winnings at poultry shows. Most of them were back-yard fanciers, and poultry for food production was only a part-time occupation which provided pin-money for the farmwife. Little by little, however, commercial exploitation of poultry as an important source of food began to develop. Following the invention of methods for storing, packing and distributing eggs, specialised egg-producing farms began to arise. Shows still dominated breeding but, in addition to classes judged on breed standards, so-called or miscalled, utility classes were introduced. In these, merit was measured on certain assumed relationships of physical appearance to production qualities, relationships which later were demonstrated to be fallacious. An even more important development was the spread of egg-laying tests in the nineteen-twenties and thirties. These existed in two patterns, based either on the production of individual birds, or on the production of pens of ten to thirteen pullets, entered by breeders in contests conducted by state agencies at central locations. Individual performance was recorded for a year by trapnesting. The breeder himself decided what birds he would enter in a test, the only qualification for entry being conformity with the breed standard. The ranking of competitors in a test was determined mainly by the number of eggs laid by a pen and replacement of birds dying in the course of the test was often permitted. Prizes were emphasised not only for pen performance but also for individual records. A "world-champion" layer would assure its owners a respectable sale of individually pedigreed hatching eggs at prices grossly inflated relative to their intrinsic breeding worth.

Although tests of this kind lasted for several decades, there was a widespread realisation that they did not necessarily reflect the quality of the commercial stock which a breeder had for sale. However, breeders themselves were not in the least interested in abandoning this type of test, since the main source of their income came from selling settings of eggs and individual offspring from test winners or, occasionally, of the test winners themselves. Great emphasis in such sales was placed on pedigrees and on even remote ancestors.

In the nineteen-twenties and thirties, recording schemes, or record-of-performance tests on the breeders' own premises under Government inspection, were started. Not only conformity to breed standards but also certain pedigree requirements were instituted as conditions for entry. Adding egg size to the qualifications for certificates of performance also characterised this stage. Disease control, especially of pullorum, carriers of which could be identified by agglutination tests, was also introduced as a condition of official recognition of breeding and production quality.

Meanwhile, the perfecting of mammoth incubators, the introduction of mail shipment of baby chicks, and the building of large plants for the commercial production of eggs supplying the increasing urban population, led to a stricter competition between breeders themselves, and also made them direct competitors of the previously existing middlemen: the multipliers and hatcherymen. Eventually both of these classes disappeared from the stage as independent operators. Their functions are now performed either by the breeding establishments themselves or by subsidiaries working on the franchise or licence system in which the immediate parents of chicks supplied to the customer are produced under the supervision of the licenser. Indeed, the majority of the breeders who survived this change and stayed in the poultry industry have now become licenced multipliers.

The next important development which occurred in the late nineteen-thirties and in the forties included the rise of population genetics, with emphasis on quantification, on objective evaluation of breeding worth, and on short-cuts and more efficient operation. With the switch in the nature of competition from an aesthetic to an economic competition, an increase in the number of traits which assumed commercial importance occurred. Such characters as freedom from blood spots, good shell quality, as well as traits of economic significance to the breeder and hatcheryman (fertility and hatchability) began to play an increasing role in the breeders' operation. An outstanding example of the latter type of character is egg number in turkeys which, of course, is of no interest whatsoever to the purchaser of baby turkeys to be raised for meat, but is of signal importance to the profit or loss that the breeder or hatcheryman can make from his flock.

These developments led to great modifications in the structure of the poultry breeding industry. Four interrelated factors may be discussed here briefly: (1) the necessity for breeding establishments to employ trained geneticists; (2) the random sample test which arose from the need for objective measurements of economic worth of the actual product sold and not of breeder-selected samples tested in the standard egg-laying test; (3) the development of the hybrid and crossbred fowl, which spelled the death of breed standards in commercial industry; (4) the self-accelerating phase in the growth of individual establishments and the accompanying reduction in their number.

The stimulus to employ professional geneticists arose from the desire of a few pioneers who wished to put their selection and mating procedures on a firm scientific footing. It was John Kimber of California who over a third of a century ago foresaw the future of commercial poultry breeding enterprises as big business with a scientific approach in every step of operation. The success and rapid expansion of his firm, in part due to the methods he adopted and in part, of course, to his personal managerial ability, has led others to accept his philosophy. In the early years of this development, commercial firms raided universities and experimental stations for staff. Certain changes in postgraduate training to supply the new market for the industrial geneticist (openings heretofore restricted to plant breeders) also occurred. At present an equilibrium probably exists. Since the number of independent breeding establishments has been drastically reduced, the total number of geneticists employed by the industry is probably no longer increasing rapidly.

The rise of the random sample test deserves notice. The principle is to test for production qualities large samples of the actual stock that is being offered for sale by a breeder, rather than to evaluate selected small groups chosen by the breeders themselves, groups which may have no genetic relation whatsoever to the birds being supplied to the customer. Random sample tests had been advocated by Hagedoorn (1927) for nearly twenty years before they came into existence in America. Shortly after the last world war an experimental random sample test was organised in California through the initiative of poultrymen, breeders and experimenters. The system has gradually spread throughout the United States and Canada. Whatever the original purpose of the random sample test, it had a considerable effect on the structure of the poultry industry by leading to the virtual extinction of the small breeder. What happened was that customers tended to take the published results of the early random sample tests too literally and accepted the ranking as measuring the exact order of genetic merit of stocks offered for sale by the various entrants. This meant that, if, by chance, a small breeder did not gain a very high place in the test (even if this reflected only an environmental component in the performance of his stock), he could lose so many sales to his larger competitor that he could not survive. In contrast, a large breeder with capital reserves could afford to contract his business temporarily if he did not succeed in a test in one year, and later expand rapidly so as to capitalise on a chance win. The process of breeder extinction in the manner that statisticians call a random walk was thus initiated.

Attempts to reform the tests so as to make them less of a competition and more a quality control measure (for example by issuing labels to breeders meeting certain standards without necessarily publicising their actual performance or rankings) were only partially successful. Dickerson (1962a) has reviewed the changes which have occurred in random-sample tests since their inception with respect to their avowed purposes, their numbers, their reliability, and the eventual use of information obtained from them. These tests are expensive to carry out, and the justification for them must vary with the stage of evolution of a poultry industry. When there are numerous breeders and breeding stocks of unknown merit, the tests perform a valuable service in comparing them under standard conditions. As time passes and the surviving stocks become closely similar in estimated total economic returns (less feed and chick costs), it grows progressively more difficult to identify superior lines, notwithstanding larger samples and repeat tests in several states. Furthermore, influences other than genetic merit, such as transport costs, advisory services, and personal contacts, play an important but imponderable part in determining a buyer's choice. The random sample test helped to establish the crossbred chick and the large scale breeder, but having done so, it needs a new purpose if it is to survive in North America. Elsewhere, its value is for the time being more apparent.

The increased utilisation of crossbred or hybrid (a cross between two inbred lines of the same or different breeds) has had a tremendous impact on the poultry breeding industry. Byerly (1964) estimates that two-thirds of all broiler chicks produced are crossbred, while 70% or more of all breeding flocks in the United States are either hybrids or crossbreds. It is impossible to tell now whether the same amount of effort expended in the improvement of large closed flock populations by selection would have produced better or poorer results, but the exceedingly skilful exploitation of the hybridisation techniques by their early developers has forced most surviving breeders of poultry for competitive reasons to apply the same methods.

All of these factors have been influenced by and contributed to the fourth one, the increased size and drastically decreased numbers of breeding establishments. Apart from the simple fact that the successful firm ordinarily expands at the expense of its less fortunate competitors, a critical mass is needed in order to obtain a certain rate of improvement. Only a large enterprise can sustain widespread advertising and permit the breeder to develop the expensive inbred lines and to employ professional planners and managers. Air transport of chicks and the franchise system have made this possible. Lerner (1962) noted that some thirty-five years ago in the Province of British Columbia with a total human population of under three quarters of a million there were more than one hundred breeders of egg-laying stock. This is to be contrasted with the twenty-nine members in 1962 of the Poultry Breeders of America, which is the association covering both the United States and Canada and includes not only breeders of egg-laying stock but also chicken meat and turkey breeders. It may be added that the 1963 membership roster of the same organisation was reduced by some 10% to a total membership of twenty-six. Because of the high capital investment which is now required to enter competition in this field, it seems unlikely that the number of independent breeders will ever increase.

There is probably no future for a company which is not big enough to support the research and development to maintain its position. The application of genetics and the results of laying trials have reduced the breeds and strains to a very small number. The laying birds now available are all very similar, so that success or failure of a business will depend, in part, on how efficient the breeding system is for making further gains, but increasingly on the sales organisation.

Having forced out the small breeder, the few remaining large establishments are now engaged in competition with each other, and are beginning to depend more heavily on foreign markets to maintain their financial soundness. The picture in many ways is very similar to that of the motor car industry in the United States in which only four companies are now left out of the hundreds that came into being since the time the automobile was invented, and in which one corporation produces more than half of all the cars made. There is a further analogy with the motor industry. Just as both Ford and General Motors had to diversify their products and put out cars, trucks and tractors, so do the breeders of one kind of poultry apparently need to expand by breeding other types of stock. Thus many of the breeders of egg-laying stock are also attempting the improvement of turkeys and some participate in the broiler industry. Indeed, in some ways, Great Britain has already outpaced the United States: at least one poultry breeder has entered both the sheep and the pig industries.

That there is an opposition to the trend from many sources is undeniable. Truths, many half-truths and a variety of myths (some examined by Robertson, 1964) about allegedly anti-humanitarian aspects of modern agriculture complicate the issues. The opinion is often expressed that the intensively produced foodstuffs are tasteless. Whether this is the case or not remains to be demonstrated by objective tests which are now very difficult or impossible to devise. Rapid growth and early killing probably do result in more tender but less tasty products. But the uniformity and reliability of quality in such products as pigs or broilers or eggs has been clearly improved under the intensive system, and from the standpoint of the welfare of society at large there is very little doubt that the increase in efficiency of production of these commodities is a desirable accomplishment. It may also be true that certain types of foodstuffs have become uneconomical to produce and, because of that, scarce. Thus, the roasting fowl has practically disappeared from most American markets. Similarly, should a consumer have a predilection for dark brown eggs, he would be hard pressed to find them at the ordinary grocery counter in California, because compulsory candling is now done by electronic scanners which cannot detect bloodspots in such eggs.

It is fair to say that this situation is a reflection of the small demand for such speciality products. Undoubtedly they could be produced and marketed, as they are in Britain, should enough people be willing and able to pay for them after they have become a luxury. To say that brown eggs should be available to all who want them now may be the same as saying that grouse and lobster should be available to all. The fact that at one time such eggs were a common enough commodity is not particularly relevant. Patterns of supply and demand change. Not long ago liver was fed only to pets and was available at very little cost. Today calf liver is not a cheap meat. Turkeys at one time were only festive fare. Today they are often one of the cheapest animal proteins available in accepted form.

IV. The Consequences of Artificial Insemination

Artificial insemination has been both a result and a cause of the revolution in large animal breeding. Its development was made possible, in the face of much criticism, by the efforts of scientists spurred on by wartime shortages and given their opportunity firstly by research administrators and then by dairy farmers. Many experts found it necessary to oppose its use for reasons they will now prefer not to recall, such as the notion that artificial insemination leads to the production of monstrosities and degeneration, inbreeding and reduced fertility. "Unnatural" the process certainly remains, but it proved itself so economical, effective and advantageous to so many herd owners that it became firmly established in all countries with advanced dairying. For general discussions of techniques and application, Perry (1960), Salisbury and Van Demark (1961) and Maule (1962) can be consulted.

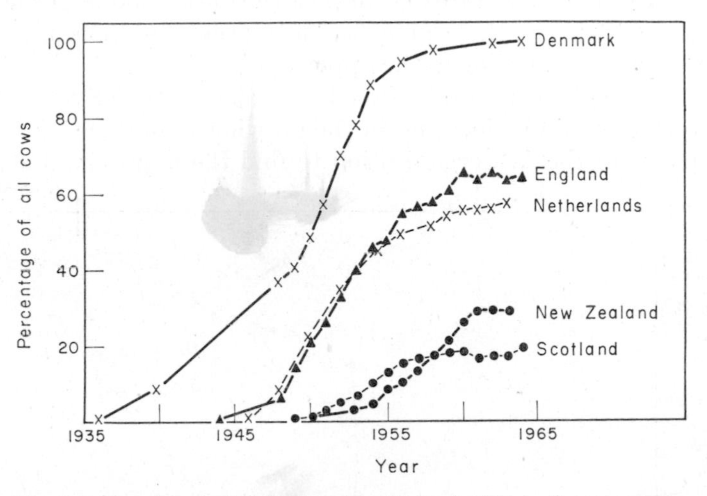

Fig. 6.2. Growth of artificial insemination of cattle in five countries.

The numbers of cows inseminated have expanded in a typical S-shaped growth curve (Fig. 6.2 shows some examples) but there are some interesting exceptions, for instance, Czechoslovakia, United States and France (Fig. 6.3). For these, it may be speculated that the usual growth process is incomplete or obscured by political and other considerations. Where countries have differed is in the point where the rate of growth has fallen away to a low value. The approximate percentages of all cows artificially inseminated in a few countries with large dairy industries are shown in Table 6.1. These few are sufficient to show the wide range of average herd sizes and their relationship to usage of artificial insemination. Distributions of herd sizes would be more informative. In the United States with its diversity of conditions for dairying, a national average herd size is of little informational value.

The appeal of artificial insemination to a dairyman varies and those with large herds, or in thinly populated or remote areas, will often prefer natural service. Others with disease problems, or needing several breeds of bull, or with small herds can be expected to favour artificial insemination. Advertising, salesmanship (especially in the U.S.) and official propaganda will also affect their choice. That there is more to it than this, however, is suggested by the fact that Denmark, where all services are closely integrated, has replaced natural mating almost entirely by artificial breeding. The point is not without significance, since the power of artificial insemination to realise its potential depends on its scale and its independence. Both are likely to be adversely affected by the trend towards better health control and larger herd size through the strength these give to the market for bulls for natural service. Influences working in the opposite direction can be found in the demand for rapid changes in breeding policy especially towards or away from beef production, and in the growing confidence of dairymen in the power of the A.I. organisation to find the improving bulls.

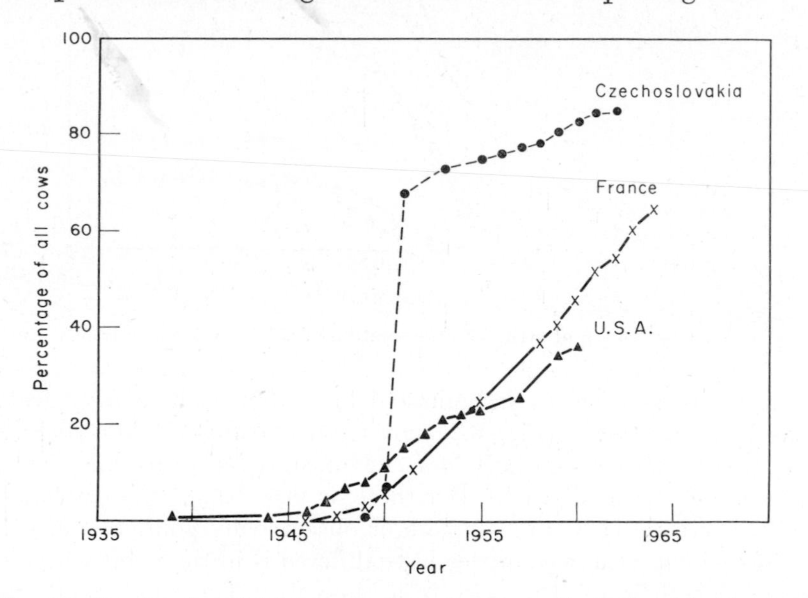

Fig. 6.3. Growth of artificial insemination of cattle in Czechoslovakia, France and U.S.A.

The effects of artificial insemination have impinged very unevenly upon those who keep cattle and breed them. Owners of cattle who have little interest in breeding have been able to use artificial insemination to reduce costs, avoid venereal disease and take advantage of the care which A.I. organisations put into bull selection. But for the smaller number interested in the sale of bulls, the market has contracted. It is not easy to discover for a particular breed how much of the change in the number of bulls sold can be attributed to artificial insemination and how much to changes in demand arising for other reasons. The production and sale of dairy bulls has no doubt been substantially reduced in total but not necessarily for a specific breed. The prices received for those that are still produced must be affected by the cost of the alternative to natural service. Furthermore, when the number of herdbook registrations declines, financial stringency for a breed society follows.

TABLE 6.1

Percentage of all cows artificially inseminated (from Adler, 1964, and other sources)

	%	Average herd size (all herds)
Denmark	100	12
Holland	60	13†
Sweden	30	11
New Zealand	31‡	70
U.S.A.	30	—
England and Wales	66§	24
Scotland	19§	44

† Tested herds only.
‡ From James (personal communication, 1965).
§ From Melrose (in Maule, 1962).

Artificial insemination has also brought powerful influences to bear in other ways. Farmers are now able to change their breed by grading up and to switch from dairy to beef-type calves very quickly. This resource in farmers' hands can lead to chasing the market and thereby cause over- or under-production of both types. Another result is to bring rapid changes in breed numbers. For instance, only 36% of Ayrshire cows in England in 1963 were bred to Ayrshire bulls. Of the rest, one-third were inseminated by Friesian semen and two-thirds by beef semen. A change in popularity can be given expression much more promptly than formerly.

The search for bulls for artificial breeding, guaranteed to improve performance, made it necessary to refine the techniques of progeny and performance testing (Carter, 1962), and to develop the recording of information in such a way as to provide ready access to the data. Figure 6.4 shows how increased family size made possible by artificial insemination improves the accuracy of the progeny test. Whereas the tests on bulls in the upper part of the figure were made on an effective average of 11·1 daughters, those in the lower part had 36·7 daughters

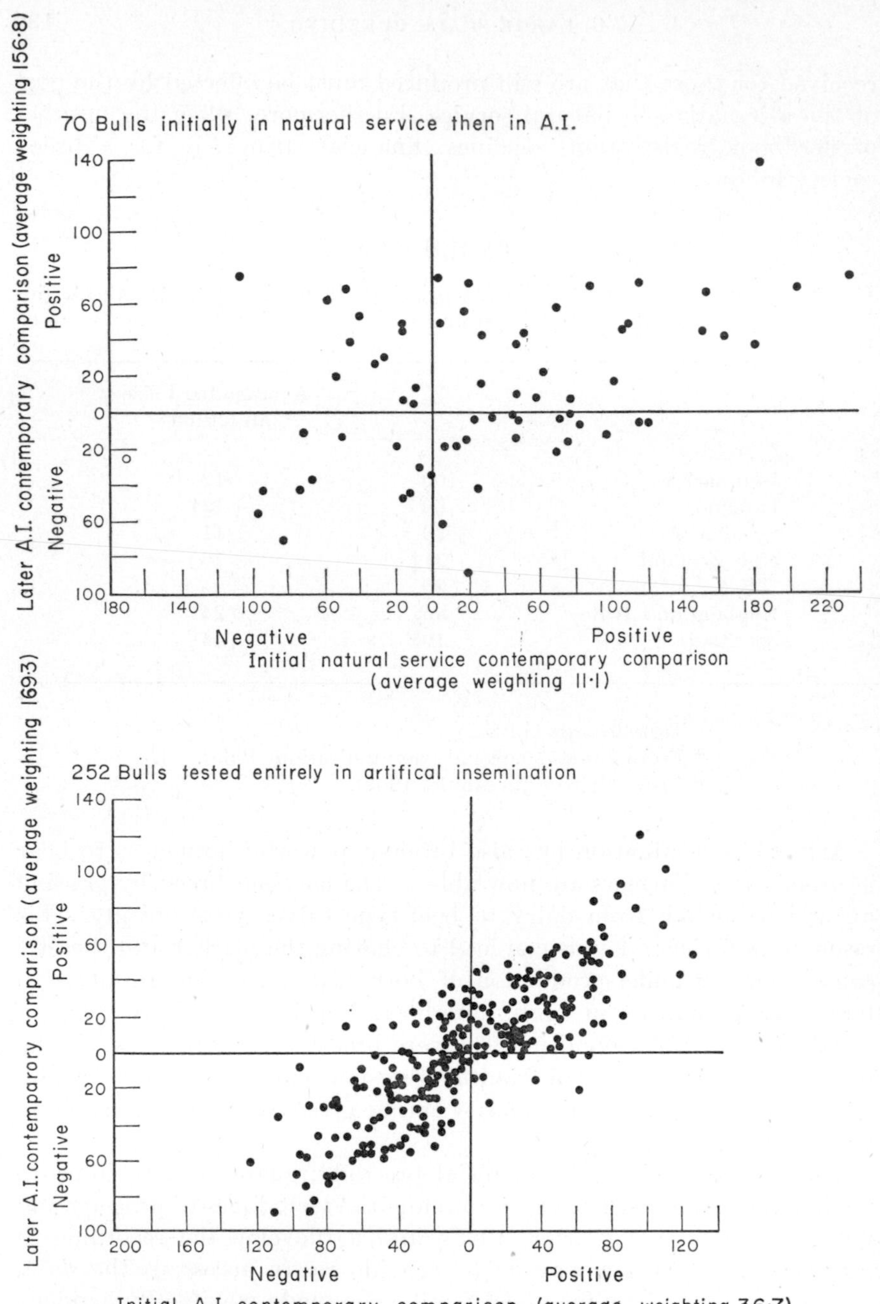

Fig. 6.4. Comparison of the accuracy of progeny testing in relation to numbers of daughters (from "Twenty-five million", a promotional pamphlet issued by the A.I. Organisation of Great Britain and reproduced by permission of the Milk Marketing Board).

on test. The figure embodies the advances made in physiology, genetics and data handling, which are adopted in different parts of the world with varying degrees of collaboration from breed societies. While such collaboration is helpful, it cannot be regarded as essential. The organisation which possesses the information on performance and has the staff capable of analysing it, is in a strong position. Indeed the emergence of a body of people with statistical machinery at their command and with the means of publicising their studies and conclusions has altered the status of the private breeder substantially. It is not surprising that sometimes he wishes to retain the importance of eye judgment and the value of winning show classes on visual appraisal, but he has now to contend with the fact that the economic characteristics are recorded and interpreted by scientists. One of the first things these people did was to reveal that genetic differences between herds were very small. In so doing, they undermined the foundations of pedigree breeding (Pirchner and Lush, 1959).

By virtue of their scientific and field staffs and central data processing, A.I. organisations are capable of carrying out operational research of value to dairymen. Witness to this is given by the annual reports of the New Zealand Dairy Production and Marketing Board and the Milk Marketing Board of England and Wales. The same publications are also stimulating to research workers and point the way to closer contact between research institutes and A.I. organisations which would benefit both parties. Indeed, as the collaboration at Cornell and Edinburgh of research workers and A.I. centres has shown, the fruit may be enjoyed over the whole world. It also suggests a means by which modern constructive breeders and those adventurous spirits who always are willing to try something new can obtain the use of up-to-date techniques and skills.

The pioneer work carried out with dairy cattle will be available to guide and hasten the development of artificial insemination with pigs, as the physiological limitations are overcome. In the U.S.S.R. the use of artificial insemination in sheep is already more extensive than in cattle. Pigs and sheep have yet to go through the stages of searching for and finding animals of superior performance to use in artificial insemination. There seems no reason to doubt that such animals exist but the finding of them will require arrangements at least as refined as those applying to dairy cattle. It can be anticipated also that the controversy between the proponents of type and the proponents of performance in terms of growth rate and fertility and other quantitative characters, will be revived. As evidence accumulates, it seems, however, unlikely that any pursuit of type for type's sake will last much longer.

Where type has a proven relationship to economic output, it will, no doubt, be retained as a criterion of selection.

The technique of deep freezing of semen makes it possible in theory to test long-dead ancestral sires against their modern descendants (see section on gamete preservation in Chapter 8). Whether any animal breeding institution would be willing to embark on long term deep freezing of sufficient semen in sufficient variety and then test it, remains to be seen. It would run the risk that ideas on conformation and on the proper objectives for breeding have so altered that no farmer would willingly agree to the use of the semen on his animals. Furthermore, there might be some uncertainty as to whether any gametic selection had taken place during the long freezing.

V. Improvement through Artificial Insemination

That there is scope for improving methods of selection among dairy sires is shown in an analysis by the United States Department of Agriculture (1964) of 1046 A.I. sires and 1508 naturally mated sires. Seven breeds were studied. Of the A.I. sires, only 46% maintained or increased milk yield and, of the others, only 51% did so. To estimate the amount by which yields of dairy cows have been raised by artificial insemination has so far proved a rather intractable problem. No doubt the merit of commercial cattle has been raised by progeny testing and preferential use of the superior sires, but there are few firm calculations to show by how much. For discussion of this problem, Robertson (1960a), New Zealand Dairy Board (1961) and Van Vleck and Henderson (1961) should be read. Table 6.2 has been extracted from more detailed information in a recent report of the New Zealand Dairy Production and Marketing Board (1964). Although only a minor fraction of the national herd is as yet bred from A.I. bulls (including young ones being tested), the calculated increase in earnings, if realised, would be substantial and well worth having in addition to the other useful services provided. The monetary cost of securing these earnings will be of great interest. At 3 shillings a pound for butterfat, the increased production of artificially bred cows is at present worth to a New Zealand dairy farmer about 60 shillings ($8.40) in each lactation.

Outlays on recording yields and testing bulls are in part an investment in the breeding of more profitable animals. To decide how large that part is, will probably engage economists for some time, but the effort should be made. Poutous and Vissac (1962) have made a start by considering the progeny test of bulls used for artificial insemination as a financial operation on behalf of the members of a centre. Profitability

TABLE 6.2

Estimated increases in numbers of artificially bred cows and sales of butterfat from a population of about two million dairy cows in New Zealand (from New Zealand Dairy Production and Marketing Board, 1964)

Year	No. of 2-yr old cows in milk bred by A.I. (in thousands)	Superiority of A.I. stock (lb fat)	Increase in earnings (in thousands of £)
1952–53	0·4	+10	0·6
1957–58	21·0	+15	81
1961–62	81·0	+19	584
1962–63	103·8	+20	822
1963–64	124·0	+20	1056
1966–67	147·3	+25	2093

will be affected by the purchase and maintenance costs of bulls, the amount of semen sold, the cost of collecting the data, and the benefit of the extended use of the best of them. Upon the answer depends a rational decision about the amount of money that should be devoted to this activity. It is not necessary to wait for the answer, however, before deciding that a breeding programme designed to maintain a balance of opposed objectives (such as more meat and more milk from a cow of fixed size) will not pay dividends on the investment.

So far, A.I. operators appear to have acted on their own or their customer's faith in the ultimate returns on outlays for improvement. Until they set out their breeding aims and show that their expenditures on performance testing are consistent with these aims and economically justifiable, there is no way of deciding whether such work should be expanded or contracted. By degrees the logic of events is pressing them slowly towards a decision on breeding aims. Unless each generation of bulls for artificial insemination is the product of the previous generation on which money has been spent, the profit from all the work on improvement through breeding is limited to expanding the use of good bulls in commercial herds and reducing the use of poor bulls. Rate and direction of change are still influenced by the breeders from whom A.I. bulls are bought. In a large business for which cattle are only a means to making money, the question "does it pay" has to be answered for the whole concern. It is directed to the kind of cow to be bred, to the policy of pure breeding, and to the amounts invested in improvement. Whether or not to bother with numerically weak breeds is another unhappy problem.

Where the breed associations are already closely integrated with artificial insemination and other services, as in Sweden and Holland, the conflict of interest is less acute and less apparent than in countries where they are not. In these, it seems that those who control artificial insemination will sooner or later be forced for economic reasons to use its potential to breed the kind of cattle customers want as fast as possible. They are not likely to do this until they have secured the support of dairymen, technical staff and equipment, and courageous leaders. If they reach this point they will feel strong enough to settle the issue about breeding objectives and if necessary act independently of the breed associations. The latter, however, are likely in this event to modify their ideas, rather than lose the market, as well as the technical help that artificial insemination can give. When this happens, the way will be clear to follow one or more jointly pursued breeding policies with new stratifications of animals and men based on new definitions of performance. Only then can the combined strength of the A.I. organisations and the breeders be brought to bear on (a) efficient purebred selection, (b) the production of distinct breeding lines within breeds, (c) developing cross-breeding merit, and (d) new breeds.

VI. Large Scale Breeding of Cattle, Sheep and Swine

The phrase "large scale breeding enterprise" deserves a closer examination than it has had yet. "Large" is a relative term. In Finland 100 ewes might be regarded as a large flock, in Scotland 1000, in the U.S., 10,000, and in Australia 100,000. In each case the flock might be managed as a unit for a single owner, and it may have an internal organisation for breeding improved sheep, but it will not be a large scale operation in the sense implied here. Neither finance, nor management, nor size relative to the modal flock size is crucial. The point can be pursued conveniently in Australia where there is a highly organised sheep industry (Short and Carter, 1955, Turner, 1956). Analysis of the Merino Stud flocks there showed that they were stratified in four main subtypes within which there were twenty-four parent studs with 164,000 ewes dominating the 100 million or more sheep through daughter studs and general studs providing the bulk of rams used in non-registered flocks. A further stratification segregates sheep into top, single, or double stud sheep and flock ewes and rams within flocks. This is an Australian version of the pedigree system.

The Danish version for swine (Lush, 1936; Jonsson, 1965) is also large scale in that it applies to the whole of the nation's stock. The concept is of elite herds breeding improved animals with the aid of progeny testing and technical supervision of breed type and performance. Swine breeding is based on 250 accredited breeders. In order to qualify for the title of "accredited", breeders' herds must not only perform to certain standards, both at home and in the testing station, but the breeders themselves are subjected to careful personal and technical selection. Their efforts are coordinated by an integrated system of advisers, four large testing stations, and recognised breeding centres or farms which have a total of about 3000 sows and 700 approved boars. The advisory service is run by Farmers' Unions.

Since the Danish bacon industry is highly dependent on an export trade which is very selective in its requirements, it is essential that the aims of breeding should be carefully aligned with the requirements of the export market and that the aims should be faithfully pursued by all the accredited breeders. Under such a system, the advisers should be well-informed, their statistics should be up-to-date, their interpretations of them should be correct, and prices should be favourable to official policy.

So far, this system seems to have worked well, at least in the absence of equally well organised competition. Naturally, there are problems and, naturally, the statistics also tend to be out of date when they become available. A task of the decision makers is to recognise deficiencies before they become serious and to take the right corrective measures. At the present time it appears that the success of the Danes in increasing growth rate and meatiness has brought a problem with it. There is evidence that some of their swine are becoming genetically liable to show a pale and watery flesh after slaughter, but there is also central authority able to direct research to the difficulty, and to see that the solution is promptly applied.

Although Danish swine breeding approaches large scale breeding in the sense used here, it does not fully qualify, because decision making in regard to breeding, although limited, is still in the hands of the individual accredited breeder.

A large scale breeding enterprise differs in one or more respects from this and other examples of the pedigree system:

1. breeding is a means to efficient production (and is not an end in itself);
2. breeding is based on modern scientific techniques;
3. breeding is closely integrated with centralised financial, managerial and marketing control.

A large scale breeding enterprise is not primarily concerned with breeding but with production and marketing. As the beef lots of California show, some relatively large businesses produce animal products without an integral breeding scheme. Where varieties of livestock of superior merit are bred for the purpose in hand, it is necessary to have sufficient resources to finance the project. These usually come from current profits on production and sale, so that there is a direct connection between the end-product and the amounts expended on research and development. For the most profitable use of these funds, breeding must be carried out on an adequate (but no greater) scale, have clear-cut aims, and apply the most effective techniques.

From the purely genetic point of view, the advantages of large scale breeding do not follow automatically from large numbers. Artificial insemination of dairy cows, for instance, is usually an extensive operation in terms of numbers of cows inseminated but it has often been conducted with no serious attempts to breed improved cows. Because it will pay for itself without doing this, because operators are often not subject to competition, and because opposition is raised by breed associations, some of the essential features of the poultry breeding revolution are missing. In Western Germany, France, and Holland the organisation of artificial insemination is much sub-divided so that the individual centres tend to be relatively small and have neither the scientific staff nor the cow numbers to mount long-term breeding programmes based on sire testing.

Regarded in a limited sense as machinery for breeding, the large organisation has or could have several considerable advantages. A large corps of fieldmen and salesmen can produce evidence of what is wanted and can sell it when it has been produced. This corps is likely to have the benefit of an extensive and competent advertising campaign. Enough suitably qualified technical staff with equipment and breeding stocks can be provided to secure results within a space of time relevant to production needs. Since a variety of projects can be undertaken simultaneously, it is possible to take risks that would deter a small business unable to withstand the results of a mistake. Finally, decision making is concentrated in the hands of a few highly-selected people able to see that action is taken quickly and effectively. This is highly important in matching production with consumption.

One effect of super-marketing is to create a demand for graded and precisely defined products. These are not easily supplied by small-scale producers. Another is to modify the law of supply and demand in a free economy in order to take account of the fact that super-markets compete in terms of total mixed purchases instead of specific items (hence "loss-leaders"). Both can be turned to advantage by vertical integration because the several steps in production and marketing can be coordinated. The hazards of over-production by enterprises seeking cost efficiency through large scale production, look as if they are being countered by forward contracting for supplies of fixed quantity and quality. Adapting breeding stock to meet the requirements is therefore becoming more urgent.

How far the combination of finance, husbandry and breeding skills, ease of transport and storage, and uniform consumer demand, that set the stage for the events in the poultry industry, can be matched with large livestock remains to be seen. It could happen, as in the past, that some of the economic or productive advantages of large units will be sacrificed in order to gain political and social ends. If this does happen, it will not necessarily happen to all kinds of activity. Small co-operatives although begun by groups of comparatively small scale farmers for buying, selling, or operating artificial insemination services, may turn out to be more of a burden than their founders hoped. Without sufficient scale, they cannot reap the advantages they seek and they may often be glad enough to throw in their lots with larger enterprises.

The present methods of breeding chickens are financed out of the profits flowing from the conjunction of large numbers, advanced technique, and unified direction. All these components are necessary for the rapid production of improved breeding lines, for extensive trials, as well as for the exploitation of the answers on an industrial scale. As far as artificially inseminated dairy cows are concerned, all are present in the U.K. and U.S.A. except the unified direction. Beef cattle are not poised in quite such an advanced position but are rapidly approaching it in the U.S. on three fronts. Artificial insemination techniques are being adapted to the circumstances of beef cattle, and organisers now have the great advantage of being able to obtain performance-tested sires. When these two techniques are allied to those of intensive dry-lot feeding, all the conditions for large scale breeding enterprises could be met.

Swine production seems in some ways ripe for development along these lines. Intensive husbandry offers scope for technical expertise in breeding and marketing, but organising it on a sufficiently large scale without the help of artificial insemination has still to be accomplished. Artificial insemination is of course possible with swine, but as yet hardly economic, except where large populations of sows are already concentrated.

In spite of its popularity in the U.S.S.R. and Eastern Europe, artificial insemination of sheep has not yet found itself a useful purpose elsewhere. Nor is it likely to, while the aims of breeding are ill-defined and the methods by which superior sires might be found are ignored. As with swine, therefore, the building up of large breeding units must be founded on something else. Many flocks quite large enough numerically to support a scientific breeding programme already exist but are not doing so. A few of the reasons can be imagined. Sometimes the profits are too small to support developmental costs; sometimes they are too large to stimulate ambition. These reasons always exist. Additionally, however, the prospects of breeding and exploiting superior sheep for an extensive husbandry (under which the large flocks tend to be kept) are not as attractive for the time being as they are for intensively managed flocks. For the latter, there would appear to be opportunities for raising output substantially by applying new technical knowledge in husbandry and breeding.

The paths towards the integration of breeding operations, as a comparison of the U.S. with Denmark shows, are not a simple function of the size of a country or its form of government. Pigs and dairy cows are bred in Denmark towards economic objectives within a form of co-ordination quite unlike the system which has evolved in the U.S. for poultry. Furthermore the paths are not mutually exclusive, either within an industry or between industries within countries. In practice an established system can present obstacles for any other. The public financing of teaching and research committed to the pedigree system, subsidies for performance testing, and official attitudes about breeding can hinder, if not prevent, new and possibly more effective organisations from arising.

Alongside the poultry industry, there exist both in the U.S. and in the U.K. co-operative ventures in dairy cattle breeding designed to overcome the disadvantages of relatively small herd size by subjecting the co-operating herds to a single breeding plan. With the aid of deep-frozen semen and artificial insemination, a group of institutional herds mustering some 700 cows in North Carolina and centred on the State College at Raleigh has adopted a concerted policy of testing young sires and exploiting the best. Semen is available from four proven sires. These are selected from an intake of three or four recruits each year which are laid off after test matings in a young sire proving programme.

A similar scheme has been organised in England around the Cambridge A.I. centre. This one involves about 2000 cows in co-operating herds and has a similar breeding plan. In both countries private breeders are learning to combine to achieve more suitable numbers for giving effect to their own ideas of type. It must be left to the future to show whether these groups can be held together. Some at least are likely to dissolve on account of dispersals, retirements, and disagreements. In Denmark, the process of co-operative breeding has been carried to the logical end point, where practically all breeding is based on local A.I. centres guided by results from the progeny testing stations.

Inefficiency of selective breeding in all its versions is paid for eventually by the nation, but more immediately by commercial producers and processors. The direct beneficiaries are a small proportion of all pedigree breeders, and those who are employed by them. Whether the maintenance of the pedigree system is a satisfactory protection against real or imaginary anti-social consequences of large scale businesses must depend on the rate at which it can evolve into a more efficient system. In its present diffuse form it cannot attract either the capital or the highly skilled, scarce and expensive organisation men capable of exploiting both scientists and salesmen.

VII. Contract Farming

Smooth flow is important in keeping costs down for the packer who prepares foods for an exacting retail market. Supplies by small farmers are too irregular, and, hence, they must combine to obtain contracts. A great variety of contracts is being tried out. Much depends on who is to do the risk taking: a banker, feed merchant, hatcheryman or final producer. Contracts increasingly specify the breeding, dates of delivery, and quality standards. The planning and discipline required are offensive to some producers.

In the United States broiler industry, a producer usually does not buy the stock or the food for it. Some contracting company will pay for all this, supervise his operations, and pay him a flat rate to cover his labour costs and depreciation. If there is a profit the company will share it with him. Depending on the contract, he may also share any losses.

Contract farming is not incompatible with good stockmanship. In fact, the converse is true. The poor stockman and manager is unreliable and an unsatisfactory partner for a contract. Efficient producers, however, acquire a powerful ally to help them with capital and technique and they stand to gain from integration.

In an industry as diverse as livestock production, it is not helpful to think of either the horizontal integration of small enterprises engaged in the same activity, or the vertical integration of adjacent stages in the process of production as if they were exclusive and unable to co-exist along with traditional farming organisation. Contract farming is a response to economic pressures and will grow where these are strong enough. Just as it may replace existing arrangements or add to them, so may it invade old markets or develop new ones. It may appeal to some farmers while repelling others who would sooner accept lower returns than be bound. So far, opportunities for contract farming seem to be greatest where intensive controlled-environment production means comparative freedom from fluctuation in supply due to weather, and easy access to retail markets.

VIII. Growth of Big Business

A reduction for all to see is now taking place in the number of independent breeders of most species of animals, and with it a strong tendency towards integrating all phases including, breeding, feed supply, equipment manufacture, medication, operation of a producer's plant, and processing as well as merchandising. Furthermore, this process of integration, at least in the poultry industry, is evolving on an international level. Production of both broiler and layer stock in Western countries is controlled by relatively few companies interlocked financially or affiliated in other ways with companies in other countries and continents. It may be a long time, if it happens at all, before a single giant company attains a monopoly and thus stifles the rivalry of which the consumer is a beneficiary. A decline in the number of independent firms controlling production of all food of animal origin could lead to reduced competition between different kinds of meat for the consumer's favour which keeps down the price he now pays for the product of his choice.

The rate of increase in the size of businesses, however, shows much variation. The largest firms in each industry with their supposedly superior efficiency do not always overwhelm their rivals completely, and incidentally gain whatever advantages there may be in having a monopoly (Florence, 1964). Some of the differentiation of industries lies in their distribution. For one thing, they may have to be near their raw material. For another, the degree of uniformity of the product and the conditions of demand will affect output. In communist countries demand is made uniform and, therefore, the advantages of large-scale in manufacture are achieved but these are denied where customers insist on a lot of variation. The less uniform the conditions of supply, the smaller the plant or firm tends to be. The outstanding examples are to be found in types of farming subject to the behaviour of weather, so that the performance of livestock is not reliable. If both farm enterprises and retailers are small scale, large wholesalers will usually be interposed between them. The size of an organisation may also be related to the ability of its employees to run it. This, no doubt, applies not only to farms but to public corporations and even governments.

Arguments both for and against the monopolistic trend which may be developing can be made. The good comes from the production of better stock faster and more efficiently while there still exists competition between the growing giants of the industry. The bad comes from the eventual control of a sector of agriculture by a few individuals whose interests may not correspond with those of the public. Whatever the case may be in manufacturing industries or in agricultural production, in animal breeding monopoly could lead to a biologically irreversible state. The capital investment required for establishing a new breed, a new artificial insemination service, or a new source of supply of breeding stock for multipliers and hatcheries, becomes incommensurate with the risks and expected returns when an established monopoly is being challenged. Spontaneously arising competition with breed monopolies is not to be expected. The break-up of trusts by government action may not be helpful if in the meantime stocks are reduced to a single and relatively homogeneous gene pool (or single sources when at least two are needed for crossbred production). Some loss of genetic variation is implicit in progress by breeding, but a further loss due to trimming the costs of maintaining unrelated stocks in a monolithic organisation might mean reduced adaptability. Fear of this, however, may be misplaced and, in any case, is not a sufficient reason to insist on retaining outmoded methods of selection.

A successful organisation, successful in the sense of having survived and grown, tends to become committed to policies based on ideas that have proved sound in the past. No reason suggests itself for exempting business firms, breed associations, dairy boards or government agencies from this generalisation. Largeness tends to encourage qualities of maturity, a desire for stability, avoidance of risks, a tendency to lapse into routine (De Carlo in Ginzberg, 1964). Too much of this kind of commitment may cause an enterprise or industry to miss opportunities for applying new methods through fear of the risks of deviating too far from past practices. For the managers of the large organisations there is a problem of adaptation to face. To add to their troubles in assessing scientists, systems experts and accountants, as well as the welter of incoming data, there is the duty to design new relationships between people within enterprises instead of new products themselves. For broader responsibilities of this kind, men are needed who will not retreat into technicalities of a lower order (for references to books on the developing management science, see Wearne, 1965).

There may be a danger that the emerging giant enterprises of animal breeding will suppress all other smaller forms of business, by being more efficient and competitive, and then relapse into stagnation. Perhaps there is a new kind of establishment growing up with great promise but in truth destined to become intolerant of domestic critics and independent reformers alike. Such fears may be ill-founded or concern only a distant future, but it may pay to devise some means of discouraging complacency or reaction in the management of the powerful new devices for breeding livestock.

The arguments in favour of bigness apply to larger animals as well as to poultry, but, as already pointed out, it does not follow that the re-organisation which has taken place in that industry will be the exact pattern for other industries. The size and capital value of a single animal, its adaptedness to intensive management or automation, its susceptibility to precise grading and contract deliveries, all will affect the size and design of production units. Stock other than swine and poultry, that is, stock more closely dependent on the kind of land they occupy, will provide opportunities for units varying widely in size. Some co-operative ventures are based on specific and limited aspects of farming, such as syndicated groups buying calves or young pigs. These kind of activities, which imply no large outlays on offices and administration and do not tax the skills of the farmers running them, may be all the bigness that some circumstances will support.

Although the tendency to increase the size of the productive unit has been shown in all industries including livestock production, it is still uncertain how far small businesses are doomed to be swallowed up by big ones. In agriculture, individual enterprises are relatively small (although growing in size). Somewhere in the whole process of production of which farms are a part, bigness may be necessary. Necessary or not, it is a fact already in many aspects such as the slaughtering, shipping, and storing of meat, manufacture of fertilizers, herbicides and pesticides, and artificial insemination. Florence (1964) interprets this trend towards large scale operations as a result of the law of increasing return. An increase of capital and labour generally tends to create an improved organisation which raises the efficiency of capital and labour.

As breeding becomes industrialised, it is increasingly subject to considerations that are quite foreign to those who practice the traditional art. In this context it is worth pondering Korach's four laws of technology (in Goldsmith and Mackay, 1964), since much of what is happening may stem from them:

1. The law of the cost variable: each process has a maximum allowable cost determined by the market price of the product.

2. The law of the great number of variables: in technology only a limited number of variables, the dominating ones, can be considered.

3. The law of the scale effect: at a certain degree of quantitative change, a qualitative change becomes necessary.

4. The law of automatisation: the range of variability can be held between determined limits only in automatic processes.

CHAPTER 7

BREED ASSOCIATIONS

The conscious creation of breeds of livestock is a comparatively recent development. As breeds go, Corriedale and Columbia among the sheep, Santa Gertrudis cattle and Lacombe pigs have a well-documented but brief history. By comparison, the origins of the Shorthorn, Hereford, Friesian, and the Merino recede vaguely into the distant past. Local varieties that were to provide the basic material for the breeds that established themselves in the nineteenth century must have been numerous, unstable, subject to war and pestilence, and crossed with the cattle of traders on stock routes. Blood group studies, like the morphological studies before them will, no doubt, endeavour to trace migrations and relationships amongst breeds, but it is uncertain whether they will ever succeed in unravelling the tangled skein.

It seems (Lush, 1945) that a likely sequence of events in the history of a breed would start with relatively high market values for some local variety and continue towards meeting a demand for pedigrees. This would narrow the sources of supply and maintain values. By no means all local varieties got that far. In his book on pigs, Davidson (1953) mentions twenty-two of them in England alone that failed to attain breed status. Shorthorn cattle, in contrast, became popular, which led to the compilation of the first cattle herdbook by George Coates in 1822. Export markets often stimulated formal breed establishment. Later, the influence of shows made it worthwhile to follow fashions with respect to type. The formation of breed societies in due course (to produce the herdbook and to promote the interests of breeders) brought about the emergence of men who were confident in their judgments and opinions. It led them to make assertions about correct type, and to explain why it was the correct type. For many breeds, export markets were very important in fixing breeding aims. Just how far demand from overseas was itself conditioned by notions of type obtaining in the home market would now be difficult to find out. Ideas about correct conformation, colour and quality of coat were very likely shared, though the reasoning behind them might have been quite wrong both at home and abroad.

I. What Is a Breed?

A population of farm animals is a breed, for practical purposes, (a) when it has some identifying features, or (b) when it has a formal association of breeders, or (c) when certain government officials say it is. These definitions are clearly not independent of each other, but they have nothing to do with performance, local adaptation, market demand or fertility, in spite of the fact that existing breeds are often described in such terms. In their early stages they all have a geographical focus that made it possible for breeders to develop joint breeding and selling policies. Although not so obvious during the growth phase of a pedigree population, the focus can usually be located by the concentration of long-established flocks and herds and, in bad times, by the shrinkage towards it of the area colonised.

Most breeds are easily recognised by their size and colour, but the genetic definition of a breed, superficial characters excepted, is not easy. Going up the biological groupings from family to strain, to local variety, to breed, a student will find no objective way of knowing when he has passed from one to the next. This is also true of the taxa in the classification of natural populations, but the concept of breed includes more than the purely genetic notion of an interbreeding population of animals characterised by gene frequencies conferring some specifiable means and variances, and isolated to an effective degree from other populations. Matings within a breed of livestock are subject to restrictions, rather like those applied to human matings, which establish hierarchical partial isolates and a strongly directional gene flow. There is a stratification in each case which encourages matings within the limits of adjacent strata without actually prohibiting them outside those limits. The net effect is a force generated by the attractions of the elite class, and holding a breed together. This class is to be thought of as an amalgam of 10–20 breeders with their associated stockmen and their animals.

Other forces superior to this one and centrifugal in character can develop within a breed. Inside a breed association, disagreement about type, and, therefore, about the main purpose of the breed, may arise and become the first stage in a new definition of breed standards. The population then may split into two or more sub-groups. This is likely to happen to a breed that is widespread, so that local varities appear within it and develop a cohesion of their own. The sub-division of Shorthorn, Jersey, and Dutch Friesian cattle are well known examples of this process.

It has often been urged (in fact, since Culley, 1794) that there are too many breeds, a view provoking the question: too many breeds for what? The numerically weak ones make no perceptible difference to the economics of farming or to the range of products. The owners may be animated by sentiment rather than by economics, but their sentiments could be tolerated by the majority with good grace. An improvement of 1% in the economic merit of Suffolk or Hampshire sires of crossbred lambs would increase quality or production more than eliminating a few relic sheep breeds. Some commentators seem to overlook the point that countries with a wide range of environments for livestock cannot escape from a substantial amount of variation in market products simply by weeding out a few unimportant breeds or crosses. Much of the variation in the size and shape and fatness of carcass meat comes from environmental causes, and much of that from within farm and breed.

The main practical guides to a breed are the breed points which are specific details of colour or conformation mostly of minor economic importance. They have come in for much criticism but there is something to be said for them. Advocates of functional beauty may be underrating the importance of breed points like colour. They confer the uniformity and the trade marks by which breeds can be easily recognised. It is, after all, a very human characteristic to be attracted by appearances, and it is likely that, so long as the users and raisers of the larger farm animals are very numerous, there will, for a long time, be a market which is influenced in favour of attractively produced animals. No doubt functional beauty will gradually count for more, and no doubt production and breeding will fall entirely into the hands of those who rate performance higher than appearance. But in the meantime, there is some way to go before breed points, in all but poultry, cease to be of significance.

II. Breed Purity

Pedigree breeders usually set considerable store by breed purity, of which the visible manifestation is an accurate register of pedigrees. Yet the value of an unblemished pedigree for export purposes is problematical and is, in any case, of little interest to the great majority of livestock farmers. There is no means of knowing now how often gene migration occurred in the past from breed to breed by accident or surreptitious crossing. Some breeds may be more important in maintaining heterosis in other "pure" breeds than is recognised. Indeed, early pedigree breeders often kept two breeds.

Two reasons suggest themselves for trying to isolate breeding populations and prevent migration of genes. Firstly, a breeder may wish to maintain genetic identity by avoiding mixing with other stocks. This applies particularly to inbred lines, novelties, or highly selected lines which would be much reduced in value by dilution or by rising too rapidly in numbers. Secondly, there may be a commercial value in the mere idea of purebred, or "well-bred" animals or in the stock of particular breeders.

Neither of these reasons has much genetic validity in a long established pedigree population, barely distinguishable in performance from common stock. The latter, in fact, if bred by successful artificial insemination sires, may be superior in some respects. Making it easy for such animals to have descendants eligible for registration in open herd books would seem to be sensible. Practice in this matter varies among breed associations. Those which insist on unsullied pedigrees may be doing themselves harm. In addition to the expense of keeping and checking the necessary records of acceptable animals, registration fees from those which are unacceptable are lost, and some possibly useful genetic variation and plasticity excluded. Since sales of breeding stock are determined mainly by the breeder's name and, to a growing extent, by performance records, rigid maintenance of breed purity could be unprofitable. With a more liberal attitude to grading-up, based on the finding that animals entering an open herdbook through a grading-up scheme do so at the bottom of the hierarchy, and, therefore, rarely leave descendants in the top herds (Robertson and Asker, 1951, Wiener, 1957), breed societies could enlarge their populations of animals with benefit to themselves. Shortening the time necessary to achieve full herdbook status, relaxing the standards applied to entrants to the first stage, and reduced fees, might increase the number of interested breeders.

To adopt a more realistic policy, however, in the matter of breed purity is difficult for a thriving association and, probably, ineffective for a declining one. What is required is a change of attitude: less emphasis on hollow claims of animal ancestry and increased emphasis on a progressive outlook among members. As interest in tests of performance grows, it seems possible that unregistered animals which have done well will sometimes become influential in a breed. They may be accepted into top herds like imported animals or they may help to provoke the constant review of breeding objectives.

The time for a breed association to encourage grading-up is when a breed is increasing in popularity. With the aid of the fees and membership dues thus added to the finances, the growth impulse is stimulated and will carry the breed further and faster in numbers than it would otherwise have done. In the past, there may have been less force than there is now to this argument, owing to the greater weight formerly attached to breed purity. The problem for breed associations is to judge where their interest lies. It is a question of timing changes in the rules to suit the economic changes in the market for pedigreed animals.

The rate at which these changes take place varies with country, with the kind of livestock, the attitudes of officials, teachers, and the press, but most important with the progress of artificial insemination and performance testing. Since the logical result of the latter activities is to establish a hierarchy of performance, there is likely to be a loss of customers for the nostrum of pedigree and purity. Purity, however, is too useful a word to be abandoned altogether, and it will probably be appropriated by geneticists for highly selected lines of predictable performance and some degree of inbreeding, which, superficially, is just what breeders meant by it. The essential difference lies in the diverse interpretations of the word "performance". In the new performance hierarchy, the value of an animal will be the greater, the higher it is. Its position will be determined by the performance of itself and its close relatives and not by its owner or its name.

Well-bred animals will need more than long pedigrees. General Lord Haig said in 1925: "Some enthusiasts to-day talk about the probability of the horse becoming extinct and prophesy that the aeroplane, the tank and the motor-car will supersede the horse in future wars. . . . I feel sure that as time goes on you will find just as much use for the horse—the well-bred horse—as you have ever done in the past" (Hart,

1959). On this issue he lacked imagination. Breeders today have less reason than he did to read the future poorly.

III. Type and Conformation

Of the vast literature on this topic only a few papers can be mentioned here. In particular, in the writing of this section Lush (1945), Touchberry (1951), Bayley *et al.* (1961), Taylor and Rollins (1963), Smith, King and Gilbert (1962), Johansson (1964), Taylor and Craig (1965), and James Biggar (in personal discussions), have been drawn upon. A useful review of the subject has been made by Nichols and White (1964).

The aims of pedigree breeding are usually based on the concept of breed type. Type is concerned with general appearance in relation to function as, for instance, it is in beef and dairy type cattle. Within a type, there are an infinite number of local differences in shape of legs, udder, head, or hind-quarters. To some extent these differences reflect variation in health and plane of nutrition, since, obviously, animals vary in shape with age and growth rate. Because of the fact that most genes influencing the shape of an animal are not local but general in action, the conformation of one part is closely correlated with the conformation of another. Muscular hypertrophy or double muscling, for instance, alters the conformation of all muscular parts, and not just where it may be most obvious (for cattle see Raimondi (1962) and Lauvergne, Vissac, and Perramon (1963); for pigs see Ollivier and Lauvergne (1964). The summation of all these changes constitutes a new type. In what follows therefore type and conformation will be regarded as interchangeable terms.

Whilst it is easy to measure many body parts and discover that they are fairly highly heritable and, hence, responsible to selection, type is as yet conveniently assessed only by subjective judgment. Heritability estimates are scarcer for type than for its components, but on the evidence of the rapid divergence of numerically large breeds into subtypes and then into distinctly different types, it may be assumed that type has at least a moderately high heritability.

A. Judges

That some breeders are prominent and skilful judges is manifest, although it is unlikely that they have any supra-genetical means of recognising genetical merit. If allied to a retentive memory for relatives and the circumstances of their rearing, a breeder's cultivated skill in choosing animals that will develop as desired can modify in his favour the chances of rearing an attractive animal. It would be interesting to know which men come to be regarded as good judges and what are their qualifications. Are they people who come to the right, meaning popular, decisions in rating show animals? In most countries, judges seem to have been drawn principally from the ranks of successful breeders, that is to say, breeders whose stock is bought by other breeders. In the United States, they are augmented by animal husbandry professors and extension workers. Attaining success as a breeder tends to make a judge to a slight degree independent of popular opinion and, to that extent, able to try to modify it.

As shows and fairs have been losing influence on breeding, the judges have naturally been losing theirs at the same time. They have lost it to those who wield the power of artificial insemination or advertising, and who may be indifferent to some traditional beliefs about conformation. However, it is usually a purpose of breed societies and show societies to commend good husbandry and good health, to attract the interest of townspeople to livestock, and to provide pleasure for country people. Judges are a means to this end which they combine with the task of maintaining and encouraging what they regard as correct external form.

B. Type and production

Breed type may well be based to some extent on characters of no known economic importance except to sellers and buyers of breeding stock, but this exception is enough to secure attention to those characters. Some breeds have, doubtless, been handicapped in this way, for instance by the requirements of colour pattern in Berkshire pigs and Shorthorn cattle. Breed type has long been highly regarded in Ayrshire cattle although this did not prevent Holstein Friesians from expanding in numbers much more rapidly than Ayrshires in Britain, Canada and the United States.

Among sheep there are extreme wool and meat types, such as the Merino and the Southdown which are not notably good milk producers. Any number of breeds with intermediate combinations of meat and milk and wool exist but none of them seems to have solved the problem of producing a large amount of milk at one time and a large amount of meat at another, or of switching from meat to another form of protein, namely wool. This suggests that the output of any product may be, primarily, a function of total body size. In addition, there seem to be certain specialisations which are incompatible with each other. From a big sheep like the Lincoln it is possible to have large amounts of meat, milk and wool, all produced by the same animal. But none of these traits is exceptionally good when the size of the sheep, in comparison to other breeds, is taken into account. It seems to be a general principle that, in all kinds of animals, extremes cannot be bred for without a relative loss of performance in some other respects.

Visually assessed type is apparently of little proven significance within breeds, so far as milk yield, fertility, carcass quality, wool production, or growth rate go. But over the broader spectrum of breeds of a given species, type is predictive of production of, at least, milk or meat in cattle and sheep, and of slaughter weight in pigs. The dangers of adopting dogmatic attitudes about the relation between type and performance are shown by a comparison of Charollais, Herefords and Holstein Friesians. Each of these breeds is an important beef producer but they are obviously dissimilar in type.

Examples that show how conformation can be related to performance are provided by show-type dogs. In some breeds, development of extremes in special traits has been carried to the point at which biological fitness and the health of the animals suffer, as in bulldogs. In agriculturally important animals, selection is not carried to such lengths, although harmful ideas about conformation have sometimes been held. Thus, hot-blooded Poland China pigs, very small teats on Ayrshire cattle, over-large heads and shoulders on Hampshire lambs which caused ewes trouble at parturition, woolly-faced Shropshires, and comprest Herefords were all favoured at one time or another. A further example is provided by the recent trend towards polled cattle which is removing a once popular but often harmful aspect of type.

Ideal conformation as defined by show judges is not necessarily ideal for commercial purposes, although judges may intend it to be. The ideal for commercial purposes is that which helps to maximise profit. Consequently, it has two main components: (a) the minimum standards necessary for sustained production, reproduction and health, as in feet, udder, mouth; (b) any features proven necessary to develop economically important traits, such as size, shortness of leg, plumpness of hams, and small heads.

Within these restrictions there is scope for much minor variety of form. A striving after refinements leads to increased costs which arise from culling and expensive purchases of breeding stock inconsistent with the aim of maximising profits from commercial stock. In pedigree herds different considerations apply if there is a profitable market for breeding stock of specific conformation. A true but trifling illustration of this is provided by those sheep breeds with "correct" and "incorrect" settings of the ears and horns.

Although it may still pay a breeder to attend more closely to conformation than to various other aspects of performance, either because his selection of breeding animals is more effective for this aspect than for the others or because buyers are willing to pay him for it, in the long run, any unnecessary diversion of effort is a handicap to the breed. Scope for selection is limited and if too much is allotted to details of conformation, more important characters are correspondingly neglected. In discussions of this topic many breeders regard relaxation of selection for type as merely the first step in a consistent series of moves in the "wrong" direction. It is true that in a small breeding population some erratic changes of type might follow suspension of selection pressure for it. But the deviations would not grow and accumulate unless "good" type is inimical to the desired performance, in which case the change might be quite noticeable and beneficial.

C. Genetic aspects

Rice *et al.* (1957) state that "while there is no genetic antagonism between good type and high production, selection for type alone will have little direct influence on production, and conversely selection for production alone will have little direct influence on over-all type rating." This group of experts can call on others, for instance, O'Bleness, Van Vleck and Henderson (1960), Bayley *et al.* (1961) and Mitchell, Corley and Tyler (1961), to support them. A breeder who reads such a statement out of its context, and who did not know what "genetic antagonism" meant could be forgiven for having his doubts about it. He would be thinking of the fact that cattle can range from extreme dairy type to extreme beef type and would not believe that milk production is not influenced by type. This is a good example of the way in which misunderstandings between breeders and geneticists can arise. A resolution of this particular difficulty requires some patience but is possibly worth it since a belief in the importance of type and conformation is deeply rooted in the minds of breeders and the policies of their associations.

The main sources of trouble seem to be semantic. The expressions *genetic antagonism* or *negative genetic correlation*, which mean in the present context that genetic increases in the average milk production of a dairy cattle population would be accompanied by undesirable changes in type, and vice versa, is, like many more technical terms, forced on authors by the need to condense their writing to save space. Worse still, the qualifications to the generalisations made are not always emphasised. Statistical studies on type ratings are made (a) with data collected on culled herds and (b) in pedigree herds that are trying to achieve a single officially approved type. Consequently the analysis is conducted on the basis of a highly restricted range of types and highly subjective observations (Van Vleck and Albrectsen, 1965). Correlations that are not significant mean only that within the limits of the material, the data and the statistical methods, no credible association between type and performance is found. It does not mean that none exists, or would not be demonstrable in more heterogeneous populations. Within a long-established breed, however, it seems reasonably certain that a breeder would be wise to discount the importance of type differences of a relatively minor character, if he is primarily interested in breeding for more milk. Sometimes, of course, as a seller of breeding stock he is not. Then, he would be right to fear that selection for milk exclusively, or nearly so, would lead to variation in type that he might not desire. Nevertheless, whether they know it or not, milk production is far more important financially to most farmers than small variation within the dairy breed type.

Some generalisations on this topic may not be generally acceptable, but nevertheless, useful in helping to clarify it:

1. Type and conformation are two words with one meaning. They refer to the shape of an animal relative to its purpose.

2. If the main purpose of a breeder is to give his animals a certain shape, the most effective way of breeding for it is to select parents on the basis of their merits in this respect.

3. There are general and specific genetic controls of type: (a) general, genes affecting skeleton, or muscling or fat; (b) specific, genes affecting heads or udders or legs. The genes affecting specific regions may in truth be general in their action but not yet recognisably so.

4. The general factors cause variation similar in character to that produced by differences in plane of nutrition or by age. Selection, however, can exploit them.

5. The components of type are numerous and of varying degrees of heritability. They include bone and joint size (Hereford *vs* Angus), appetite (as shown by depth and thrift), mature size, and rate of maturing. Consequently, the effect of selecting for type varies with the component and the emphasis placed on it.

6. Within a breed of moderately uniform type, the minor variations in type have very little influence on milk production, longevity, economy of food use, fertility, or carcass quality of an animal or of its offspring.

7. Breeding for extreme types is effective as shown by the history of the Shorthorn and Friesian breeds.

The ability of an animal to produce meat and milk may be supposed to depend on a number of characters including those listed below. Selection causing changes in these characters would not always be equally beneficial for meat and milk production. If the symbols + for beneficial and — for neutral or adverse can be applied, these changes might be listed as follows:

	beef	*milk*
better appetite	+	+
later or less fat deposition	+	+
intramuscular fat	+	—
greater thickness of muscle	+	—
blockiness of build	+	—
mature size of frame	+	+
more efficient udder	—	+

Could it not then be that the attempt to breed a dual purpose animal will have varying degrees of success, according to which of these characters is being favoured in breeding? Some will enhance both kinds of production, others will perhaps tend to increase the amount of milk but be neutral or be negatively correlated with meat production. If there is or could be an effective physiological mechanism for switching food energy to meat in males and milk in females, there would seem no reason preventing a high performance dual purpose type. A part-way illustration of this situation is provided in chickens in which hormonal implantations in males of primarily egg-laying breeds made them at one time an acceptable poultry meat producer. In mammals, however, such useful sexual dimorphisms seem to have been achieved neither by man nor by nature to more than a very limited degree. Consequently emphasis on meat is regarded by many breeders as inimical to milk in cattle as well as in sheep and pigs. This negative correlation may be obscured to some extent by selective breeding for characters such as appetite and mature size which favour both.

In general it would seem worthwhile to discuss the problem in terms of the various components of meat character. This phrase is used to cover combinations of size, muscling, fat to lean ratio, proportion of cuts and a traditional blocky carcass. The genetic relationships of all these with milk yield are at present unknown. Even if there is a negative genetic correlation between any of these traits and milk yield, it may be possible to reduce it or break it.

Until fashion changes, most breeders are rather wary about concentrating on a few characters to the exclusion of all others. After sixty years of performance testing, Danish pig breeders still exercise much less selection on the test data than is theoretically possible (Smith, 1963), and the same is true in Britain. Just how much dissipation of selection differential is due to a shortage of suitable animals when they are needed, how much to high cost or negligence, and how much to preserving conformation is not known.

Deciding on the proper balance of type and production is not easy for a breeder. Extreme positions carry risks, but there are breeders who appear to believe that type is of overriding importance. Current show standards for yearling beef cattle, for instance, attract the support of breeders though to others they seem to verge on the ridiculous. It has to be conceded, however, that breeders who take risks of supplying the markets of the world with beef bulls must be allowed to use their own judgment of what is the best way of breeding. The time has come, as the growth of the performance registry in the United States shows, when their assertions about merit in beef cattle are being put to the test of economy and rate of gain and carcass quality. Should it be found that they are mistaken, the usual penalties for such mistakes will be demanded, that is a loss of buyers and the favouring of competitive breeds.

The other extreme position (paying no attention at all to conformation) is impossible, since natural selection removes the worst misfits, but it can be closely approached. Those who advocate or practise selection of breeding stock solely on the basis of one or two characters of high economic importance do so. If these characters have no correlation with conformation, the selected animals, as a group, should be of average conformation, but among them will be some rather poor looking specimens. To breeders accustomed to using breeding stock above average in conformation, even average animals will be repugnant. Those who try to keep a strong selection differential for economic characters will have to overcome their desire for "excellent" conformation and be prepared for an immediate, but not progressive, deterioration of their stock in this respect. Fashions in conformation change, however, and what is "good" now may be superseded by a "better" that is more relevant to economic demands or more adapted to parturition. For the new cattle breeds in Australia and United States, containing a proportion of zebu genotype, acceptance of a novel type may become crucial to their success.

In the field of large-scale commercial poultry breeding where no aesthetic or sentimental considerations for specially attractive looking birds exist, it is significant that the number of traits over which selection is spread is not small but large. The sixteen-item selection index used in breeding for high egg production (Dickerson, 1962) does not include conformation. All the other traits are objectively recorded and given an importance appropriate to their economic value, heritability, and correlation with other useful characters because chickens have been bred to fit their purpose with an exactness not remotely approached by larger livestock.

A.I. sires with the best proofs in milk or butterfat yield are apt, in the view of the breeders, to be of undesirable type. So far, large breeding organisations have not been consistent in their attitudes to type. In cattle breeding, the controllers of artificial insemination seem unwilling for the most part to declare policies relegating type to a minor role in bull selection. An exception is the American Breeders' Service (Prentice, 1963), which arranges the insemination of one-and-a-half million cattle and which has decided to ignore type classification. In Britain, type is still important and inhibits the full exploitation of the most outstanding sires as judged by milk yields of daughters. And the costly data produced by progeny and performance testing of swine is not yet put to full use (Smith, 1964).

IV. The Natural History of Breed Associations

Livestock Record Associations and Breed Societies are a part of the cultural inheritance of many countries. Their origins and development came in response to needs that were felt, and their activities, like those of many other institutions before them, tend to become habits essential to their continued existence but no longer serving the original purposes.

As in other social groups, conflict is an essential ingredient of an association's existence. Radicals must always do battle with conservatives but they are bound together by acceptance of a history of judgments on method and beliefs, and agreement to serve the needs of the association so that it shall survive.

In an association of breeders with a common interest in a particular breed the status of an individual has two elements: firstly, the rights pertaining to his membership of the group (for example, to register animals), and secondly, competition among members for position in the social order within the group. Partly as a cause and partly as an effect, high rank in this order is associated with extensive social and business contacts, and with privilege in breeding. This stratification, which is so characteristic of human and lower animal societies, evolved because it was an effective means to survival. It combines the advantages of a large group acting in unison with leadership by a few. The role of those lowest in the hierarchy is to contribute to the funds, buy stock from their superiors, resign in periods of depression, and provide a reservoir of men and animals in case of need. For the highest, in addition to material rewards, there are the rewards of office, the benefits of distinction in breeding, the right to opinions, and the obligation to defend the group. In all this, the objects or the methods of breeding have been irrelevant. The basic urge is to survive and, if possible, to grow. Beyond that there is no group philosophy.

The concept of breed is intentionally restrictive with a view to benefiting the registered owners who make the rules. The concept is restrictive in several respects, not equally applicable to all breeds:

1. discouraging the immigration of genes by overt or covert crossbreeding;
2. spending selection effort on breed points not essential to the main purpose;
3. concentrating selection on a narrow definition of breed type;
4. resisting emphasis on performance data;
5. discouraging the planned obsolescence which is implied by a policy of improvement.

Wilson (in Finlay, 1925) said that "herd books have had little effect on the methods of the stock breeder. . . . But herd and stud books have made unimportant characters important, handicapped breeds with characters they had better have been without, prevented them obtaining characters which would have increased their value, and debarred them from all improvement excepting from within." With the passage of time some of the force has gone from these damaging remarks. Wilson was, however, regarding it as the duty of a breed association to improve, though this was not and is not its primary object.

How much restrictiveness is in the interest of the livestock industry now, is the question. Assuming that breeds (or other isolates) are necessary, some degree of restriction on the immigration of genes is required. How much of it is desirable depends on the function of the breed or strain. An inbred strain involved in a crossbreeding routine would be destroyed if it were badly contaminated. An outbred population of purebred animals, by contrast, owes its future to the genetic variability which enables it to progress. Breeders, therefore, with a regard for their own interests, must strike a balance between the risks of becoming out of date by progressing too little and the risks of letting their flocks and herds lose acceptability by moving too fast.

Unless a breeder has a most unlikely amount of capital to invest in the future, he will not breed animals for which there is no current market. Innovations are hard to get started. The same thing is to some extent true of an artificial insemination service. It may have objectives towards which it is working but cannot get too far ahead of those whom it serves. It is still the case that sires with popular prefixes can be most easily tested, since dairy farmers prefer them to bulls with unknown prefixes. Part of the activities of a large artificial insemination service however, can be devoted to progressive policies for the progressives among its clientele.

The methods of breeders of the larger animals are basically pure-breeding, mass selection, eye judgment, and, as opportunity offers, an assessment of progeny. In the sense intended here, these methods have done service since Varro. Without doubt, a long future is in front of them. The burden of the geneticist's complaint is that they are inefficiently applied, and, inevitably so, within the context of small independent breeding herds.

A. Rights and duties

No rights or duties have been placed on breed associations by legislation requiring them to improve livestock. If they claim responsibility for improvement in the past, by virtue of the skill of a small minority of their members, they must also accept responsibility for failures. Neither, however, can properly be laid on them. Amelioration of livestock is an exceedingly complex process, in which breed societies and associations have played a relatively minor role. What breeders spend, individually or collectively, on attempts to produce a more saleable article, is entirely at their own risk and wish, and entails no obligation on others to meet the cost. Until recently, this statement would have been regarded as contemplating the obvious. A decaying enterprise, however, can be expected to hope that it will be regarded as essential and enabled to operate on a cost plus profit basis. It could be that in due course this argument might be found acceptable in some form. If so, this form might require certain duties to be carried out in return for financial support, and would effectively alter the character of breed associations. Their basic function would have to be transformed into a public service.

B. Mental attitudes

Although the seeds of change had been sown long ago, breed societies and livestock registry associations did not have any serious premonitions about their future until artificial insemination began to expand rapidly after the end of the Second World War. They were used to booms and depressions. Breeds competed and varied in popularity, while types within breeds came and went. But breeders did not doubt that they were an essential and respected element in the livestock industry. If they had heard of genetics, they had also read in their breed journals that its exponents were unpractical people who knew nothing of breeding animals. If they had heard of the performance trials steadily carried out in agricultural experiment stations, they thought them remote from the breeding of pedigree stock.

Considering that there have been and still are many thousands of pedigree breeders in the United States, the United Kingdom and elsewhere, it is surprising at first glance that not one of them has made a significant contribution to knowledge of genetics. Their associations have a history of resistance to scientific advances dating from their establishment. Under pressure they beat a slow retreat marked by milsetones such as milk and litter recording, performance testing and selective registration. No other behaviour should be expected. Members of breed associations are essentially individualists, combining together for mutual advantage, and working collectively only for bare necessities such as herdbooks, advertising and showing, and such other activities as have also become obligatory. They are producers and sellers of livestock, not research workers or evangelists. They are responsive to economic pressures but not to advice or moralising. They are doers rather than thinkers.

Since there are a great many animal geneticists in the world, it is also surprising at first glance that so few of them personally breed high priced livestock. Contrary to assertions that are sometimes made, geneticists have, in fact, bred commercially valuable stock, for instance, the modern chickens, turkeys and broilers, Columbia and Targhee sheep and Lacombe pig. Thousands of show winners have been bred and reared by college professors who were neither farmers nor geneticists. In the West, except possibly for poultry, livestock are not usually provided by universities or governments (Canada excepted) for geneticists to improve. State-supported competition tends to be resented by breeders, and freedom to advertise, or buy and sell stock, or hire skilled stockmen is restricted for geneticists. More important, success in showing or selling would signify merely that some geneticists were as capable as some breeders of practising these arts. Such a demonstration would hardly justify the trouble of making it. At the level of an artificial breeding organisation, a dairy board, or a pig industry controlling authority, whole populations of animals become subject to breeding influence, and it is in these circumstances that the geneticist is best able to apply his knowledge.

It is easy to see why breeders are unreceptive to the science of genetics. The business of breeding pedigree stock for sale is not just a matter of heredity, perhaps not even predominantly so. The devoted grooming, feeding and fitting, the propaganda about pedigrees and wins at fairs and shows, the dramatics of the auction ring, the trivialities of breed characters, and the good company of fellow breeders, constitute a vocation, not a genetic exercise.

The opposition of the breeder untrained in quantitative genetics springs from a variety of causes:

1. Loss of income or status due to the shrinking market for stock without records of performance, or with poor records. It is better from a breeder's point of view for an animal to have no records than records of poor performance.
2. His distrust of paper records, that is, of records on a few traits which serve as bases for judgments instead of visual assessments of whole animals. It is immaterial to him that the top ten % of a large population in, for example, milk yield is nearly *average* in type. He wants to be sure that the particular animal he uses is at least average. Fundamentally, the problem here is one of objectives, not of methods. Breeders still rank type highly as an objective.
3. The importance he ascribes to ancestors and collateral relatives. To the extent that his knowledge of them does duty for missing data he may be right, but he seems at times to over-value them.
4. His difficulty in accepting the concept of heritability when he is assessing individual animals. He is inclined to attribute to heredity what is due to his husbandry.
5. His belief that in judging the breeding value of an animal by eye there is a skill which gives better results than the machinery of genetics. That there is an acquired skill in judging on appearance, and, perhaps, even an inborn capacity for acquiring it, may be conceded. The difficulty is to find the best application for it.

These are all reasonable objections raised by craftsmen overtaken by mass production methods and a computer technology. Occasionally, one of them may succeed in producing the right article when it is wanted, but the chances of his doing so, and of being recognised as having done so, are too slim to justify depending entirely on him and his fellows or on his methods. Given *the same objectives* but more extensive material the computer will be more efficient than the craftsman. As a result of the combination of artificial insemination and milk recording, breeders now have accessible to them much more information about milk and fat yield than ever before. By making an assistant rather than an enemy of the computer, they would become more competitive.

C. Population dynamics

A population of purebred animals, constituting a breed, has a characteristic structure, which is to some extent determined by the nature of the animals, but to a far greater extent by the nature of man himself. Pig numbers fluctuate more than those of cattle because they are cheaper and reproduce more quickly. Breeders of pedigree pigs wax and wane in enthusiasm correspondingly, and the turnover in membership of breed associations is high. Since farmers who decide to become breeders of registered animals vary in age, ability, and resources, there is great variation in the skill of their operations and the experience and scope they bring to them. The structure of the pedigree industry which has evolved is a workable arrangement for integrating this diverse material.

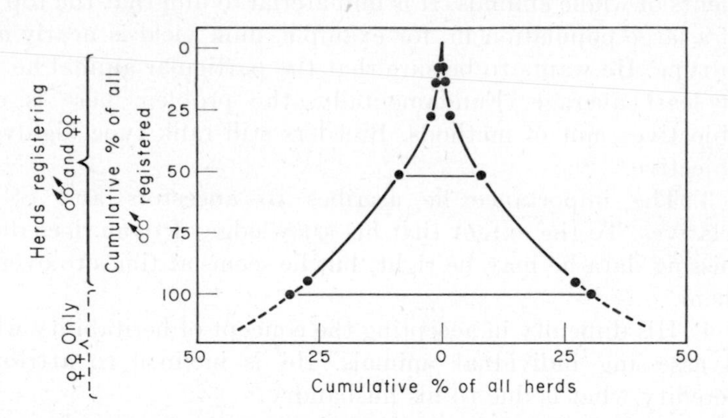

Fig. 7.1. Population structure of pedigreed Ayrshire cattle (from Wiener, 1953; reproduced with the permission of *The Journal of Agricultural Science* and the author).

The most obvious feature is the pyramidal stratification of herds illustrated for cattle in Fig. 7.1 taken from Wiener (1953). It also applies to other large animals, though in poultry the picture is now different, as has been made clear in the previous chapters. The lower ranks contain what is known as the multipliers' herds and most breeders never rise into the higher echelons. Those who do, however, tend to be the ones with older herds of greater than average size. Their output of breeding stock, male and female, increases and more of it finds its way into other pedigree herds. Eventually, a breeder may reach the peak of the pyramid where there are usually ten to twenty herds which are the source of the genetic material that will in time filter downwards. Decisions on breeding policy taken in these few herds will have a predominating influence on their breed, for their herd prefixes will occur with high frequency in the pedigrees of animals in the lower strata.

It does not matter in fact how the majority of breeders mate or cull their breeding stock, but it does matter how they buy them. They provide the market in which the principal breeders earn much of their income. At the very base of the pyramid are crossbred derivatives of the purebred layer immediately above. It seems more charitable to suppose that those who use pejorative terms like "scrub cattle" and "mongrels" to describe these animals are trying to make debating points than to conclude they do not understand what they are saying. These terms although once valid, are now descriptive of appearance and, since this is mainly determined by the owner's control of disease and nutrition together with his choice of the near ancestry (usually purebred), they ought to be applied, if at all, to his management. Some evidence bearing on the quality of the progeny of allegedly poor specimens has been offered by McLean (1952). Not surprisingly, his reject rams, although scarcely scrub animals, produced groups of fat lambs indistinguishable from those by highly priced rams. Negative measures which are based on discrimination against the "scrub" cattle are not only futile but misconceived. It might well be more difficult to identify the worst genotypes than it is to identify the best if, as is likely, disease is more serious among badly managed than well managed stock.

An hierarchical arrangement is usual where the influence of a few animals is to be made most effective, as for instance in government schemes for breeding and distributing superior stock, or in poultry breeding establishments. In the bull studs of artificial insemination centres it is gradually being developed. Being just an operative technique, there is no assumption that the results are good. It can work to raise or lower standards, and it can work to no effect at all. Although the basic principle is universally applied, there are many versions other than the pedigree systems of the United States and Britain. In Denmark (see Jonsson, 1965, for a clear description) and Israel for instance, breeding is closely integrated with other governmental and agricultural institutions. All systems, however, have the same fundamental distinction between elite and non-elite stocks. Where they differ is in the rules and qualifications of the elite category, and in the form of administration including finance.

To the extent that the elite class is absolutely isolated in a genetic sense, it will undergo a gradual process of inbreeding due in part to the very limited population size, in part to intentional linebreeding or inbreeding, and in part to co-ordinated selection for particular genes. Historical studies (summarised by Lush, 1946) mostly show a rather slow but persistent degree of inbreeding. Unless, therefore, this is counter-balanced by some outcrossing, a pure breed has this particular handicap inherent in its population structure. The characters which are susceptible to inbreeding degeneration such as milk yield, fertility, and physical vigour, will suffer a decline if enough selection effort in a contrary direction is not exerted. The loss, however, may be hidden by a rising standard of husbandry.

The expectation of life of a registered pedigree herd at its inception is rather short (Donald and El Itriby, 1946). In periods of little or no expansion in population numbers, as many as 50% of all herds started may disappear within five years of their origin. Only a minority at any time will be over ten years old, no matter whether pigs, sheep or cattle are considered. Capital gains, disillusionment, death or retirement take a heavy toll, which is more than matched by the intake of new members in a growing society and which outruns the intake in a weakening one.

Owing to the spread of artificial insemination there has been a considerable reduction in the number of multiplier herds which can find a market for young bulls. This is happening at the same time as the growing power of publicity is leading to increasing importance of the few top men in each breed. Their market for young bulls, both in artificial insemination and amongst the multipliers' herds which remain, is still quite large in most countries (see Fig. 6.2). Consequently, the shape of the pyramid representing the structure of the pedigree industry may be altering. The lower strata of the pyramid may be a little larger than before owing to the number of people who register females only, but the apex of the pyramid may be a little higher (due to the growth in the size of herds) and the waist somewhat slimmer, through the loss of some of the multipliers' herds.

Herd and flock sizes tend to be small, especially in the younger age groups. In Britain about half the registered sheep flocks are of less than fifty ewes (Wiener, 1961). Ayrshire cattle herds, which tend to be comparatively large among dairy breeds, contain, on average, about seventy cows. Most pedigree herds of other breeds will be substantially smaller. By genetic criteria, both the size and the duration of most pedigree flocks and herds are ill-adapted to the task of constructive breeding. Selection differentials are inevitably small, progeny testing is restricted and there are too few generations of animals to achieve much.

Bulls used in natural service fail, more often than not, to have ten daughters with recorded milk yields. The risks of making wrong assessments of sires that have few daughters is brought out in Table 7.1. This shows the distribution of sire estimates made by the contemporary comparison method on samples of varying size drawn from 1050 daughters of a bull used for artificial insemination which, on average, exceeded their contemporaries by 23 gallons of milk. With sample sizes of ten daughters the probability of misjudging the sire's genotype by 25 gallons or more either way is about 67%.

TABLE 7.1

Percentage distribution of sire estimates made by contemporary comparison on samples drawn from 1050 daughters of an A.I. bull with a contemporary comparison of +23 gallons (Livestock Records Bureau, 1964)

Sample size	No. of samples	Sire estimates (gallons) Minus deviations 100+	50–99	1–49	Plus deviations 1–49	50–99	100–149	150+
10	105	6	6	23	33	18	10	4
20	52	—	—	31	38	31	—	—
30	35	—	3	23	48	26	—	—
50	21	—	—	10	80	10	—	—

Although the breeding operations of most pedigree herds are of no great interest in themselves, these herds are vital to the pedigree system as a whole, since it depends on having a large number of multipliers and commercial users of registered animals. Commercial buyers contribute indirectly to breed society funds, partly by choice and, in Britain, partly as a result of official sanction against unregistered males. Multipliers are important as direct contributors of subscriptions, registration fees, and advertising costs and as buyers of stock from "superior" herds and flocks. They are financially essential and any change which reduces the social cachet of pedigree breeding or which reduces their numbers is bad for breed societies and the pedigree system.

D. Breed numbers

Stonaker (in Hodgson, 1961) has described the changes in popularity of breeds of livestock in the United States as measured by the numbers of registrations in herd books. Since 1930 the total annual registrations in all beef breeds have risen from about 200,000 to 800,000, that is by a factor of four. During the same period the total population of beef cattle has climbed from 27 to 65 million—roughly by a factor of two-and-a-quarter. Hence the proportion of all cattle that are registered has been rising. At present about one calf out of twenty-five born is registered. The Shorthorn which once accounted for 65% of all registrations now contributes a mere 5%; its place has been taken mainly by the Hereford. Latterly, the Hereford has had to make room for the expansion of Polled Herefords, Aberdeen-Angus, and half a dozen new breeds.

Dairy cattle in the United States reached their peak numbers about 1943 since when they have slowly declined. Registrations, however, have been mounting steadily, and exceed 400,000 from a total dairy cattle population of about 30 million. Most of the increase can be claimed by the Holsteins. In fact the other breeds have tended to attract fewer registrations since 1950.

TABLE 7.2

Numbers of bulls licensed in Britain since 1934 in 5-year periods (in thousands)
(compiled from Ministry of Agriculture sources)

Period ending	Ayrshire	Friesian	Shorthorn	Angus	Hereford	Total
1937†	16	9	88	11	8	133
1942	24	28	114	10	10	187
1947	37	48	81	7	8	180
1952	38	43	45	9	9	144
1957	26	38	26	13	11	144
1962	21	33	12	12	15	93

† Four years only.

Although registrations of females do not always parallel those of males, breed fortunes in Britain can be conveniently followed in terms of the numbers of young bulls officially licensed in England and Scotland. These are shown in Table 7.2 for a few selected breeds of the greatest numerical importance. Until 1944 the Ayrshire and Friesian were expanding at the expense of the Shorthorn; thereafter all three began to contract as artificial insemination made itself felt. By 1962, less than half the number of bulls licensed in 1942 were required. Not all the reduction is attributable to artificial insemination since in recent years small herds have been disappearing at the rate of 3000 to 4000 a year and their demand for bulls with them. In spite of these changes, the two chief beef breeds have been relatively prosperous.

Pigs and sheep are no less susceptible to changing popularity. Some data illustrating the point have been culled from more extensive information given by Byerly (1964) and appear in Table 7.3. Although it would be interesting to know the causes of the rise and fall of breeds, they would be difficult to trace accurately. No doubt they include changes in market demand (for example, leaner hogs), show successes or

TABLE 7.3

Purebred registrations of swine and sheep in U.S.A. (Byerly, 1964)

Swine			Sheep		
Breed	No. of registrations		Breed	No. of registrations	
	1946	1961		1946	1961
Berkshire	24,628	14,138	Hampshire	31,875	29,586
Duroc	84,413	60,542	Rambouillet	13,433	9047
Spotted Poland China	43,731	13,020	Shropshire	22,100	9466
Landrace	—	27,307	Suffolk	6489	37,910
Yorkshire	3500	28,111	Columbia	3200	10,000
Hampshire	41,663	54,675	Corriedale	11,857	15,901

lack of them, advisory and extension work, and the effectiveness of breed propaganda. Whatever may be the exact combination of causes, the effect of both rising and falling numbers on the finances of breed associations is obvious. As income grows, more publicity of all sorts can be undertaken to ensure still more growth while a shrinking budget forces on members the inactivity that will make matters worse. It is worth noticing that once past the formative stages a breed which has failed to grow in numbers or to keep pace with its principal competitor rarely recovers or notably expands its popularity. Dominant breeds tend to give way to something new, not something old.

E. Breed improvement

As Robertson (1963) has observed, breeds can be altered from within only slowly whereas buyers can change their minds quickly. A breed can be effectively dead before anyone realises that it is ill. When buyers change their allegiance, they do not leave behind the enthusiasm, the capital, or the time to make the deserted breed competitive again.

Once a breed declines in numbers to the point where it cannot match the amount of performance or progeny testing of its chief competitors, its power to compete is sapped.

Unless there were a wholly unlikely unanimity of view about the objects of breeding and about the amount to be spent on data collection, the prospects that outlays on improvement would be recovered in a small market for breeding stock seem poor. Many of the small societies can have no real interest in improvement. They are small because the breeds they support have little economic importance and until they acquire it, costly efforts to breed new models cannot be anticipated.

It would appear from past history that at any one time many breeders are investing in the wrong breed or the wrong ideal. Most breeds are either static in numbers or shrinking, and, while it is not impossible to make money in a falling market, a rising market usually provides the better investment. Although a long view is necessary for a constructive breeder, especially of slowly reproducing cattle, the short-term prospect of financial advantage from the sale of registered animals at a premium over the price of similar but unregistered animals must be the main attraction for most members of breed associations. When grade stock are equal in performance to registered animals, the ethics of this practice seem dubious. How many pedigreed animals would be required solely for breed improvement is a question that could only be answered by deciding the aims and the effort to be put into it. For artificial insemination organisations this question is by no means academic.

Left to themselves, large animal breeders would still disregard genetics for it is neither convenient nor economic for them to practise what it preaches. After 1945, however, the technological revolution made itself felt in livestock breeding. Of the numerous developments taking place, artificial insemination was the most conspicuous. In the post-war years there was a large increase in the number of men trained in genetics and physiology, who were inspired by the potentialities of artificial breeding and able to exploit the resources now in their hands. They had machinery which could cope with vast amounts of data, and they had the means to publicise their findings and opinions. Artificial breeding centres had to be efficiently and economically run, and they had to have some sort of breeding policy in which the features of the old craftsmanship could have only a minor part. In due course, the disagreeable truth came out that breeders did not know, in fact, how to select bulls able to increase yields. Their self-delusion was exposed by the progeny test. Pig breeders suffered the same fate, and sheep breeders can presumably look forward to it.

During this period, poultry breeding began its astonishing metamorphosis at the hands of a combination of geneticists and business men. In its present form, the poultry industry has little use for pedigree breeding in the sense in which this term is used for larger animals. This is not to say that the whole apparatus of pedigree breeding of cattle, sheep and pigs should be allowed to fall into decay. By and large, the turn of events seems to have caught breeders collectively unprepared for changes. Years after the warning signals were first flown, resentment still paralyses good judgment. Paradoxically, therefore, it falls mainly to those who have been the instruments of modernisation to consider carefully what role predigree breeders could play, that would give them a reasonable return for their efforts. There is no point in parading ancestral breeders, or in claiming virtue for past performances. At its crudest, the question is whether their current assets can be put to good use.

What are these assets? Numerically, pedigree breeders comprise perhaps 5% of all breeders. Financially, their societies tend to be weak and rarely able to conduct operational research bearing on their own efficiency and survival, or to hire trained scientific staff of a calibre to compel respect. Technically, they have their pedigrees and their associated systems of identification and some other records, and a market for breeding stock dependent on the diminishing goodwill of commercial breeders, extension workers and educators. Socially, they comprise loosely organised groups of farmers with an interest in breeding. This last item is possibly the most important.

These general assessments of course do not apply equally to all countries and all breeds. Danish and German organisations, for instance, differ considerably from those in the United States and United Kingdom. Furthermore, breeders of sheep and beef cattle who have not yet felt fully the disenchantment with pedigree breeding, may be more inclined to overvalue their pedigrees, their stock, and their status, than are breeders of dairy cattle and pigs. (Robertson, 1963; Rae, 1964).

Since the days when it was written into the constitutions of breed associations and societies that improvement of their breeds was an objective, much has happened to make them act as relatively conservative bodies. Improvement, as they define it, to be sought at the rate they find convenient, however, is no longer acceptable to a generation with scientists, artificial insemination, computers, publicity and finance at its command. What was once an institution leading the way to better livestock, is becoming a brake on progress. Except in the numerically large breeds, societies achieve little more, from a public point of view, than a numbering system and the preservation of some breeds, that might not be missed if they failed to survive. In the largest breed associations, the great mass of subscribers have no special significance unless the few breeders of elite herds and flocks are in fact improving their stocks. Of this, there is remarkably little evidence that does not emanate from them. Were all breed records, by some mischance, to be lost, it would not matter much, except to the herd owners concerned and perhaps not even to them. They would promptly re-establish the numbering system, appoint themselves the arbiters of new elite herds based on unregistered purebreds, and in a short space of time all would be as before.

It is probably necessary to add that these remarks arise from the fact that in countries with highly developed livestock industries, there is now little difference between registered and unregistered animals of the same breed. Furthermore, the problem of deciding what constitutes improvement should not be one for constructive breeders alone. Their ability to make slow changes in type is not in question, but their ability to define objectives, control modern resources, and cope with invisible traits, is highly suspect, unless they have power over very large numbers of recorded animals. If they can acquire this power, and remain constructive breeders, they will have no need of a breed society. From the point of view of the livestock industry as a whole, the real danger is that those who do exercise control over large numbers may saddle themselves with the inhibitions of pedigree breeders without their reasons.

Numerically small breed societies have shown in the past remarkable ability to survive on little more than the loyalty of a few members and they may well continue an unobtrusive existence especially if the stock is concentrated rather than geographically dispersed. Where, however, these small societies have a representative on joint committees or councils, derived from breed associations in general, they may exert an influence beyond their numbers or importance. Such can happen in the business of shows and fairs, and in the deliberations of government departments. Although a reactionary point of view is not necessarily to be assumed from such representatives, it is obvious that they would be tempted to resist technical developments that might diminish further their ability to compete for sales of breeding stock. It is also obvious that they might not welcome any amalgamations with other societies whereby they might lose their identities.

As in other kinds of business, there is a trend towards the emergence of a large and dominant breed association in each main class of stock. Today, in English speaking countries, this means Jersey or Holstein-Friesian type cattle in dairying, and Herefords in beef production. The process has not gone so far in pig breeding, and is not very evident as yet in sheep breeding, excepting the Merino in Australia and the Romney in New Zealand. Although economic forces may cause other breeds, at present relatively less popular, to displace the present leaders, it becomes progressively more difficult for them to do so, unless they can command sufficient modern publicity. The advantages of size and big numbers are not easily overcome. Large populations of animals support effective advertising and exhibiting and they are more likely to include some herds run by owners exceptionally gifted in some relevant way. Where a change of direction in breeding needs to be brought about quickly, they offer more opportunity of finding suitable animals. The idea that there is a critical mass which has to be exceeded before a breed generates enough activity and energy to become competitive is attractive, but measures for improvement probably depend not only on numbers but also on their rate of increase and the intangible element of enthusiasm.

Assuming a widening interest among commercial farmers in performance-tested stock, the large breed associations would be in a strong position to provide, if they wished, this kind of stock, whereas it would be practically impossible for the small ones to do so, unless the financial burden were to be carried by some other body. The same principle applies to milk recording in most countries, and it might well apply to spending money on the testing of numerically small breeds, especially the offshoots of large breeds. Indeed, testing could be the key to success for a new breed. By the same token, testing could damn a breed as inferior. Testing as a technique for improvement, however, has a value which increases with the numbers available for testing and selection, so that there ought to be some non-linear relationship between outlays on testing and the degree of improvement. Where these outlays originate in public funds, there is the additional question of the proportions which should be spent on sheep or pigs or other classes of stock. If there are official policies in the Western democracies implying answers to these questions, they do not seem to have been publicised.

When money speaks clearly enough, it is remarkable how quickly breeds can sometimes respond, notwithstanding all the handicaps of small scale operations. This follows when breeders become united in striving for a simple objective. Changes in length of bacon pigs, and reduction in fatty deposits of sheep, pigs and cattle take place rapidly when quality differentials are paid. These traits have a moderately high heritability.

Where many small selection lines are being carried (that is, individual herds) some will proceed much faster than the statistical average for a large population. If breeders are agreed on objectives, and the objectives are simple enough, recognition of superior stocks can lead to rapid advances. The usual difficulty is that the condition for such progress, a high selection differential, is not met. Superior stock does not turn up in the top herds, or is not recognised, or the objectives are too complex for rapid change even in small lines. Given a relaxation of type requirements, however, good progress could be made in spite of low heritabilities, provided the wish were there. It is not genetic theory which is at the root of controversy, but the rate at which breeds will be allowed to become obsolete. Like politics, improvement of livestock is an art of the possible.

Convenient though it may sometimes be for protagonists and antagonists of breed associations to discuss them in extravagant terms, general statements about their present or future value seem scarcely advisable. In the past there has been so much evolutionary diversity, that, both between and within countries, the functions exercised by breed associations vary widely. If they ceased to exist, some other body would have to be invented to register births and parentage; but it would not necessarily select the same animals to register, or conduct propaganda. Given the stimulus of a livelier interest in economic traits and the threat of competition from splinter groups, some breed associations will no doubt evolve further.

F. Possible reforms

Since breed associations regard themselves as essential components of a national industry, it is perhaps not out of place to recall some of the suggestions that have been made for their reform. Prosperous associations tend not to take kindly to such suggestions, but the weak and apprehensive bodies are more receptive. Obviously, those who try to be helpful are interested in the vitality of breed associations as constructive forces in animal breeding and not as private bodies pursuing private ends.

Warwick (1963) has dealt with this subject sympathetically. Although he was discussing beef cattle, his remarks apply more widely. Since there is no substitute for stock well adapted to their purpose, much advice has been given with a view to modifying practices thought to hinder this objective. Closed herdbooks are often criticised, for they are still common and where they are open are merely ajar. The admission of faults, and the recording of lethals and defective animals, would place an association in a much stronger position to rectify complaints. Protecting business interests is less important than breed improvement, yet some breeders tend to oppose rather than collaborate in the use of artificial insemination and performance testing which could help them. Discouragement of this sort eventually drives progressive breeders out of breed associations.

TABLE 7.4

Approximate operating expenses of breed associations per registration in 1963

Breed	U.S.A. No. registered	$	U.K. No. registered	£
Aberdeen-Angus	345,600	33·9	6300	5·4
Hereford	513,000	3·1	20,300	1·9
Holstein Friesian	259,000	6·4	82,000	1·7
Ayrshire	14,600	11·8	41,000	1·0
Jersey	42,400	13·3	23,100	1·7

Pedigree breeders need modern techniques if they are to stay in competition with the large corporations. They can have ready access to the modern tools they need if they are willing to pay for them. As an example of what can be achieved, the Swedish Livestock Management organisation repays study (see Fig. 5.1). The average amounts spent by breed associations for each animal registered vary widely. Publicity, field officers, prizes, research, performance records, and administrative costs show an uneven incidence and consequently the ratio of total costs to total registrations, as given in Table 7.4 for a few breeds varies correspondingly. This table may serve, however, to give some idea of its order of magnitude. The figures may be compared with the difference in value of registered and unregistered stock, provided the breeder's own costs of advertising and other expenses are not forgotten. None of the values shown are strikingly large relative to the value of the animals. They would have to be increased to finance performance testing.

There are a number of collective actions breeders can and have taken where survival requires them. They are all varieties of co-operation, which is generally difficult for such individualists as breeders, and include:

1. merging of office staffs and routines, as was done by the eight breed societies now constituting the National Pig Breeders Association of the United Kingdom;

2. getting rid of the burden of keeping and publishing long pedigrees;
3. organising breeders groups to enlarge the effective size of breeding herds and flocks;
4. putting greater emphasis on economic merit;
5. exploiting the facilities available for data collection and analysis, advice, publicity, performance testing and artificial insemination;
6. discarding the narrow approach to pure breeding;
7. constructively using crossbreeding and importation.

In addition to submerging the individual in the group, most of these actions involve parting with some independence.

If an object of a society is to help to ensure the growth of capital and interest on the investment of members in pedigree stock, members had best take an objective rather than an emotional view of artificial insemination, fairs and shows, performance testing, propaganda, purity of pedigree, type classification, herdbooks, livestock exports, breed councils, and genetics. The problem before them is, perhaps, not whether amateurs or professionals should control the apparatus of livestock breeding, but how both may use it to advantage.

V. Geneticists and Breeders

In the course of the discussion, contrasts and conflicts between scientists and practical breeders have been repeatedly pointed out. To conclude the chapter dealing with breed associations, it may be appropriate to return once more to this theme.

There are three strata of active participants in animal breeding at the decision-making level. There is first of all the professional research geneticist employed by the state or by a university. He has a vested interest, not in industry, but in the genetic concepts which he or his teachers have formulated. He may have difficulty in freeing himself from them and in becoming objective, should a total evaluation of breeding theory be required.

He has his counterpart in some industrial establishments, particularly those dealing with poultry breeding. Although the commercial professional geneticist may be more flexible in his thinking and be more readily swayed by economic facts, it is probable that industrial geneticists are more dogmatic than the academics. This arises from differences in personalities between the two groups which choose these divergent careers and even more so from the divorcing of the practising industrial geneticist from the academic atmosphere of untrammelled enquiry.

The managerial stratum is next. Here are men of both worlds such as are probably found in any other type of industry, as well as in agriculture. Although sometimes in positions of administering quasi-government agencies (and here again a Milk Marketing Board comes to mind), they may have come through a training in biology and have often more open minds than many businessmen.

Finally, there is the practical breeder. Once more, the poultry industry is the exception to the general rule. In it, geneticist-managers have taken over the direction of breeding policy in most if not all advanced countries. It is only a matter of time before the same will happen in under-developed areas. But, in the larger classes of livestock, the biologically untrained men are still in charge and the differences in background, in outlook, and in sophistication between them and the academic or research biologists, needless to say, leads to mutual antagonisms.

It is possible that geneticists underestimate many practical difficulties in their simplified theoretical concepts, but it is certain that many breeders refuse to understand and to credit the geneticists with having advanced their cause in any way. And the genetic views of small breeders, agricultural journalists, and, in many relatively backward countries, agricultural college staffs trained in the non-experimental or field-trial tradition, abound.

The usual question asked by the breeder of the scientist is what genetics has contributed to animal breeding. The accomplishments of a Bakewell, and other eighteenth and nineteenth century practitioners of the art are cited in contrast with those of the modern population geneticists, and the work of the practical plant breeders, alleged to have produced remarkable varieties of alfalfa, wheat or fruit, is often favourably contrasted with the contribution of geneticists to animal breeding.

It is essential to counter these accusations, not in order to justify the work of research geneticists or to put control of breeding policy into their hands. The importance of a rebuttal lies in placing their work in proper perspective, since future decisions, which will affect most of mankind, will depend on it. It is essential to point out what the scientific approach is, why there is a necessity for quantitative and precise measurement instead of the guessing characteristic of the pre-scientific days of breeding, and why there is a need for a rational, instead of an emotional, evaluation of the requirements and the directions of future livestock breeding.

Large scale operations of breeding, which *nolens volens* most of the world is forced to adopt for the sake of efficiency, depend now on the quantitative approach to the only three variables that anyone, practical breeder or academic geneticist, has been able to devise as the basis of genetic improvement of diploid sexually reproducing animals: (a) the choice of parents, (b) the selection of the way in which the males and the female parents are combined in mating, and (c) the number of offspring that a chosen parent is permitted to contribute to the next generation. All of these have to be rationally and quantitatively manipulated for the sake of speed and efficiency. The present population genetics theory may be a relatively poor thing and is certainly susceptible to improvement but it is vastly superior to the rules-of-thumb of the early breeders who could not do the job they did 150 or 200 years ago on the scale which is demanded by the exigencies of today.

The early breeder most certainly chose parents largely on the basis of phenotypic excellence as determined simply by his eye. When progeny testing was introduced (and this may have been of biblical antiquity), the choice of the animals to be tested still had to be based on the phenotype, and the use of pedigrees, which came in later, cannot be shown to have contributed a great amount to the efficiency of breeding practice. But the principle of using parents, themselves undistinguished in appearance but able to combine well in crosses, derives from post-Mendelian experiments. This systematic utilisation of hybrid vigour in animal improvement, although put into commercial practice by breeders was based on genetic research. Genetic theory is playing a vastly more significant part in animal breeding than many practical agriculturalists realise in spite of the changes it has wrought in the aims, the methods and the structure of their own industry. No conglomeration of individually small operators can produce the genetic improvement of which a large concern is capable, unless they band together and so become large themselves. In short, the task of breeders as well as of professional research geneticists is to adjust their operations to the needs of the present and the future so that they can contribute not only to their own personal welfare but to that of humanity.

Part IV: The Future of Animal Breeding

CHAPTER 8

OPERATIONAL PROBLEMS

Organisational changes as extensive and fundamental as those now taking place have numerous consequences. So it comes about that stockmen must learn to talk with mechanics, and auctioneers, to read statistics. Journalists have their own adjustments to make as do the nostalgic elder statesmen of agriculture. However, only a few of the problems that arise can be discussed here and then only briefly. They fall conformably into two classes: those presenting themselves as a result of the expanding demand for the prompt solution of practical problems by research, and those that come of thinking in terms of individual animals or of herds and flocks.

I. The Explosion of Scientific Information

Since the last war there has been a rapid increase in the amount of scientific effort expended by both public and private agencies. This has brought its own difficulties, for the faster progress has meant a high rate of obsolescence of research equipment. More serious still is the obsolescence of research workers and teachers which now amounts to a galloping disease. Although there is much inconclusive worrying about this problem, no radical solution to it is in sight. In the paragraphs which follow some of its salient features have been described with the aid of Price (1961, 1963), Dedijer (1962) and reports of the National Science Foundation (1961, 1963). Many other sources have also been consulted.

It is generally agreed that the human population now doubles itself approximately every thirty-five to forty years. Yet the doubling time for the number of scientific journals and for scientific papers published is estimated to be at the most fifteen years (shorter in countries only recently developing their scientific effort). The doubling time for the number of scientists and engineers (U.S. estimate) is close to ten years; that for total expenditures for research and development is about six years; that for expenditure per head varies from about two years for China and Japan to five years for the United Kingdom, the United States, and Canada. In Commonwealth Agricultural Bureaux Journals, the number of abstracts has gone up from 25,661 in 1946 to 61,474 in 1962. Since some of the increase is due to better coverage, it is safe to conclude that the rate of expansion in numbers of scientific papers in applied biology is rather less than in physics and chemistry, but it is still formidable. In animal breeding, the growth rate is less spectacular than in many other fields, but nevertheless rapid. Thus, for Animal Breeding Abstracts, the annual number of papers abstracted has risen from 1200 to 3600 over the last twenty years; but the amount of material in the *Journal of Animal Science* has increased some six-fold from the first volume published in 1942 to the last complete volume in 1964.

Among the fascinating implications of these and similar figures which have been brilliantly explored by Price is the fact that about 87½% (an estimate of 93% has been made by others) of all scientists who have ever lived are alive and working today. The number of scientists listed in "American Men of Science" has increased in the last five decades from 5500 (60 per million population) to 96,000 (480 per million). Because of the increase in the frequency of multiple authorships of articles, the total number of papers may not have multiplied at a similar rate, but its rate of growth is still staggering, especially since every author will apparently produce on the average 3·5 articles (the individual record of 995 items in one person's bibliography seems to be held by the British mathematician Arthur Cayley).

In addition, a great diversification of sources of scientific information is occurring. The unchecked flood of new textbooks, symposia, proceedings, reviews and monographs is threatening to defeat its own ends. Thus the number of scientific journals has risen in the last two centuries from 10 to about 100,000, and no slowing down of the rate of increase is evident although a ceiling somewhere must be supposed to exist.

It is true that attempts at modernising methods of information retrieval are being made and with a considerable degree of success, but they will not solve the basic problem of the high rate of obsolescence of individual scientists. The majority of practising scientists already face the dilemma either of reading books and journals or concentrating on other aspects of research. Irrespective of their choice, they are bound to fall steadily behind their younger colleagues relative to the state of knowledge and in comprehension of new ideas. Given the fact that in the realm of animal breeding research experiments may take a dozen or more years to complete, this state of affairs obviously threatens the pursuit of livestock investigations. For one thing, it may be found that by the time all the data of an experiment are in, the methods and designs used, and possibly its very objectives, are out of date. For another, recruitment of professional personnel of high competence is jeopardized, if the satisfactions that scientists obtain from successful solutions of problems are withheld in this manner from them.

A similar state of affairs might be expected to occur also among scientists working in industrial research and development, Surprisingly, this is not necessarily the case in all fields. Kovach (1960) has presented graphs showing that whereas the lag between the date of a discovery in mathematical theory and its application is increasing (fifty years in 1860 as against seventy years projected for 1960), that between scientific discovery and engineering application is dramatically being reduced (sixty years a century ago but only three or four now). Indeed, in animal breeding practice, the lag may even be negative with application, on occasion, preceding scientific formulation or explanation. The use of inbreeding in breed formation for instance, took place long before the mechanics of attaining homozygosity were understood; and so did the utilisation of heterosis, the genetic basis of which has still not been completely elucidated. This means that in contrast with their colleagues in universities and research institutes, geneticists working in industry can commonly not only reap greater financial remuneration, but are also likely to obtain more of the less tangible satisfaction of seeing the fruits of their labours early. These are factors with an adverse bearing on the incentives for productive and creative scientists to engage in basic research on animal breeding.

The fact that the bigness of science is affecting scientific methodology and the direction which research takes is indisputable. In animal breeding research specifically, accurate world-wide data showing how big the field has become are not available, but it is possible to give some figures of overall expenditures by governments and by industry for research and development.

These data have limitations, some of which arise from the definitions used to distinguish between basic research, applied research, and development. For present purposes there seems to be no need to split hairs. The broad definitions used in the various reports of the National Science Foundation can be accepted. In them, research is described as systematic, intensive study directed towards further knowledge of the subject studied. Basic research addresses itself towards increasing knowledge. Mendel's experiments would fall into this category. Applied research is directed towards practical applications of knowledge. In this sense, investigations on the use of Mendelian principles for practical ends, for example, transforming horned into polled breeds of cattle, would be termed applied research. Finally, development refers to the systematic use of knowledge for designing and producing useful prototypes, materials, devices or processes. The creation and improvement of a polled breed prior to its release to farmers would be designated as development.

TABLE 8.1

Expenditures on science in different countries (from Dedijer, 1962)

Country	Year	Research and development expenditures as % of gross national product	Research and development expenditures per caput in U.S. $
U.S.A.	1960–61	2·8	72·4
U.S.S.R.	1960	2·3 (?)	36·4
U.K.	1958–59	2·5	26·0
Sweden	1959	1·8	24·3
Canada	1960	1·2	21·9
France	1961	1·3	15·2
Australia	1960–61	0·6	8·9
Japan	1960–61	1·6	6·2
Yugoslavia	1960	0·7	1·4
China	1960	?	0·6
Ghana	1960	0·2	0·4
Egypt	1960	?	0·3
India	1959–60	0·1	0·1
Pakistan	1960	0·1	0·1

The figures at hand are only approximate and include expenditure on work dealing with many subjects not relevant to this discussion. They indicate, however, an order of magnitude. Table 8.1 shows how much variation there is among countries. The distribution of expenditures for research and development by different agencies in the United States, a rich country, which has allocated the greatest amounts is shown in Table 8.2. Although it is taken from the latest full analysis, the amounts shown have since been greatly increased as is clear from

Table 8.3, where it may be seen that the investment by the U.S. government in research and development activities went up from under $10,000 million in 1961–62 to well over $16,000 million in 1965.

TABLE 8.2

Total expenditure for research and development, 1961–62, millions of dollars (from National Academy of Sciences, 1965)

Sources of funds	Research and development performers				
	Federal government	Industry	Colleges and universities	Other non-profit institutions	Total
Federal government	2090	6310	1050	200	9650
Industry		4560	55	90	4705
Colleges and universities			230		230
Other non-profit institutions			65	90	155
Total	2090	10,870	1400	380	14,740

TABLE 8.3

United States Federal government budgeted research and development expenditures (from Walsh, 1964)

	In millions of dollars	
	Fiscal 1964	Fiscal 1965
Defence	7300+	7000
Including basic research	205	220
Aeronautics and space	4976	4900
Atomic energy operating expenses	2320	2226
Including basic research	268	292
National Institutes of Health	918	965
National Science Foundation	353	420

In Great Britain gross government expenditure on universities and scientific research was about £260 million (Barber, 1963). The distribution of different types of scientific effort is shown in Table 8.4.

The growth of research in agriculture is shown in Table 8.5. These sketchy data provide at least the general financial background against which animal breeding research may be viewed. Admittedly, space, military, atomic energy and medical research attract vastly more support than does that in agriculture. But the point to be made is that

TABLE 8.4

Distribution of scientific expenditures in Britain (from *Anon.*, 1962)

	In %			
	Basic research	Applied research	Development	Total
Government departments (military)	1	19	80	100
Government departments (civilian)	5	45	50	100
Atomic Energy Authority (civilian work only)	20	50	30	100
Research councils	40	55	5	100

TABLE 8.5

Expenditures on research by some government agencies (from National Academy of Sciences, 1965)

	Year			
	1947–48	1956–57	1960–61	1962–63 est.
U.S. Department of Agriculture, millions of $	39	88	131	160
U.K. Agricultural Research Council, millions of £	0·9		5·6	6·5
Scottish Department of Agriculture, millions of £			3·7	4·6†
French National Agricultural Research Institute, millions of NF		9·0	29·0	37·6

† Includes advisory service.

research expenditures on the scale of 1 to 3% of the total national income are possible for all but the most undeveloped countries. Whether this amount is spent for developing armament, for reaching the moon, for research on limiting population size, for a search for cancer cures, or for work on increasing food supplies, is a matter of choice for individual nations or for supra-state organisations. The policies governing expenditures of this magnitude must, of course, be considered from many angles with which the present discussion is not concerned. For a fuller account, readers are referred to a report of the National Academy of Sciences (1964).

II. The Tactics of Animal Breeding Research

The investigations of the Lush school, although taking into account non-additive genetic variability, addressed themselves primarily to the additive variation which is amenable in some degree to straightforward selection. More attention has recently been paid to the theoretical bases of other types of variation, notably those involved in heterosis. It seems, however, that the law of diminishing returns is beginning to assert itself in this approach to problems of livestock amelioration. Refinements in the statistical theory of genetics, as well as in the estimates of parameters of concern to breeders are still being made. But they seem to be mainly of a technical rather than of a conceptual kind, and it may well be asked whether or not the time has come to consider completely fresh avenues of attack on the unknown in animal breeding. The question has acquired some piquancy because suspicions that early claims implying that population genetics could solve all problems of animal breeding were exaggerated, are being entertained now even by their warmest advocates.

Only a few years ago Lush (1960) himself surveyed problems of dairy cattle improvement which require further information for solution. It is somewhat distressing to find that, with a few exceptions, they refer to the very same issues that could have been listed, say, fifteen years earlier. They are:

1. Information on values to be used in selection indexes, that is, heritabilities and economic weights for each of the traits to be considered, and genetic and phenotypic correlations between them.
2. Separation of epistasis from overdominance effects.
3. Examination of the possibilities of rotational crossbreeding.
4. Possibilities of improvement by selection for combining ability.

5. The significance and possibility of utilising genotype-environment interactions.
6. Problems of data collection and processing.
7. More accurate statistics on natality, mortality, sterility and other relevant properties of populations.
8. Techniques of creating new breeds.

Lush also refers to artificial mutagenesis, but in such pessimistic tones so far as cattle are concerned as virtually to exclude it from the list of prospective research projects.

There is little doubt that all of the topics in the list are worthy of continued attention. How many of them should fall within the province of state-supported research is, however, an issue. The list is presented here merely to highlight the point that, useful as the population genetics approach has been, and fruitful as it may continue to be, in itself it does not now offer many opportunities for threshold developments, a succession of which is necessary for research to thrive. It has been validly said that the greater part of scientific drudgery consists of mopping-up operations. All of the eight points listed are manifestly of this type and are unlikely by themselves to lead to important fundamental discoveries.

Such discoveries are not arrived at merely by wishing. The time must be ripe for them to occur. There must be the right people to recognize where they may appear, perhaps to seize on an anomaly, perhaps to attempt a new synthesis of observations or disciplines, perhaps to catalyse the work of others. The recent great advances in genetics seem to have come about in great measure by the intervention in biology of physicists, crystallographers and chemists, bringing with them fresh and uncommitted outlooks. Indeed, an historical example of such cross-fertilisation is provided by Sewall Wright's own short excursion from strictly zoological and evolutionary pursuits into animal breeding. The question then arises how a new phase in animal breeding research can be encouraged to begin and where it may be expected to take place.

The answer to the first of these questions, at least in part, depends upon that to the second since there are now several types of research agencies and for each of them a different reply may be appropriate. These include:

1. Government-supported and controlled research establishments. The Animal Breeding Research Organisation at Edinburgh, the Experiment Station of the United States Department of Agriculture at Beltsville, the Institute of Genetics of the U.S.S.R. Academy of Science in Moscow, or the C.S.I.R.O. in Australia, are examples of this type of institution.

2. Research units supported in part by state funds and eligible to receive subventions from private money-granting foundations or agencies. These are usually connected with University or Agricultural College departments, and are the most widespread type of unit for animal breeding research in Europe and America.

3. Private institutions engaged in research as a public service. The Rockefeller Foundation with its work on Latin-American agriculture is an outstanding example of this kind.

4. Statutory institutions, the primary function of which is production or marketing but not research, though part of their income may be devoted to research either directly or by supporting other agencies. The Milk Marketing Board of England and Wales is one example of such an organisation.

5. Private industry. Compared with non-agricultural and some agricultural industrial businesses (e.g. in stock nutrition, veterinary medicine or ornamental horticulture) those dealing with animal breeding still devote a pitiable amount of their expenditure to research and development. The outstanding exception is the poultry industry, which, both in the United States and increasingly so in Great Britain, is investing very heavily in this activity. It may be expected that similar developments will occur in other classes of stock in due course of time, especially if commercially sponsored research institutes are extended to this field as much as they already are in the physical sciences.

The two latter types of agencies are at present devoting their efforts primarily to development and not to basic research. At best, they can be expected to carry out research on specific methods and techniques of improvement of their product or on the principles governing the biological properties of their own stocks. There is a paradox here in that research findings are, generally speaking, of much less value to small scale producers than to large ones, and, consequently, science may even work to their personal disadvantage. Indeed, the original intention behind the public financing and administration of agricultural research was to help an industry composed of relatively small productive units. But by now, the picture has changed so that at the moment it behoves the first two kinds of organisations to devote themselves to basic research, to investigate general principles, and to tackle speculative, imaginative "far-out" problems, ideas and experiments. State-supported institutions should be encouraged to search for breakthroughs. And if they are to do so unhindered, responsibility for some of the applied research might be delegated to the other kinds of agencies as they become intellectually and financially able to support it.

It has been said by Pavitt (1963) that "in the industrially advanced countries the long-term, risky and often costly investment in research and development, and the often imperfect knowledge of industrialists as to its potential profitability, make it highly unlikely that an adequate level of expenditure on research and development for economic purposes will be achieved without governmental encouragement and aid." As far as livestock improvement is concerned, this point seems well taken.

It should be understood that, since as yet much of the agricultural production in the capitalist world is carried out by individuals, co-operatives and companies which operate on too small a scale to afford extensive applied research and development, governments (as suggested by Pavitt) might aid and advise establishments largely financed by individual companies but carrying on research and development that would benefit the industry as a whole.

Clear-cut distinctions between basic and applied and developmental research projects are notoriously difficult to make and to commit large research laboratories exclusively to one kind would usually quickly prove foolish for various reasons. For one thing, recruits, even if so cleverly chosen that all are skilled at basic work to begin with, will not all remain so. For another, a project may slip rapidly in either direction so that the original classification of either the project or the workers on it cannot be maintained. Some projects are primarily for educating the people who carry them out. Others are to maintain an active corps of able technologists who can apply and adapt advances made elsewhere. Still others may serve as an insurance against the risk that some advantage in technique or resources will accrue exclusively to a competitor. Notwithstanding all this, it is possible to place varying degrees of emphasis on the several stages of research. What is now suggested is that, as livestock breeding becomes more organised, development and operational work will fall increasingly to the industry itself to support, whilst basic and applied work will continue to depend on taxpayers.

No recipe guaranteed to result in a breakthrough is known, but a number of ingredients can be given. The first would be the abolition of specific crop-oriented research units in institutions where they exist. If departmentalisation is necessary in agricultural colleges, functional designations, such as genetics, physiology, nutrition, would be more profitable that commodity ones, such as dairy husbandry or poultry science. The principles of breeding apply to all species. Whatever the organisation of the intra-institutional units, no impediments to cross-departmental flow of staff activities of ideas and of co-operation between the units should be permitted.

There should be freedom but not licence to explore any byways that a responsible investigator may wish to follow. There have to be limits, and budgetary costs of experiments have to be kept in line with the total resources of an institution. But only broad administrative restrictions on the approaches used, the materials employed, and, indeed, upon the aims of the experiments need be exercised. In choosing the scientists who would have such scope and independence in their activities, much care has to be exercised. Imagination must be tempered by intelligence, and non-conformity by purposefulness. As a desirable counterpart to such men, there should be others of sober judgment, and scientific conformity. No applied, operational, or developmental research can flourish without a solid underpinning by fundamental work. Developments in the United States have prompted Storer (1963) to say: ". . . we must expect to find that basic research will become the tail rather than the dog in coming years. Unless knowledgeable efforts are made now to protect it we may find it being wagged right off the dog without our being aware of it . . . when applied research comes to dominate science, it will come under the guise of basic research."

Experts in non-biological subjects, or in biological disciplines other than genetics, could be encouraged to join the staffs of these research institutions. If the motivation for scientific work is either pleasure or gain, opportunity for one or the other, if not both, must be supplied for recruits. All of these conditions simply mean great flexibility in administrative matters.

III. Contract Research and Consultants

It is now possible for anybody to engage the services of experts or consultants who will advise on many aspects of business administration and organisation. They will conduct surveys of consumers, advise about advertising campaigns, and how to make the most persuasive charitable appeals. There are institutions which are well equipped and willing to carry out contract research. This is a form of activity which is somewhat more highly developed in the United States than it is in the United Kingdom, and it seems likely that it will grow for the simple reason that not all business firms would find it worth their while to establish research laboratories to carry out what might be an extremely specialised or short-term investigation. At the moment there seem to be few institutions of this kind for servicing the agricultural industry. In most countries farmers have come to depend on state-aided institutions for investigating their problems and giving them advice. It is doubtful whether either applied research or development in the field of livestock improvement can any longer be handled entirely in this way. This is especially so in the operation of breed societies or associations where the position is rather unsatisfactory. If they depend on advice from research institutions and colleges, one or other of three things is likely to happen. They may get good advice which they can understand and apply; they may get good advice which is couched in indigestible and unacceptable form and is like corn sown on stony ground; or they may get bad advice which arises from the fact that their problem has been insufficiently studied by someone who has other things to do, and is not being paid for his efforts or lacks sufficient personal incentive to finding the best solution. It may be that in the future, as university institutions occupy themselves with fundamental work, industry would be well advised to make greater use of contract research.

The concept of contract or sponsored research dates back to 1886 although the first sponsored research institute was not set up until 1911, following the establishment of industrial fellowships at the University of Kansas. There are now numerous commercial laboratories and a few large institutes such as the Battelle Memorial and the Arthur D. Little, Incorporated, which undertake sponsored research (Woodward, 1960). In some industries, chiefly those with a large content of modern physics and chemistry, contract research has expanded until it now employs thousands of scientists.

Laboratories carrying out contract research are to be distinguished from research associations which are common in Europe. These are laboratories maintained by co-operative efforts of industrial firms. Their results are available to all and they are usually partially state supported. Agricultural research institutes of various kinds are similar in character except that they tend to be mainly, if not wholly, state-supported. It is perhaps noteworthy that the biggest users of science, such as the chemical, electronics, oil and pharmaceutical industries, have no research associations. Where industries are not able or not willing to establish their own research laboratories in a specific field, the contract laboratories offer a versatile and speedy means of carrying out research on specific problems. They can usually arrange for teams of investigators with a wide variety of skills to be concentrated on a particular problem. They can be relied upon not to share their discoveries with competitors. Workers in contract research laboratories tend to have a different attitude towards research from that of workers in secure jobs in state-aided institutions. They have to select problems carefully so that there is a reasonable chance of coming to definite conclusions within the budgeted period, and within the amount of money made available. They are not, of course, without their limitations. So far, they have not been anxious to expand their activities to crops or livestock which would require large capital outlays in farms and buildings. Their research has no direct advisory or teaching value. Some laboratories, however, do provide an advisory service through which experts, especially in operational research, can be obtained.

For people and firms in the agricultural industry who are not accustomed to the real costs of research, the expense of contract research is likely at first sight to appear alarming since research projects have to carry their proper share of the overhead costs. Nevertheless, contract research is now highly developed in the United States and in spite of some disillusionment due to too optimistic views of its possibilities and its use in unsuitable circumstances, it is likely to grow much more. Meanwhile, it is expanding rapidly in the United Kingdom, in Norway and Germany and other European countries where research associations have often failed to thrive.

A prerequisite for the successful use of contract research is that there must be somebody with a problem. It is not necessary that the person or institution should be aware that there is a problem, but should be capable of becoming aware. That person or company must also be interested in having the problem solved and in acting on the results. The solution, however, may turn out to have consequences which management had not foreseen and does not wish to accept. It goes without saying that whoever wishes to benefit from contract research, must be willing to pay for it.

Business enterprises, government departments and large scale organisations of farmers are already using contract research for many purposes. Agriculture is moving gradually away from the small individual enterprises towards organisations big enough to finance contract research on problems that directly affect their running. Its growth starts slowly however, not only because the demand for it has to develop, but also because its success depends on acquiring temperamentally suitable scientists of great skill. Normally, such people can find very attractive occupations either in industrial laboratories attached to large corporations or in universities. Hence, the success of contract laboratories is likely to be measured by the rate at which they can acquire the men needed.

There seems no doubt that in the newer types of agriculture there will be plenty of room for solving *ad hoc* problems in biology and in applying operational research techniques. Producing, distributing and packaging of farm products seem to be no less and are probably far more complicated than those arising within other industries. Office administration, data handling and co-operatives also provide a very fruitful field for the unbiased investigator. Furthermore, the shortage of good scientists in agricultural research institutions who are experienced in these matters on conjunction with the difficulties of switching them from one activity to another when they have teaching and administrative duties to perform, will place an increasing strain on the machinery for agricultural research.

Operational research was developed during the war as a result of an attempt to apply numerical thinking to wartime problems. Twenty years later it is possible to say that it has developed extensively in industry but is still in the process of discovering the appropriate scientific methods for combining the social sciences with the others whenever required to so in the solution of practical problems. The exponential acceleration of research means that each generation's life and uncertainties differ more and more from those of its forbears. Precedent is a less and less useful guide and consequently operational research faces an expanding volume of work.

Another method which is rapidly developing in the United States to meet difficulties is the use of consultants. Where these are completely independent they operate like contract research laboratories but where the consultants are members of the universities' staffs it is necessary to have some institutional agreement to their acting as consultants and to receiving the necessary rewards of their work. This is common practice in the United States (though not in agriculture) but in the United Kingdom, while the employees of universities are usually able to accept substantial fees, this is not encouraged in the state-aided research institutions.

Academic consultants, who are nowadays highly specialised, are not likely to be of much help in resolving industrial complexities (Bragg, 1964). Practical animal breeding does not generate ideal questions which have ideal answers. It is perhaps nearer to being a social than a natural science for it involves economics, psychology, and other disciplines besides genetics. For its problems there are "best" solutions, but not unique ones. Consultants however can make major contributions in two sets of circumstances: when an enterprise launches into a new field where it has no experience, and when ideas and methods from another field have direct applications.

The American Society of Agricultural Consultants, which consists of several dozen commercially-oriented agencies providing a variety of services to the agricultural industry has recently been founded. The group recognises five main areas in which advice as well as contract research will be provided: nutrition, physical plant, economics and management, animal health and engineering. Despite the omission of genetics and animal breeding, at least one enterprise specialising in this field is represented in the Society. It is a consulting firm which will process data, set up breeding programmes and carry out specific analytic assignments for private individuals, associations of breeders and government research institutes which are attempting to solve some short-term problem. The species of animals in which the clients of this organisation are interested include chickens, turkeys, cattle, dogs, rabbits and several kinds of fish.

Should this type of sponsored research prove to be as satisfactory in the field of animal improvement as it has in chemistry and physics, it may be possible to organise small scale breeders who are willing to submerge their independence in a consortium large enough to compete successfully with the emerging giant poultry and livestock breeding companies. Little attention so far has been paid to biological problems and less still to social and motivational ones but contract research and consultation is a developing industry and it is likely to move into these fields more and more as time goes on. The discovery that business can conduct industrial research at a profit is an economic advance of the first order and it is by no means certain that the profit from better physical and chemical processes will be any higher than that from applying it to the biological and social sciences.

IV. International Cooperation

The post-war years have witnessed a great development of international scientific co-operative efforts. They range all the way from administratively imposed joint programmes of research between scientists in different countries to informal co-operation on projects arising from research workers at the grass-root level. There are many international agencies that support, supervise, conduct or aid research and development. They include the International Atomic Energy Agency with headquarters in Vienna, other establishments concerned with atomic energy (CERN in Geneva, CEN in Brussels), as well as establishments dealing with public health (WHO in Geneva) or agriculture (FAO in Rome). Table 8.6 shows a number of further examples. In addition, international military treaty organisations such as NATO support some basic and much applied research in different countries, and United States government agencies such as the Atomic Energy Commision, the National Science Foundation and the U.S. Public Health Service provide subventions for scientific endeavours in many countries.

Research programmes on a world-wide scale, including the International Geophysical Year, the Year of the Quiet Sun and the

International Biological Programme have been in operation or are being planned. Finally, a great number of international scientific congresses and symposia, largely for exchange of information, are being held, often under the auspices of international scientific unions or councils.

TABLE 8.6

Types of international co-operative enterprises (adapted from King in Goldsmith and MacKay, 1964)

Organisation	Number of member countries	System
CERN	13	Large centralised institution based on high cost of equipment
EURATOM	5	Separate laboratories in different member countries
ESRO	11	Co-operation between different laboratories (individual initiative)
ELDO	7	Division of labour between countries (politically organised)
ENEA	—	Independently managed and financed projects with central supervision
OECD	21	Common programming and voluntary collaboration but locally financed

It is not at all certain that all of these efforts are always worthwhile. Clearly there are some general advantages to approaching the solutions to fundamental scientific questions in this way. These may include mutual intellectual stimulation and interaction between individuals of different scientific and national backgrounds, the avoidance of unnecessary duplication of effort by different countries, and the possibility of undertaking projects which single small nations could not handle either because of their high cost or because of shortages of specialised talents or training. But there are possible disadvantages in devoting much of the research effort of a given country to enterprises of this sort. International institutes with extensive facilities, with possibilities of co-operation with colleagues and with higher standards of pay, provide a temptation for the best brains of each country to leave their homelands. However, the results of such research efforts, if made public, as they usually are, may counterbalance this objection. Participants and non-participants in them stand to benefit from joint undertakings. But it has to be recognised that expatriation of food research workers may adversely affect the training of the younger scientists by removing founts of inspiration. This aspect of the matter could be partly remedied by the introduction of a rotational scheme of assigning all of the researchers in an international laboratory to teaching duties for periods of time in various member countries, independently, of course, of national origin. These teaching duties should not necessarily be in the form of lectures or university courses but could be adapted to the special talents of a given individual so that seminars, conferences, the writing of books, and the preparation of other instructional devices could be assigned depending on his special abilities.

Breeding research appears to be adaptable to an international scheme of operation. The costs of fundamental work with large animals, if it is to be carried out on an adequate scale, are prohibitive for many of the smaller countries which have recently acquired their independence. And yet these are the countries that need much research of this type on which to base their own applied work and development. Mere reliance on work done in the United States or the United Kingdom has adverse psychological effects both on the political morale and, in the long run more importantly, on the morale of the people who are responsible for development. Establishment of international research centres could provide for them a sense of participation. It would support a systematic exchange of visiting teaching personnel and would go a long way towards inculcating the scientific attitude and a sense of responsibility. But it might be exceedingly important not to permit such institutions to become toys for bureaucrats, a dangerous enough failing on the national scale, and possibly fatal when international co-operation is concerned.

There is no dearth of subjects on which research projects, either centrally established or carried out in several locations, could be undertaken economically and profitably. Principles derived from basic research must underlie development, and therefore the special directions that basic research takes should, if only in part, be a feed-back response to the needs of development. By pooling resources the have-not countries would have a better opportunity of obtaining the information on which development of their livestock breeding must depend.

Beyond this, there are many applied research problems that could be more readily solved on such a co-operative basis. For example, if genotype-environment interactions are, as many believe, an important limiting factor in the improvement of our livestock, the possibility of testing genotypes over a wide variety of environments is greatly enhanced when experimental stations working on the same problem are scattered, for instance, throughout the continent of Africa, instead of located in only one country. There is also the possibility of using partial isolates or sub-populations with occasional interchange of germ plasm. This was considered by Wright to be the most efficient way of producing long range evolutionary change, and it could be exploited by pooling the resources of several countries. The introduction of exotic varieties and creation of unrelated inbred lines for crossing might be facilitated in the same way.

Finally, not the least of the benefits that may be expected would come from improvement of documentation, standardisation of data-publishing methods, increased reliance on a common language, especially for abstracts and, perhaps, the stemming of the uncontrolled tide of scientific journals by internationalising them.

V. Decay of Genetic Variability

It may soon become one of the implied responsibilities of any organisations or institutions which control the genetic destiny of a whole species to maintain a reserve of variation for further improvements and for unforeseen shifts in the environment or in demand. Indeed, it may be said that each generation has an obligation to see that genetic variation, like soil fertility, is not handed on to its successors in an exhausted state. The argument, however, that such reserves may be needed in case new diseases spread is only partially valid. In the past the advances made in veterinary science appear to have come more quickly than truly disease resistant individuals could be bred. The length of a generation in the larger animals means that pathogens could probably outpace selection for resistance (an argument first advanced by Haldane against the possibility of successful natural selection for infectious disease resistance). But it is, of course, possible that changes in environment, or extensions of the existing range of environments (and hence in the most efficient genotype), may be necessary. Such could arise, for instance, from mechanisation or intensification. Similarly, shifts in demand can be rather rapid and require rapid responses. An example of this type is provided by depth of backfat on pigs. Had there been no genetic variability in this character, pig breeders would not have been able to reduce the amount of fat by breeding when the market dictated this.

In closed populations some decay of genetic variability is inevitable. Indeed, selection advances must normally always be made at the expense of variability. Restoring genetic variance by artificial mutagenesis is a theoretical possibility, but as yet no demonstration that it is practical in domestic animals has been given. Furthermore, until directed mutations can be induced, the undesirable changes simultaneously produced might make this too expensive to put into practice.

Breed or strain crossing, that is, introgression of genes from one population to another, presents more immediate means for renewal of variation. But its use involves the maintenance of separate strains, breeds, or gene pools. Whether or not a monopoly which has found one particular population to be the most successful money-maker would necessarily keep in storage other populations simply with an eye to the future, depends on the position of the horizon in its planning. It is readily conceivable that exigencies of a purely economic kind may require such organisations to discard their stocks even though originally their managers had fully intended to keep them indefinitely. This argument should not be construed as advocating the maintenance of all available and newly arising genetic variants. In the future genetic variation may be contrived as needed, if not by mutagenesis then by the use of exotic breeds, and hence it would be wasteful to regard every variety of livestock as a potential treasure house of genes to be preserved at all costs. Given Holstein and Jersey cattle, it is likely that every other known dairy breed could be reconstructed as closely as need be, and many more. For some time yet, much genetic variation will be preserved because of the innate slowness of evolutionary processes and, in the absence of regimentation, because of the wishes of breeders who pursue breeding policies of their own independently both of breed societies and artificial insemination organisations.

Some play is made from time to time with the idea that unpopular breeds may be useful stores of genes. If anyone knew what genes would be useful in the future (but not now), if anyone could say whether a specific breed had them, if anyone could be nominated to undertake the task of extracting and exploiting them, the idea might have a better chance of leading to action. Although few will raise a finger to save redundant breeds, except for research, much effort has been devoted to making or introducing additional breeds. There may be a time coming when qualities such as resistance to metabolic and bacterial diseases, or efficiency of food use, may have to be rapidly developed by crossing with exotic breeds instead of by the slow process of modifying current populations.

In poultry breeding the question of variability may be more acute. Here dependence of industrial producers on a relatively few inbred lines, and the spread of these lines to many countries poses a threat to the reserves of genetic variation. If there is a case for preserving old breeds it is probably strongest for poultry. Where larger animals can pay their way to some extent by providing research material, it is only sensible to preserve some that have become superseded as has been argued by Rowlands (1964) for wild species of animals in zoos.

Those who are interested in numerically small breeds would often like support to keep them alive, through financial assistance if necessary. A much more pertinent way in which the problem of maintaining genetic variance can be met is to subdivide large gene pools into sub-populations. Organising gene pools to which moribund breeds contribute (particularly if they are moribund on account of their small numbers and not because of their lack of merit), might be worthwhile in spite of being a highly improbable undertaking. A simpler solution would lie in banking frozen sperm, since genes stored as haploids are as useful for the purpose as those stored as diploids (see next section).

The prime issue in the whole matter is the question who is to initiate any scheme of preservation of genes or breeds, who is to direct it, who is to carry it out, and, above all, who is to pay for it. Fortunately, there is no immediate urgency in this connection, except, perhaps, for chickens. The overall philosophy of responsible decision in a capitalist society is a much more weighty matter than its application to animal breeding, and the probability is that the solution to this particular problem will have to be found from general principles. Further comments on the subject of genetic variability appear in the next chapter.

VI. Gamete Preservation

Much attention has been devoted recently to methods of preserving gametes for long periods of time. A good many of them have been found successful. Thus offspring have been produced from frozen sperm of various species of animals, frozen eggs have been successfully thawed, fertilised and in a certain percentage of cases developed into embryos following transplantation. VanDemark (in Hodgson, 1961) can be consulted for a review of the history of experimentation on this topic, beginning with the first successful artificial insemination of dogs performed by Spallanzani in the 18th century. When techniques for long range and large scale storage of frozen sperm and eggs (*in vitro* cultures may provide another way for preserving gametes) are perfected, considerable use of them might be made in livestock production and improvement. Indeed, proposals to establish extensive facilities for acquiring, preserving and distributing gametes of a variety of species have been under consideration by various agencies for some time. Although possibilities for doing so seem on the face of it to be exciting and although many purposes to which the use of such facilities could be directed have been suggested, a sober examination of the contribution that storing gametes can make to animal improvement tends to dampen enthusiasm. Among the uses to which repositories of gametes could be put, the following may be noted.

1. *Storage of germ plasm.* This is in part a matter of regulating supply and demand. Should a type of stock be produced in excess of the needs, gametes may be saved for future expansion. Similarly, various types of germ plasm which may not have any current economic value could be stored in case they are required, and reserves of genetic variability maintained.

2. *Long range control populations.* Activation of frozen gametes after many years could be undertaken to provide populations to serve as controls on the progress achieved by selection. This idea requires the assumption that no gamete selection among the frozen and successfully reactivated gametes will occur. Whether this research tool would be useful for industry may be questioned. Other ways for maintaining controls can perhaps be devised at a lower cost and with greater assurance that they would serve the purpose (see the following section on control lines).

3. *Back-crossing successive generations of daughters to a single sire in order to establish highly inbred lines.* This is a likely usage of gamete preservation. But it is also likely that alternation of son × dam and sire × daughter matings would be satisfactory enough to avoid the need for long-term storage.

4. *Increasing the numbers of contemporary offspring of single individuals.* Undoubtedly, egg preservation could be of value in the progeny testing of females. However, it is probable that transplantation of ovarian segments or egg transfer could be used in a better and simpler way. For male progeny testing, more effective usage could be made of fresh semen. The only remaining possibility under this rubric is the increase in the number of offspring from progeny-tested individuals, but the considerations advanced for progeny testing also apply here.

5. *Obtaining offspring from individuals after carcass evaluation.* At present, this may be one of the most important uses of gametic preservation in terms of large animal breeding practice. However, the utility of this process for this purpose is becoming less and less, as techniques of live evaluation of carcass quality are being developed.

6. *Preservation of sperm from young animals before mutations have accumulated with the ageing of the gametes.* This procedure has been advocated in order to reduce the mutation load in man. By storing

sperm in lead-shielded containers, the number of mutations would be reduced, since exposure of gamete-producing cells in a live organism to various mutagenic agencies would be avoided. There may be some biological justification for this practice in the human species but its social, psychological and economic aspects involve controversial issues which this is no place to discuss. For livestock, the matter would appear to be of trivial significance.

7. *Maintenance and introduction of new or exotic stocks.* Transportation of gametes may be a cheaper as well as a more hygienic way than introduction of live animals. Nevertheless gamete storage for this purpose does not seem at the moment to be a particularly important technique.

Thus, while sperm and egg banks of livestock may become commonplace in some distant future, no important functions for them can be discerned at this time.

VII. The Use of Controls

The difficulties of verifying the results of selection experiments and of genetic trends in the quantitative characters of domestic animals have already been alluded to in Chapter 3. But difficult or not, operating managers of artificial insemination schemes, economic planners, and other citizens and officials need information about what genetic changes are occurring in populations of animals designed to supply food for the world. Indeed, if investigators and breeders are to provide critical interpretations of experiments, more general applications of techniques which will separate genetic from non-genetic changes is imperative, as has been argued by Goodwin, Dickerson and Lamoreux (in Kempthorne, 1960). While at first thought the cost of measurements needed to accomplish this end with livestock may seem prohibitive, more careful appraisal suggests that selection experiments without such measurements, and thus not susceptible to critical analysis, may be even more costly. The use of controls has been adopted in poultry breeding and appears to be already paying dividends to one commercial breeder employing it (Dickerson, 1962).

Although they would probably not wish to call their selection policies "experiments", those who formulate these policies in large enterprises might also be expected to be in a position, or put themselves in a position, to say whether their policies were bearing fruit. It may be easy enough to secure agreement to this as a general principle, but when it comes to the allocation of funds to carry it out there may be more room for disagreement. No conventional proportion has ever been decided upon for guiding companies and institutions in making decisions on this point. It may then be suggested, if only for the sake of argument, that perhaps 10% of the estimated gain in performance converted into financial terms should be devoted to discovering whether or not an expected gain has taken place.

In the past a great deal of effort has been invested in the so-called improvement of livestock. This has been done as a matter of faith rather than on any assurance that the improvement as hoped for will actually take place. The effort of trying to estimate the cost of running control lines might even have the incidental benefit of drawing up a balance sheet which would show on the one side the costs incurred in attempting to obtain improvement and on the other side the value of the improvements achieved.

Table 2.2 shows how the milk yield of a cow has risen in recent years. In no country do those responsible for research, administration, economics, or breeding in the dairy industry know why it rose. Save for hopeful extrapolations of the trends they can make no predictions nor tell where to put more effort. They have no measure of the results of actions they did take. In these circumstances, control groups to act as a check on breeding or other processes may be less expensive than ignorance. As far as breeding is concerned they can be visualised as checks on the efficacy of progeny and performance testing, of breeding plans, of artificial insemination and any other improvement scheme; and they could be incorporated in all trials of breeds and crosses, new or old.

Four main types of control group have been devised. The first is a breeding line which is selected in a direction opposite to that of another. Being contemporary the two lines experience the same environmental changes and therefore the responses to selection can be estimated by the difference between them. The second method is to make parallel comparisons among lines starting from the same population at the same time but subjected to differing selection systems. These are also contemporary and presumably equally affected by environmental changes, of either a temporary or permanent character. This method does not, however, enable the effects of genetic changes and of environmental changes to be estimated independently. The third method relies on a group bred at random or under relaxed selection (several such populations have been established; see Gowe and associates, 1959, and King, Van Vleck and Doolittle, 1963). Like the main breeding line itself, such a control group may be affected not only by changes in gene frequency due to natural selection or to chance fluctuations, but also by inbreeding due to the small population size. The fourth type of control consists of matings repeated in a pattern that will allow genetic, environmental and maternal effects to be isolated (for a statistical analysis of this method, see Giesbrecht and Kempthorne, 1965). It is probably best suited to poultry where truly contemporary comparisons between sufficiently large populations are feasible.

All types of control so far devised are open to various kinds of objections, but it is to be doubted whether any of the objections are more serious, or any combination of them are more serious, than to have no controls at all. The main expense of keeping control lines, either of the independent closed population or of the repeat mating types advocated by Goodwin and his associates (*loc. cit.*), arises from the cost of maintaining a sufficiency of males, either to prevent inbreeding or to enable them to be used in successive years when they would normally be culled after one year.

In cattle, pig and sheep populations, the maintenance of control lines might be less expensive than in a poultry breeding concern where this operation has to occupy space that could be devoted to the production of chickens for sale. Recording of the necessary information and subsequent analysis are, however, major sources of expense in all animals.

The type of control adopted in any circumstances should be determined first of all by the specific questions which it is to answer, and secondly, by the permissible number that shall be available for it. This means in effect that the costs of maintaining the control animals determine their number and the volume of records made on them.

One of the complications with mammals is that maternal effects often affect the performance of repeat matings. Whenever this might happen, it would be important to have good estimates of them. One way of avoiding this complication would be to have repeated matings on the male side, either by keeping the males for several generations of breeding, or by using frozen semen. Variations of this method have been exploited by Van Vleck and Henderson (1961) for dairy cattle and by Smith (1962) who found suitable material in a pig herd for his purpose. Although valuable as a demonstration of what is possible, this method is perhaps too uncertain in operation to be acceptable as a permanent policy. It is probably only a matter of time, however, before improvements in technique will allow wider use of control lines or other methods of estimating real genetic changes, a consummation in both public and private interest.

VIII. Animal Improvement in the U.S.S.R.

Nearly all that has been so far said regarding the operational problems of animal breeding refers to the situation in the Western world. In the

Communist countries the political situation and economic structure of agriculture may make much of the previous discussion irrelevant. However, it is by no means clear what the situation is even in the Soviet Union, literature on which is abundant. The difficulties arise firstly from the rapid changes in instruction, research and development policies which seem to be unpredictable. Secondly, there is the paradox that in a monolithic state in which it is avowed that the only purpose of science is to serve the economic needs of mankind, agricultural research has, in fact, been divorced from fundamental biology.

As is well known, Lysenko was a virtual dictator with respect to agricultural genetics between 1948 and 1964. Mendelian theory and especially its statistical aspects were derided and considered to be philosophically unsound. Although Lysenko's own background was an agronomical one, he had recently turned to problems of dairy cattle improvement. His own results did not represent anything startlingly unorthodox and, even before his fall, it seemed that some of his associates were moving into the direction of ordinary population genetic approaches to animal improvement.

Since November 1964, a veritable revolution on the genetic front has occurred. Not only has Mendelism become respectable, but a drive to update textbooks and research programmes along the lines of modern molecular biology and biophysics was initiated. The great difficulty with respect to livestock genetics is the lack of informed technologists. A gap of a whole generation of genetically oriented teachers, research workers and breeders presents a formidable handicap to the prompt revival of the genetic work of the kind carried out by Serebrovsky (before Lysenko's rise to power) which in many of its features resembled that of Lush. Only the future will show how the problems discussed in this book will be tackled in the U.S.S.R.

Meanwhile, it may be instructive to give a translation of the summary of a report on the Eleventh International Congress of Genetics held at the Hague in 1963 by Lysenko's deputy director, and later acting director of his former laboratory (Kushner, 1963). Appearing, as the report did, in a belligerently anti-Mendelian journal, it gives some notion if not of the state of affairs, then at the least of the state of mind about a year before Lysenko's views fell into official disrepute.

The summary reads:

> "In deliberating upon the results of the Congress and comparing them with the condition and problems of our research in the area of genetics and selection of agricultural animals, the following basic conclusions must be reached.

"1. The time has come to pass a special decree regarding the utilisation of heterosis in the different branches of animal husbandry analogous to the degree which in its time ensured the mass production of hybrid corn seed in this country.

"2. A systematic study of blood groups in the leading breeding populations of cattle must be organised.

"3. For a cheap mass organisation of the analysis of milk for protein content it is essential to arrange the manufacture of modern equipment and reagents similar to those widely used in Holland.

"4. It is highly desirable that feed cost in the growth and finishing of swine for pork should become a basic criterion of selection.

"5. It is necessary to commission networks of zootechnical scientific establishments which have poultry departments to create specialised strains of chickens, which upon crossing would give highly economic broilers that, at the time of slaughter (at 60–70 days of age), would expend not more than 2·5–2·6 kilograms of food per kilogram of gain.

"6. In zootechnical scientific establishments, it is necessary to broaden investigations on the problem of correlations between the most important economically useful traits of animals, so that rational selection methods for combining in the organism the most desirable traits could be worked out in different species of animals.

"7. The question of the possibility of organising studies on hereditary improvement of livestock and birds with respect to their resistance to certain dangerous diseases should be discussed by zootechnical and veterinary scientific establishments.

"8. Measures should be taken to improve the teaching of genetics in agricultural colleges and to introduce into their curricula questions of population genetics."

As the reader will recognise, the recommendations made would have been *mutatis mutandis* conservative in the West twenty years ago, albeit a good many of them still remain to be implemented. It will be of considerable interest to see whether these reforms will be introduced faster into Soviet agriculture than into Western.

CHAPTER 9

GENETIC RESEARCH

The genetic amelioration of animals parallels the evolutionary process in nature in its three basic aspects: *genetic divergence*, or the creation of new and distinct forms either by mutation (spontaneous or induced) or by hybridisation, both usually followed by selection; *improvement*, or the selection of increasingly more useful variants (mutations are needed for this process to supply the variability); and *persistence*, or the multiplication and expansion of the types in use.

Selection, control of mating (including that between species) and mutagenesis, therefore, are the methods of animal breeding. The first two have been discussed in Chapters 3 and 4. Artificial mutagenesis has apparently been attempted in domestic animals (apart from the silkworm) only once (with chickens) and that unsuccessfully (Abplanalp *et al.*, 1964). But all of these methods may acquire broader uses than their names imply today. Direct manipulation of genes at the molecular level either for inducing new variability or for changing or "falsifying" instructions leading to undesirable phenotypes in order to obtain desirable ones (genetical engineering), could eventually become part of mutagenic techniques. Similarly, tissue culturing, including exchange or combination of genetic information between cells of different species, a technique already attained (Ephrussi and Weiss, 1965; Harris and Watkins, 1965), or production of mosaics by combining cells from different embryos (Mintz, 1965), might be eventually subsumed under the term "mating". It may not be at all preposterous to suggest that, before long, direct production of meat for human consumption may be carried out in vats by tissue culture, and selection techniques will certainly be refined in ways unimagined today.

It is a main objective of animal breeding research to make these methods more efficient and more effective, which implies that this research must, in the ultimate sense, be directed towards understanding the nature of the biological phenomena underlying them. Basic research results in the public domain can benefit the world at large, whereas developmental investigations may have only local value. Furthermore, profit making industrial establishments cannot be expected to invest in excessively speculative research although public interest demands that somebody does. A whole spectrum of research projects ranging from strictly operational problems and within the ambit of industrial establishments, to fundamental biological studies likely to be undertaken only by exclusively research organisations, can be readily outlined. The limit of such a catalogue of possibilities in animal breeding research is the limit of imagination, courage and space. Carrying out the research itself is a matter of technical competence and money.

The have-not nations must conserve their resources and employ them primarily for increasing capital investment and for meeting immediate consumers' needs. Long range research which is a gamble, then must clearly fall to the have-nations. In many ways both the United States and United Kingdom have risen to the challenge, albeit still at an inadequate level (see Chapter 8). But, for a reason difficult to fathom, in the socialist countries, only now does there seem to be an awakening to the need for a fundamental approach to agricultural research.

The perusal of agricultural science journals from the Soviet Union and Eastern Europe suggests that the main occupation of their experiment stations and agricultural colleges is to engage in empirical field tests without the least concern for the fundamental biological significance of the results obtained. While proper statistical analysis is being found in an increasing number of papers, proper designs are not. Furthermore, the objectives of most of the experiments reported seem to be to try something without rhyme or reason and see what happens. The alleged transformations of chickens by blood transfusion from other breeds, however, has a sort of theoretical basis in Lysenko theory, if it may be called that. But experiments of irradiating chicken eggs with small doses and observing the laying performance of the birds hatched from them seem to be based on no preconceived idea of any sort. Scientific methodology presumably involves testing of hypotheses. An experiment like this does not seem to be testing any hypothesis. In the Western world also there is the risk that too many such experiments may be attempted merely because there are not enough good ideas to go around. There is, therefore, a danger that continuous spending of public funds on trivial and pointless research will rebound and result in lowered support of science. This must not be allowed to happen. As more control of operational research is placed in the hands of commercial enterprises, it is imperative to preserve the quality of fundamental research in agriculture, including animal breeding. The bigness of science is very relevant here. The growth of scientific endeavour brings in its wake an emphasis on expensive and fashionable equipment, digital computers, electron microscopes, amino acids analysers, and the like. And the costliness of experimentation with large animals makes it important to ensure that the available money and time must be spent on projects that spring from good thinking.

Lerner (1962) has reviewed a variety of research problems specific to poultry. Much of the following chapter is an extension of his article. The topics dealt with are only a sampling, and they are presented in no particular order. Although the emphasis is on basic questions, a good many topics considered have an immediate bearing on breeding practice.

I. Operational Research

Breeders of all kinds of animals feel the need for early recognition of superior genotypes. The earlier the genetic endowment of an animal could be identified, the less expensive breeding would be. If, instead of having to test several dozen daughters of a dairy bull in order to decide whether his genes should be widely disseminated, his superiority could be identified by a blood test at the time of his birth, the economies in running a breeding programme would be considerable, and about five years of waiting would be avoided.

Whether or not specific adaptation of animals to specific environments contributes a significant amount to the variance in performance is a pressing question. The problem goes beyond the issue of locally adapted breeds (discussed in Chapter 5) in a situation where country-wide and international distribution of stock is concentrated in a few hands. Should interactions between genotypes and environments (considered in Chapter 4) be of only slight importance, it would be possible to carry on selection at one location for performance in any other location. If, however, interactions are of great importance (and Dickerson, 1962, thinks that in poultry they are), then it may be necessary to develop special stocks for each location. The problem may disappear in time should production be carried out under a constant environment everywhere. It must be understood that what is meant here is not necessarily the constancy of environment throughout the year or throughout an animal's lifetime, but, rather, a uniformity of environment imposed on all herds or flocks wherever they are kept, even if the environment itself is fluctuating or cycled.

Allied to this general problem is its complement. In other words, selection has until recently entailed trying to produce or find genotypes suitable for a particular environment. Emphasis could be reversed by determining the kind of particular environment, including lighting, feeding, and temperature regimes, or hormone treatments, that would be suitable for specific genotypes already in existence. If new species of animals are to be utilised in food production, this approach may be the only practical one.

On the border line between operational and basic research lies the search for highly heritable markers which might be linked with or might depend on genes controlling the phenotypic expression of economic traits. There has already been work on the determination of the levels of certain enzymes in blood or blood group as an indication of production qualities. At this stage, it does not seem clear whether the blood-grouping laboratories associated with many American poultry breeding enterprises are used for selection or have other reasons behind them, such as advertising or policing franchised dealers. Nevertheless, there are many developments in the field of immunology and biochemical polymorphism which would suggest that it may be worthwhile to continue examining blood groups, blood proteins, enzymes, metabolic breakdown products, for correlations with economic traits (see a subsequent section on biochemical and immunological variation).

Some way of circumventing the biological inefficiency of production, perpetuation and multiplication of superior genotypes is needed. It is bad enough to have sexual reproduction break up an optimal genotype because of Mendelian segregation, but added to this is the fact that production of heterotic types, dictated by the economics of the situation, is often less efficient than closed population selection. If it were possible to induce cloning or asexual reproduction, both complications of this sort would be overcome. How soon this millenium will come cannot be said at the moment, but it is the essence of an enquiring mind not to set bounds to the possible, improbable as their transgression may seem.

Much of the mathematical theory of selection and selection limits is based on simplifying assumptions many of which have already been discussed. It seems rather important to be able to verify these assumptions, and, further, to be able to remove restricting conditions. This clearly means extending biological knowledge of each species. It may be granted that general notions about the expected genetic changes under selection which have been derived from experiments with laboratory animals can provide a partial guide to practice when larger animals are concerned. But each species has its own biology and may have peculiarities not present in the laboratory animals used in pilot studies. It is well known that under natural selection a great variety of mechanisms for attaining the same basic ends has been produced and it is hence important to find out exactly which ones have been adopted by each species. How general, for example, are segregation abnormalities in cattle, sheep and turkeys? What role does cytoplasmic inheritance or virus transmission through the cytoplasm play in chickens, swine or sheep? Is the number of alleles the same per locus in cattle as it is in bees? How important is the fact that in most domestic animals there is a large number of chromosomes instead of the few found in *Drosophila*? Is there much non-randomness in crossing-over with the consequent tendency to build up polygenic blocks? Is the estimate of mutation rate characteristic of *Drosophila* or mice applicable to pigs or beef cattle?

From what is known of genetic theory, it can be predicted that no system of selection may be expected to work efficiently for an indefinite period of time. Sooner or later (and the more efficient the system, the sooner), exhaustion of genetic variation, induction of negative genetic correlations among favoured traits, and reduction in Darwinian fitness as a correlated response, are bound to enter the picture. It is, therefore, small wonder that selection for annual egg production may after several generations lose its initial effectiveness. Reports of this have been provided by Abplanalp (1962) and Morris (1963). It is one of the great deficiencies in current breeding programmes that no signals to show when a selection or breeding system should be changed have been developed. The difficulties of measuring selection progress, owing to the great environmental variation from year to year and the costliness of maintaining adequate controls (discussed in Chapter 3), are at present too great to permit such signals, but a search for markers, or biochemical indicators of some sort, which would measure changes in the gene array from generation to generation as an index of selection progress, should undoubtedly be instituted.

II. Genetic Variability

Another broad subject which has both applied and basic ramifications is genetic variability. Present notions of selective improvement are rightfully based on the extent of genetic variation now available. However, there is no reason why creation of new variability and beyond that, of new genotypes and phenotypes, should be out of the question. Artificial mutagenesis is one technique that may be used to create new variation. Mutation breeding has been tried with various degrees of success in many plants. To mention only a few: barley, peanuts, soya beans, rice, potatoes, fruit trees, subterranean clover, tobacco, ornamental plants have all been subjected to radiation in order to produce variability. But the attempt, already cited, to extend this sort of study to poultry has been unsuccessful. The only project dealing with useful animals which has indications of reasonable success is one dealing with mutagenesis in the silkworm (Tazima, 1964).

Alternative methods of creating new variability include distant hybridisation which is likely after a period of selection to result in the production of new combinations. Contrary to mutation breeding, where the improvement in the genotype may be expected to occur by gene substitutions at relatively few loci, distant hybridisation may be best attempted for the production of complete novelties. Introduction of single genes from wild populations in the manner done by plant breeders is probably of limited utility in livestock. But an avenue for creating new types may lie in the selection for crossability of different species. For example, considerable variation in fertility and hatchability is found when different strains of chickens are crossed with the Japanese quail. It should be possible to select for breakdown of isolation barriers between these species and thus originate new forms of animals which might be useful for food production.

A still further possibility lies in subjecting populations to stress either by manipulating environment or by introducing foreign genes, not for their own value but for unmasking cryptic genetic variation that is present but not manifested. Examples of this approach are provided by the study of Dun and Fraser (1958) of whiskers in mice, and by Abplanalp's (1962) investigation of shock breeding in chickens.

III. Sex Control

An intriguing challenge of long standing is sex control in mammals. Many attempts of varying degrees of sense have been made to influence sex ratios at conception or birth, but so far compelling evidence for success is lacking. The most hopeful procedures are attempts to separate X- and Y-bearing sperm in the semen. Differences between them are of immunological, morphological or electrical nature. Immunological work has not so far helped much although claims for serological identification of even smaller antigenic differences than those to be expected between whole chromosomes have been advanced.

Thus, Gullbring (1957) states that the A and B blood group antigens can be distinguished in human sperm. However, Owen (in a discussion of Braden's paper, 1960) has cast considerable doubts on this interpretation of Gullbring's results, even though he thinks that the possibilities in this direction are open.

Morphological separation of the two kinds of sperm by taking advantage of the unequal amounts of chromatin in them was suggested long ago (Lush, 1925). The recent investigations along these lines are not particularly encouraging (Lindahl, 1958, 1960; Andersen and Rottensten, 1962), although some separation of rabbit X and Y sperm by differential sedimentation was reported by Bhattacharya (1962). The long standing attempts at influencing sex ratios by dietary control of pH of blood serum do not seem to very successful (Weir and Haubenstock, 1964).

Claims that there are differences in electrical charge between the X and Y kinds of sperm were first made some thirty years ago and reiterated recently on the strength of experiments with rabbit semen. They have also been criticised in a review of the evidence on the subject by Rothschild (1960). Thus, at the present time it may be said that no reliable methods of influencing sex ratios of livestock are available, but the search for dissimilarities in the two kinds of sperm, be they chemical or physical, has by no means been exhaustive and is worthwhile pursuing further with restrained optimism.

Because in birds it is the female sex that is heterogametic, similar work in the chicken, the turkey or any other species of bird, is somewhat less promising. There the mechanics of influencing sex ratio would be to identify, if possible, the X- or Y-carrying eggs, although there are no techniques for achieving this in immediate prospect. An alternative lies in the feminisation or masculinisation, as desired, of embryos whose sex has already been genetically determined. Work on the endocrinological aspects of this problem has been in progress for many years, but so far it does not seem to have led to fruitful results. In fact, the only example of an useful animal in which experimental control of sex ratio has been attained is the silkworm (Astaurov, 1962 and Astaurov and Ostriakova-Varshaver, 1957). Nevertheless, the reward for solving this problem is so great that continued research, at least on a pilot scale with laboratory animals, should be encouraged.

IV. Parthenogenesis

Studies on this subject in birds and in mammals have also been in progress for some time. A useful summary, which curiously omits

Olsen's work with turkeys and chickens (see below), is provided by Beatty's (1957) monograph. Of the various kinds of parthenogenetic processes listed by him, those which lead to the production of diploid forms seem to be of greatest potential. There are at least three: (a) polar body fertilisation, (b) doubling of the chromosome numbers of the egg, (c) tetrad formation with subsequent reduction in chromosome numbers. Although much effort has been expended, conspicuous success in finding spontaneous or repeatably producing experimental parthenogenetic foetuses brought to term has not been achieved with mammals. Most of the treatments induce parthenogenetic development but the embryos usually do not survive.

Yet there are many techniques which appear sufficiently promising to be pursued, especially since survival in at least the early stages of life is compatible with some of these methods. Evidence suggesting the possibility of producing parthenogenetic mammals is available on many invertebrates and some lower vertebrates. In turkeys, Olsen (1960 and earlier) has found a considerable proportion of individuals of spontaneous parthenogenetic origin, as well as some conditions under which their incidence is enhanced (e.g. certain virus infections of the dams). Unfortunately, in his experiments, all individuals that lived long enough to allow sex identification were found to be males. This might be expected if they arose by doubling of chromosome numbers and if X-less birds are inviable. More than twenty-five of such parthenogenetically produced males were so far found to be capable of siring offspring (Olsen, 1965). Furthermore, selection for increased incidence of parthenogenesis has been remarkably successful. Although such males are of great biological interest, it is difficult to see the other than experimental purposes for which they could be used.

It may be hoped that research on mammals in this field will be continued. Among them, it would be females that would result from a similar parthenogenetic process. And in mammals, unlike birds, practical application of a method of producing such individuals is much more promising. Rapid inbreeding, virtual self-fertilisation or cloning (depending on the nature of the apomictic process) could be practised. It has even been suggested that parthenogenetic techniques may make the male completely obsolete. Not only could females be self-propagating, but they would virtually be assured of a sort of immortality. Since no immunological incompatibility would exist between mother and daughter, a ready supply of daughters could be maintained to furnish the mother with replacements for worn-out organs.

This type of research requires the investigation of successful methods of inducing parthenogenetic development and determination of the intra-uterine and post-natal regimens and environments in which such animals could survive. There is no doubt that the search will be a long and expensive one, but, should it be successful, some of the great problems of utilising heterosis in the larger animals would be overcome.

V. Maternal Effects

There is a great deal to be learned about maternal effects. The subject bears not only on population genetics, but also on a great many other aspects of animal production. For present purposes, a list of the types of maternal effects which need to be investigated is sufficient. Such a list will demonstrate their widespread significance:

1. *Genetic* (a) The effects of common environment of litter mates and, to a lesser degree, of non-contemporary sibs on the variance between unrelated individuals; (b) the accumulation of spontaneous mutations in oocytes of the dam as a function of her age.

2. *Cytological* (a) Cytoplasmic or plasmon inheritance; (b) chromosomal abnormalities due to irregularities of segregation in egg formation.

3. *Embryological* (a) Congenital malformation due to insults to the dam or to other causes, such as ageing and changes in the physiological or nutritional state; (b) growth of embryos *in utero* or in eggs of birds.

4. *Immunological* Mother-foetus incompatibility.

5. *Physiological* (a) The possibility of increases in the variability of metric traits of offspring due to a progressive loss of the dam's ability to maintain the proper intra-uterine environment as she grows older; (b) variation in post-natal growth of mammals as a residual of influences on the embryo or of pre-weaning nutritional effects; (c) haemodynamic and placental influences on the growth of the foetus.

6. *Pathological* (a) Endocrine disturbances of the dam; (b) cytoplasmic transmission of viruses; (c) milk transmission of viruses or other infective agents; (d) transfer of passive immunity.

This list is by no means exhaustive (for instance, it does not include studies on intra- and inter-specific transfers of eggs which might be listed under several of the above categories). Nor is it certain that all of the items are of significance. Nonetheless their exploration is necessary before all the different kinds of the relation between mother and offspring can be fully understood. Since so much of animal production depends on this relation, the importance of this topic of research can hardly be overestimated.

VI. Disease Resistance

There is very little doubt that Gowen's (1937) dictum that "No investigator who has adequately sought inherited host differences in disease response has failed to find them" is as valid today as when it was first pronounced. Hutt (1958) or Fredeen (1965) may be consulted for examples ranging from oysters and bees through the whole gamut of laboratory and domestic animals, illustrating the ubiquity of a genetic component in the ability to withstand disease. Although Hutt recognises that under certain circumstances other methods of checking disease may be used, he is probably the leading advocate of selection for disease resistance in farm animals. At the other extreme, stand the proponents of the view that control of most diseases can be attained much more effectively by environmental rather than by genetic means.

The case for selection for resistance at the expense of reducing selection pressure for other traits is very weak when diseases of nutritional origin are considered. However, conversely, selection for high performance may be, in essence, selection for the correction of metabolic errors disguised as susceptibility to nutritional deficiencies.

For infectious diseases, at least in some cases, the argument is more open. Thus, until veterinarians are able to prevent or treat such a disease as leucosis in fowls more successfully than they can now, some selection for resistance may sensibly be incorporated into breeding programmes, even though no breeder has yet fully solved his problems in this way. Bacteria and viruses can, no doubt, evolve faster than chickens or cattle, and reliance on resistant genotypes to keep livestock free from disease is weak insurance. It appears that much of livestock production is now moving towards disease-free environments. Both gnotobiotic (germ-free) as well as specific-pathogen free herds of swine have already been made available (Trexler, 1960; see also Pollard's 1964, article on the use of gnotobiotes in biological research, and Luckey's 1963 book). Indeed, Ross (1960) has computed that 20 million head of hogs of this type could be produced by 1967 in the United States. Whether or not this would be economically sound is another matter, but it does seem that before long, disease-free farming will be possible technically, at least, for swine and for chickens.

Interdisciplinary research on this subject is called for and the cooperation of pathologists and immunologists is vital, but there is a long tradition of geneticists approaching it on their own. The line between geneticists and immunologists is growing very thin, and it hardly matters in which section of this chapter the matter is brought up. But it should be kept in mind that the whole issue of disease control by genetic or environmental means is rendered exceedingly thorny, not only by its relation to animal improvement but also because in it the safeguarding of public health is at stake.

VII. Biochemical and Immunological Variation

Biochemical variation in livestock involves at least three related topics: (a) intra-specific serological differences in red blood cell antigens; (b) histo-compatibility antigens; (c) serum, albumen and enzymatic variants.

For the first of these, there have been recent literature reviews by Stormont (1958), Briles (1960), Rendel (1961) and Gilmour (1962). In addition, a comprehensive investigation of the relationship of blood groups and production traits of dairy cattle has been made by Neimann-Sørensen and Robertson (1961). The basic facts which hold for all species are:

1. Polymorphism (existence of genetic variants in a population at levels that cannot be accounted for by recurrent mutation) is common.
2. Inbreeding does not seem to lead to the expected increase in degrees of fixation at many of these loci.
3. In some instances, non-inbred populations have more heterozygotes than expected from Hardy-Weinberg equilibrium frequencies.
4. In spite of these facts, no compelling evidence for a strong association between fitness or production characters on the one hand, and genotypes for blood group antigens on the other, has been produced either in cattle or in sheep (Stansfield *et al.*, 1964). Chickens are a possible exception, though in them the situation does not seem to be quite clear (see for instance Morton *et al.*, 1965).

The problem, naturally, arises as to what the circumstances are that maintain this extensive polymorphism, since it seems unlikely that blood-groups which have succeeded in perpetuating themselves at a high frequency in a population would prove to have very detrimental effects on any character for which livestock had been strongly selected over many generations. This question is not only relevant to domestic animals, but also to many other species, including man. Although various selective mechanisms, involving either heterozygous advantage or fluctuating advantage of one or another of the homozygotes depending on age or on environment (in particular, exposure to disease-producing agents of various kinds), have been suggested, the field may be considered to be an open one deserving further close study.

Whether any of the useful domestic animals, except the chicken, is suitable for pursuing generalised investigations on this subject may be open to doubt. Yet, because of the widespread practice of blood-typing of cattle for other purposes, attempts to link the data thus obtained with information on fitness or productivity should certainly be fostered.

Polymorphisms at histocompatibility loci raise many important biological problems going much beyond the bounds of this discussion. In the realm of livestock improvement, some studies are only in their initial stages as is the investigation of Schierman and Nordskog (1961) on the relationship of blood groups to transplantation antigens in chickens (now confirmed by Craig and McDermid, 1963). The value of this kind of research may be greater than that of other polymorphisms, since full understanding of ways of bypassing immune reactions can lead to successful organ transplantation.

This technique is of great interest from the standpoint of therapeutics and rejuvenation, and, even more so, when specifically used for genetic improvement. Small sections of ovaries from superior dams could be transplanted to any number of potential foster mothers after their own ovaries had been removed or destroyed so that they could serve as incubators. This would permit a great number of descendants of a single female to be produced. In chickens, perhaps one thousand to two thousand offspring of a single hen could be obtained in a short period of time. The recent advances in immunology imply that projects of this sort are by no means visionary, although many procedural difficulties remain to be overcome.

Polymorphisms have been observed in serum proteins, egg proteins and in various enzymes. Development of the technically simple but powerful methods of electrophoresis has led to a blossoming of studies of genetic differences between and within species. The comprehensive review by Ogden (1961) on biochemical individuality of larger animals mentions a great many polymorphisms in protein composition. Particularly numerous are haemoglobin variants in cattle. In some other species of domestic animals, different types of haemoglobin exist (adult and foetal) but they are usually present in all animals and, therefore, do not belong to the class of polymorphs. A recent report by Manwell *et al.* (1963) suggests that control genes for switching off production of early embryonic haemoglobin may not be the same in bantams as in the larger breeds of fowl.

Among other serum proteins, the transferrins show many polymorphisms in cattle, sheep. pigs and goats. A suggestion has been made by Ashton, Fallon and Sutherland (1964), that there is a relationship between genotype for this serum constituent and some useful traits in dairy cows although Datta *et al.* (1965) were not able to confirm this fully. More recently, genetic differences in serum transferrins of chickens have also been found by Ogden *et al.* (1963).

In the proteins appearing in milk and eggs, there also seem to be distinct polymorphisms. The different types of β-lactoglobulins of cows' milk are controlled by genes with frequencies varying from breed to breed (see Ogden's review, *loc. cit.*). In chicken eggs, independent studies by I. E. Lush (1961) and by Feeney *et al.* (1963), utilising starch-gel electrophoresis, indicate that there are genetic variants in the kinds of protein constituents a hen secretes in the production of egg albumen. Evidence for linkage of egg protein loci (Buvenendran, 1964), as well as of genes controlling casein variants in cows' milk (Grosclaude *et al.*, 1964, and King *et al.*, 1965) has been produced.

Finally, genetic variability in such enzymes as catalase, esterases and phosphatase have been reported in dogs, rabbits, cattle and pigs (Ogden, *loc. cit.*), in mice (Petras, 1963), and in chickens (Wilcox, Van Vleck and Shaffner, 1962). In chickens it has also been reported that there is corresponding variation in egg production. If this finding is confirmed, selection on the basis of the serum alkaline phosphatase of young birds might be useful in attempts to improve egg production.

In the light of current discoveries in molecular biology confirming the fact that genes are essentially determinants of protein structure, it is not surprising to find so many genetic biochemical markers distinguishing individuals within a species. In human beings, it is known that certain physical and mental properties are associated with biochemical departures from the normal. Common examples of this are phenylketonuria, galactosemia, and a great number of other inborn errors of metabolism (Harris, 1959), some of which can be corrected when identified early in life by providing proper environment. The important problem which concerns a geneticist dealing with stock improvement lies, firstly, in explaining how variants of this kind come to be maintained in relatively inbred populations (a concern he shares with the student of evolution), and, secondly, in attempting to harness them for livestock improvement.

Of less immediate practical interest is the biochemistry of inter-specific differences. However, this subject provides such a radical departure from the classical methods of studying the genetics of higher organisms that it is one of the most exciting and interesting new fields of evolutionary studies. In the past, genetic investigations could be carried out only on animals which interbreed. It is true that attempts at phylogenetic reconstructions have been made on the basis of degree of similarity in morphological, physiological or immunological traits of different species, but the newer techniques present opportunities undreamed of only a few years ago. One step in this development lay in discovering the procedures for identifying amino acid sequences in the structure of various proteins ("finger-printing"). Ingram's (1963) extensive work on haemoglobins provides a good example. Studies on various proteins in different species, including comparison of the structures of adreno-corticotrophin, insulin, cytochrome-*c*, and many others have already been carried out (for a recent study involving lactic dehydrogenase, see Wilson *et al.*, 1964).

This type of information is superimposed on knowledge of the manner in which the amino acid sequences are determined by codons, which are the intra-genic units that instruct the cellular apparatus what amino acid is to occupy a given place in a protein chain (see Woese, 1963, and Lanni, 1964, for reviews of "genetic coding"). As a result, it becomes feasible to determine the number of mutational steps which have occurred in the course of differentiation of species from a common distant ancestor. Essentially this amounts to the study of the genetics of non-interbreeding animals. Probable phylogenies of different species can be constructed, and the historical past thereby illuminated. To a scientist it feels aesthetically and intellectually satisfying to learn the likely steps by which horse and human haemoglobin diverged in the evolutionary past. It is a great achievement to be able to establish with some certainty the origin of man from African rather than Asian ancestors. In due time, perhaps full understanding of the relationships among domestic or potentially domestic species of animals may help in some way in food production. From the standpoint of adding knowledge to our insight of evolution, these biochemical discoveries have already proved exceedingly valuable. Their use in animal breeding is, for the time being, not easy to foresee, but Faraday's classical riposte, "what use is a new-born baby?", comes to mind. To acquire knowledge for knowledge's sake is a compulsion for human beings. The problem, however, in defining the functions of organisations specifically set up to study animal breeding is to justify why they should be charged with a given kind of research. More attention to molecular biology should probably be paid in animal breeding and animal husbandry departments. It is clear, however, that if they were organised on a substantial scale to do so, they would have to compete for staff and facilities with other units which are devoted to fundamental research. Peaceful coexistence suggests itself, but there may not be enough first-class talent to go around and this may prove to be the limiting factor in the advancement of basic knowledge in animal breeding.

VIII. Cytogenetics

Attention should be drawn to the recent rekindling of interest in mammalian cytology. With the invention of new methods for chromosome study, the discovery of sex chromatin, the identification of the Y-chromosome as the sex-determining factor in mammals, the accumulation of many instances of defects and diseases produced by chromosome abnormalities, the hypothesis regarding inactivation of one of the sex chromosomes in somatic tissues of female mammals, and with many other stirring developments, the field of cytology has become exceedingly active. Not only have clinicians become greatly interested in it, but also students of evolution, who have once more started investigations of the phylogenetic relationships of primates and other higher organisms from chromosomal evidence. So far, activity of this sort in domestic animals is limited, but the field is one worthy of fresh attack, especially since it is one in which the researcher escapes from the limitation of long inter-generation intervals on the rate of scientific advance. The reviews by Chu (1963) and Welshons (1963) are recommended as an entry to the literature on this subject.

Some excitement in human genetics has been caused by the discovery of variation in chromosome number and its connection with various human disorders of mind and body. There can be little doubt that chromosomal aberrations, similar to those found in man, occur in domestic mammals, and possibly, birds. It seems to be a fertile field for fundamental investigation, since experimental breeding in them is feasible, whereas in man it is not. Healthy animals that are homozygous for translocations are conceivable. Should these become a reality, copyrighted stocks of hybrids produced from crosses of homozygotes for the standard and the translocated arrangements, respectively, could become an exceedingly important item in stock production, because it would be impossible to breed from them. One apparently successful attempt at making a translocation homozygous in chickens (Bernier, 1960) has been reported, although unfortunately this spontaneous translocation studied could not be seen under a microscope.

Contrariwise (N. Inouye, unpublished) tried in vain to make homozygous an X-ray induced and cytologically identifiable translocation (also in chickens).

IX. Euphenics and Genetical Engineering

In the past, it has been often debated whether nature or nurture, the genotype or the environment, was the more important in shaping the phenotype. Today the argument is no longer a matter of subjective opinion, but can be readily solved in plants and animals (though not so easily in man) by experimental determination of the degree of heritability. Hence it is possible to make reasonably intelligent guesses as to how much improvement is to be expected from changing the average genotype of a population by selection (for a theory of selection limits, see Robertson, 1960, and the discussion in Chapter 4).

It is certain that much more is to be learned about improvement of performance by changing the environment. Only relatively recently the discovery was made that feeding antibiotics increases growth rate was made. The last word has probably not been said on such topics as vitamins or trace elements, optimal lighting or temperature regimens for different animals, and sub-clinical infections; but it is clear that in the future, improvement of genotype and environment must go hand in hand. A changed genotype may perform better in a changed environment, and a changed environment may call for selection to change the genotype.

Beyond these tried and tested methods of increasing food production, recent developments in molecular biology open vistas of still other techniques to make the most of our animal resources. These are Lederberg's (1964) euphenics, and, what Tatum (1964) calls genetical engineering.

Euphenics by analogy with eugenics refers to the improvement of phenotypes based on defective genetic constitutions. For example, in man there is an inherited metabolic error leading to the disease phenylketonuria. It is caused by the inability of the liver to produce an enzyme needed in the conversion of the amino acid phenylalanine into tyrosine. This block causes phenylalanine and phenylpyruvic acid to accumulate in the blood stream and in various tissues including the central nervous system. These substances have toxic effects and produce a variety of unpleasant symptoms, including severe mental retardation. Euphenic correction of the disease might be brought about by transforming liver cells by treatment with genetic material (DNA), that includes instructions how to manufacture the missing enzyme. Transformation of this kind has been achieved in tissue culture, and it is probably only a matter of time before it can be accomplished *in vivo*.

Naturally, the first application of euphenic measures, except for preliminary experiments, would be in man. Indeed, the costs of applying this idea to domestic animals may forever remain prohibitive, but the possibilities of transformation, grafts, organ transplants, and other corrective methods should be kept in mind.

More remote in time, but probably much more applicable to domestic animals, is genetical engineering, that is, artificial manipulation of the chemical information which determines the phenotypes of the generations to be produced. Here there would be only the initial cost: once a change is induced, the information will continue to be passed on to successive generations through the germ cells. In Tatum's (1964) opinion, there are three important aspects of the subject of which man would need to acquire mastery in order to make genetical engineering more than a dream. The first of these is control of mutational processes. The second is the development of methods of designing or synthesising the genetic determinants, be they similar to the classical concepts of the gene or of some other nature. The third is the perfection of methods of introducing into living systems genetic determinants tailored to order.

As in Lederberg's vision of euphenics, Tatum's concern was primarily with human beings. However, before this stage is reached, a similar approach to animals will have been accomplished. It may be debatable whether or not centuries or merely decades will provide enough time for biologists to accomplish the feats which genetical engineering of this precision would require. There may be some doubting Thomases to question whether the goal is attainable at all. Be that as it may, it would be foolish to be dogmatic to the point of denying the possibility of these and even more startling innovations.

If directed mutations or even beneficial artificial mutations, which are not directed, become available to technicians in animal breeding, a vast array of new genotypes of livestock will emerge. Similarly, if genetic determinants can be manufactured synthetically, and then introduced into the germ cells or early zygotes, completely novel phenotypes will be obtained. Genetical engineering, distant as it may seem, would make much of what has been said in this book obsolete; but it would also go a long way towards solving the many problems that have been discussed.

X. Interdisciplinary Studies

Most research projects require the combined efforts of various specialists, and would be unlikely to be solved soon by specialists in genetics working alone. To conclude this chapter, some very broad cooperative efforts between different disciplines will be referred to, merely as an indication of the potential in breeding research opened up by modern biology. The opportunities here are so numerous that only a few can be mentioned and even these without details.

The vital problem of the nature of heterosis requires the collaboration of enzymologists and geneticists. If it is assumed that proteins and especially enzymes are responsible for rates of biological processes and their buffering, it seems reasonable to expect that the investigation of the major enzymes found in hybrids could throw light on the physiology of this obviously important aspect of animal production.

The fields of embryology and developmental genetics have not yet contributed extensively to the much needed synthesis between them and population genetics, at least as far as animal breeding is concerned. Yet there are many exciting and challenging ideas in the marketplace of these disciplines which show promise of developing. So far, biological aspects of such concepts as redundancy, developmental noise, and canalisation (see the section on information theory in Chapter 10), are far from practical animal production. But at least an attempt to disrupt canalisation (equivalent to experiments of Dun and Fraser, 1958, which involved changing the background of a gene to unmask cryptic variation) is planned for sheep (S. S. Y. Young, personal communication).

Physiology of growth, physiology of reproduction, and physiology at large also promise an important contribution to the amelioration of animals. Not only is understanding of the elements of many basic processes that enter the production of useful foodstuffs and fibre deficient, but even at the purely empirical level, little is known about the optimum way of handling animals. Circadian rhythms have not been studied in large domestic animals. Is cycled environment a necessity for efficient production of meat, eggs, milk or wool? Joint studies on the larger animals should lead to answers to such questions.

It would be helpful to know, to return to the subject of hybrid vigour, whether the manifestation of heterosis is rooted in the same physiological pathways in different species. There is reason to suspect that egg production in chickens depends on a series of interrelationships between different endocrine glands, and the heterotic phenomena observed in this character are probably tied up with optimal levels, balances, or sensitivities of these endocrine organs. Heterosis as expressed in growth rate or, perhaps, in hardiness, or disease resistance, may be based on entirely different physiological mechanisms in different species.

The contribution that animal pathologists could make to genetic improvement of animals is obvious. No matter what has been said earlier regarding disease resistance, should specific bases for the ability of organisms to withstand invasion or damage by pathogens be discovered, it is clear that such knowledge could be put to useful ends by breeders.

Psychologists and ethologists could combine with geneticists in the study of animal behaviour. Indeed, behaviour genetics is becoming an exceedingly important field in the eyes of psychologists. From the standpoint of animal breeding, very little exploration has been carried out. And yet such questions as the genetic basis of docility, the peck order in birds and mammals, the relationships between twins, the influence of social contacts on performance, will have a significant bearing on the level of production that can be obtained by suitable management.

The catalogue of projects of this type could be made interminable. But enough examples have been given in this chapter to show how vast the arena of research yet to be done is.

CHAPTER 10

INFORMATION AND DECISION

The ability of an organisation or farm business to succeed by adapting itself to changes, or by taking advantage of them, lies in the hands of its management. Although a manager is commonly thought of as a doer, in modern management decision-making is his main function. The pace of events now means that wrong decisions can have disagreeable consequences very quickly, and management cannot afford costly trial and error methods.

In recent years statisticians and experts in business administration have been increasingly preoccupied with a quantitative approach to decision-making. Not only highly intricate mathematics, information theory, and the statistical theory of gains, but philosophy, sociology, and psychology also enter into the complex discipline which is evolving. All this is highly relevant to animal breeding. Population genetics aims to set out the rules of variation in populations but it does not set them out to the last detail, nor, indeed, to the last significant detail. Its very comprehensiveness and power to accommodate any regular forces controlling the segregation or multiplication of genes makes it unsuitable for coping with the behaviour of individuals. The same difficulty obstructs those who would apply decision theory to the practice of law, probably the most ancient decision-making institution. While geneticists generalise, lawyers, and judges try to deal with specific cases in a specific manner (Cowan, 1963).

The processes of arriving at decisions therefore vary from law to politics to science. The originally gradual, and lately rapid, metamorphosis of scientists from being observers of nature to being decision-makers on a national scale has created a need for lawyers and politicians to understand not only how they arrive at decisions themselves but also how scientists arrive at them. It is up to scientists to study and understand the workings of their own minds and to make them clear to others.

In a political democracy, it is impracticable for everyone to take part in decision-making except on the largest and most general issues. Where governmental decisions are needed in technical matters, they are made by a few people who are supposed to be armed with all the relevant information and the ability to use it properly to come to decisions. As far as animal breeding is concerned, and it is a large industry, this "intelligent layman" technique manifestly does not work.

In this field it sometimes happens that the decision-maker somehow decides what he wants and then supposes it is the duty of the scientist to tell him how to get it. A significant part of management, however, consists in trying to find out what results are wanted, and the scientist is thus put in the position of having to suggest what decisions ought to be taken. This leaves the decision-maker with the task of dealing with incomplete data and of taking the risks. Furthermore many decisions arrived at on the basis of scientific counsel given in one situation may have entirely unforeseen consequences when implemented in a changed situation of which the advisors are not apprised.

In animal breeding it is now impossible for those unfamiliar with genetic science to criticise its assertions after only brief consideration, no matter how penetratingly intelligent they may be. If advantage is to be taken of genetics, some reliance must be placed in its principles and chief exponents. Churchman (1961), in a different connection, puts the matter as follows: "The true 'facts' are the assertions accepted by a few whom our society trusts. As a culture we tend to trust our physical scientists, and tend not to trust our social scientists. We trust no one if we can see the direct relevance of his assertions to changes in our national policy and our personal lives." Although the difference between the policy-maker, the administrator, and the scientist may well disappear in a future blessed with an accepted science of values, the present seems to be characterised by distrust.

I. Types of Enterprise

There is little doubt that what the future promises is integration not only within animal breeding but embracing food production and distribution. The paths by which the eventual structure of agriculture is reached are uncertain but at least four levels of integration may be visualised:

1. The breeding, manufacture and distribution of the products of a single species.
2. Breeding operations involving different species but independent of production and distribution.
3. International co-ordination with a single company or an inter-locking organisation controlling the breeding of a given species in different countries.
4. Complete integration of all processes of animal food production from breeder to retailer covering all species, and internationally organised.

The highest level depends on political and economic relationships among nations and is unlikely to develop in the immediate future. Meanwhile, although no complete monopolies have developed within the other three stages, a trend towards them may be discerned and it has naturally increased the amplitude of decision-making in animal breeding.

Decisions by governments regarding prices or subsidies for agricultural products, or research, or marketing and development boards, can change the fortunes of breeds very suddenly. Then there are actions leading to the growth of business empires with incidental effects on breeding and breeders. At the other extreme are the decisions by some individual breeders to do nothing to adapt themselves to changed circumstances and so to acquiesce by default in the changes which will make them redundant as breeders.

Viewed from the standpoint of control and decision-making, there are four categories of enterprises in animal breeding.

1. Completely nationalised industries like those in communist countries.
2. Quasi-governmental organisations depending on economists, scientists and technicians for management on behalf of farmers or farmer co-operatives. The Milk Marketing Board of England and Wales is an example.
3. Individuals or corporations employing professional managers to operate the business for profit. American Breeders Service, Curtis Candy Company, and Kimber Farms in the United States, and the Ross group in Britain are cases in point.
4. Unorganised small scale private enterprises. As a class this category has rather poor prospects in industrialised countries, except in minor types of food production not worthwhile nor easy to organise (such as bee-keeping). Many such undertakings, for

instance, those dealing with ducks, geese, rabbits or goats, may still contribute substantially to the agricultural economics of some under-developed nations, but they have little place in an industrialised society.

II. Profits and Values

How is the success of a breeding enterprise to be judged? As Churchman (1961) says, the idea of maximising profits in private enterprises is too vague. If it is enquired *when* they are to reach a maximum, the uncertainties arise at once. The goal of immediate profits rules out research and investment for the future. At the other end of the time range, there is horizon planning, that is, planning specifically for the point in the future beyond which no useful information is available. The concept of horizon, however, in essence begs the question, since creating information, and hence defining where the horizon is, must be part of the plan itself. Difficulties would arise for breeding establishments in both types of planning, especially when dealing with long-lived animals. Rapid genetic gains by a competitor who abandons long range goals may prove financially fatal to a horizon planner. Yet it is essential to carry on long term development and operational research in view of the need to raise national productivity, and to cope with changes in markets and husbandry. Among these, secular changes in the environment of herds and flocks, hitherto inexperienced diseases, unpredictable climatic variation, new developments in food processing and merchandising, changes in consumer preferences, and many others come to mind. One of the reasons for the integration in agriculture lies in the need for maximising controls over the direction in which some of these factors move. In the poultry industry, for example, the breeding programmes as well as production operations with birds laying brown eggs, would probably have been cushioned against the economic set-back they experienced in California had they been under the control of the groups which developed the highly efficient apparatus for candling only white eggs.

Most pedigree herd owners in practice fix their horizon near at hand, for the good reasons that they can neither see ahead very far nor wait long for profits. For larger private enterprises, such as poultry firms or artificial insemination concerns, the problem of allocating current resources to genetic research and development seems to have no logical solution. Profit and loss accounts say nothing about neglected opportunities for securing new markets, or breeding better strains, and overhead costs cannot be related to any objective measure of efficiency or skill in organisation. A business which does not strike a satisfactory compromise in the conflict between long and short term considerations will sooner or later be at a disadvantage in a competitive economy.

Troublesome as the problem is for a large private concern, it is more complex still for a public institution such as the English Milk Marketing Board. Are the rules for its administration the same as for a private business? How much should it spend on research? Should it follow policies favoured by its constituents, or government policies? Should it encourage or try to destroy competitors? Should it try to make a large profit, a small profit, or no profit? Is it expected to help uneconomic small scale farmers, and ailing breed societies? What kinds of cattle ought it to select and breed for? Are its interests to be limited to dairying, to agriculture, to Britain, to Europe, or to the world? Each of these questions is political, but each has a scientific undertone.

III. Information

Decision-makers will have much information on which to base their decisions, but they will still be required to exercise foresight and judgment. This applies as much to those who are operating public or semi-public artificial insemination schemes, as to the managers of poultry enterprises. More efficient methods will lead to a faster rate of change, but it does not follow that the direction will be permanently correct. However, as more information about markets and production becomes available, the risk of persisting in wrong policies should in fact be progressively reduced. Yet, since economic and agricultural statistics have an authoritative air about them, it is well to remember that they often suffer from three defects. They may refer to a relatively distant past and therefore be irrelevant to some degree; where based on estimates or samples, they can be inaccurate; and, owing to varying assumptions and meanings, they can mislead if used for comparisons between countries or regions. In rapidly changing circumstances, statistics open to any of these criticisms will not be helpful, with or without a computer.

Since decision-makers cannot know of all current and future events that would influence them, some selection of data has to be made. It is a reasonable guess that, had different selections been made in the past, decisions would also have been different. Facts available are often those which have been collected for some other purpose, and subject to various biases that may have been irrelevant to the original purposes. If, instead of being based on lactation milk yields because there is an apparatus for collecting them, dairy bull selection had been based on body weight, cow populations might now have been different and perhaps superior. If breed societies had fully recognised the value of artificial breeding as an aid to progeny testing, they might have tried harder to exploit it from the beginning. Once decisions have been taken and effects produced, there can be no return to the original position. They can be reversed but the new decisions now apply to a changed population of animals and farmers in new circumstances.

Data cost money and there must inevitably be a tendency to make do with the cheapest or dispense with them altogether. What information to collect and how much of it are two of the first questions to be answered by operational research. Even when the answers are available, however, their value is short-lived. To be influential, therefore, statistics and performance records of all kinds must be available when the moment for making decisions is at hand. Speed must be allied with accuracy and relevance.

Animal breeding can never be the same as it was before the computer. Data handling that would require an inconvenient army of clerks is now necessary in the poultry and dairy cattle organisations, and will be necessary for the pig and sheep organisations of the future. In any conflict of views, information is a powerful weapon and it is in the hands that control the computer.

Information is not just a question of market intelligence or prompt reports of progeny tests. It includes a constant weighing of alternative products and procedures and an awareness of the results of research. Communication, however, even among scientists now verges on anarchy (Coblans in Goldsmith and MacKay, 1964), and for others interested in research findings, the position is worse. At its best, it has been likened to wind-pollination. In some ways, the process of making known the results of research appears extravagant and wasteful, as is the process of getting the results in the first place. Many workers and papers of no great significance have to be accepted in order to make sure of obtaining the few that are important. Van Vleck and Henderson (1965), who incidentally provide an admirable discussion of the use of statistics in reports of experiments related to animal production, think that negative results should be even more widely publicized to help in the planning of research.

Publication in a scientific journal serves three purposes: those of the author, the reader of current issues, and the future inquirer for recorded facts and hypotheses. Although these conflicting interests have not yet been reconciled, the spate of literature is forcing some changes. Coblans (*loc. cit.*) considers that it is gradually bringing nearer to realisation proposals made from time to time that short pithy resumes be widely distributed, while long papers with data be centrally stored and be available on film for those who need detail.

Some such separation will have to be achieved for decision-makers if they are to have time to expose themselves to new ideas and put them promptly into action. Meantime, the gulf between research and its application tends to widen for those managements which cannot or do not wish to keep in touch with research. This has happened with animal genetics, except where enterprises are large enough to support the employment of geneticists.

IV. Information Theory

Like the theory of games and systems analysis, information theory is far too complex and mathematically technical to permit an exposition here (for a discussion of its relation to biology see Yockey, 1958). Indeed, it has not been developed at all in relation to animal improvement and it is to be hoped that specialists in this field will before long explore this approach to breeding. As a foretaste of the possibilities, the speculations of King (1961) on information theory as related to the topics of inbreeding and heterosis may be referred to.

In brief, two important concepts entering information theory need to be considered: redundancy and noise. The first of these measures the amount of information, expressed in "bits" that is, a logarithmic function of the number of alternative choices in conveying a certain specification (for instance, the specification that the King of Hearts is desired from a pack of playing cards might involve five choices: black or red suit; Hearts or Diamonds; figures or numbers; males or female, King or Knave) which is given in a message in excess of the minimum required to ensure that the proper communication is delivered. The second, noise, refers to the interference by random events simulating the instructions of choosing one or the other alternative and thus distorting the message. Thus, a great amount of redundancy makes it certain that, by repetition of the same message in different or alternative ways, the intent of the message conveyed is preserved when it arrives at the receiver's end. On the other hand, the existence of noise may lead to the wrong alternative choice at every stage of selection (for instance, specifying Diamonds instead of Hearts) and thus result in a garbled or erroneous communication.

An illustration related to selection may be given here. When it is proposed to choose a sire of the most desired genotype, redundancy can be provided by incorporating in a selection index duplicating information on a variety of traits which are genetically correlated with the desired ones or on the phenotypes of related animals. Noise, in this instance, is produced by environmental effects which tend to obscure the real genotype of the animals. There is an important relationship between these concepts. To quote King: "Just as the ratio of amount of information in a message to the minimum necessary to express it determines the redundancy; the ratio between the amount of information sent and the amount received determines the 'equivocation', the amount lost because of noise. Therefore, for a given message transmitted over a given channel, whenever redundancy is greater than the equivocation, the probability will be high that the message received will be accurate."

Now, to provide the necessary amount of redundancy and to minimise noise, costs time, labour and, of course, money. The numbers of measurements on individuals and on their relatives that could be made is, naturally, unlimited. The question is which measurements permit the most accurate evaluation of the genotypes when the costs of increases in accuracy of the message relative to the amount of extra gains obtained are taken into account.

Thus the application of information theory to selection would require, among other things, the determination of the most economical levels of excess of redundancy over equivocation. This is equally true of the methods of the old-fashioned breeders, whose evaluation of economic merit may have been based on the examination of individuals for pedigree, or for breed, or on individual type points, or on type as an unquantified entity, and of the elaborate selection indexes which call for measurement of a great number of traits.

In the first case, instead of redundancy, it may be noise which is being added by extra measurements or impressions which the breeder uses in selection. In the second, the required redundancy may be too costly for the degree of accuracy which can be obtained, especially if an index, despite its comprehensiveness, fails to lead to genetic improvement in a population. Conversely it is also possible that the genetic theory of selection indexes may in fact produce low equivocation but may be failing to keep redundancy at the optimal level.

There is no doubt that not only selection but many other points involved in the organisation, theory and practice of animal improvement could also benefit from drawing on information theory. It may be hoped that it is only a matter of time, and not a long time, before extensive use of the considerable arsenal of techniques available to the experts in this subject will be introduced into animal breeding.

V. Cybernetics

Cybernetics, or the science of communication and control, has in recent years begun to pervade biology. At the molecular level, modern theories of gene action are based on the existence of intra-cellular self-regulatory devices (Jacob and Monod, 1961). The development of individual organisms has similarly been treated in cybernetic terms (Waddington, 1957). And on the population level, organic evolution itself has been viewed as a feedback process between genotype and phenotype, and between phenotype and environment (Schmalhausen, 1960). For a comprehensive review of the subject, Ashby (1956), as well as Norbert Wiener's original work (1948), should be consulted.

The idea is applicable to domestic populations of animals and the the manipulation of the contents of their gene pools by man. Essentially, whether breeders realise it or not, feedback processes of various kinds have been continually dominating their operations. For example, when selection for increased fat in pigs was so successful as to make lard a drug on the market (there were, of course, other reasons than successful selection), the feedback to the breeder led him to reverse the goals of selection in the direction of leanness.

A more automatic process of this type can occur when selection for some trait carries in its wake reduction in fitness and thus puts a brake on further gains in the desired character (see section on correlated responses, in Chapter 4). Thus, selection for size and conformation of turkeys was accompanied by loss of fertility.

Both the pig and the turkey examples show negative feedback, that is to say, the cutting-off of the stimulus when the response has reached a certain level. There are, however, also examples of positive feedbacks. One of these is the growth of breeding monopolies: the more efficient operation of the bigger breeding establishments makes them more successful, allows them to take over a bigger share of the market, permits them to acquire more resources, which in turn leads to even greater efficiency of their breeding programme, and therefore accelerates the whole process of their expansion. Computer selection, which has advantages over selection by eye, ear, or imagination, or even that carried out with the aid of a hand calculator, is justifiable only for large populations. But adequate size can be reached only when a breeder already has superior stock. It is, however, possible that, when the only breeders left are those of a sufficient operating size to use computers, some other factor will emerge which would set an upper limit to their expansion. In other words a negative feedback may eventually appear.

Effective breeding programmes depend on feeding genetic and economic parameters into calculating machines or computers in order to work out selection indexes or other selection criteria. But (as noted in Chapter 4) every time that there is response to selection, the values of these parameters change. Re-evaluation has then to be undertaken at frequent intervals. In time, no doubt, a fully automated circuit will make such adjustments without specific instructions at each step. It is interesting to speculate whether or not the lag between selection and marketing of a product, which in large animals must occupy a considerable period of time, will prevent such automatic procedures from becoming useful.

A single breeding establishment responsible for propagating not one but many species of animals may be common in the future. Cybernetic connections between all of the breeding programmes under the control of such an establishment will have to be made. And here it is perhaps the economic parameters that would play a more important role than the biological ones. A breeder who has a virtual monopoly of the improvement of both turkey and chicken broilers may find that he is competing with himself. He may therefore have to shift the emphasis in his selection programmes so as to make an economic differentiation between the two species under his control.

It may be a long time before economic planning of livestock production on a nation-wide basis is introduced in free-enterprise countries. Indeed, it appears that even in socialist states decentralisation of planning is being tried. Nevertheless, since such questions as the number of cattle or the number of broilers or turkeys that would meet the foreseeable demands will need to be answered, linear programming for arriving at an optimum structure of the breeding industry will have to be introduced, if only as a guide-line for private entrepreneurs. The computing services for working out maximum profit plans for individual farmers, which are already available, would merely have to be expanded on, first a regional, later national, and perhaps eventually, world-wide basis. The complications of horizon-planning would be greatly magnified but there is nothing intrinsically improbable about this kind of future.

As with many other topics considered in this book, only a passing reference, to cybernetics can be made. It takes, however, but little imagination to extend the ideas both to broader areas and to more specific points than has been done here. There is little doubt that cybernetic theory, like information theory, is destined to play an increasingly important role, not only in research, but also in commercial production based on animal breeding.

VI. Decision Theory

Mathematical and statistical tools for evaluating courses of action are now being brought together under the label of decision theory. It is really not a single theory but a collection of techniques for weighting numerous factors in a logical and systematic fashion. Where all the factors are known and predictable, decisions are made with certainty. Other methods, such as linear programming, cope with cases where there are chances that can be accurately measured or calculated or where there are only uncertainties. These are merely aids for there is no way to be sure of coming to the right decision. Such questions as the highest safe stocking rate, the best breed or cross to use, and the best ratio of crops to stock, are not capable of exact answers for any one farmer, nor for all farmers. Assumed biological constraints of density-dependent character are out of date for the most part. Furthermore, economic conditions change, and farmers vary in skill, energy and resources. The newer aids to decison-making are intended to be an advance on intuition and rule of thumb but carry no guarantee of correctness.

According to Alfandary-Alexander (1965), who has discussed this subject in general terms, one of the most useful services a decision-theorist can perform is to list alternative courses of action and spell out the consequences associated with each. He might also encourage preliminary trials or samplings so that as in classical statistical theory, probabilities based on experience can be used.

The choice of one or other of the various criteria is a matter of temperament: pessimists favour one kind and optimists another. What the decision maker may do after he has made his decision to prejudice the chances of his being right is, doubtless, also a personal trait.

Decision-makers who are bad losers can attempt to minimise the regret experienced after results are known. The amount of regret would be measured by the difference between the actual results and what might have been had the future been accurately foreseen. To minimise regret they should choose that course of action which has the minimum of all the maximum regrets. This criterion is called minimax regret. For executives with more sanguine temperaments, there are other ways of approaching decisions.

One of the most interesting results of decision theory has been the discovery that there is no one best criterion for selecting a strategy. Some decision theorists used to cherish the hope that mathematics alone might produce the perfectly automatic decision. However, managerial intuition is still quite important. Indeed, decision-theorists are now looking for more effective ways to integrate expert judgment and mathematics. They want to take full advantage of the comprehensiveness of the human mind to augment the analysis. Something might then be done to encourage probing or sampling the environment in order to get better estimates of the conditions likely to prevail. Managers or decision-makers, in whatever occupation, will from now on neglect to apply analytical methods to their operations only at their own risk. For the foreseeable future, however, they have no reason to fear that they will be completely replaced by machines producing decisions untouched by human brains.

Those responsible for livestock breeding enterprises will find a relevant process in the ramifications of the screening by which superior animals are found and multiplied. Given a predetermined outlay on facilities and resources for testing (an assumption which by-passes a major problem), the disposal of these resources in the most effective way makes a complex study. Robertson (1957) and Skjervold and Langholz (1964a) among others, have made important contributions to the specific problem of testing artificial insemination sires, King (1955) to the use of pig testing stations, and Young (1961) and others, to the general theory. These are examples from the greater field that includes the search for new varieties of plants, drugs, antibiotics and herbicides. Federer (1963) has briefly reviewed the subject and lists over 500 references. The fundamental question in this whole area is one that artificial insemination sire testers are well acquainted with: at what point in the progression of testing, terminating the test, and starting a new test will the expected gain be at a maximum? Finney (1958) opened up the subject of the relationship between the resources allocated to the screening process and the importance to the national economy of the improvements thereby obtained. If today there is any correlation here, it might well be coincidental. To discover how decisions about resources came to be made would require much research. It is virtually certain, however, that they were not based on a careful calculation of expected benefits. Yet there must be a theoretical optimum amount of effort distributed in an optimum way among the several stages of selection. It will depend in part upon the relative importance of the animal, and in part on the cost in time and money of each extension of the testing programme. For the larger livestock, a clarification of this topic is called for, but meantime it is worth remembering that any scheme for the selection, testing and approval of new varieties which proceeds too cautiously, can be a handicap if the breeding programme on which it is based is to be effective.

VII. Farmers as Decision-makers

Farmers as a class include some who are able to manage large enterprises. At this stage of history, a sifting is taking place. A few will succeed in organising big units. They are unlikely to be those who are constitutionally incapable of appreciating the opportunity science creates or of providing attractive work for the technologists who could serve them. The class known as scientists also contain a proportion temperamentally able to harness skill and energy in the building up of large production and marketing organisations. Farmers' unions and co-operatives are wise to recognise that the ability which makes a good research worker depends in part on brains and resourcefulness in the face of variations and obstacles, qualities that have their value in other occupations.

In all the recent history of animal breeding, covering a period of flux in methods and objectives, it is not easy to think of any example, in any country, of organised pedigree breeding in the forefront of developments now obvious to all. This was to be anticipated since all evolution proceeds at a variable pace. Advances lead to periods of comparative stability while forces are marshalled for further progress. Bureaucracies, whether of breed associations, research administration or government, are the marks and the symbols of those temporary ends. Reformers have to remember that the rapid rate of change now means that a new and apparently forward looking institution can become before anyone realises it, a fortress of reaction. To concentrate exclusive power over some important phase of animal breeding in a single institution, such as a government department, a centralised pedigree registry, or a national artificial insemination service, could be an error of judgment.

There have, however, been agriculturalists who were outstanding in the history of plant and animal production and the present is no exception. All research institutes are familiar with the practical man who follows closely on the heels of research workers and is sometimes ahead of them in applying their ideas. Such men are known to be influential in guiding the thoughts and actions of fellow farmers, and it would probably be advisable to encourage them by making technical assistance more readily available.

At a time when animal breeding is in a turmoil, it is useful to consider how breeders arrive at decisions affecting their business. They face many questions regarding their breeding objectives, attitudes to be adopted towards performance testing, or pedigree breeding, gains to be expected from co-operative breeding projects and many others. Emery and Oeser (1964) have made an interesting study of the way in which Australian farmers develop their conceptual range. They believe that the adoption by a farmer of a new practice he has heard of depends on (a) understanding, (b) accepting its relevance, and (c) deciding whether or not to adopt it. Among Australian farmers, the attitude to knowledge was different from that of town dwellers. Thinking, reading and planning were not thought of by farmers as work. Knowledge achieved by practical experience or personal communication was valued more than book-learning or ideas tested remotely by others. The investigation showed further that readiness to read about and absorb new scientific ideas, and to adopt new practices was positively correlated with skill in coordinating and planning, suitable environmental opportunities, and an unbiased outlook on information. Where a topic is ill-defined (such as "science", or "improvement", without specific definition) individual judgments tend to be vague and much subject to the influence of those with prestige. Consequently, it is natural that a traditionalist group of leaders in a farming community will encourage the rejection of a new idea because they are anchored to their traditional views. Progressives, on the contrary, tend to be wedded to the mobile concept of progressive farming. Although not exactly a surprise, this finding has its interest in the present context. When the adaptability of breed associations, the likelihood of new breeding organisations, and the performance of producer co-operatives is being assessed, the behaviour patterns of elected representatives of farmers and breeders are very pertinent.

Farming is an occupation pursued by men of great variety of age, temperament and circumstance. They cannot be characterised in a word; yet to them decision-taking is an almost hourly occurrence. To a particular challenge, farmers will react at different speeds and in different ways. An interesting survey has been made in New Zealand of the reasons why farmers did not adopt herd testing and artificial insemination (Table 10.1). There, artificial insemination, after a late and slow start taking about four years, experienced a rapid expansion for nine years, at the end of which time rather more than one-third of all cows were artificially inseminated. During the last three years, the rate of growth has become much slower. Herd testing has a comparatively long history of over forty years and covers about one-quarter of all cows. In New Zealand, as elsewhere, the value of herd-testing depends heavily on the attitudes of mind of farmers who test and on their willingness to apply the lessons learned from the figures. Artificial insemination is similarly affected, although in New Zealand the evidence that it can raise yields is clearer than in most other countries (see

Table 6.2). On the basis of the rather small samples obtained from one area, it would not be wise to place much trust in the details of Table 10.1, but they will serve well as illustrations.

TABLE 10.1

Reasons for not using dairy herd testing and artificial insemination given by New Zealand dairy farmers (from New Zealand Dairy Production and Marketing Board, 1964)

Reason	In % Herd testing	Artificial insemination
Not interested in increasing butter fat production	12	14
Not rearing replacements	6	8
No scope for culling	12	—
Testing a nuisance	16	—
A.I. stock or C.R. not good enough	—	10
Likes to select own bulls	—	28
Expense	34	22
Apathetic	5	7
Sundry	15	11
Number of farmers surveyed	215	140

In a country which encourages its agriculture, which showed an increase in average herd size (in herds over ten cows) from fifty-three to seventy cows in the last ten years, and which has increased butterfat production per cow by about 20 lb a year in the same period, it is significant that such a high proportion of farmers remain unimpressed by the two main improvement techniques. In Holland and the Scandinavian countries the proportion is much smaller, but in Britain and the United States it is about the same as in New Zealand.

It is instructive to compare these New Zealand observations with the analysis Florence (1964) gives of the motives he thinks might prompt the owner of a business or his salaried manager to decide for or against enlarging the business (Table 10.2). Mixed though they may be, they are sufficiently diverse to guarantee that opportunities will not be grasped with equal fervour by everyone. What the owners of pedigree herds will do when faced with the option of expanding operations, submerging themselves in co-operatives, or becoming agents of large firms, will vary in a manner hinted at by Tables 10.1 and 10.2. What the executives of establishments that are already large will do to exploit further opportunities will likewise vary. At one extreme may be those who are young and energetic enough to think on an international scale; at the other extreme are those who will be happy just to retain their posts.

TABLE 10.2

Motives for and against expanding the size of business (adapted from Florence, 1964)

Motive	Entrepreneur for	Entrepreneur against	Salaried manager for	Salaried manager against
		Economic		
Cash	Greater profits	Profit not greater	Bigger job	No personal gain
Real cost		Cannot be bothered Tax burden		Expansion might fail
		Psychological		
Power	Empire building	Dependence on outside funds	Power-seeking	Fear of others' power
Hobby	Love of work	Pre-occupied with other work	Love of work	
		Sociological		
Fame	Prestige	Preference for amateur status	Prestige of belonging to the business	Professional loyalty
A-logical		Own money to play with		No initiative

This variation is not just of personal or private significance. As the individual dynamism of some few men is overthrowing the resistance to change of existing institutions, they are creating a new context within which animal breeding will have to work. New institutions of their designing are growing out of the ruins of the old. The pace is set by those who, fascinated by growth, are determined to make a profit and will serve no other cause. Such men are to be found in the poultry and artificial insemination industries. Few of them have arisen through farming organisations, for the qualities they display are not the same as those required for democratic leadership in farming or politics.

The prehistoric distinction between nomadic herdsman and settled agriculturalist can still be traced in the behaviour patterns of the stockman and the grower of crops when confronted with change. Animals are an extension of self for the stockman in a way impossible for less directly responsive plants. But the surviving traces are disappearing, as the arts of animal busbandry give way to the techniques of mass production. The livestock breeders who aim to secure for themselves a share in the ownership and management of large breeding enterprises will have to adapt themselves accordingly. They cannot remain livestock breeders in the usual sense of this phrase.

VIII. Scientists as Decision-makers

Scientists cannot escape decision-making: they decide on problems to study, experimental design, data to collect, and interpretations to accept. Since every true decision, as distinct from an inference, involves an element of choice, the constraint imposed by logic and mathematics rules out, according to Cowan (1963), the study of creative decisions. In spite of the fact, however, that the sources of creative impulses such as desire for knowledge, inspiration and faith are but poorly understood, their practical consequences are evident enough.

Although formerly pursued largely as a cultural activity with the aim of revealing and contemplating the truth, science has become for the most part an unabashed striving for mastery over nature. The great edifice of knowledge has been built mainly to prosecute war, defeat disease, prolong life, and produce more of life's necessities and amenities. In the course of events, scientists may have become personally involved (much as Professor Higgins in Shaw's " Pygmalion "). They will now have to learn to carry ever-increasing responsibility for their decisions, as those who participated in the creation of the atomic bombs became fully aware.

Scientists must be judged less by their intellectual or physical equipment than by the effectiveness of their actions in discovering or applying knowledge (Cox, 1964). There are no pre-ordained ways of making discoveries. If, in the same sense, decision-makers heading a breeding enterprise are to be judged by results, there must be some acknowledged aims. But, to be sure of progress, the methods must be chosen imaginatively and applied skilfully as in successful art.

Yet, organised science in the shape of scientific societies and research councils has not been notable for enterprise (Welsh, 1965). It has, in fact, turned out to be the function of politicians, a few controllers of firms and industrial boards, and occasional farmers to have the courage of their scientists' convictions. As a class, scientists are not power-seekers like politicians, notwithstanding some well-known empire-builders among them. Nevertheless, science and technology, being both a means and an end of public policy, involve all scientists in ambivalent situations. Not unnaturally, some leave themselves open to the suspicion that they are more interested in pursuing a policy for the benefit of science than a scientific policy for the benefit of society.

The stereotype picture of a scientist is of a man driven by curiosity, whose equipment is intelligence, integrity, observation, thinking and creativity. Very few have this equipment developed to a high degree. Most could lay claim, like other human beings, to some degree of ignorance, bias, poor judgment, vanity and snobbery. Few scientists are non-conforming, original thinkers. The outstanding fact about the motivation of scientists according to Price (1965) is the highly competitive quality of their employment so that they must have rather special inducements to enter it in the first place and even more special ones to succeed in it. As he sees it, the 7% annual increase in their numbers comes through a "birth" rate of some 17% and a "death" rate of 10%—rather like a human population in primitive stages of development.

Policy making, even when it concerns science, has its own problems and techniques, of which neither scientists nor economists, as groups, can presume to possess special or exclusive understanding (Mesthene, 1964). Exceptional individuals, whatever their technical labels, may have personal gifts, which enable them to exert moral responsibilities, and to deal with human beings of irrational faith, tradition and prejudice. The pursuit of science does not confer these gifts on all teachers or research workers. Nevertheless when policy-makers have to grapple with problems and arrive at balanced judgments, they disqualify themselves if they do not insist on the participation in the act of policy-making of those who are specially trained in distilling the essence of technical matters, and who understand the implications of scientific knowledge and new discoveries. For this purpose, it is not enough that an administrator should have taken a degree forty years ago, and it is not enough that he should supplement this by occasional acceptable advice from "experts" on advisory committees, willing to give off-the-cuff answers to complex and important questions. He needs to incorporate outstanding and active scientists in the apparatus of decision-making and insist that their share of executive responsibility is publicly understood.

This is of great importance in problems of research financing. Thus, the machinery of allocation of public funds for research purposes by such agencies as the National Science Foundation and the U.S. Public Health Service is based on decisions made on the advice of panels of practising scientists by administrators, who as a rule are no longer active in the scientific profession. In spite of the increasing budgets that these agencies have, competition for funds is becoming progressively more severe. For instance, in one unpublished example, while the available funds rose 70% in a recent period of three years, the requests for subvention received during the same time more than doubled. Since, as may be seen from Table 8.2, American colleges and universities depend for 75% of their research and development costs on Federal government agencies, it is clear that scientific administrative decisions may come perilously close to determining the course and scope of the science to come. There is no evidence, so far that programme directors of the fund-granting agencies have become dictators, but the possibility is inherent in the situation.

Few scientists have declared themselves in favour of a scientific autocracy, and many are especially wary of what is called "informed scientific opinion." This does not mean that administrators of entirely non-scientific origins and attitudes would be any more acceptable. Geneticists have reason to regret the time lag in the application of their science caused by decision-makers in the livestock industry who, in the light of subsequent events, seem to have over-valued their own untutored opinions.

It is characteristic of recent years throughout much of the world, that concern about participation of scientists in national policies is manifested in all levels of society (Gilpin and Wright, 1964). The use of scientists as advisors to ministries and departments is not at all new. In the United States, the National Academy of Sciences was founded some hundred years ago by President Lincoln to provide the government with scientific advice. In the field of agriculture and animal breeding, government-supported fact-finding and fact-dissemination agencies are also of long standing in the shape of experiment stations, research institutes, extension and advisory services. But scientific advice has been nearly always on the technical level. The scientists, on the one hand, presented their findings for action to government offices, and, on the other, distributed technical information to farmers and breeders. The notion, however, that scientists should sit in the inner councils of a nation and participate in decision making and policy formulation is much more recent. At the moment, it seems that the extent of this practice depends largely on the political ideas and personalities of high elected officials. It is, however, clear that the existence of high level advisory committees and councils carries no guarantee that proper study will be made of major problems such as priorities for research expenditure, or that the scientists who advise will bear any burden of responsibility for implementing decisions.

There are undoubtedly many scientists who are anxious to contribute their knowledge and wisdom to decision-making. But it must be recognised that natural scientists are sometimes lacking in social tradition, in comprehension of political constraints, and in understanding of the economic facts of life. Many of them are ill-equipped for a role in high-level decision making and allocation of priorities owing to pride in their scientific morality and a distaste for compromises. The accommodation of politics and the restrictions on openness of thought and word necessary in Big Science in industry and government (especially, in military developments) are foreign to them. Nevertheless, there are those who rise superior to these handicaps. It seems that, eventually, a new breed of administrator combining the multifarious talents required in this position will be needed. Just as the ancient Chinese scholar-bureaucrat, the mediaeval viziers and agents, and the Victorian literary dilettanti have served their turn, so, perhaps, in the future, a class of modern technocrats adapted to recent changes in science-based industries will arise. Whether such a breed can be made to order, as the Shorthorn breed of cattle has been, remains to be seen. The problems of shortage of scientific manpower so far as experimenters and teachers are concerned is probably less acute, even in the face of the tremendous increase in these activities, than that of accomplished and dependable decision-makers.

Readers may have some reservations about the scientific bureaucrat as a decision-maker, and may express some apprehension about the extension of Parkinson's Law to science. Nonetheless in the area of animal breeding, rational and scientifically based decision-making on a national and international scale is going to be called for very soon. Indeed, it may become a rather critical issue as to where and what kind of improvement of livestock should be carried out by whom. It is not sufficient to rely on economists, simply because the biological facts are so intimately connected with the decisions to be made. A fuller understanding of the mechanics of animal breeding rather than simple reliance on technical advice would certainly be needed by managers responsible for plans to be put into effect.

Manifestly, the strong case for harnessing genetics more effectively to livestock production would be weakened by claiming more than could be delivered. There is much more to be learned about genetics and about its interactions with other forces. As an example, the question of control of breeding policy will serve well.

No matter what the respective merits of breeding procedures recommended by geneticists or developed and practised by breeders may be, the assumption that any one system or method is the only right one appears to be too dangerous to accept. Granted, that if in evaluating the operational merits of competing systems of organisation and breeding, one system will appear to be superior, it does not follow that it would be wise to eliminate all other systems.

The amount of information on which any decision can be based is inadequate to provide an answer which will always remain valid. Accepting that a solution now may interfere with, or make it impossible to adopt later another that might have been better still, there is no doubt that one or more of such better ones exist but are unknown today. To adopt a single system does not allow for the adjustability which will be needed for future changes in environment, and the new objectives towards which animal improvement may be directed.

A geneticist might well be more cautious than a pedigree breeder in making claims for his preferred breeding system, but in that case his decision-making would be influenced by considerations difficult to specify exactly and he would be acting as an informed layman. Faced with the common situation in which the breeding population is too small to sustain two or more different breeding policies, the geneticist will not be able to fall back on his theory for guidance. He will know that in natural evolution, species that have effective numbers too small to avoid specialisation and rigidity run great risks of becoming extinct.

If, in the computations underlying the determination of the optimum system of breeding or breeding organisation, a factor for flexibility, anticipating future developments, could be fed into the computer along with heritability values, genetic correlations and other known parameters, the situation might be different. One might venture a guess that what would come out of the computer would be a series of multiple answers, much like solutions of equations, containing unknowns raised to the *n*th power.

Where possible, it might be sensible to manage one part of breeding enterprises in a way which appears to be most efficient today and another part as a flexibility reserve. This could be maintained in a variety of ways that would ensure a potential for later changes and modifications should they be necessary, and its size would be determined by the optimism or pessimism of the decision-makers.

IX. Group Values

The chief concern of individuals is with the values which they recognise to be the values of the many groups to which they belong (Churchman, 1961). A "group" of people, such as the membership of a breed association, has identifiable actions and objectives and it functions within a certain teleological framework.

If a group of, say, farmers or scientists works towards certain ends and has certain operational theories, its members tend to think in predetermined ways, collect information favourable to its views, interpret observations subjectively, and constitute a vested interest. Heritability is an instance of a concept which is much conditioned by group values set on it, in this case, by population geneticists who know its meaning and implications, its strength and weaknesses.

Group values are like net genotypic values for complex traits in having much variation of individual components of which the resultant is the group value. This need not correspond to the value of any one component. The more the individuals resemble each other, however, the more distinctive the group value. Where there are conflicting views, they destroy each other and the group value is weak and unobtrusive. Within a group, all individuals are not equal and hence group values tend to approximate the values of persuasive or powerful individuals. As such, they are not necessarily correct or beneficial.

The importance of group values, when they are taken in conjunction with personal characteristics, arises from the various kinds of decision-making situations in which breeders and scientists may find themselves. A rough classification can be made following Churchman and using his eight categories (Table 10.3). It is a condensed version which might be described inadequately as an over-simplification. As a first attempt to use this classification will probably show, real people do not fit conformably into it, but it will also make clear that the group values and attitudes of geneticists and breeders may be far more important in decision-making than academic or social distinction.

It is natural that the majority of the present exponents of population genetics should belong to the category ABC. By virtue of their own genotypes and environments, they have chosen for themselves occupations and activities which are essentially deductive and conventional. It follows that the best use which can be made of them is in work requiring these qualities, namely, development of new breeds and strains by methods in which these workers are expert.

Management and control of large breeding populations has fallen, at various times and places, to: (a) government departments; (b) tycoons in business; (c) boards representing farmers and others; (d) technocrats.

TABLE 10.3

A classification of scientists (after Churchman, 1961)

A—a = conventional —non-conventional
B—b = formal, precise—non-formal, less precise
C—c = deductive —inductive

A	B	C	Forecast the unexpected; originality. Mathematics.
		c	Arbitrary "facts" as prelude to study of implications. Science fiction.
	b	C	Speculators by profession.
		c	Arbitrary "facts" as prelude to study of implications. Science fiction.
a	B	C	Precise, principled, e.g. mathematical economics. Sparing in invention of principles.
		c	Precise formulation of theory and experiment. Usefulness a minor consideration.
	b	C	Poets. Use vague terms, but general and original—common in economics—significant rather than trivial.
		c	Generalisers from facts—data collection is primary aim—limited precision and theorising.

A. Conventional means follows the rules of the group; non-conventional applies to judgments of a group that are not determined by its rules.
B. Formal applies to an inferential judgment when the group is aware of the rule.
C. Deductive means finding commitments *from* primal judgments.

The success or failure of their operations is bound up to some extent with the personalities of the managers and of the people they employ to provide information. These people are likely to be different in these four categories, although they should, perhaps, not be. Whereas the tycoon is likely to be attracted to type ABC, because it pushes things to logical conclusions or grapples with complex situations by original methods, the technocrat tends to be a strong believer in his own judgment working on data produced by type abc. This type seems likely to commend itself also to boards and government departments whose collective tastes for originality and enterprise are usually rudimentary. To some extent, the selection of "scientists" for service with breeding enterprises is likely to influence the nature of the "facts" on which they operate and, thereby, on their natures and activities. The most effective place for types ABC and abc would appear to be university departments and research institutes where freer rein can be given to originality.

X. State Responsibilities

Whatever the case in socialist countries, in non-socialist states the weighty question of state obligations arises in a variety of forms. Thus, responsibility for conducting or supporting research is an important state function. It has been discussed in some detail in Chapter 8.

Among other responsibilities, there is also advisory or extension work. The more advanced the technology of breeding is, the less significant does this function become, largely because breeders or cooperative breeding enterprises have their own technical staff. Whereas, no doubt, an agricultural advisor could be very helpful to a Scottish mountain sheep breeder, it is fair to say that in the whole of the United States not one extension man capable of improving the genetic operations of the breeders of egg-laying stock could be found. Even in commercial poultry production much extension work both in the United States and in Britain is being taken over by fieldmen for hatcheries as a competitive service.

Promotion of superior breeding methods has been a recognised function somewhat distinct from straightforward advisory work. Not only advice, but financial contributions in the form of assuming costs of record keeping, testing, inspection, providing publicity and making stimulus payments for successful stocks are included among the various methods used. It seems that this function of the state is withering away. Some subsidy payments for agricultural products eventually find their way into the breeders' hands. Promotion of superior breeding methods is then indirectly encouraged by subsidies to the producer.

Non-genetic regulatory activities have probably been in existence for a longer time than other government activities. This refers not only to enforcement of fair trade practices and honest advertising, but even more so, to regulation of disease. In many states it is illegal for breeders and hatcheries to sell chicks from flocks which have not been tested for pullorum disease. This a matter of protecting the producer. There is also the problem of protecting humans from diseases carried by livestock, though this may be of limited concern in breeding operations.

Official regulation of breeding practices is decreasing in some countries and increasing in others, and may be the most controversial of all of the government activities considered. Licensing of males on basis of conformation or the introduction of regionalisation fall into this category. In California at one time, an attempt was made to enforce poultry breeders' participation in federally-operated improvement plans. Indeed, disease-regulating features of such plan did become part of the law, though other aspects (for instance, restriction of use for breeding purposes to cockerels from dams which have laid some minimum number of eggs) were frustrated by an exceedingly strong opposition from free-enterprise-minded breeders.

Afterword

Afterword

Most citizens pay very little attention to the provenance of their meat, milk and eggs, but there are nevertheless strong links between the consumer and the breeding of the animals the products of which he eats. His influence makes itself felt through his preferences and discriminations among the goods offered. In areas where the standards of living are high, the economic pressure he exerts is towards higher quality and standardisation, the exact meaning of these descriptions being determined locally according to taste. By degrees, therefore, farmers are becoming obliged to try to deliver uniform products through a closer control of breeding and husbandry, and the subsequent grading of output. In less fortunate areas the need for protein from an acceptable and usually conventional source is paramount, and there is less attention to taste, tenderness and fat. Everywhere, however, the consumer is interested in cheapness and therefore in low cost methods of production. Universal aims of breeding and rearing livestock can hardly be defined except possibly in terms of maximum output of protein. Whether this should be a maximum for a unit of land, a unit of labour or a unit of capital investment, will depend on which component is in least supply. Ultimately, perhaps there will be a suitably weighted combination of the three. Where protein deficiency is unknown, breeding aims may be unacceptable over the short term unless they are closely related to quality.

In the last analysis, it is still the consumer, actual or potential, who brings about the use of science to alter any stage of the process of production and distribution. But the consequences hardly ever affect only the consumer. While his food habits are being changed by hygiene, freeze-drying, prepackaging, and other marketing devices, the added complications and skills required for producing and transporting foods tend to favour the more adaptable, the better financed, and the larger farmer or breeder.

Each new concept arising out of research leads to others. It thus gathers impetus, although the freedom of thought which engenders it always encounters opposition, for it is a human trait to resist any change which adversely affects the skills and status of people. As the rate of scientific development accelerates, it is to be expected that more people will become aware that they are not just spectators of agricultural evolution. Animal breeding is everyone's concern. Those most intimately affected are the breeders themselves, their employees, and the producers who depend on them for breeding animals. There are, at one remove, numerous suppliers of goods and services, extension personnel, journalists, bankers, administrators and research workers. And beyond these are the processors, wholesalers, distributors, retailers and, finally, consumers.

Over most of the world, animal production is firmly rooted in a close, and even personal, relationship of man with his livestock that is the result of an age-long process of reciprocal adjustments. For a century or so, they have been driven unceasingly to make greater demands on each other but society is still not satisfied. In some places there is hunger and in others cheaper products of higher quality are wanted. Each country strives to improve its competitiveness and achieve the economic growth which will finance its social aspirations. Agriculture must make its contribution and it must do so, as always in the past, by exploiting land, labour, capital and brains. Where technical developments have gone far enough, the advantages of scale accrue to animal production just as they do to other industries, but they are almost impossible to realise while maintaining the traditional husbandry that required a farmer to know and care for his animals. Growing along side it, therefore, is a new science-based industry with economic goals that allow of no concessions to sentiment.

This new industry, and to a lesser extent the old one, are at once the creations and the creators of the technology which serves them. Large sums of money have gone into education, research and development for industry. Agriculture has been allotted only a small share, yet enough of it has been expended to provide, among other things, a theoretical basis for breeding the animals that produce meat, milk, wool and eggs, and a set of techniques more effective than those formerly available. Until recently, however, two necessary conditions for the use of the new methods have usually been lacking: the objectives of breeding policy have not been defined in economic terms, and there have been no enterprises large enough to seek them independently of the established system of breeding. As is well-known, one or the other, and, occasionally, both of these conditions are now being met in some countries, so that the process of industrialising animal protein production can be expected to continue and eventually to embrace all forms of livestock.

Once started, the fusion of economic efficiency with technical expertness is likely to continue at a rising pace. Under the circumstances, the attack on traditional methods, wherever they are deemed to be restrictive, has proceeded from matters such as cultivations and herbicides which a farmer may adopt at discretion, to fundamentals such as land tenure, crop rotations and co-operative marketing which are accompanied by legal sanctions against non-conformists. The idea that animal production can be thought of and organised after the manner of factory production is steadily gaining ground in spite of the fact that it generates a strong opposition from those who are emotionally or financially engaged. It is at least as radical a departure from man's most ancient occupation as the automated production line is from a cottage craft. Where a political balance will be struck between the supporters of traditional methods and the proponents of cheap food no-one can tell. But in view of the growing human population and expanding scientific research, it seems unlikely that any one answer will finally dispose of the problem. As long as agrarian squalor and hunger afflict the world, the cry for more and better food will be heard.

Constructive breeding in future, especially for intensive production in a science-based industry will probably necessitate the subdivision of animal populations into breeding groups of operationally adequate size and a fast service of performance testing and data-processing. These will permit selection aims to be energetically pursued and inferior varieties discarded. Inter- and intra-breed rivalry will be encouraged by regional autonomy (where breeding is organised on a national scale) by adapting breed associations to the techniques of breeding for economic merit, by having performance testing adequate in extent and more effective in method, and by systematic importation. Selection will occur not only within herds or flocks but between them and between breeds as it does now. The ultimate goal is the mutual adaptation of genotypes and environments.

The success of a breeding programme can be measured in several ways. In some circumstances the rate of change may be of overriding importance; in others, the kind of change, or the cost of obtaining a unit of change. But whichever is taken as the criterion, the number of breeding animals in a population is of critical importance. It does not follow that control over large numbers, which is necessary for rapid advances from selection of breeding stock, will automatically result in progressive improvement. The other advantages of size will secure its economic future. Auctioneering, artificial insemination and sales promotion sometimes depend heavily on the continued prosperity of the pedigree breeding system. For them, improvement by breeding may well be irrelevant and even a threat to their existence.

Large-scale operations do not have to be modelled exactly on those now becoming characteristic of poultry, but by some means the power of fixing policy and securing action must be concentrated in a few capable hands. How to achieve this, while ensuring that those who wield the power remain accountable for their use of it, is a fundamental social and political question common to most human activities. Put in a form appropriate to animal breeding, the problem is to evolve policy-making bodies whose activities are consistent with the national or world interest. Without such institutions there can be no acceptable aims of breeding, and without aims there can be no exploitation of the power of science to increase the productivity of the livestock industry through breeding.

Pedigree breeding, as it is commonly understood, has in the past contributed a great deal to livestock production, but it must now expect its services and costs to be critically examined. These costs have to be borne by the industry and ultimately by the consumer, so it is reasonable to ask what services are rendered and what they are worth, questions that commercial enterprises have to ask themselves. Public interest sometimes requires restraint on the pursuit of efficiency. Without going so far as to shield inefficiency absolutely, a country may decide to offer partial or temporary protection, with a view to encouraging an established institution to improve its competitiveness against its rivals. There may also be such a strong public distaste for monopolies that some loss of efficiency is tolerated in order to avoid them. In other situations, the strength and political influence of those who support the existing breeding system may be great enough to suppress the knowledge that it could be improved. An acceptable solution to conflicting views may be found in allowing or encouraging two or more independent breeding organisations that may differ in aims and methods but are subject to performance tests under neutral supervision. This stage has yet to be reached for the larger livestock.

Means of measuring performance need to be constantly developed and refined for the comparison of breeds and crosses and for testing new varieties. Although performance testing of pigs, sheep and cattle is still inadequate, even rudimentary, the remedies lie in a striving for perfection, not in discouragement and the assumption of a superior attitude. There will be no going back to unaided eye judgment. It is easier to imagine international arrangements for testing livestock, especially pigs and dairy cattle in Northern Europe where conditions of husbandry are rather similar.

Making sure that varieties produced abroad can be imported and adequately tested is the best insurance that complacency will not develop in those responsible for breeding. In many countries, there may be little prospect of a government establishing an agency for the express purpose of making available the material from which more advanced biological machines could be developed, yet there is much to be said in its favour. Such an agency would be charged with making available the best livestock, no matter what their origins. Practically all varieties of crops and livestock in Western countries are of imported origin (and man himself is no exception), but no attempt has yet been made to carry out systematically a process that occurred sporadically in the past. Whether or not this comes to pass, there seems to be a good case for an intelligent policy of importation and testing. Even in the unlikely event of a country having nothing to gain from foreign breeds themselves, importations may provide a stimulus to home breeders. Appropriate veterinary precautions must be taken and some way found of discouraging projects that have little hope of being carried out. This is far from easy, since successful importations have very often been due to breeders prepared to fight for their ideas against the establishment and the orthodox masses. In future, however, the enterprising breeder will often be a large corporation or consortium which is clear about what new genetic material it needs, and is technically equipped to test stocks quickly and adequately.

As the methods of agriculture change and the craftsman gives way to the technician, the nature and financing of research call for re-appraisal. What passes muster in agriculture based on comparatively small farms will not do for a highly competitive economy with big businesses. Intensive production cannot be served in the same way as an extensive pastoral husbandry. But existing arrangements tend inevitably to be better adapted to the more out-of-date or undeveloped aspects of agriculture than to the most intensive and highly capitalised. Many aspects of this complex activity of a modern society that apply to animal breeding justify some constructive thinking now. In the past, much effort has been put into studying questions arising in the pursuit of agriculture by farmers and much more into trying to convey the results to them. While this effort still goes on, different research with a high order of urgency is required as a result of the organisational changes in farm industries. Operational problems are seen in far greater financial clarity than before, and it is worth paying much for speedy solutions to them. Slow deliberate attack over a broad front, inadequate facilities for applied and developmental studies and lack of financial incentives to students and research workers of ability create dissatisfaction in an active and progressive industry. Since time is short, it should not be squandered. The horizon of a scientist interested in a basic biological problem may be as much as ten years away, while for another concerned with adapting a new process to industry it may be only one or two, or even less, if the problem is operational in character. As a result, geneticists who are devoted exclusively to research, tend to take a longer view of breeding plans than do breeders pre-occupied with economic survival from year to year. But this incompatibility is becoming academic.

Decision makers responsible for a large enterprise are obliged to address themselves at first to developmental or operational research but as the enterprise grows and they begin to plan for a confident and receding future, basic problems attract more of their attention. Somehow present needs must be balanced against long-term hopes. For a business which has to finance research out of profits, there are several factors that determine what is spent. The size of the business and its financial strength exert a weighty influence, and so does the stimulus provided by powerful and aggressive competitors. How much the research programme will cost and what advantages it will confer on the marketing of the product are questions that have to be answered no matter how awkward they are. When the temperaments of research workers, their interests, qualifications, and age are added to this fluid situation, the public and private supporters and administrators of research in agriculture are saddled with a task which can be neither neglected nor finally accomplished. How much to invest in livestock improvement is, however, a question that is not susceptible of an exact answer by any kind of decision maker. Added to the genetic uncertainties about achievable rates of progress, there are economic and psychological imponderables that influence enthusiasm and willingness to take risks or mortgage the future. No rigid or doctrinaire attitudes can be supported when the justifiable investment can rise or fall with time or from one class of stock to another. This applies with as much force to the pedigree system as to any other method of breeding an elite population, for in the last resort the incidental costs and profits of all methods must be met by someone. Unless there is a change for the better in the genetic and economic merit of each breed, it is difficult to see how any outlays on breeding for improvement can bring returns to the ultimate financial source. Constructive breeding therefore must have positive aims, anticipate the future correctly, and temper the rigours of genetic theory with a proper regard for costs and returns.

Animal genetics stands ready to help in reaching the objectives of society. It is for society to decide where its ambitions lie and what are its purposes. If there is nothing so important as the production of foods to allay hunger, mitigate poverty, or support luxury, there is no-one more important than producers. The economic goals of agriculture are paramount. Institutions are devised that will bring them closer, and men are made to fit the institutions. Like many other sciences, genetics has been nourished mainly in Western countries where orthodox economic theory shelters public apathy and bureaucratic indifference to the hastening ills of mankind. Accordingly geneticists can apply themselves without moral distinctions to improving the poor man's pig, or the rich man's conspicuous beef. They have also long known that the power of their science can be frustrated by vested interest and a distaste for controversy. Individualism flourishes, and old habits die hard, the spectre of overpopulation notwithstanding. In time, society may be obliged to become more explicit about its purposes. When that day comes, it may be easier to define how science can best contribute to the quality of life of all the earth's inhabitants.

References and Author Index

The numbers on the right-hand margin refer to the page in the text where mention of a given work or person is made.

Abplanalp, H. (1962). Modification of selection limits for egg number. *Genet. Res.* **3**, 210–225. [65, 219, 220]

Abplanalp, H. [227]

Abplanalp, H., Asmundson, V. S. and Lerner, I. M. (1960). Experimental tests of selection index. *Poult. Sci.* **39**, 151–160. [86]

Abplanalp, H., Lowry, Dorothy C., Lerner, I. M. and Dempster, E. R. (1964). Selection for egg number with X-ray induced variation. *Genetics, N.Y.* **50**, 1083–1100. [215]

Adler, H. C. (1964). Artificial insemination of cattle. *V Congresso Internazionale di Riproduzione Animale e Fecondazione Artificiale, Trento* **4**, 40–61. [139]

Ahlgren, H. L. (1962). An extension program in a changing agriculture. *J. Dairy Sci.* **45**, 438–442. [9]

Albrectsen, R. [164]

Alfandary-Alexander, M. (1965). Decisions for the doubtful. *New Scient.* **25**, 437–439. [245]

Allen, C. P. (1962). The contribution of the plasmon to specific reciprocal cross differences in poultry. *Poult. Sci.* **41**, 825–839. [49]

Allison, W. S. [228]

Andersen, H. [105]

Andersen, H. and Rottensten, K. (1962). [Experiments with control of sex ratio on rabbits.] *Arsberetn. Inst. Sterilitäts-forsch., Copenhagen* (in Danish, cited from 1963 *Anim. Breed. Abstr.* **31**, 245). [221]

Andrewartha, H. G. and Birch, L. C. (1954). "The Distribution and Abundance of Animals", 782 pp. Univ. Chicago Press, Chicago. [80]

Andrews, F. N. [125, 163]

Ankorion, Y. [42]

Anon. (1962). Scientific research and development under government auspices. *Nature, Lond.* **193**, 710–714. [194]

Anon. (1964). Decisive calculations. *New Scient.* **24**, 145. [109]

Aristotle. [123]

Aschaffenburg, R. [227]

Ashby, W. R. (1956). "An Introduction to Cybernetics", 295 pp. Chapman and Hall, London. [243]

Ashton, G. C., Fallon, G. R. and Sutherland, D. N. (1964). Transferrin (β-globulin) type and milk and butterfat production in dairy cows. *J. agric. Sci., Camb.* **62**, 27–34. [227]

Asker, A. A. [158]

Asmundson, V. S. [86]

Astaurov, B. L. (1962). The present state of problems of artificial parthenogenesis in the silkworm. *Symp. genet. Biol. Italy* **9**, 20 pp. [221]

Astaurov, B. L. and Ostriakova-Varshaver, V. P. (1957). Complete heterospermic androgenesis in silkworms as a means for experimental analysis of the nucleus-cytoplasm problem. *J. Embryol. exp. Morph.* **5**, 449–462. [221]

Aylward, F. [6]

Babcock, S. M. [78]

Baker, C. M. Ann. [227]

Bakewell, R. [17, 123, 124, 185]

Barber, A. (1963). Civil Estimates 1963–64. Class VII. Universities and Scientific Research. 41 pp. H.M. Stationery Office, London. [193]

Bates, Lynn, S. [7]

Bayley, N. D., Parker, J. B., Heidhues, T., Plowman, R. D. and Swett, W. W. (1961). Dairy type: its importance in breeding and management. *U.S.D.A. Tech. Bull.* **1240**, 26 pp. [34, 160, 163]

Beatty, R. A. (1957). "Parthenogenesis and Polyploidy in Mammalian Development", 132 pp. Camb. Univ. Press, Cambridge. [222]

Belajev, D. K. and Trut, L. N. (1963). The ways of reorganization of the reproductive function in seasonly reproducing mammals. Genetics Today (*Proc. XI Int. Congr. Genet., The Hague*) **1**, 250. [95]

Benoit, J., Leroy, P., Vendrely, Colette and Vendrely, R. (1958). Phenotype du bec des canetons provenant de première et deuxième générations des canard Pekin antérieusement traités a l'ADN de canard Khaki Campbell. *C.r. hebd. Séanc. Acad. Sci., Paris* **247**, 1049–1052. [50]

Berg, R. (1961). The ecological significance of correlation pleiades. *Evolution, N.Y.* **14**, 171–180. [93]

Berge, S. (1961). The historical development of animal breeding. Scientific problems of recording systems and breeding plans of domestic animals (Max-Planck-Inst. Tierz. Tierernähr, Mariensee/Trenthorst), 109–127. [123]

Berge, S. (1963). Protein/fat in milk from different species of domestic animals. *Acta agric. scand.* **13**, 220–226. [31]

Bernal, J. D. (1939). "The Social Function of Science", 482 pp. Routledge, London. [13]

Bernier, P. E. (1960). A spontaneous chromosome aberration in a S.C. White Leghorn. *Poult. Sci.* **39**, 1234. [229]

Bhattacharya, B. C. (1962). Die verschiedene Sedimentations-geschwindigkeit der X- and Y- Spermien und die Frage der willkürlichen Geschlechtsbestimmung. *Z. wiss. Zool.* **A166**, 203–250. [221]

Bigalke, R. C. (1964). Can Africa produce new domestic animals? *New Scient.* **21**, 141–146. [41]

Biggar, J. [160]

Billett, F. S., Hamilton, L. and Newth, D. R. (1964). A failure to transform metazoan cells by DNA. *Heredity, Lond.* **19**, 259–269. [50]

Birch, L. C. [80]

Blackwell, R. L. [225]

Blaxter, K. L. (1960). Efficiency of feed conversion by different classes of livestock in relation to food production. *Proc. V. Int. Congr. Nutr. Washington D.C.* Panel VI, 8–14. [31, 32]

Bodmer, W. F. and Parsons, P. A. (1962). Linkage and recombination in evolution. *Adv. Genet.* **11**, 1–100. [51]

Bohren, B. B. [65]

Bohren, B. B. and McKean, H. E. (1964). Relaxed selection in a closed flock of White Leghorns. *Genetics, N.Y.* **49**, 279–284. [65]

Bose, N. K. (1961). "Cultural Anthropology", 140 pp. Asia Publ. House, New York. [13]

Boyazoglu, J. G. (1963). Aspects quantitatifs de la production laitiere des brebis. I. Mise au point bibliographique. *Annls. Zootech.* **12,** 237–296. [34]

Boylan, W. J. [95]

Braden, A. W. H. (1960). Genetic influences on the morphology and function of the gametes. *J. cell. comp. Physiol.* **56** (Suppl. 1), 17–29. [221]

Bradford, G. E. [225]

Bragg, S. L. (1964). Using academic consultants in industry. *New Scient.* **23,** 383–385. [202]

Briggs, H. M. 1958. "Modern Breeds of Livestock", 754 pp. Macmillan, New York. [125]

Briles, W. E. (1960). Blood groups in chickens, their nature and utilization. *Wld's. Poult. Sci. J.* **16,** 223–242. [225]

Brinks, J. S., Clark, R. T. and Kieffer, N. M. (1965). Evaluation of response to selection and inbreeding in a closed line of Hereford cattle. *Agric. Res. Service, U.S.D.A. Tech. Bull.* **1323,** 36 pp. [67, 74]

Buffon, de, G. L. L. [123]

Burger, R. E., Shoffner, R. N. and Roberts, C. W. (1961). Treatment of fowl sperm and developing embryos with deoxyribonucleic acid extracts. *Poult. Sci.* **40,** 559–564. [50]

Buschinelli, A. (1961). Injeções periódicas de sangue em aves e suas implicações geneticas. Thesis, Faculty of Sciences and Letters, University of Saõ Paolo, 142 pp. [50]

Buvenendran, V. (1964). Evidence of linkage between two egg albumen loci, in the domestic fowl. *Genet. Res.* **5,** 330–332. [227]

Byerly, T. C. (1962). Research for the future. *J. Anim. Sci.* **21,** 142–145. [7]

Byerly, T. C. (1964). "Livestock and Livestock Products", 422 pp. Prentice-Hall, Englewood Cliffs, N.J., U.S.A. [18, 77, 134, 177]

Campbell, Rosa M. [43]

Candolle, de, A. P. [123]

Carter, H. B. [144]

Carter, H. W. (1962). Effectiveness of artificial insemination in dairy cattle improvement. *J. Dairy Sci.* **45,** 276–281. [130, 139]

Cayley, A. [190]

Chapman, A. B. [59, 60]

Chu, E. H. Y. (1963). Mammalian chromosome cytology. *Am. Zool.* **3,** 3–14. [229]

Churchman, C. W. (1961). "Prediction and Optimal Decision", 394 pp. Prentice Hall. Englewood Cliffs, N.J., U.S.A. [14, 236, 238, 256, 257]

Clark, C. [4]

Clark, Joan R. [227]

Clark, R. T. [67, 74]

Clary, J. J. [227]

Clayton, G. A., Morris, J. A. and Robertson, A. (1957). An experimental check on quantitative genetical theory. I. Short-term responses to selection. *J. Genet.* **55,** 131–151. [66]

Clayton, G. C. and Robertson, A. (1957). An experimental check on quantitative genetical theory. II. The long-term effects of selection. *J. Genet.* **55,** 152–170. [66]

Coates, G. [107, 125, 156]

Coblans, H. [240]

Colburn, O. (1963). Report of the sheep panel. British Livestock Breeding, 76–94, H.M. Stationery Office, London. [125]

Cole, H. H., (ed.) (1962). "Introduction to Livestock Production", 787 pp. Freeman. San Francisco. [77, 88]

Commoner, B. (1961). In defense of biology. *Science* **133**, 1745–1748. [10]

Comstock, R. E. (1960). Problems and evidence in swine breeding. *J. Anim. Sci.* **119**, 75–83. [116, 127]

Comstock, R. E. [68, 95, 119]

Corley, E. L. [163]

Cowan, T. A. (1963). Decision theory in law, science, and technology. *Science* **140**, 1065–1075. [101, 235, 251]

Cox, G. (1964). Personal responsibility and technocracy. *Pharm. J.* **193**, 389–390. [251]

Craft, W. A. (1953). Results of swine breeding, *U.S.D.A. Circ.* **916**, 51 pp. [74]

Craig, J. V. and McDermid, E. M. (1963). Prolonged skin homograft survival and erythrocyte (β-locus) antigens in young chicks. *Transplantation* **1**, 191–200. [226]

Craig, J. V., Norton, H. W. and Terrill, S. W. (1956). A genetic study of weight at five ages in Hampshire swine. *J. Anim. Sci.* **15**, 242–256. [68]

Craig, Jean. [160]

Crittenden, L. B. [65]

Cruden, Dorothy M. [86]

Culley, G. (1794). "Observations on Live Stock", 222 pp. Robinson, London. [124, 157]

Cuthbertson, D. P., (ed.) (1963). "Progress in Nutrition and Allied Sciences", 452 pp. Oliver and Boyd, Edinburgh. [21, 43]

Cuthbertson, D. P. [43]

Darwin, C. (1872). "The Origin of Species by Means of Natural Selection", 6th Ed. 458 pp. Murray, London. [93]

Darwin, C. R. [41, 56, 125]

Dassman, R. F. (1964). "African Game Ranching", 75 pp. Pergamon Press, Oxford. [41]

Datta, S. P., Stone, W. H., Tyler, W. J. and Irwin, M. R. (1965). Cattle transferrins and their relation to fertility and milk production. *J. Anim. Sci.* **24**, 313–318. [227]

Davidson, H. R. (1953). The production and Marketing of Pigs", 2nd Ed. 537 pp. Longmans, Green, London. [155]

De Carlo, C. R. [151]

Dedijer, S. (1962). Measuring the growth of science. *Science*, **138**, 781–788. [189, 192]

Dempster, E. R. [55, 215]

Dempster, E. R. and Lerner, I. M. (1947). The optimum structure of breeding flocks. I. Rate of genetic improvement under different breeding plans. *Genetics, N.Y.* **32**, 555–566. [81]

Dempster, E. R. and Lerner, I. M. (1957). Multiple shifts for testing cockerels. *Poult. Sci.* **36**, 143–146. [65]

Dempster, E. R., Lerner, I. M. and Lowry, Dorothy C. (1952). Continuous selection for egg production in poultry. *Genetics, N.Y.* **37**, 693–708. [93]

Dettmers, A. E., Rempel, W. E. and Comstock, R. E. (1965). Selection for small size in swine. *J. Anim. Sci.* **24**, 216–220. [68]

Dickerson, G. E. (1955). Genetic slippage in response to selection for multiple objectives. *Cold Spring Harb. Symp. quant. Biol.* **20**, 213–224. [93]

Dickerson, G. E. (1959). Techniques for research in quantitative animal genetics. "Techniques and Procedures in Animal Production Research." (Am. Soc. Animal Prod.), 56–104. [83]

Dickerson, G. E. (1962). Experimental evaluation of selection theory in poultry. *Proc. XII Wld's Poult. Congr. Sydney* (*Symposia*), 17–25. [66, 85, 105, 166, 210, 217]

Dickerson, G. E. (1962a). Random sample performance testing of poultry in the U.S.A. *Anim. Breed. Abstr.* **30**, 1–8. [134]

Dickerson, G. E. (1962b). Implications of genetic-environmental interaction in animal breeding. *Anim. Prod.* **4**, 47–63. [79]

Dickerson, G. E. [210, 212]

Dickerson, G. E. and Hazel, L. N. (1944). Effectiveness of selection in progeny performance as a supplement to earlier culling in livestock. *J. Agric. Res.* **69**, 459–476. [81, 109]

Dillon, J. L. [109]

Dobzhansky, Th. (1962). "Mankind Evolving", 381 pp. Yale Univ. Press, New Haven, Conn., U.S.A. [12]

Donald, H. P. (1959). Evidence from twins on variation in growth and production of cattle. *Proc. X Int. Congr. Genet. Montreal* **1**, 225–235. [49, 72]

Donald, H. P. [71]

Donald, H. P. and El Itriby, A. A. (1946). The duration of pedigree herds in three breeds of cattle in relation to selective breeding. *J. Agric. Sci., Camb.* **36**, 100–110. [174]

Donald, H. P., Read, J. L. and Russell, W. S. (1963). Heterosis in crossbred hill sheep. *Anim. Prod.* **5**, 289–299. [129]

Doolittle, D. P. [211]

Downs, J. H. [211]

Dun, R. B. and Fraser, A. S. (1958). Selection for an invariant character—"vibrissa number"—in the house mouse. *Nature, Lond.* **181**, 1018–1019. [220, 232]

Dunn, L. C. [51]

Durham, R. H. [88]

Edwards, D. L. [227]

El Itriby, A. A. [174]

Ellis, N. R. (1964). New trends in pig nutrition. "Livestock Nutrition 1964." (Graham Cherry Org. Ltd., London), 29–37. [29, 34]

Emery, F. E. and Oeser, O. A. (1958). "Information, Decision and Action", 132 pp. Melbourne Univ. Press. [247]

Ephrussi, B. and Weiss, Mary C. (1965). Interspecific hybridization of somatic cells. *Proc. natn. Acad. Sci. U.S.A.* **53**, 1040–1042. [215]

Epstein, H. (1965). Regionalisation and stratification in livestock breeding with special reference to the Mongolian People's Republic (Outer Mongolia). *Anim. Breed. Abstr.* **33**, 169–181. [127]

Ernle, Lord (1961). "English Farming", 6th Ed. 559 pp. Heinemann and Frank Cass, London. [17]

Everson, D. O. [129]
Ewart, J. C. [125]

Falconer, D. S. (1960). "Introduction to Quantitative Genetics", 365 pp. Oliver and Boyd, Edinburgh. [49, 72]
Fallon, G. R. [227]
Faraday, M. [228]
Farthing, B. R. and Legates, J. E. (1957). Genetic covariation between milk yield and fat percentage in dairy cattle. *J. Dairy Sci.* **40**, 639–646. [84]
Federer, W. T. (1963). Procedures and designs useful for screening material in selection and allocation, with a bibliography. *Biometrics* **19**, 553–587. [246]
Feeney, R. E., Abplanalp, H., Clary, J. J., Edwards, D. L. and Clark, Joan R. (1963). A genetically varying minor protein constituent of chicken egg white. *J. biol. Chem.* **238**, 1732–1736. [227]
Finlay, G. F. (ed.) (1925). "Cattle Farming." Proceedings of the Scottish Cattle Breeding Conference. 495 pp. Oliver and Boyd, Edinburgh. [168]
Finney, D. J. (1958). Plant selection for yield improvement. *Euphytica* **7**, 83–106. [246]
Finney, D. J. (1962). Genetic gains under three methods of selection. *Genet. Res.*, **3**, 417–423. [83]
Fisher, R. A. [125, 126]
Florence, P. S. (1964). "Economics and Sociology of Industry", 258 pp. Watts, London. [150, 152, 249, 250]
Foght, Martha [227]
Food and Agriculture Organization (1963). "Emerging Diseases of Animals." FAO Agric. Studies, 61. 241 pp. [37]
Fraser, A. S. (1962). Simulation of genetic systems. *J. Theoret. Biol.* **2**, 329–346. [109]
Fraser, A. S. [220, 232]
Fredeen, H. T. (1965). Genetic aspects of disease resistance. *Anim. Breed. Abstr.* **33**, 17–26. [224]
Fuller, J. L. and Thompson, W. R. (1960). "Behavior Genetics", 396 pp. Wiley, New York. [36]

Galton, F. [125]
Garnier, J. [227]
Gaul, H. (1965). Induced mutations in plant breeding. Genetics Today (*Proc. XI Int. Congr. Genet. The Hague*) **3**, 689–709. [91]
Gibson, R. [211]
Giesbrecht, F. G. [67]
Giesbrecht, F. and Kempthorne, O. (1965). Examination of a repeat mating design for estimating environmental and genetic trends. *Biometrics* **21**, 63–85. [212]
Gilbert, N. [29, 160]
Gilmour, D. G. (1962). Current status of blood groups in chickens. *Ann. N.Y. Acad. Sci.* **97**, 166–172. [225]
Gilmour, D. G. [225, 227]
Gilpin, R. and Wright, P. (eds.) (1964). "Scientists and National Policy Making." 315 pp. Columbia Univ. Press, New York. [253]
Ginzberg, E. (ed.) (1964). "Technology and Social Change", 158 pp. Columbia Univ. Press, New York. [151]

Goldsmith, M. and Mackay, A. (eds.) (1964). "The Science of Science" 235 pp. Souvenier Press, London. [13, 152, 204, 240]

Goodwin, K. [210, 212]

Gowe, R. S. [88]

Gowe, R. S., Johnson, A. S., Downs, J. H., Gibson, R., Mountain, W. F., Strain, J. H. and Tinney, B. F. (1959). Environment and poultry breeding problems. 4. The value of a random-bred control strain in a selection study. *Poult Sci.*, **38**, 443–462. [211]

Gowen, J. W. (1937). Contributions of genetics to understanding of animal diseases. *J. Hered.* **28**, 233–240. [224]

Gregory, W. C. (1961). The efficiency of mutation breeding. Symposium on mutation and plant breeding (NAS-NRC, Washington D.C. Publ. 891), 461–486. [91]

Grosclaude, F., Garnier, J., Ribadeau-Dumas, B. and Jeunet, R. (1964). Etroite dependance des loci controlant le polymorphisme des caseins. *C.r. hebd. Séanc. Acad. Sci., Paris* **259**, 1569–1571. [227]

Gullbring, G. (1957). Investigation on the occurrence of blood group antigens in spermatozoa from man, and serological demonstration of the segregation of characters. *Acta med. scand.* **159**, 169–172. [221]

Gustafsson, Å. (1947). Mutations in agricultural plants. *Hereditas* **47**, 1–100. [91]

Hafez, E. S. E. and Lindsay, D. R. (1965). Behavioural responses in farm animals and their relevance to research techniques. *Anim. Breed. Abstr.* **33**, 1–16. [80]

Hagedoorn, A. L. (1927). An improved method of testing the quality of a breeder's entire flock. *Rep. Proc. Wld's Poult. Congr. Ottawa*, 95–98. [133]

Haig, D. [159]

Haldane, J. B. S. [96, 125, 126, 206]

Hamilton, L. [50]

Hanson, W. D. and Robinson, H. F. (eds.) (1963). Statistical genetics and plant breeding. Nat. Acad. Sci.—Nat. Res. Counc. Publ. 982. 623 pp. Washington D.C. [59]

Hardy, G. H. [53, 225]

Harris, H. (1959). "Human Biochemical Genetics", 310 pp. Camb. Univ. Press, Cambridge. [101, 227]

Harris, H. and Watkins, J. F. (1965). Hybrid cells from mouse and man: artificial heterokaryons of mammalian cells from different species. *Nature, Lond.* **205**, 640–646. [215]

Hart, B. H. L. (1959). "The Tanks", Vol. 1. 462 pp. Praeger, New York. [159]

Haubenstock, Harriet S. [221]

Hayman, R. H. [69]

Hazel, L. N. (1943). The genetic basis for constructing selection indexes. *Genetics, N.Y.* **28**, 476–490. [83, 86]

Hazel, L. N. [81, 109]

Hazel, L. N. and Lush, J. L. (1942). The efficiency of three methods of selection. *J. Hered.* **33**, 393–399. [83, 86]

Heady, E. O. and Dillon, J. L. (1961). "Agricultural Production Functions", 667 pp. Iowa State Univ. Press. Ames, Iowa, U.S.A. [109]

Heidhues, T. [34, 160, 163]

Henderson, C. R. (1964). Selecting the young sire to sample in artificial insemination. *J. Dairy Sci.*, **47**, 439–441. [82]
Henderson, C. R. [104, 142, 163, 212, 240]
Herskowitz, I. H. (1965). Genetics. 2nd Ed. 623 pp. Little Brown, Boston. [50]
Hickling, C. F. (1962). "Fish Culture", 295 pp. Faber and Faber, London. [42]
Higgins, H. [251]
Hodgson, R. E. (ed.) (1961). "Germ Plasm Resources", 381 pp. American Association for Advancement of Science, Washington D.C. [59, 60, 120, 175, 208]
Hoogschagen, P. (1958). Present and future importance of animal production in small holdings (a) in lowland areas. European Association of Animal Production, Publication C.f., **7**, 29–55. [9]
Hubbard, C. (1954). "The Complete Dog Breeders' Manual", 353 pp. Sampson Low, London. [97]
Hutt, F. B. (1958). "Genetic Resistance to Disease in Domestic Animals", 198 pp. Constable, London. [119, 224]
Hutt, F. B. (1964). Animal Genetics. 546 pp. Ronald Press, New York. [70]

Ingram, V. M. (1963). The Hemoglobins in Genetics and Evolution", 165 pp. Columbia Univ. Press, New York. [228]
Inouye, N. [230]
Irwin, M. R. [227]

Jacob, F. and Monod, J. (1961). Genetic regulatory mechanisms in the synthesis of proteins. *J. molec. Biol.* **3**, 318–356. [243]
James, J. P. [139]
James, J. W. [80]
Jeunet, R. [227]
Johansson, I. (1959). The genetics of milk yield. *Proc. X Int. Congr. Genet., Montreal* **1**, 236–243. [72]
Johansson, I. (1960). Progeny testing methods in Europe. *J. Dairy Sci.*, **43**, 706–713. [78]
Johansson, I. (1961). "Genetic Aspects of Dairy Cattle Breeding", 259 pp. Univ. Illinois Press, Urbana, Illinois, U.S.A. [77, 90, 104, 105]
Johansson, I. (1961a). Züchterische Massnahmen zur Leistungsteigerung beim Milchvieh. *Z. Tierzücht. ZüchtBiol.* **75**, 221–237. [84]
Johansson, I. (1964). The relation between body size, conformation and milk yield in dairy cattle. *Anim. Breed. Abstr.* **32**, 421–435. [160]
Johnson, A. S. [211]
Jonsson, P. (1965). ["Analysis of Characters in the Danish Landrace Pigs."] 490 pp. Komm. Landhusholdningsselskabets. Copenhagen (in Danish). [145, 173]

Kaplan, N. O. [228]
Kato, M. [50]
Kelleher, Therese. [59]
Kempthorne, O. (1957). "An Introduction to Genetic Statistics", 545 pp. Wiley, New York. [83]
Kempthorne, O. (ed.) (1960). "Biometrical Genetics", 234 pp. Pergamon Press, London. [210]
Kempthorne, O. [212]

Kiddy, C. A. [227]
Kieffer, N. M. [67, 74]
Kimber, J. [133]
King, A. [204]
King, J. C. (1961). Inbreeding, heterosis and information theory. *Am. Nat.* **95**, 345–364. [241, 242]
King, J. W. B. (1955). The use of testing stations for pig improvement. *Anim. Breed. Abstr.* **23**, 347–356. [109, 246]
King, J. W. B. [29, 160]
King, J. W. B., Aschaffenburg, R., Kiddy, C. A. and Thompson, M. P. (1965). Non-independent occurrence of β_{s1} and α-casein variants of cow's milk. *Nature, Lond.* **206**, 324–325. [227]
King, S. C., Van Vleck, L. D. and Doolittle, D. P. (1963). Genetic stability of the Cornell randombred control population of White Leghorns. *Genet Res.* **4**, 290–304. [211]
Kojima, K. I. [59]
Korach, M. [152]
Kosin, I. L. and Kato, M. (1963). A failure to induce heritable changes in four generations of the White Leghorn chicken by inter- and intra-specific blood transfusion. *Genet. Res.* **4**, 221–239. [50]
Kovach, L. D. (1960). Life can be so nonlinear. *Am. Scient.* **48**, 218–225. [191]
Kunkel, B. W. (1938). The fat of the land. *Sci. Monthly* **46**, 47–58. [6]
Kushner, H. F. (1963). [The XI International Congress of Genetics.] *Zhivotnovodstvo, Mosk.* (1963) **2**, 76–83 (in Russian). [213]

Lahman, A. [42]
Lall, H. K. (1956). "Breeds of Sheep in the Indian Union ", 39 pp. Government of India Press, New Delhi. [128]
Lamoreux, W. F. [210, 212]
Langholz, H. J. [82, 109, 246]
Lanni, F. (1964). The biological coding problem. *Adv. Genet.* **12**, 1–141. [228]
Lauvergne, J. J. [160]
Lauvergne, J. J., Vissac, B. and Perramon, A. (1963). Étude du caractère culard. I. Mise au point bibliographique. *Annls. Zootechn.* **12**, 133–156. [160]
Lederberg, J. (1964). A crisis in evolution. *New Scient.* **21**, 212–213. [230]
Lederberg, J. [231]
Legates, J. E. [84, 125, 163]
Lerner, I. M. (1950). "Population Genetics and Animal Improvement", 342 pp. Cambridge Univ. Press., Cambridge. [11, 83]
Lerner, I. M. (1951). Natural selection and egg size in poultry. *Am. Nat.* **85**, 365–372. [67]
Lerner, I. M. (1954). "Genetic Homeostasis", 134 pp. Oliver and Boyd, Edinburgh. [56, 57, 75]
Lerner, I. M. (1958). "The Genetic Basis of Selection", 298 pp. Wiley, New York. [11, 49, 55, 56, 65, 72, 73, 93, 95, 97, 109]
Lerner, I. M. (1959). The concept of natural selection: a centennial view. *Proc. Am. phil. Soc.*, **103**, 173–182. [55]
Lerner, I. M. (1962). Perspectives in poultry genetics. *Proc. XII Wld's Poult. Congr., Sydney* (*Symposia*), 9–16. [135, 217]
Lerner, I. M. [65, 81, 86, 87, 93, 215]

Lerner, I. M., Asmundson, V. S. and Cruden, Dorothy M. (1947). The improvement of New Hampshire fryers. *Poult. Sci.* **26**, 515–524. [86]

Lerner, I. M. and Dempster, E. R. (1951). Attenuation of genetic progress under continued selection in poultry. *Heredity* **5**, 75–94. [55]

Leroy, P. (1962). Observations faites sur des poules "Rhode Island Red" génétiquement contrôlées et sur leur descendants de 1re et 2e générations après injections répétées de sang de pintade. *C.r. hebd. Séanc. Acad. Sci. Paris* **254**, 756–758. [50]

Leroy, P. [50]

Levine, L. [228]

Lewontin, R. C. (1964). The interaction of selection and linkage. I. General consideration; heterotic models. *Genetics N.Y.* **49**, 49–67. [51]

Lewontin, R. C. and Dunn, L. C. (1960). The evolutionary dynamics of a polymorphism in the house mouse. *Genetics N.Y.* **45**, 705–722. [51]

Li, C. C. (1955). "Population Genetics", 366 pp. Univ. Chicago Press, Chicago. [49]

Lincoln, A. [253]

Lindahl, P. E. (1958). Separation of bull spermatozoa carrying *X*- and *Y*-chromosomes by counter-streaming centrifugation. *Nature, Lond.* **181**, 784. [221]

Lindahl, P. E. (1960). Experimental influence upon the distribution of the sexes in mammals by separation of male and female determining spermatozoa. *Z. Tierzücht. ZüchtBiol.* **74**, 181–197. [221]

Lindsay, D. R. [80]

Livestock Records Bureau (1964). L.R.B. Newsletter 25, 4 pp. Edinburgh. [175]

Lowry, Dorothy C. [93, 215]

Luck, J. M. (1957). Man against his environment: the next hundred years. *Science* **126**, 903–908. [31]

Luckey, T. D. (1963). "Germfree Life and Gnotobiology", 512 pp. Academic Press, New York and London. [224]

Lush, I. E. (1961). Genetic polymorphisms in the egg albumin proteins of the domestic fowl. *Nature, Lond.* **189**, 981–984. [227]

Lush, J. L. (1925). The possibility of sex control by artificial insemination with centrifuged spermatozoa. *J. Agric. Res.* **30**, 893–913. [221]

Lush, J. L. (1936). Genetic aspects of the Danish system of progeny testing swine. *Iowa St. Agric. Exp. Sta. Res. Bull.* **204**, 105–196. [77, 145]

Lush, J. L. (1945). "Animal Breeding Plans", 3rd Ed. 443 pp. Collegiate Press, Ames, Iowa, U.S.A. [49, 72, 109, 155, 160]

Lush, J. L. (1946). Chance as a cause of changes in gene frequency within pure breeds of livestock. *Am. Nat.* **80**, 318–342. [173]

Lush, J. L. (1960). Improving dairy cattle by breeding. I. Current status and outlook. *J. Dairy Sci.* **43**, 702–706. [31, 195]

Lush, J. L. [83, 86, 109, 120, 125, 126, 141, 195, 213]

Lysenko, T. D. [50, 213, 216]

Mackay, A. [13, 152, 204, 240]

Mangelsdorf, P. C. (1961). Biology, food and people. *Econ. Bot.* **15**, 279–288. [7]

Manwell, C., Baker, Ann C. M., Roslansky, J. D. and Foght, Martha (1963). Molecular genetics of avian proteins. II. Control genes and structural genes for embryonic and adult hemoglobins. *Proc. natn. Acad. Sci. U.S.A.* **49,** 496–503. [227]

Marlowe, T. J. (1964). Evidence of selection for the Snorter dwarf gene in cattle. *J. Anim. Sci.* **23,** 454–460. [91]

Mason, I. L. (1957). "A World Dictionary of Breeds", 272 pp. Commonwealth Agriculture Bureaux, Slough. [125]

Mason, I. L. (1964). Lethals and reproduction. *V Congresso Internazionale Riproduzione Animale e Fecondazione Artificiale, Trento* **2,** 58–69. [90]

Mather, K. (1943). Polygenic inheritance and natural selection. *Biol. Rev.* **18,** 32–64. [56]

Mather, K. (1949). "Biometrical Genetics", 162 pp. Methuen, London. [49]

Maule, J. P. (ed.) (1962). "The Semen of Animals and Artificial Insemination." 420 pp. Commonwealth Agriculture Bureaux, Slough. [137, 139]

Mayer, Jean (1964). Food and population—a different view. *Nutr. Rev.* **22,** 353–357. [4]

Mayr, E. (1963). "Animal Species and Evolution", 797 pp. Harvard Univ. Press. [54, 125]

McBride, G. (1964). Social behaviour of domestic animals. II. Effect of the peck order on poultry productivity. *Anim. Prod.* **6,** 1–7. [80]

McBride, G., James, J. W. and Wyeth, G. S. F. (1965). Social behaviour of domestic animals. VII. Variation in weaning weight in pigs. *Anim. Prod.* **7,** 67–74. [80]

McDermid, E. M. [225, 226, 227]

McKean, H. E. [65]

McLean, J. W. (1952). Progeny testing in sheep. Lincoln College (N.Z.) Techn. Publ. 8, 45 pp. [173]

Melrose, D. R. [139]

Mendel, J. G. [10, 47, 48, 125, 126, 192]

Merritt, E. S. and Gowe, R. S. (1960). Combining ability among breeds and strains of meat type fowl. *Can. J. Genet. Cytol.* **2,** 286–294. [88]

Merritt, E. S. and Slen, S. B. (1963). Response to selection for body size at two ages in the fowl. Genetics Today (*Proc. XI Int. Congr. Genet. The Hague*) **1,** 261–262. [86]

Mertz, E. T., Veron, Olivia A., Bates, Lynn S. and Nelson, O. E. (1965). Growth of rats fed on opaque-2 maize. *Science* **148,** 1741–1742. [7]

Mesthene, E. G. (1964). Can only scientists make government policy? *Science* **145,** 237–240. [252]

Mintz, Beatrice (1965). Genetic mosaicism in adult mice of quadriparental lineage. *Science* **148,** 1232–1233. [215]

Mitchell, H. H. (1964). "Comparative Nutrition of Man and Domestic Animals", Vol. 2. 840 pp. Academic Press, New York and London. [32]

Mitchell, R. G., Corley, E. L. and Tyler, W. J. (1961). Heritability, phenotypic and genetic correlations between type ratings and fat production in Holstein-Friesian cattle. *J. Dairy Sci.* **44,** 1502–1510. [163]

Moav, R. [42]

Monod, J. [243]

Morris, J. A. (1963). Continuous selection for egg production using short-term records. *Aust. J. Agric. Res.* **14,** 909–925. [65, 219]

Morris, J. A. [66]

Morton, J. R. [227]

Morton, J. R., Gilmour, D. G., McDermid, E. M. and Ogden, A. L. (1965). Association of blood-group and protein polymorphisms with embryonic mortality in the chicken. *Genetics N.Y.* **51**, 97–107. [225]

Mountain, W. F. [211]

Mudd, S. (ed.) (1964). "The Population Crisis and the Use of World Resources", 562 pp. Junk, The Hague. [3]

National Academy of Sciences (1964). Federal support of basic research in institutions of higher learning. 98 pp. *Nat. Acad. Sci. Washington D.C.* [195]

National Academy of Sciences (1965). Basic research and national goals. 336 pp. *Nat. Acad. Sci., Washington D.C.* [193, 194]

National Science Foundation (1961). Investing in scientific progress. 30 pp. *Nat. Sci. Found. Washington D.C.* [189]

National Science Foundation (1963). Research and development in industry. 118 pp. *Nat. Sci. Found., Washington D.C.* [189]

Neimann-Sørensen, A. and Robertson, A. (1961). The association between blood groups and several production characteristics in three Danish cattle breeds. *Acta agric. scand.* **11**, 163–196. [225]

Nelson, O. E. [7]

Newth, D. R. [50]

New Zealand Dairy Board (1961). Population genetics. 37th Annual Report, 102-107. Wellington, N.Z. [142]

New Zealand Dairy Production and Marketing Board (1964). 3rd annual report. 103 pp. Wellington, N.Z. [142, 143, 249]

Nichols, J. E. (1957). "Livestock Improvement", 4th Ed. 240 pp. Oliver and Boyd, Edinburgh. [127]

Nichols, J. R. and White, J. M. (1964). Correlation of meat and milk traits in dairy cattle. *J. Dairy Sci.* **47**, 1149–1155. [160]

Nordskog, A. W. [226]

Nordskog, A. W. and Giesbrecht, F. G. (1964). Regression in egg production in the domestic fowl when selection is relaxed. *Genetics, N.Y.* **50**, 407–416. [67]

Norton, H. W. [68]

Novitski, E. [51]

O'Bleness, G. V., Van Vleck, L. D. and Henderson, C. R. (1960). Heritabilities of some type appraisal traits and their genetic correlations with production. *J. Dairy Sci.* **43**, 1490–1498. [163]

O'Connor, L. K. (1962). The use of milk records in the selection of male and female breeding cattle. 121 pp. *European Ass. Anim. Prod.* (Mimeo.) [113]

Oeser, O. A. [248]

Ogden, A. L. (1961). Biochemical polymorphism in farm animals. *Anim. Breed. Abstr.* **29**, 127–138. [226, 227]

Ogden, A. L. [225]

Ogden, A. L., Morton, J. R., Gilmour, D. G. and McDermid, E. M. (1962). Inherited variants in the transferrins and conalbumins of the chicken. *Nature, Lond.* **195**, 1026–1028. [227]

Ollivier, L. and Lauvergne, J. J. (1964). Étude du déterminisme heréditaire de l'hypertrophie musculaire du porc de Pietrain: premiers résultats. 5 pp. *European Ass. Anim. Prod.* (Mimeo.) [160]

Olsen, M. W. (1960). Nine year summary of parthenogenesis in turkeys. *Proc. Soc. exp. Biol.* **105**, 279–281. [51, 222]

Olsen, M. W. (1965). Twelve year summary of selection for parthenogenesis in Beltsville small white turkeys. *Brit. Poult. Sci.* **6**, 1–6. [51, 222]

Oppenheimer, J. R. (1963). Communication and comprehension of scientific knowledge. *Proc. natn. Acad. Sci. U.S.A.* **50**, 1194–1200. [12]

Organisation for Economic Cooperation and Development (1964). The problem of low incomes in agriculture. *O.E.C.D. Observer*, 1964 (10), 27–33. [9]

Ostriakova-Varshaver, V. P. [221]

Ovington, J. D. (ed.) (1963). The better use of the world's fauna for food. *Symp. Inst. Biol.* No. 11. 175 pp. Institute of Biology, London. [6, 41, 42]

Owen, R. D. [221]

Parker, P. A. [34, 160, 163]

Parkinson, C. N. [254]

Perramon, A. [160]

Parsons, P. A. [51]

Pavitt, K. (1963). Research, innovation and economic growth. *Nature, Lond.* **200**, 206–210. [198]

Perry, E. J. (1960). "The Artificial Insemination of Farm Animals", 3rd Ed. 430 pp. Rutgers Univ. Press, New Brunswick, New Jersey, U.S.A. [137]

Pesce, A. [228]

Petras, M. L. (1963). Genetic control of a serum esterase component in *Mus musculus*. *Proc. natn. Acad. Sci. U.S.A.* **50**, 112–116. [227]

Phillips, R. W. [120]

Pirchner, F. and Lush, J. L. (1959). Genetic and environmental portions of the variation among herds in butterfat production. *J. Dairy Sci.* **42**, 115–122. [141]

Pirie, N. W. (1958). Leaf protein as a human food. *Nutrition* **12**, 17–22. [43]

Pirie, N. W. (1962). Indigenous foods. *Adv. Sci.* **18**, 467–475. [43]

Pirie, N. W. [41, 42]

Plowman, R. D. (1964). World meat supply—its distribution and outlook. *J. Dairy Sci.* **47**, 1135–1137. [27]

Plowman, R. D. [34, 160, 163]

Pollard, M. (1964). Germfree animals and biological research. *Science*, **145**, 247–251. [224]

Portal, M. and Quittet, E. (1950). Les races ovines françaises. 90 pp. *Fed. Nat. Ovine Parcs.* [128]

Poutous, M. and Vissac, B. (1962). Recherche theorique des conditions de rentabilité maximum de l'épreuve de descendance des tareaux d'insémination artificielle. *Annls. Zootechn.* **11**, 233–256. [142]

Prentice, R. (1963). "Utility is Beauty", 12 pp. American Breeders Service, Chicago. [167]

Prevosti, A. (1955). Geographical variability in quantitative traits in populations of *Drosophila pseudoobscura*. *Cold Spring Harb. Symp. quant. Biol.* **20**, 294–299. [93]

Price, D. J. de S. (1961). "Science Since Babylon", 149 pp. Yale Univ. Press, New Haven, Connecticut, U.S.A. [189]

Price, D. J. de S. (1963). "Little Science, Big Science", 119 pp. Columbia Univ. Press, New York. [189]

Price, D. J. de S. (1964). The scientific foundations of science policy. *Nature, Lond.* **206**, 233–238. [11, 252]
Prunster, R. W. [69]
Purser, A. F. [69, 70]

Quittet, E. [128]

Rae, A. L. (1964). Genetic problems in increasing sheep production. *Proc. N.Z. Soc. Animal Prod.* **24**, 111–128. [33, 179]
Rahnefeld, G. W., Boylan, W. J., Comstock, R. E. and Singh, M. (1963). Mass selection for post-weaning growth in mice. *Genetics, N.Y.* **48**, 1567–1583. [95]
Raimondi, R. (1962). I bovini Piemontesi a groppa doppia. 92 pp. Paravia, Turin. [160]
Read, J. L. [129]
Reed, C. A. (1959). Animal domestication in the prehistoric Near East. *Science* **130**, 1629–1639. [42]
Reichlin, M. [228]
Rempel, W. E. [68]
Rendel, J. (1961). Recent studies on relationships between blood groups and production characters in farm animals. *Z. Tierzücht. ZüchtBiol.* **75**, 97–109. [225]
Ribadeau-Dumas, B. [227]
Rice, V. A., Andrews, F. N., Warwick, E. J. and Legates, J. E. (1957). Breeding and improvement of farm animals. 5th Ed. 537 pp. McGraw-Hill, New York. [125, 163]
Roberts, C. W. [50]
Robertson, A. (1955). Selection in animals. Synthesis. *Cold Spring Harb. Symp quant. Biol.* **20**, 225–229. [55]
Robertson, A. (1956). The effect of selection against extreme deviants based on deviation or on homozygosis. *J. Genet.* **54**, 236–248. [56, 66]
Robertson, A. (1957). Optimum group size in progeny testing and family selection. *Biometrics* **13**, 442–450. [104, 109, 246]
Robertson, A. (1960). A theory of limits in artificial selection. *Proc. roy. Soc.* **B153**, 234–249. [97, 230]
Robertson, A. (1960a). The progeny testing of dairy bulls—a comparison of tests on father and son. *J. Agric. Sci.* **54**, 100–104. [130, 142]
Robertson, A. (1963). Summing up and proposals for future action. British Livestock Breeding, 95–102, H.M. Stationery Office, London. [177, 179]
Robertson, A. (1963a). Influence of breeding on production efficiency. World Conf. Animal Prod. Rome (*European Ass. Anim. Prod.*) **1**, 99–110. [97]
Robertson, A. (1964). A dispassionate view of those factory farms. *New Scient.* **22**, 568–569. [136]
Robertson, A. [66, 109, 225]
Robertson, A. and Asker, A. A. (1951). The genetic history and breed-structure of British Friesian cattle. *Emp. J. exp. Agric.* **19**, 113–130. [158]
Robertson, A., Waite, R. and White, J. C. D. (1956). Variations in the chemical composition of milk with particular reference to the solids-not-fat. II. The effect of heredity. *J. Dairy Res.* **23**, 82–91. [31]
Robinson, H. F. [59]

Rollins, W. C. [160]
Roslansky, J. D. [227]
Ross, G. J. S. [83]
Ross, O. B. (1960). Specific pathogen-free swine. Proc. 1960 Meeting Nutritional Council (*Am. Feed Mfrs. Assoc.*), 13–15. [224]
Rossi, J. [50]
Rothschild, Lord (1960). *X* and *Y* spermatozoa. *Nature, Lond.* **187**, 253–254. [221]
Rottensten, K. [105, 221]
Rowlands, I. W. (1964). Rare breeds of domesticated animals being preserved by the Zoological Society of London. *Nature, Lond.* **202**, 131–132. [208]
Royal Statistical Society (1963). "Food Supplies and Population Growth", 85 pp. Oliver and Boyd, Edinburgh. [3, 41, 42]
Russell, E. J. (1954). "World Population and World Food Supplies", 513 pp. Allen and Unwin, London. [18, 21]
Russell, W. S. [129]

Salisbury, G. W. and Van Demark, N. L. (1961). "Physiology of Reproduction and Artificial Insemination of Cattle." 639 pp. Freeman, San Francisco. [137]
Sandler, L. and Novitski, E. (1957). Meiotic drive as an evolutionary force. *Am. Nat.* **91**, 105–110. [51]
Sager, R. 1964. Nonchromosomal heredity. *New Engl. J. Med.* **271**, 352–357. [50]
Schierman, L. W. and Nordskog, A. W. (1961). Relationship of blood type to histocompatability in chickens. *Science* **134** 1008–1009. [226]
Schmalhausen, I. I. (1949). "Factors of Evolution", 327 pp. Blakiston, Philadelphia. [93]
Schmalhausen, I. I. (1960). Evolution and cybernetics. *Evolution* **14**, 509–524. [243]
Searle, S. R. (1961). Phenotypic, genetic and environmental correlations. *Biometrics* **17**, 474–480. [93]
Searle, S. R. (1964). Review of sire proving methods in New Zealand, Great Britain and New York State. *J. Dairy Sci.* **47**, 402–413. [110]
Sen, B. K. and Robertson, A. (1964). An experimental examination of methods for the simultaneous selection of two characters, using *Drosophila melanogaster*. *Genetics, N.Y.* **50**, 199–209. [66]
Serebrovsky, A. S. [213]
Shaffner, C. S. [36, 227]
Shaw, G. B. [251]
Sheldon, B. L. (1963). Studies in artificial selection of quantitative characters. *Austral. J. biol. Sci.* **16**, 490–515. [59]
Shoffner, R. N. [50]
Short, B. F. and Carter, H. B. (1955). Analysis of Merino stud flocks. *C.S.I.R.O. Bull.* **276**, 35 pp. [144]
Sidwell, G. M., Everson, D. O. and Terrill, C. E. (1962). Fertility, prolificacy and lamb livability of some pure breeds and their crosses. *J. Anim. Sci.* **21**, 875, 879. [129]
Šiler, R. [118]

Simpson, G. G. (1953). "The Major Features of Evolution", 434 pp. Columbia Univ. Press, New York. [93]
Singh, M. [95]
Skårman, S. (1963). Crossbreeding experiments with sheep. *LantbrHögsk. Annlr.* **29**, 63–98. [88]
Skårman, S. (1965). Crossbreeding experiments with swine. *LantbrHögsk. Annlr.* **31**, 3–92. [88]
Skjervold, H. (1963). The optimum size of progeny groups and optimum use of young bulls in A.I. breeding. *Acta agric. scand.* **13**, 131–140. [82]
Skjervold, H. and Langholz, H. J. (1964). Factors affecting the optimum structure of A.I. breeding in dairy cattle. *Z. Tierzücht. ZüchtBiol.* **80**, 25–40. [82, 109]
Skjervold, H. and Langholz, H. J. (1964a). Der optimale Einsatz der Nachkommenprüfung-Stationen in der züchterischen Arbeit. *Z. Tierzücht. ZüchtBiol.* **80**, 197–207. [82, 246]
Slater, W. (1963). Man and his food. *Nature, Lond.* **199**, 1225–1226. [4]
Slen, S. B. [86]
Smith, C. (1962). Estimation of genetic change in farm livestock using field records. *Anim. Prod.* **4**, 239–251. [212]
Smith, C. (1963). Genetic change of backfat thickness in the Danish Landrace breed of pigs from 1952 to 1960. *Anim. Prod.* **5**, 259–268. [166]
Smith, C. (1964). Pig progeny testing in Great Britain. The amount of selection on test results. *Anim. Prod.* **6**, 254. [109, 167]
Smith, C. (1964a). The use of specialised sire and dam lines in selection for meat production. *Anim. Prod.* **6**, 337–344. [89]
Smith, C. (1965). Results of pig progeny testing in Great Britain. *Anim. Prod.* **7**, 133–140. [119]
Smith, C., King, J. W. B. and Gilbert, N. (1962). Genetic parameters of British Large White bacon pigs. *Anim. Prod.* **4**, 128–143. [29, 160]
Smith, C. and Ross, G. J. S. (1965). Genetic parameters of British Landrace bacon pigs. *Anim. Prod.* **7** (In press). [83]
Smith, H. F. (1936). A discriminant function for plant selection. *Ann. Eugen.* **7**, 240–250. [83]
Sonneborn, T. M. (1964). The differentiation of cells. *Proc. natn. Acad. Sci. U.S.A.* **51**, 915–929. [50]
Spallanzani, L. [208]
Spector, W. S. (ed.) (1956). "Handbook of Biological Data", 584 pp. Saunders, Philadelphia. [83]
Stanford Research Institute (1959). Possible nonmilitary scientific developments and their potential impact on foreign policy problems of the United States. 100 pp. Government Printing Office. Washington, D.C. [3]
Stansfield, W. D., Bradford, G. E., Stormont, C. and Blackwell, R. L. (1964). Blood groups and their associations with production and reproduction in sheep. *Genetics, N.Y.* **50**, 1357–1367. [225]
Stent, G. S. (1963). "Molecular Biology of Bacterial Viruses", 473 pp. Freeman, San Francisco. [50]
Stonaker, H. H. [175]
Stone, W. H. [227]
Storer, N. W. (1963). The coming changes in American science. *Science* **142**, 464–467. [199]

Stormont, C. (1959). On the applications of blood groups in animal breeding. *Proc. XI Int. Congr. Genet., Montreal* **1**, 206–224. [107]
Stormont, C. [225]
Strain, J. H. [211]
Stroun, J., Stroun-Guttières, L., Rossi, J. and Stroun, M. (1963). Transfer to the progeny of alternatives induced in the White Leghorn by repeated injections of heterologous blood. *Archs Sci. Genève* **16**(2), 1–16. [50]
Stroun, M. [50]
Stroun-Guttières, L. [50]
Sutherland, D. N. [227]
Swett, W. W. [34, 160, 163]
Sykes, G. (1963). "Poultry as a Modern Agribusiness", 242 pp. Crosby, Lockwood, London. [130]

Tatum, E. L. (1964). Genetic determinants. *Proc. natn. Acad. Sci. U.S.A.* **51,** 908–915. [230, 231]
Taylor, St. C. S. [29]
Taylor, St. C. S. and Craig, Jean (1965). Genetic correlation during growth of twin cattle. *Anim. Prod.* **7**, 83–102. [160]
Taylor, St. C. S. and Rollins, W. C. (1963). Body size and conformation in identical twin cattle. *Anim. Prod.* **5,** 77–86. [160]
Tazima, Y. (1964). "The Genetics of the Silkworm", 253 pp. Logos Press. Academic Press, London and New York. [92, 220]
Terrill, C. E. (1958). Fifty years of progress in sheep breeding. *J. Anim. Sci.* **17,** 944–959. [35]
Terrill, C. E. [129]
Terrill, S. W. [68]
Thompson, M. P. [227]
Thompson, W. R. [36]
Tinney, B. F. [211]
Touchberry, R. W. (1951). Genetic correlations between five body measurements, weight, type and production in the same individual among Holstein cows. *J. Dairy Sci.* **34**, 242–255. [160]
Touchberry, R. W., Rottenstein, K. and Andersen, H. (1960). A comparison of dairy sire progeny tests made at special Danish testing stations with tests made in farmer herds. *J. Dairy Sci.* **43**, 529–545. [105]
Trexler, P. C. (1960). A new approach to the control of infectious disease in laboratory and farm animals. Proceedings of the 1960 Meeting of the Nutrition Council (*Am. Feed Mfrs. Assoc.*), 13–15. [224]
Triffitt, L. K. [69]
Trut, L. N. [95]
Turner, Helen Newton (1956). Breeding plans for sheep—past and possible progress. *Proc. Aust. Soc. Anim. Prod.* **1**, 100–116. [144]
Turner, Helen Newton, Hayman, R. H., Triffit, L. K. and Prunster, R. W. (1962). Response to selection for multiple births in the Australian Merino: a progress report. *Anim. Prod.* **4**, 165–176. [69]
Tyler, W. J. [163, 227]

Udall, S. L. (1963). "The Quiet Crisis", 209 pp. Holt, Rinehart and Winston, New York. [8]

United States Department of Agriculture (1964). (Sire Evaluation). Dairy herd improvement letter 40(5), 7 pp. U.S.D.A., Washington, D.C. [142]

Van Demark, N. L. [137, 208]
Van Valen, L. (1963). Haldane's dilemma, evolutionary rates, and heterosis. *Am. Nat.* **97**, 185–190. [96]
Van Vleck, L. D. (1964). Sampling the young sire in artificial insemination. *J. Dairy Sci.* **47**, 441–446. [82]
Van Vleck, L. D. [36, 163, 211, 212, 227]
Van Vleck, L. D. and Albrectsen, R. (1965). Differences among appraisers in the New York type appraisal program. *J. Dairy Sci.* **48**, 61–64. [164]
Van Vleck, L. D. and Henderson, C. R. (1961). Measurement of genetic trend. *J. Dairy Sci.* **44**, 1705–1710. [142]
Van Vleck, L. D. and Henderson, C. R. (1963). Bias in sire evaluation due to selection. *J. Dairy Sci.* **46**, 976–982. [104]
Van Vleck, L. D. and Henderson, C. R. (1965). Statistics in the design and analysis of physiology experiments. *J. Anim. Sci.* **24**, 559–567. [240]
Varro, M. T. [169]
Vendrely, Colette. [50]
Vendrely, R. [50]
Veron, Olivia A. [7]
Vissac, B. [142, 160]

Waddington, C. H. (1957). "The Strategy of the Genes", 262 pp. Allen and Unwin, London. [243]
Waite, R. [31]
Wallace, L. R. (1955). Factors influencing the efficiency of feed conversion by sheep. *Proc. Nutr. Soc.* **14**, 7–13. [33]
Wallace, L. R. (1958). Breeding Romneys for better lambing percentages. *N.Z. J. Agric.* **97**, 545–550. [68, 69]
Walsh, J. (1964). Federal R. & D.: Congress continues to boost budget, but increases are on last year's reduced scale. *Science* **146**, 623–624. [193]
Warwick, E. J. (1963). Role of a breed association in modern beef production. 20 pp. Santa Gertrudis Breeders International, Kingsville, Texas, U.S.A. [182]
Warwick, E. J. [125, 163]
Watkins, J. F. [215]
Wearne, S. H. (1965). Towards a science of project management. *New Scient.* **27**, 162–163. [151]
Weinberg, W. [53, 225]
Weir, J. A. and Haubenstock, Harriette S. (1964). Independence of sex ratio and diet-induced changes in blood-pH of mice. *J. Hered.* **55**, 187–192. [221]
Weiss, Mary C. [215]
Welsh, E. C. (1965). (quoted by D. S. G. under News and Comment). *Science* **147**, 381. [252]
Welshons, W. J. (1963). Cytological contributions to mammalian genetics. *Am. Zool.* **3**, 15–22. [229]
White, J. C. D. [31]
White, J. M. [160]
Wiener, G. (1953). Breed structure in the pedigree Ayrshire cattle population in Great Britain. *J. Agric. Sci.* **43**, 123–130. [172]

Wiener, G. (1957). The significance of grading-up in the Ayrshire cattle population of Great Britain. *J. Agric. Sci.* **49**, 313–318. [158]

Wiener, G. (1961). Population dynamics in fourteen lowland breeds of sheep in Great Britain. *J. Agric. Sci.* **57**, 21–28. [174]

Wiener, N. (1948). "Cybernetics", 194 pp. Technology Press, Wiley, New York. [243]

Wilcox, F. H., Van Vleck, L. D. and Shaffner, C. S. (1962). Serum alkaline phosphatase and egg production. *Proc. XII Wld's Poult. Congr.* (*Section papers*), 19–22. [36, 227]

Williams, R. J. (1956). "Biochemical Individuality", 214 pp. Wiley, New York. [101]

Wilson, A. C., Kaplan, N. O., Levine, L., Pesce, A., Reichlin, M. and Allison, W. S. (1964). Evolution of lactic dehydrogenases. *Fed. Proc.* **23**, 1258–1266. [228]

Wilson, J. [168]

Woese, C. R. (1963). The genetic code—1963. I.C.S.U. *Rev. Wld. Sci.* **5**, 210–252. [228]

Wohlfarth, G., Lahman, M., Moav, R. and Ankorion, Y. (1965). Activities of the Carp Breeders Union in 1964. Bamidgeh, **17**(1), 9–15. [42]

Wolstenholme, G. (ed.) (1963). "Man and His Future", 410 pp. Churchill, London. [4]

Woodward, F. N. (1960). Sponsored research. *Chem. & Ind.* (1960), 33–36. [200]

Wright, P. [253]

Wright, S. (1921). Systems of mating. *Genetics, N.Y.* **6**, 111–178. [73]

Wright, S. (1923). Mendelian analysis of the pure breeds of livestock. II. The Duchess family of Shorthorns as bred by Thomas Bates. *J. Hered.* **14**, 404–422. [95]

Wright, S. (1955). Classification of the factors of evolution. *Cold Spring Harb. Symp. quant. Biol.* **20**, 16–24. [53]

Wright, S. [54, 109, 125, 126, 196, 206]

Wyeth, G. S. F. [80]

Wynne-Edwards, V. C. (1962). "Animal Dispersion in Relation to Social Behaviour", 653 pp. Oliver and Boyd, Edinburgh. [80]

Yamada, T., Bohren, B. B. and Crittenden, L. B. (1958). Genetic analysis of a White Leghorn closed flock apparently plateaued for egg production. *Poult. Sci.* **37**, 565–580. [65]

Yockey, H. P. (ed.) (1958). "Symposium on Information Theory in Biology", 418 pp. Pergamon Press, London. [241]

Yoon, C. H. (1964). Bases for failure to induce transformation in vivo with exogenous, homologous DNA. *J. Hered.* **55**, 163–167. [50]

Youatt, W. (1834). "Cattle; Their Breeds, Management and Diseases", 600 pp. Baldwin, London. [123, 124]

Young, G. B. [29]

Young, S. S. Y. (1961). A further examination of the relative efficiency of three methods of selection for genetic gains under less-restricted conditions. *Genet. Res.* **2**, 106–121. [83, 86, 246]

Young, S. S. Y. [232]

Zeuner, F. E. (1963). "A History of Domesticated Animals", 560 pp. Hutchinson, London. [42, 98]

Subject Index

D

E

F

G

H

I

Genders and Sexualities in the Social Sciences

Titles include:

Jyothsna Belliappa
GENDER, CLASS AND REFLEXIVE MODERNITY IN INDIA

Edmund Coleman-Fountain
UNDERSTANDING NARRATIVE IDENTITY THROUGH LESBIAN AND GAY YOUTH

Niall Hanlon
MASCULINITIES, CARE AND EQUALITY
Identity and Nurture in Men's Lives

Brian Heaphy, Carol Smart and Anna Einarsdottir (*editors*)
SAME SEX MARRIAGES
New Generations, New Relationships

Sally Hines and Yvette Taylor (*editors*)
SEXUALITIES
Past Reflections, Future Directions

Meredith Nash
MAKING 'POSTMODERN' MOTHERS
Pregnant Embodiment, Baby Bumps and Body Image

Meredith Nash
REFRAMING REPRODUCTION
Conceiving Gendered Experiences

Barbara Pini and Bob Pease (*editors*)
MEN, MASCULINITIES AND METHODOLOGIES

Victoria Robinson and Jenny Hockey
MASCULINITIES IN TRANSITION

Yvette Taylor, Sally Hines and Mark E. Casey (*editors*)
THEORIZING INTERSECTIONALITY AND SEXUALITY

Thomas Thurnell-Read and Mark Casey (*editors*)
MEN, MASCULINITIES, TRAVEL AND TOURISM

S. Hines and Y. Taylor (*editors*)
SEXUALITIES: PAST REFLECTIONS, FUTURE DIRECTIONS

Yvette Taylor, Michelle Addison (*editors*)
QUEER PRESENCES AND ABSENCES

Kath Woodward
SEX POWER AND THE GAMES

Genders and Sexualities in the Social Sciences
Series Standing Order ISBN 978–0–230–27254–5 hardback
978–0–230–27255–2 paperback
(*outside North America only*)

You can receive future titles in this series as they are published by placing a standing order. Please contact your bookseller or, in case of difficulty, write to us at the address below with your name and address, the title of the series and one of the ISBNs quoted above.

Customer Services Department, Macmillan Distribution Ltd, Houndmills, Basingstoke, Hampshire RG21 6XS, England

Also by authors:

Richardson, D., Mclaughlin, J. and Casey, M. (2006/12) (eds) *Intersections Between Feminist and Queer Theory*. Palgrave.

Taylor, Y., Hines, S. and Casey, M. (2011) (eds) *Theorizing Intersectionality and Sexuality*. Palgrave.

Men, Masculinities, Travel and Tourism

Edited by

Thomas Thurnell-Read
Senior Lecturer in Sociology, Coventry University, UK

and

Mark Casey
Lecturer in Sociology, Newcastle University, UK

palgrave
macmillan

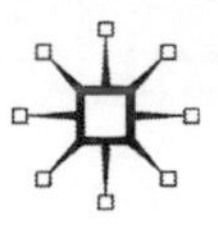

First published 2014 by
PALGRAVE MACMILLAN

Palgrave Macmillan in the UK is an imprint of Macmillan Publishers Limited, registered in England, company number 785998, of Houndmills, Basingstoke, Hampshire RG21 6XS.

Palgrave Macmillan in the US is a division of St Martin's Press LLC, 175 Fifth Avenue, New York, NY 10010.

Palgrave Macmillan is the global academic imprint of the above companies and has companies and representatives throughout the world.

Palgrave® and Macmillan® are registered trademarks in the United States, the United Kingdom, Europe and other countries

ISBN: 978–1–137–34145–7

This book is printed on paper suitable for recycling and made from fully managed and sustained forest sources. Logging, pulping and manufacturing processes are expected to conform to the environmental regulations of the country of origin.

A catalogue record for this book is available from the British Library.

Library of Congress Cataloging-in-Publication Data
Men, masculinities, travel and tourism / [edited by] Thomas Thurnell-Read, Mark Casey.
pages cm.—(Genders and sexualities in the social sciences)
ISBN 978–1–137–34145–7
1. Tourism – Social aspects. 2. Men – Travel. 3. Men – Identity.
4. Masculinity. I. Thurnell-Read, Thomas, 1982– II. Casey, Mark E.

G155.A1M434 2015
305.31—dc23 2014028179

Contents

Notes on Contributors

Hazel Andrews is Reader in Tourism, Culture and Society at Liverpool John Moores University, UK. Her research explores the social and symbolic constructions of national, regional and gendered identities in the context of British package tourism in the Spanish resort island of Mallorca. She is the author of *The British on Holiday: Charter Tourism, Identity and Consumption* (2011). Hazel is a co-founder and an editor of the *Journal of Tourism Consumption and Practice*. In addition, Hazel is chair of the tourism committee of the Royal Anthropological Institute and chair and founder of the Tourism and Embodiment Special Interest Group of The Association for Tourism and Leisure Education. Her work has appeared in numerous international journals including *Journal of Material Culture* and *Tourist Studies*, and she is the editor of *Liminal Landscapes* (with Les Roberts, 2012) and of *Tourism and Violence* (forthcoming).

Mark Casey is Lecturer in Sociology at Newcastle University, UK. His interest in sexuality is reflected in his current British Academy research grant examining sexuality, identity and space in the north-east of England. He has written about gay male identity and mental health (2009, 'Addressing Key Theoretical Approaches to Gay Male Sexual Identity', in *Critical Public Health*, 19(3–4): 293–306). He has researched and published on gay male travel in Australia and is currently undertaking research funded by the School of Geography, Politics and Sociology in the island of Mallorca, examining the effects of the current economic downturn on British residents there. He teaches across the three stages of the sociology degree at Newcastle including the modules 'The Sociology of Tourism' and 'Regulating Sexuality'.

Rosalina Costa is a sociologist and holds the position of Assistant Professor at the University of Évora in Portugal where she teaches across the three stages of the sociology degree. She is also a researcher affiliated with the Research Centre for the Study of Population, Economy and Society (CEPESE). She completed her doctoral studies in 2011 at ICS-UL with a thesis titled 'Small and Big Days: The Rituals Constructing Contemporary Families'. Her research interests cover a range of subjects, including family and personal life, social time and ages of life, ritualisation, consumption, memory and the imaginary. While currently publishing work within these topics, she is also enthusiastic about social research methodologies and has extensive experience teaching data analysis and research methodologies to both sociology and tourism students at the University of Évora.

Monica Gilli is Assistant Professor at the University of Milano-Bicocca where she teaches 'Sociology of Territory', and Tourism. Her research interests are tourism as a factor in urban regeneration and local development, cultural and heritage tourism, museums, the relationship between tourism and identity construction, and landscape and sustainability. Among her publications are *Authenticity and Interpretation in Tourist Experience* (2009, 'Autenticità e interpretazione nell'esperienza turistica', FrancoAngeli, Milano), *The Voyage Out: Sociological Studies* (2012, The Voyage Out. Studi sociologici, Scriptaweb, Napoli) and with E. Ruspini, M. Decataldo and M. del Greco *Tourism Gender Generations* (2013, 'Turismo Generi Generazioni', Zanichelli, Bologna).

Helen Goodman is a current doctoral candidate in the English department at Royal Holloway, part of the University of London. Her thesis addresses the linked topics of madness and masculinity in English literature and culture between 1850 and 1910. Her interdisciplinary research draws on a range of sources including fiction, psychiatric journals, asylum photographs, personal diaries and legal and medical reports.

Nigel Jarvis is Senior Lecturer at the University of Brighton, UK. His teaching interests include the impacts of tourism and sport tourism. He completed his PhD in 2006; his thesis examined the meaning of sport in the lives of Canadian and British gay men. The research critically aids understanding about how the lived experiences of gay men taking part in sport relate to and inform relevant hegemonic and queer theoretical debates. He has a keen interest in gender and sexuality issues in leisure and tourism. He is currently working on a series of research papers related to lesbian and gay men's perceptions of the cruise industry in general as well as the gay cruise sector specifically. He is also investigating how heterosexual staff working on gay cruise charters interact with customers. Nigel competed at the 2014 Cleveland Gay Games and his publication on the legacy of the event, titled 'Masculinity and the Gay Games: A consideration of hegemonic and queer debates', is forthcoming in 2014.

Yasmina Katsulis is Assistant Professor of Women and Gender Studies at the School of Social Transformation at Arizona State University, United States. Prior to this, Yasmina received her doctoral degree in anthropology from Yale University in 2003 and undertook further post-doctoral training at the Center for Interdisciplinary Research on AIDS in the School of Epidemiology and Public Health, also at Yale University. She is a medical anthropologist who specialises in research on the structural determinants of gender and sexual health disparities and is author of *Sex Work and the City: The Social Geography of Health and Safety in Tijuana, Mexico* (2009).

Chun-Yu Lin completed her PhD at Lancaster University in 2014. Her doctoral research project is titled '"From strangers to spouses?": International marriage immigrants in Taiwan'. It focuses on international marriage between marriage migrant women from Southeast Asian countries and Mainland China and Taiwanese men, and explores immigrant women's experiences of their transition from 'brides-to-be' to Taiwanese wives. Her research interest in gender as it relates to immigrant women began in graduate school and was part of her research, exploring Southeast Asian women's learning experiences in literacy programmes when she was studying at the National Taiwan Normal University. She specialises in the area of gender and transnational migration, with particular interest in transnational marriage, matchmaking businesses and bodywork.

Álvaro López López holds BA, masters and PhD degrees in geography from the National Autonomous University of Mexico (UNAM) and a post doc in tourism and environment from the University of Waterloo, Canada. He is currently a researcher at the Economic Geography Department of the Geography Institute and a lecturer in undergraduate and graduate programmes in the Faculty of Arts, UNAM. His main research areas are tourism geography, gender-sexuality geography and regional geography. In 2008, he was awarded the Social Sciences Research Award by the Mexican Academy of Sciences and the Young Researchers in Economic and Administrative Area Award by the UNAM. He was the secretary (2007 to 2009) and president (2010 to 2012) of the Mexican Academy of Tourism Research (AMIT). He is currently the academic secretary of the Geography Institute, UNAM.

Kristin Lozanski is Associate Professor of Sociology at King's University College in London, Ontario, Canada. She holds adjunct positions with the Department of Women's Studies and Feminist Research and the Migration and Ethnic Relations Program, both at Western University, Canada. Her scholarship focuses on the intersections of gender, racialization, globalization and mobilities. She has previously written about the intersections of gender, travel and colonialism (2007, 'Violence in Independent Travel to India', in *Tourist Studies*, 7(3): 295–315). Her work in critical travel and tourism studies has also appeared in the *International Journal of Qualitative Methodology*, *Social Identities*, and *Critical Sociology*.

Kevin Markwell is Associate Professor in the School of Tourism and Hospitality Management, Southern Cross University. He has published widely on the intersections of homosexuality and tourism and has co-authored a number of publications with colleague Gordon Waitt, including *Gay Tourism: Culture and Context* (2006). His interest in the social science of tourism extends beyond sexuality, and his most recent book is the collection *Slow Tourism: Experiences and Mobilities* (2012), edited with Simone Fullagar and Erica Wilson. He is currently editing a collection titled *Birds, Beasts and Tourists: Human-Animal Encounters in Tourism*, to be published in late 2014.

J. Carlos Monterrubio received his PhD in tourism from Manchester Metropolitan University, UK. He is a lecturer and tourism researcher at the Autonomous University of the State of Mexico, Mexico. His research interests include the sociocultural aspects of tourism, tourism and sexuality, and tourism as a field of study, particularly in Mexico. He is the author of three books: *Host Community Attitudes: A Perspective on Gay Tourism Behaviour, Tourism and Social Change: A Conceptual Perspective* (Spanish), and *Non-Conventional Tourism: Sociocultural Impacts* (Spanish). He has written several articles in international journals in English, Spanish and French. He was the secretary of the Mexican Tourism Research Academy and is currently an active member of the Mexican Research System.

Elisabetta Ruspini is Associate Professor in Sociology at the University of Milano-Bicocca and coordinator of the AIS (Associazione Italiana di Sociologia – Italian Sociological Association) Research Section 'Studi di Genere' (Gender Studies). She is a board member of the ESA RN33 (Women's and Gender Studies). She is the main editor of the book series 'Generi, Culture, Sessualita' (Gender, Cultures, Sexualities). She has extensive teaching and research experience, and has published widely on gender issues. Among her recent publications: with Jeff Hearn, Bob Pease and Keith Pringle (eds), *Men and Masculinities Around the World: Transforming Men's Practices* (2011) and *Diversity in Family Life: Gender, Relationships and Social Change* (2013).

Thomas Thurnell-Read is Senior Lecturer in Sociology in the Department of International Studies and Social Science at Coventry University,

UK. Having received his PhD from the University of Warwick in 2010, Thomas held an Early Career Fellowship at the Institute of Advanced Studies, also at the University of Warwick, before joining Coventry University in the autumn of 2010. His work explores social constructions of masculinities in relation to leisure and consumption practices. His research interests also include the sociological study of alcohol and drinking cultures, embodiment and emotion, and qualitative research methods. He teaches on a range of degree level modules relating to sociology of gender, leisure and consumption, the media, and social divisions and identity. He has published work in a range of international journals including *Annals of Tourism Research, Men and Masculinities, Sociological Review Online* and *Sociology* and has recently been contracted to write a text book titled *Tourism: Sociological Perspectives*.

James Treadwell is Criminology Lecturer at Birmingham Law School at the University of Birmingham, UK. He researches and teaches in the areas of criminology and criminal justice and has particular expertise on topics of professional and organised crime, violent crime and victimisation. He is the author of the best-selling textbook *Criminology* (2006) and the revised and updated *Criminology: The Essentials* (2012). James is best known as an ethnographic and qualitative researcher and has published articles in a number of leading national and international ISI-ranked criminology and criminal justice journals.

Gordon Waitt is Professor of Human Geography at the Department of Geography and Sustainable Communities, University of Wollongong. His recent publications in *Annals of Tourism Research, Leisure Studies, Social & Cultural Geography* and *Continuum* explore the importance of affective and emotional relationships. He is currently undertaking research on household sustainability, 'voluntourism' and bird watching. With Christine Metusela, he recently co-authored *Tourism and Australian Beach Culture: Revealing Bodies* (2012). With Kevin Markwell, he co-authored *Gay Tourism: Culture and Context* (2006). With Chris Gibson, Carol Farbotko, Nicholas Gill and Lesley Head, he co-authored *Household Sustainability: Challenges and Dilemmas in Everyday Life* (2013). At the University of Wollongong, he teaches across the honours, third-year and first-year curriculum of the geography and land and heritage management degrees.

Michael R. M. Ward is a research assistant at the Faculty of Health and Social Care at the Open University. He is currently working on an ESRC project titled 'Beyond Male Role Models', which explores gender identities and practices in work with young men. His doctoral research was an ethnographic study that centred on the lives of a group of young working-class men (aged 16–18) in a community in the South Wales Valleys and will be published in late 2014. His other work has examined the transport needs of older people in rural Lincolnshire. Alongside colleagues at the University of Lincoln, he wrote a report seeking to develop community transport in the county. He has taught sociology at both further and higher education institutions to students of all ages and is a tutor at the Lifelong Learning Centre at Cardiff University and an associate lecturer at the Open University.

1
Introduction

Mark Casey and Thomas Thurnell-Read

With its roots in the Grand Tours of the seventeenth and eighteenth centuries (Withey, 1997; Inglis, 2000; Littlewood, 2001) and its development entwined with the expansion of European colonialism (MacKenzie, 2005), modern tourism – particularly the international tourism industry – is an inescapably gendered phenomenon (Swain, 1995), arising from profoundly gendered societies and from the global interconnections among them (Enloe, 1989). Just as, in the most general sense, 'tourism has become a metaphor for the way we lead our everyday lives in a consumer society' (Franklin, 2003: 5), so, too, has it emerged as an important context in which contemporary relationships based on both entrenched and emergent positioning of gender, ethnicity and class can be studied. As noted by Cara Aitchison (1999: 61), 'tourism needs to be considered not just as a type of business or management but as a powerful cultural form and process which both shapes and is shaped by gendered constructions of space, place, nation and culture'.

Although there is some existing work that has focused upon gender and tourism (Sinclair, 1997; Aitchison, 1999; Swain and Momsen 2002; Pritchard et al., 2007) the focus upon men, masculinities and tourism has been rather limited (e.g., Knox and Hannam, 2007; Noy, 2007). While the implicit masculine position of the tourist gaze has been questioned, understandings of specific male tourists have often failed to engage with the gendered notions of independence, adventure, embodiment and risk that underpin much tourism experience and practice. Indeed, considerable overlap can be identified between the markers of successful travel and notions of toughness, independence and resilience in the face of risk or adversity foregrounded in hegemonic masculinity (Connell, 1995; Connell and Messerschmidt, 2005). Further, many of the freedoms and benefits of travel and tourism are only open to men or, at least, only available to women via a more complex negotiation of problematic assumptions about the nature of danger, risk and (in)dependence when travelling. This book aims to provide the first comprehensive exploration of the interdependencies of masculinities and tourism by bringing together a diverse range of empirical studies.

Both tourism studies and the critical study of men and masculinities only developed as coherent fields relatively recently. While early seminal texts (MacCannell, 1976; Urry, 1990; Hearn, 1992; Connell, 1995) outlined important theoretical and conceptual foundations, in both fields it was not until the later years of the 1990s that studies of sufficient diversity allowed earlier paradigmatic developments to be applied and observed in a range of empirical investigations. At the same time, important developments in the conceptualisation of tourism from sociology (Urry, 1990), anthropology (Selwyn, 1996) and geography (Crouch, 1999) in particular, developed new ways of looking at tourist phenomena and gave rise to what has become a genuinely interdisciplinary field of study. It is to the work of recent years in both fields that we look for important and progressive conceptualisations of tourism. Thus, tourism is experienced through embodiment (Crouch and Desforges, 2003), tourism practices are often spatially antagonistic (Edensor, 1998; Mordue, 2005), and tourism spaces and activities are often characterised by their liminality (Shields, 1991; Selänniemi, 2003; Andrews and Roberts, 2012). Similarly, the critical study of men and masculinities has sought to highlight the ways in which the social construction of masculinity gives rise to an increased understanding of masculinity as plural (Aboim, 2010), as embodied (Kehily, 2001a, b), and as open to renewed formations based around new articulations of masculinity specific to changes in the socioeconomic context of late modernity (Nayak, 2003; Anderson, 2011).

In light of this, the book sits at the intersection of two different but complementary fields of study. It is therefore offered both as a useful introduction to scholars of gender and sexuality who have little familiarity with developments in the sociological study of tourism and as a chance for academics and researchers of tourism to return focus to the notion that masculinity is central to much tourist practice and experience.

The centrality of highly gendered notions of hospitality and sociability (Hochschild, 1983) to the tourism industry frequently gives rise to social situations configured around the service of men by women. The expectation is that women in such settings will readily and willingly engage in a range of support roles, all based on gendered, classed and frequently racialised notions of hospitality. Many, but not all, tourist encounters (Crouch et al., 2001) are mediated by monetary exchanges, although tourism industries and workers work hard to conceal this from clients. The economic underpinnings of many highly gendered social interactions taking place as part of tourism cannot therefore be underestimated. This appeal to patriarchal benevolence appears in work on male sex tourists (Sanchez Taylor, 2000) who worryingly justify, condone and, indeed, even valorise their acts of paying for sex in terms of the economic assistance offered to a gendered, classed and racialised 'Other'.

Sex is central to tourism (Littlewood, 2001; Bauer and McKercher, 2003); just as its study readily reveals great gender inequalities, it also exposes complex social constructions of sex, sexuality and desire (Frohlick, 2010). The fantasies of tourist escapism have long been built upon deeply entrenched inequalities between individuals and between countries. Erik Cohen identified this, in his influential essay on sex tourism in Thailand, as the myth of 'instantly available women' that pervades the tourist imagination (Cohen, 1982: 407).

The disinhibition associated with many tourism destinations may encourage traits associated with hegemonic masculinity to be pursued with greater vigour, and with a different array of associated senses, feelings and emotions, than they would at other times. Guilt, shame or embarrassment may be mitigated by the temporary entry into a social milieu that endorses the pursuit of hedonism, excess and indulgence. Notable, then, are the male tourists who appear to seek out experiences and encounters that specifically simplify, compartmentalise and de-problematise a version of hegemonic masculinity and its relationship with masculine and feminine 'others'. While we do not wish to reinstate the home-away binary of early tourism studies, which has been shown to be far more porous and hazy than first envisaged, we do suggest that the contributions to this book demonstrate that the ways in which masculinity is enacted whilst 'away' can readily increase our understanding of how gender subjectivities shift whilst at 'home'. Thus, changes to the dominant form of masculinity might result in reconfigurations of how tourists and travellers experience and enact gender whilst 'away'. Whether or not and to what extent these changes in the constitution of gender roles and identities are carried over into tourism practice is the subject of chapters by Costa (Chapter 9) and Gilli and Ruspini (Chapter 14).

Studies that have highlighted the problematic masculinities of many young men in a deindustrialised late modernity (Winlow, 2001; Blackshaw, 2003; Nayak, 2006) also indicate that the leisure sphere can provide a space for the assertion of a masculinity clinging to the values of 'traditional' hegemonic masculinity. While individual men may enact a masculinity no longer available to them at 'home', their performances are part of a wider and deeply entrenched structuring of the tourist industry along heavily gendered lines.

Written through many of the studies in this book is, if not the actual and complete transformation of masculinities for the individuals involved, then the continued allure and the sense of possibility for tourism to offer a retreat from or an escape to certain experiences of masculinity. The sometimes fantastical and visceral – and at other times mundane and quotidian – imagining of potential masculinity as fantasy pervades the tourist imagination. For all this fantasy, the experiences of male tourists and travellers also remain rooted in the pragmatics of negotiating risky spaces and attendant dangers (Katsulis, Chapter 11) as is travel and tourism for many women, for whom adventurous and risky travel is only problematically achieved whilst remaining stubbornly associated with the masculine (Lozanski, Chapter 3). Through a familiar narrative of risk, adventure and activity, the gendered nature of tourist experience becomes apparent in many of the studies in this book.

Since Connell's influential *Masculinities* (1995), the critical study of men and masculinities has developed a greater appreciation of the variety of masculinities that are pluralised (Aboim, 2010) and commonly vary across the life course (Spector-Mersel, 2006). This plurality is evident throughout this book, as the chapters illustrate the diverse experiences of men who, through their engagement with tourism and travel, engage with some elements of gender, class and ethnic positioning whilst disowning others. While the mobility that tourism offers may allow men to reassert, and to some extent recreate, masculinity, they also carry with them sometimes-burdensome markers of class locality (Ward, Chapter 7) and ethnicity (Waitt and Markwell, Chapter 8).

The concepts of reinvention associated with tourism and travel are readily mapped onto this more flexible, plural conception of masculinities. The empirical works collected in this book speak to this diversity; gender performance is relational and contingent. A recurring theme is that men in various empirical studies enact and construct their masculinity in relation to the gender of others. The supposed masculinity of some Polish men, perceived by 'stag tour' participants as overly aggressive and threatening, is used to frame the male stag tourist's own gender subjectivity as playful and spontaneous (Thurnell-Read, Chapter 4). Likewise, in Treadwell's contribution, a perceived 'other' of respectable middle class values hangs in the air as a foil to which the working class men can contrast their excesses.

We are therefore, to deploy Tim Edensor's (1998) dramaturgical analogy, interested in the diversity of actors/tourists who frequent, traverse and often contest these tourist stages/spaces. Tourism spaces are constructed and experienced relationally. Across the hotel swimming pool (Casey, Chapter 6) or the nightclub dance floor (Thurnell-Read, Chapter 4; Waitt and Markwell, Chapter 8), tourists gaze upon (Urry, 1990) and encounter (Crouch et al., 2001) each other in ways that are at times mundane and at others spectacular. These spaces are never solely tourist spaces. Both Lin (Chapter 10) and Waitt and Markwell (Chapter 8) demonstrate the coming together of various actors within the tourist setting where tourists, non-tourists and staff present us rich imbrications of biographies and subjectivities of gender, sexuality, class and ethnicity. It is through such intersections that tourism spaces are potential sites for dominant gender codes to be challenged and reconfigured and, from a methodological point of view, for sociological insights to be drawn, which often have a bearing far beyond the tourist setting.

The contributions fall broadly into four key areas of theoretical engagement that reflect the themes in this book. While the chapters that form section one ground the book in key debates about the nature and form of hegemonic and non-hegemonic masculinity, the remaining three sections relate identity, sexuality and embodiment respectively. While our intention is not to detract the reader's attention from the considerable overlaps among these themes and the work of the contributing authors, the division of the book into these sections, we hope, will guide the reader through both general and specific connections between masculinity and tourism. In particular, the sections reflect the advances made in the fields of tourism studies and of men and masculinities detailed above and highlight the need to envisage gendered hierarchies and structures of unequal power whilst not losing sight of the embodied and subjective experiences of individual tourists and travellers.

Forming the first section, the initial four chapters theorise hegemonic masculinity and its performance within, and relationship to, travel and touristic experiences. In doing this, the chapters offer the reader insight into theoretical debates around how men 'do' hegemonic masculinity whilst travelling, and the emergence of recent challenges to its dominance. Helen Goodman's chapter reflects the longstanding association between masculinity and travel and reflects on the ease of men's global mobility, empowered through their claims to hegemonic masculinity. Providing a useful historical and cultural context for the following chapters, Goodman draws from both diaries and fiction to examine the relationship between masculinity and newly emerging travel during the Victorian era. By engaging with the dominance of hegemonic masculinity then, her chapter reflects the primacy given to the middle and upper class white male body within such travel accounts. Following this, Kristin Lozanski explores meanings and experiences of risk for both male and female travellers in India. In her chapter, Lozanski examines how narratives of risk frame women's holiday experiences and meanings in negotiation with hegemonic masculinity in Rajasthan, Goa, and Kerala in India. Male tourists' negotiation of risk (and females' perceived vulnerability to it) represents the privilege attributed to the male body and masculinity as men move through different cultural and geographic regions. Similarly, through his focus on the European stag weekend phenomenon, Thomas Thurnell-Read explores one of the key markers of successful heterosexual hegemonic masculinity. His chapter presents the centrality of heterosexual masculinity to the touristic experiences desired and explored during the stag weekend. In so doing, he examines more nuanced accounts of how masculinities are increasingly fragmented and pluralised, with a need to reconcile this with the hyper-hegemonic masculinity as presented and performed through the increasingly commodified stag weekend. Nigel Jarvis' chapter draws on the Gay Games and the event's potential to challenge the assumed relationship between hegemonic heterosexual masculinity, the sporting body and global mobility. The advent of the Gay Games, for Jarvis, exposes residents and tourists to multiple gay male identities and diverse ways of doing masculinity that may challenge hegemonic masculinity and its cultural and spatial domination. Taken together, the chapters of this first section ground debates relating to hegemonic and subordinate masculinity in empirical examples of tourism practices and illustrate how the social construction of masculinity is revealed to be central to many of the principle discourses of travel and tourism.

In the next section, again comprising four chapters, the book moves the reader into discussions surrounding masculinities, tourism and men's identities. Mark Casey's chapter theorises about the way working class masculinities are presented on holiday in the fictional television show, *Benidorm*. Casey asks if some performances of masculinity whilst on holiday are extensions of how men 'do' their gender and class 'back home', or if the liminality of holiday experiences allows men to explore multiple masculinities and other ways of doing their social class? Developing upon the themes present in Casey's chapter, Mike Ward explores the classed and gendered identities of young working class men from areas that have witnessed significant social, cultural, political and economic change in South Wales. By temporarily moving away from home in search of holiday experiences, these men are marked as 'flawed' tourists because their gendered, economic and geographical identities position them as out of time and place in touristic settings associated with the globally successful and mobile body. Notably, Ward's chapter illustrates that such markers of identity and belonging – at times stigmatizing – are not shrugged off with ease; they travel with the young men into and through tourist spaces. In their chapter, Gordon Waitt and Kevin Markwell explore the ethnocentrism of gay tourism by examining the experiences of local gay men and their daily lives as they intersect with global gay tourists in Bali. Similar to some of the themes in Ward's analysis of the (im)mobile identities of his young male participants, the chapter explores the spatialised negotiation of local/global men's gay and ethnic identities and the complex interaction of the diverse masculinities of Balinese and Western gay men. In the final contribution to the section, Rosalina Costa offers insight into how family holidays may allow some men and women to renegotiate or suspend traditional gender roles, as men are offered the freedom to explore diverse ways of doing masculinity away from the rigid demands of hegemonic masculinity in everyday life. Costa's contribution, as with the other three chapters in the section, opens to the reader the potential transformative nature of the holiday and of tourism times and spaces. All four chapters also indicate how the gendered identities of individual tourists negotiate a wider array of connections and (inter)relationships linking the performance of masculinity to that of femininity as well as to class, ethnicity and sexuality.

Sexualised tourist encounters have long been a means to reveal complexities relating to gender and power inequalities and, with this in mind, the three chapters in the third section explore sex and sexuality and their place in men's lives and travels. Chun-Yu Lin explores the centrality of the heterosexual relationship and marriage for cross-cultural mobility between Taiwan and other Asian countries, such as Vietnam and China. Her chapter moves the reader away from the assumed relationship between travel and the search for relaxation, escape, or the potential liminal experiences which travel can offer. Instead, Taiwanese men utilise their economic, cultural, gendered and ethnic privilege, and their global mobility, to seek future wives through the use of international matchmaking agencies. Following this, Yasmina Katsulis offers a critical analysis of the sex industry within Tijuana, Mexico and the role of online communities in allowing heterosexual American men to plan, map and compare their holiday plans and sexual practices. The centrality of risk, adventure and desire to discourses deployed in such virtual communities represents a sobering insight into exploitation in Tijuana and the deeply rooted power imbalances between the United States and Mexico. Carlos Monterrubio and Álvaro López-López draw the reader's attention to the diverse economic realities of gay men in Mexico and their access to, and experiences of, travel. Through a focus on the diverse social, political and cultural issues distinct to Mexico, they present the implications of identity development through travel for individual gay men and the wider gay community. As with the work of both Lin and Katsulis, their chapter examines the place of economic privilege in successful performances of male identity and expressions of sexual desire through tourism and travel.

Although embodied tourist practice and experience is present in all of the chapters in this collection, the final three chapters pay particular attention to how masculinity is embodied within travel and tourism practice. James Treadwell, in his analysis of travelling English football fans, exposes the importance of football fandom and an associated array of hedonistic practices in defining some men's masculinity, heterosexuality and national identity. His chapter critiques earlier approaches to football-associated violence and moves the reader to an analysis of the embodied performances of drink, drugs, sex and violence in and around the 2006 World Cup hosted by Germany. Treadwell's chapter has echoes of Lozanski's, where risk and danger are negotiated as part of a perceived 'authentic' travel experience, and where more mundane, 'safer' forms of tourism might be deemed contemptible. Finally, Monica Gilli and Elisabetta Ruspini, through their analysis of Italian holiday brochures, examine how images utilised in such publications, although still privileging the heterosexual male gaze, are presenting different ways that men may embody changing notions of masculinity on holiday. Similar to Costa's earlier chapter, they examine how changing notions of masculinity may be borrowing from attributes once considered 'feminine', and in so doing, allowing men to move away from restrictive norms of how masculinity should be embodied.

In concluding this collection, Hazel Andrews contributes an afterword that brings together these varied themes and reflects on her own experiences of researching tourism in the heavily gendered environment of the Mallorcan resort towns of Palmanova and Magaluf. Bringing into clear focus the performative, embodied and sensory aspects of how tourism and travel offer up spaces and practices that are informed by – and in turn inform – understandings of masculinity (invariably interwoven with sexuality, class, ethnicity and national identity), the book closes by indicating that tourism remains a significant context in which masculinity is both maintained and challenged.

References

Aboim, S. (2010) *Plural Masculinities: The Remaking of the Self in Private Life*. Farnham: Ashgate.

Aitchison, C. (1999) 'Heritage and nationalism: Gender and the performance of power', in D. Crouch (ed.) *Leisure/Tourism Geographies: Practices and Geographical Knowledge*. London: Routledge, pp. 59–73.

Anderson, E. (2011) *Inclusive Masculinity: The Changing Nature of Masculinities*. Abingdon: Routledge.

Andrews, H. and Roberts, L. (2012) *Liminal Landscapes: Travel, Experience and Spaces In-between*. Abingdon: Routledge.

Bauer, T. and McKercher, B. (2003) *Sex and Tourism: Journeys of Romance, Love, and Lust*. Binghamton, NY: Haworth.

Blackshaw, T. (2003) *Leisure Life: Myth, Masculinity and Modernity*. London: Routledge.

Cohen, E. (1982) 'Thai girls and Farang men: The edge of ambiguity', *Annals of Tourism Research*, 9(3): 403–28.

Connell, R. and Messerschmidt, J. W. (2005) 'Hegemonic masculinity: Rethinking the concept', *Gender and Society*, 19(6): 829–59.

Connell, R. W. (1995) *Masculinities*. Cambridge: Polity.

Crouch, D. (1999) *Leisure/Tourism Geographies: Practices and Geographical Knowledge*. London: Routledge.

Crouch, D. and Desforges, L. (2003) 'The sensuous in the tourist encounter: Introduction: The power of the body in tourist studies,' *Tourist Studies*, 1(3): 5–22.

Crouch, D., Aronsson, L. and Wahlstrom, L. (2001) 'Tourist encounters', *Tourist Studies*, 1(3): 253–70.

Edensor, T. (1998) *Tourists at the Taj: Performance and Meaning at a Symbolic Site*. London: Routledge.

Enloe, C. (1989) *Bananas, Beaches and Bases: Making Feminist Sense of International Politics*. London: Pandora.

Franklin, A. (2003) *Tourism: An Introduction*. London: Sage.

Frohlick, S. (2010) 'The sex of tourism? Bodies under suspicion in paradise', in J. Scott and T. Selwyn (eds) *Thinking Through Tourism*. London: Berg, pp. 51–70.

Hearn, J. (1992) *Men in the Public Eye: The Construction and Deconstruction of Public Men and Public Patriarchies*. London: Routledge.

Hochschild, A. (1983) *The Managed Heart: Commercialization of Human Feeling*. Berkeley: University of California Press.

Inglis, F. (2000) *The Delicious History of the Holiday*. London: Routledge.

Kehily, J. (2001a) 'Understanding heterosexualities: Masculinities, embodiment and schooling', in G. Walford and C. Hudson (eds) *Gender and Sexualities in Educational Ethnography*. Bingley: Emerald, pp. 27–40.

Kehily, M. (2001b) 'Bodies in school: Young men, embodiment, and heterosexual masculinities', *Men and Masculinities*, 4(2): 173–85.

Knox, D. and Hannam, K. (2007) 'Embodying everyday masculinities in heritage tourism(s)', in A. Pritchard, N. Morgan, I. Ateljevic and C. Harris (eds) *Tourism and Gender: Embodiment, Sensuality and Experience*. Wallingford: CABI, pp. 263–72.

Littlewood, I. (2001) *Sultry Climates: Travel and Sex Since the Grand Tour*. London: John Murray.

MacCannell, D. (1976) *The Tourist: A New Theory of the Leisure Class*. New York: Schocken Books.

MacKenzie, J. (2005) 'Empires of travel: British guide books and cultural imperialism in the 19th and 20th centuries', in J. Walton (ed.) *Histories of Tourism: Representation, Identity and Conflict*. Clevedon: Channel View, pp. 19–38.

Mordue, T. (2005) 'Tourism, performance and social exclusion in "Olde York"', *Annals of Tourism Research*, 32(1): 179–98.

Nayak A. (2006) 'Displaced masculinities: Chavs, youth and class in the post-industrial city', *Sociology*, 40(5): 813–31.

Nayak, A. (2003) '"Boyz to men": Masculinities, schooling and labour transitions in de-industrial times', *Educational Review*, 55(2): 147–59.

Noy, C. (2007). 'Travelling for masculinity: The construction of bodies/spaces in Israeli backpackers' narratives', in A. Pritchard, N. Morgan, I. Ateljevic and C. Harris (eds) *Tourism and Gender: Embodiment, Sensuality and Experience*. Wallingford: CABI, pp. 47–72.

Pritchard, A., Morgan, N., Ateljevic, I. and Harris, C. (eds) (2007) *Tourism and Gender: Embodiment, Sensuality and Experience*. Wallingford: CABI.

Sanchez Taylor, J. (2000) 'Tourism and "embodied" commodities: Sex tourism in the Caribbean' in S. Clift and S. Carter (eds) *Tourism and Sex: Culture, Commerce and Coercion*. London: Pinter, pp. 41–53.

Selänniemi, T. (2003) 'On holiday in the liminoid playground: Place, time, and self in tourism', in T. Bauer and B. McKercher (eds) *Sex and Tourism: Journeys of Romance, Love, and Lust*. New York: Haworth Press, pp. 19–31.

Selwyn, T. (1996) *The Tourist Image: Myth and Myth Making in Tourism*. New York: Wiley.

Shields, R. (1991) *Places on the Margin: Alternative Geographies of Modernity*. London: Routledge.

Sinclair, T. (1997) *Gender, Work and Tourism*. London: Routledge.

Spector-Mersel, G. (2006) 'Never-aging stories: Western hegemonic masculinity scripts', *Journal of Gender Studies*, 15(1): 67–82.

Swain, M. (1995) 'Gender in tourism', *Annals of Tourism Research*, 22(2): 247–66.

Swain, M. and Momsen, J. (2002) *Gender/Tourism/Fun*. New York: Cognizant (MC) Communication Corp.

Urry, J. (1990) *The Tourist Gaze: Leisure and Travel in Contemporary Societies*. London: Sage.

Winlow, S. (2001) *Badfellas: Crime, Tradition and New Masculinities*. Oxford: Berg.

Withey, L. (1997) *Grand Tours and Cook's Travel: A History of Leisure Travel 1750–1915*. London: Aurum Press.

Part I

Hegemonic Masculinity, Travel and Tourism

2

Masculinity, Tourism and Adventure in English Nineteenth-Century Travel Fiction

Helen Goodman

Modern conceptions of masculinity may be traced to the nineteenth century, when a combination of socioeconomic factors provoked a reformulation of manliness. Policies such as the breadwinner's wage combined with an ideological shift towards an opposition between domestic femininity and travelling masculinity highlighted gender difference, contributing to a distinctive new masculinity in the mid-nineteenth century. Towards the fin-de-siècle, fears about the degeneration of the British population prompted renewed interrogation of the desirable (and undesirable) qualities of manliness. This chapter explores masculinity specifically related to Englishness as opposed to Britishness, since the authors and characters discussed here are English and explicitly refer to themselves as such. These individuals also define themselves in middle- to upper-class heterosexual terms, although there is certainly scope for future studies to interrogate these categories. The focus of this chapter is the period from 1865 to 1905, examining work by Anthony Trollope alongside the later fiction of H. Rider Haggard, Joseph Conrad, Rudyard Kipling and W. H. Hudson. While previous studies by Dixon (1995) and others have tended to focus on adventure fiction from the late 1880s to 1910, this slightly earlier focus enables analysis of shifting trends towards the end of the century in the context of mid-Victorian conceptions of masculinity.

This chapter frames international tourism as a masculine activity, reserved primarily for the middle and upper classes in the nineteenth century. A distinction is made between two modes: firstly, earlier narratives in which travel to paradisiac lands by young, upper-class men reinforces masculinity, as a rite of passage for personal development upon entering manhood, following education and preceding marriage, as for the protagonist in Trollope's *He Knew He Was Right* (1868–9); and secondly, later writing in which harsh landscapes threaten masculinity (a tradition dating back to Homer), as in Haggard's *King Solomon's Mines* (1885) and *She* (1887), as well as Conrad's *Heart of Darkness* (1902). Although focussed on tourist adventure rather than military travel, the chapter makes connections with imperialist masculinities. These texts depict a masculine identity that both constitutes and is constituted by imperial instincts, in exploration of exotic British territories (the fictional Mandarin Islands in Trollope) and a masculine desire to control and subdue hostile environments (the central African jungle in Conrad). Both types of travel represent masculine endeavours to conquer the feminine: Trevelyan marries a girl from the Mandarins, while in *King Solomon's Mines*, a group of men explore a mountainous landscape referred to as 'Sheba's Breasts'.

Adventure narratives are analysed here within the context of normative Victorian expectations of male behaviour, focussing on central tenets including independence, freedom and good health. Recent research on muscularity has highlighted associations between the idealised male physique and the moral gentleman. Muscularity could indicate to readers a worthy character to be both admired and revered. Haggard and Conrad's fiction depicts imperial masculinity outside an explicitly military context, describing Africa as a dark, amoral but exciting place, where both the native population and the landscape itself must be subdued and overcome. Their male characters embody an intrinsically colonialist identity, becoming role models for new generations of middle- and upper-class British boys.

Masculinity in crisis

By the 1880s, fears about the mental and physical degeneration of the British population were gaining pace, and mid-Victorian representations of the British male as a supremely rational creature began to wane (Pick, 1996). Darwin's *Descent of Man* (1871) contributed to a network of anxieties about heredity and racial purity, which increased in the following decades. The eugenics movement is testament to the lengths to which many were willing to go in an attempt to halt the rate of this perceived degeneration. A preoccupation with such concerns is revealed by much popular literature of the period. Writers including Conan Doyle envisioned a future in which evolution led not to progress, but to degeneration. Sherlock Holmes explains such a possibility to Watson:

> When one tries to rise above Nature one is liable to fall below it. The highest type of man may revert to the animal if he leaves the

> straight road of destiny...There is danger there – a very real danger to humanity. Consider, Watson, that the material, the sensual, the worldly would all prolong their worthless lives. The spiritual would not avoid the call to something higher. It would be the survival of the least fit. What sort of cesspool may not our poor world become? (Conan Doyle, 1927: 196–7)

Studies of nineteenth-century gender have often made a sharp distinction between mid-century roles (considered to reinforce a strict separation of gentlemanly masculinity from the female ideal, described in Patmore's famous 1854 poem as the 'Angel in the House') and fin-de-siècle transgressive eccentricity (including the new woman and the dandy). Such readings risk assuming that mid-century masculinity was a stable construct. Kiely (1964: 21) observes that 'in the decade of the 1880s there were signs of a small but vigorous movement in Great Britain' reacting against novels in the tradition of Eliot and Trollope. Showalter (1990: 9) has argued that a crisis of masculinity emerged later, in the 1890s, and Gagnier (1986: 98) has written that this decade saw a crisis 'of the male on all levels – economic, political, social, psychological, as producer, as power, as role, as lover'. More recent studies by Kestner (2010) and others have not interrogated this observation by searching for earlier reappraisals of masculinity.

I suggest that this crisis was already well under way in the 1870s, and may be observed in some 1860s texts such as Trollope's *He Knew He Was Right*. In this novel, Louis Trevelyan's masculinity is subject to incessant scrutiny, from his social behaviour and his relationship with his wife, to his ability to sustain rational thought, before toppling entirely during a period of jealous monomania, ending in death. His wife Emily constantly expresses her wish that her husband was more 'manly'. Although she does not define this, her main concern is that, in jealously suspecting her of conducting an affair (particularly with someone as unlikely as her ageing godfather), he does her a great dishonour. Trollope demonstrates the failures of attempts to impose the male will. Having demanded that Emily not receive Colonel Osborne at home, Louis soon notes that 'so far he had hardly gained much by the enforced obedience of his wife' (p. 54). He is acutely aware that his behaviour threatens his status as a gentleman, and berates himself for it: 'He had meant to have acted in a high-minded, honest, manly manner [...] on looking at his own conduct, it seemed to him to have been mean, and almost false and cowardly' (p. 55). In contrast to broader contemporary ideas of manliness that either turned a blind eye to or actively encouraged the enforcement of wifely obedience, this particular mode of *gentle*manliness posited by Trollope requires the opposite. It is desirable in patriarchal terms for Emily to be obedient to her husband, but crucially, she must want to be so, and voluntarily.

It has been observed that 'the nineteenth-century reading public was led from the stress on Christian manliness that marked the fictional works of Charles Kingsley and Thomas Hughes to Henty and H. Rider Haggard's depiction of robust heroes driven by brutal energy and dominating wills' (McLaren, 1997: 33). The subject of 'muscular Christianity' has undergone a recent critical reappraisal, identifying it as an important focus of nineteenth-century masculinity. Wee (1994) has explored this theme in relation to national identity, tracing Kingsley's 'commitment to the possibility of an English racial regeneration' back to Lamarck's faith in acquired characteristics (p. 74–5). Haggard's presentation of previously unexplored African regions, like Conrad's, is predicated on an assumption that 'natives' are fundamentally inferior in moral, if not necessarily muscular, terms. Language emphasising Africa's 'darkness' is used to preclude the possibility that it could rejuvenate England except by its colonial subjugation. Despite the masculine bravado that adventure fiction exudes, characters' masculinity is perpetually tested, often falling short and judged to be degenerate. Novels that conclude abroad tend to present masculine triumph, and those describing the traveller's return to British soil and society, failure. Such men are sometimes psychically or psychologically damaged by the trials of travel, as in Kipling's *The Man Who Would be King* (1888) (in which Carnehan becomes insane and dies in an asylum of sunstroke) and R. L. Stevenson's *Treasure Island* (1883). This may result from traumatic experience abroad or indicate a dreaded case of a British man 'gone native', as in Conrad's short story, 'Karain: A Memory'. In a sense, the latter was the most troubling, representing a counter-imperial triumph of the foreign 'Other' over the English. *Lord Jim* (1889–1900) explores the guilt-wracked psychological implications for a young man, brought up to dream of heroism, who cowardly jumps ship. Narratives of this kind, in contrast to Haggard's bombastic and triumphant masculinity, mark a vital crisis of imperial manliness.

Transitions in gender and genre: domestic realism and adventure fiction

By bridging the gap between domestic realism and adventure fiction, traditionally critiqued entirely separately, a productive new area for research may be established. There is a danger for literary critics and cultural historians to think and write in terms of artificial boundaries such as different decades, different ideologies and different genres of fiction. Although such neat categorisation may be tempting, particularly in discussions of the Victorian period, which was itself obsessed with classification, it tends to produce a somewhat reductionist representation of the era, which belies the reality of a complex network of influences. Fin-de-siècle and modernist literature ought not to be analysed without reference to what had preceded it, or with the assumption that a particular time (considered by many to be the late 1880s) marked a complete watershed, after which mid-Victorian ideas were entirely overthrown. While the style of Haggard and Conrad differs greatly from earlier writers such as Dickens and Trollope, their concerns about a declining standard of masculinity are shared.

Carlyle's published lecture series, *On Heroes, Hero-Worship, and the Heroic in History* (1841) promoted key tenets of masculinity, such as adventure, action and bravery, alongside strong moral character: the same characteristics embodied by Allan Quatermain and his companions. Salmon (1886, in Kestner, 2010: 17) observed that boys' fiction was part of a 'vast system of hero-worship'. The word 'system' implies deliberate ideological teaching that may be considered child indoctrination – part of the dominant pro-empire English ideology of the period and a vital instrument of what Althusser (1971) termed the 'Ideological State Apparatus'. Katz (1987) considers that 'the imperial hero, whether a soldier, an adventurer, or simply an embodiment of "manliness", has no ideological dispute with his society – although he may believe that his society is losing its integrity, getting soft, or somehow straying from its true course', being 'a traditional man of action who expresses the politically conservative aspirations of his society' (p. 61). I suggest that Haggard's protagonists perform a more active function than that implied by Katz's view that he merely has 'no ideological dispute' with his society. Figures such as Allan Quatermain serve to reinforce imperial support and conservative models of masculinity at a time when the former was lagging in the wake of revolts in India and Ireland, and the latter was challenged on home soil by alternative styles of masculinity such as dandyism.

Mad dogs and Englishmen: gentlemanliness and Englishness at home and abroad

Englishness and manliness are inextricably linked, not only in the more explicitly imperial narratives of Haggard, but also in Trollope's largely domestic fiction. The eponymous protagonist in Haggard's *Allan Quatermain* (1887) pins down what he considers to be the core of English masculinities:

> That is what Englishmen are, adventurers to the backbone; and all our magnificent muster-roll of colonies, each of which will in time become a great nation, testify to the extraordinary value of the spirit of adventure which at first sight looks like a mild form of lunacy [...] The names of those grand-hearted old adventurers who have made England what she is, will be remembered and taught with love and pride to little children. (p. 101)

The adventurous spirit is lauded for its exciting possibilities for travel, appealing even to, or perhaps especially to, those least likely to go abroad. However, the primary motivating factor for men's travel during this period was colonisation. To modern postcolonial readers, the exhilaration that imbues Haggard's fiction and inspires his young male readers is far from innocent. Instead, such plots are indelibly marked by the (often militarily or politically aggressive) dominance of the British Empire. Green (1979: 23) defines the adventure genre as 'a series of events, partly but not wholly accidental, in settings remote from the domestic and probably from the civilised (at least in the psychological sense remote), which constitute a challenge to the central character'. However, I suggest that by examining Trollope's fiction, written slightly earlier, a significant continuum from domestic to adventure fiction may be revealed.

English gentlemanliness figures in Trollope's novel as a goal to which the protagonist aspires, but struggles to reach. Two decades later in Haggard's novels, Quatermain achieves this golden standard, while another two decades on, Conrad's characters deconstruct it, questioning both its attainability and its desirability. *Allan Quatermain* bears a dedication this time to Haggard's son who died four years later:

> I inscribe this book of adventure to my son Arthur John Rider Haggard in the hope that in days to come he, and many other boys whom I shall never know, may, in the acts and thoughts of Allan Quatermain and his companions, as herein recorded, find something to help him and them to reach to what, with Sir Henry Curtis, I hold to be the highest rank whereto we can attain – the state and dignity of English gentlemen. (1)

Conrad's *Heart of Darkness,* written from 1898–99, questions not only the nature of English gentlemanliness, but also the validity of the imperial project as a whole:

> The conquest of the earth, which mostly means the taking it away from those who have a different complexion or slightly flatter noses than ourselves, is not a pretty thing when you look into it much. What redeems it is the idea only. An idea at the back of it; not a sentimental pretence but an idea; and an unselfish belief in the idea – something you can set up, and bow down before, and offer a sacrifice to. (p. 9)

This statement by the character Marlow is, in various ways, a telling one, weighing up the validity of the imperial project. For the purposes of this particular context, it encapsulates perhaps the largest foundation of British masculinity: the need to define male selfhood in terms of duty and sacrifice. Trevelyan, Quatermain and Marlow are united by their imperative drive to live up to such expectations.

Said (1994) suggests that *Heart of Darkness* 'beautifully captured' a particular imperial attitude still highly visible in modern Western foreign policy. 'The thing to be noticed', Said argues, 'is how totalising is its form, how all-enveloping its attitudes and gestures, how much it shuts out even as it includes, compresses, and consolidates'. In its modern manifestations (American conflicts in Vietnam, Iran and Iraq), 'we suddenly find ourselves transported backward in time to the late nineteenth century' (p. 24). Conrad's position on imperial matters differs considerably from those of his contemporaries in making direct connections with 'the redemptive force, as well as the waste and horror, of Europe's mission in the dark world' (Said, 1994: 25). Conrad's position as a Polish expatriate as well as an employee of the imperial system placed him well to interrogate imperial masculinities. While Haggard aimed to inspire future generations of individual, heroic, imperial male explorers, Conrad's framing of Marlow's narrative emphasises that the listeners are men of business. This, Said explains, is 'Conrad's way of emphasising the fact that during the 1890s the business of empire, once an adventurous and often individualistic enterprise, had become the empire of business'.

Conrad wishes his readers to reconsider concepts such as empire and masculinity in the light of new perception. In his preface to *The Nigger of the Narcissus* (1897), two years before writing *Heart of Darkness,* Conrad explained: 'My task which I am trying to achieve is, by the power of the written word to make you hear, to make you feel – it is, before all, to make you *see*. That – and no more, and it is everything'.

While Stevenson and Haggard seem to endorse a certain brand of masculine travel, Conrad's writing has no such agenda. Rather than encouraging his readers to follow a certain example, Conrad illuminates new ways of seeing and questioning. His version of African 'darkness' implies moral uncertainty rather than absolute barbarism and amorality, in which masculine heroism is challenged but does not emerge triumphant.

From boys to men: masculine travel as initiation

Haggard's fiction explicitly represents and encourages the intergenerational transfer of masculine ideals, dedicating *King Solomon's Mines* to 'all the big and little boys who read it' (p. v). Similarly, the author's note accompanying the 1898 edition expresses his hope to 'fall into the hands of an even wider public, and that it may in years to come continue to afford amusement to those who are still young enough at heart to love a story of treasure, war, and wild adventure' (p. vii). The novel maintained a relatively narrow readership, engaging almost exclusively the middle-class white male, albeit of varying age. Such a dedication for a fictional narrative illustrates the direct ideological line of influence that writers like Haggard sought to maintain with their readers. By appealing to young boys as well as men, adventure fiction provided a vehicle for the transfer of the imperial values. As well as imbuing the young with support for the British Empire, often based on representations of other nations as barbarous, this kind of literature was foundational in establishing early ideals of masculinity. Dawson (1994: 1–2) has examined the workings of this process, observing that 'masculinities are lived out in the flesh, but fashioned in the imagination'. The masculinity propagated and glamorised by adventure fiction is that which has 'proved efficacious for nationalistic endeavour'. Dawson recalls wondering, 'Might there be a relation between the fantasies of boyhood, the reproduction of idealised forms of masculinity, and the purchase of nationalist politics?' (p. 5), describing his own Gramscian method of 'examining the self as a jumbled repository of unsorted traces from the past, and the compilation of an historical inventory; and so meshed with my wider enquiry into the historical origins, use and effects of the adventure form' (p. 5).

Literature since Homer's *Odyssey* has often presented travel as a key rite of passage in establishing the strength and endurance that have been central to the majority of Western dominant adult male identities. (Bookending the nineteenth century, Romantic masculinity and dandyism are two notable exceptions). Travel and adventure provide opportunities for male characters to be tested for their psychological and physical strength, their ability to exercise reason while taking brave risks, and their stamina to achieve greatness in the face of adversity. Jim in *Treasure Island* begins his travels as a naïve boy, gradually negotiating different forms of masculinity, so that Stevenson is able to instruct his young readers in the nature of the genuine masculinity they ought to strive for, in opposition to inferior or false shows of manliness. Nelson (1990: 130) has argued that masculine morality in the novel is highlighted by its contrast with androgyny: 'if Smollett is *Treasure Island*'s forbidding father figure, Silver is its enticing mother'. Armstrong (1987: 52) observes that in Dickens's *Oliver Twist* (1838), 'Fagin's true villainous nature is initially cloaked behind a maternal exterior of sizzling sausages, schoolroom games, and terms of endearment. But his simulation of benign authority disintegrates as the profit motive comes into conflict with his feminine virtues and cancels them out' (p. 52). Drawing on this analysis of a similar effect, Nelson contends that 'Silver's motherliness cloaks its mercenary opposite' (p. 130).

Preceding Stevenson's classic, there was a strong tradition of this masculine mode in sea voyage literature on both sides of the Atlantic. Melville's *Moby Dick* (1851) was a major contribution to this genre, and like Stevenson and Haggard's fiction, it was explicitly directed at a male readership. In a letter to Mrs Sarah Morewood, a friend with whom Melville had a close and, as far as we know, entirely platonic, relationship (see Schultz and Springer, 2006: 30), he warned, 'Don't you buy it – don't you read it [...] it is by no means the sort of book for you. It is not a fine feminine piece of Spitalfields silk – but of the horrid texture of a fabric that should be woven of ships' cables and hausers. A Polar wind blows through it, and birds of prey hover over it. [...] Warn all gentle fastidious people from so much as peeping into the book – on risk of a lumbago and sciatics' (1851: 206).

As early as 1868, Trollope was questioning the effectiveness of travel as initiation into manhood. Emily Trevelyan's determination to put quarrels with her husband aside is interpreted by Louis as 'being treated as a naughty boy, who was to be forgiven', perceiving another threat to his manliness, this time in the guise of an affront to his maturity. As Tosh has observed, masculinity required manly maturity, distinct from boyishness, in a way that did not translate to the maturity from girlhood to womanhood. Trevelyan's world tour, having graduated from Cambridge or Oxford (Trollope, presumably accidentally, cites one university early in the novel and the other later), is the next logical step in his development of upper-class English gentlemanliness, before settling in London for married life.

Travel undertaken as tourism is repeatedly framed in alternative language to incorporate masculinised ideals of duty and honour, implying a greater ordeal than holidaying would suggest. On his second expedition, Trollope's Trevelyan considers himself to be driven abroad to escape his wife, who might attempt to reclaim custody of their young son. Quatermain's endeavours in Africa are not under military orders, or to continue his elephant-hunting profession, but are an entirely voluntary retirement activity. He and his two travelling companions apparently search for the gold and diamond mines out of a perceived duty to Sir Henry's brother, Neville, who has disappeared. Fortune hunting as their primary motive would contradict the particular gentlemanliness that the novel idealises, itself predicated on a social ethical system in which money was considered vulgar. That Haggard's heroes leave Africa with countless diamonds is presented as a reward bestowed upon them, rather than a goal actively sought – enormous wealth is a by-product of perfect masculinity. As Monsmon (2000) has bluntly declared in his opening statement on the novel, 'Everyone knows that the trio in *King Solomon's Mines* went to Kukuanaland for its diamonds; the rescue of Sir Henry's brother served only as a convenient pretext for recovering Solomon's treasures [...] The wisest and richest of Biblical rulers, Solomon would have provided the strongest moral sanction for the Victorian resumption, as it were, of mineral extraction on the dark continent' (p. 280). Haggard obscures the exact nature of the trip's purpose by using language of empire and duty, describing the 'civilising' European influence, in order to endow his characters' mission with nobleness vital to the mode of masculinity he promotes.

Masculinity and femininity

Paradoxically, nineteenth-century efforts to recodify masculinity relied on persistent reference to femininity. Creating gender difference, seeking to construct the masculine and feminine as opposite poles, involved repeated reference to femininity as the most effective entity against which masculinity could be contrasted. As McLaren (1997: 33) has observed, '"Manly", which in the eighteenth century was used to mean the opposite of boyish or childish, was in the Victorian age increasingly employed as the antonym of the feminine or effeminate'. There are, however, exceptions to this. Trollope presents Louis Trevelyan's lack of manliness in terms of the protagonist's immaturity. Trevelyan's wife finds him childish rather than effeminate. The kind of initiation into adult manhood discussed here is often understood to mark a complete separation and independence from the feminine. For example, Zweig (1974: 70–1) contends that 'The initiated adult undoes his original birth from woman, replacing it by a wholly masculine birth [...] He becomes not only his own father – that represents the "Oedipal" side of the initiation – but also his own mother'. However, I suggest that the masculine identity interrogated and reaffirmed by travel reaches the climax of manliness only when contrasted against femininity. Odysseus's impressive adventures are rendered more masculine by their contrast against Homer's image of Penelope at home, waiting for her husband to return. Furthermore, Odysseus's encounters with femininity in the course of his travels, most notably with the Sirens, serve to heighten his manliness. His strength to avoid the temptations of their beautiful sound marks his achievement of key tenets of masculinity, such as self-control and the inability to be overpowered by women.

Masculinity as depicted in Victorian travel fiction may at first appear to stand alone, without reference to femininity. Besides accounts of honeymoon tours and some exceptions such as Wollstonecraft's *A Short Residence in Sweden* (1796), the vast majority of travel texts from the long nineteenth century depict the journeys of men, either individually or in groups, but without women. I suggest that this apparent absence of femininity can, perversely, serve to draw attention to gender contrast, and that the feminine persona, where not manifest as a human character, takes an alternative inanimate form. In Haggard's *King Solomon's Mines*, this form is the physical landscape of the African continent. The mountains that characters must climb to prove their masculinity are overtly gendered – described as shaped like a woman's breasts. In constructing such a direct equation of femininity and land, Haggard extends metaphors of the masculine conquering the feminine by subduing the female body. Millman (1974:22) argues that in adventure modes of writing, and specifically that of Haggard, 'The product was the "male novel", written by men, for men, and about the activities of men'. While it may remain easy for such critics to ignore the subtexts of femininity against which masculinity is constantly distinguished, such clear representations of the masculine ruling the feminine, as in Haggard, ought certainly to be acknowledged. When placed in contrast to femininity, heterosexual masculinities in Haggard's fiction become more sharply defined.

In *Green Mansions* (1904), Hudson accentuates gender division and its implications for heterosexual masculinity. This less well-known novel depicts another female native whose supernatural beauty threatens adventuring white men by distracting them from danger. Rima's melodious sounds draw Mr Abel in, following Homer's depiction of the Sirens, suspending masculine power. Mr Abel repeatedly finds himself 'standing still, as if in obedience to a command, in the same state of suspense' (p. 58), rendered impotent in his explorations. Like Haggard's explorers, Mr Abel's masculine success depends on his ability to overcome the adverse challenges represented by the feminine.

Contemporary criticism occasionally alluded to femininity in what was otherwise a deliberately masculine genre. Stevenson (1884: 143) wrote that the 'novel of adventure [...] appeals to certain sensual and quite illogical tendencies in man'. Literature and art more broadly was, for Stevenson, inherently feminine, not aligned with the broadly masculine realities of life: 'Life is monstrous, infinite, illogical, abrupt and poignant; a work of art, in comparison is neat, finite, self-contained, rational, flowing and emasculate' (p. 142). This contemporaneously controversial connection between the rational and the emasculate marks what I suggest was a vital shift in Victorian conceptions of masculinity towards the end of the nineteenth century.

More recently, critics such as Kestner (2010: 142) have argued that 'unlike *King Solomon's Mines* or *Allan Quatermain*, where the erotic is not particularly stressed, the subject of *She* is eroticism'. In line with this reading, Haggard's references to femininity and sensuality are understood to become more overt in his later writing. In 1887, the year in which he published *She*, Haggard publicly asserted that 'sexual passion is the most powerful lever with which to stir the mind of man' (p. 176). However, I suggest that in *King Solomon's Mines* 'the erotic' certainly is 'particularly stressed' on occasion. Most obviously, Haggard's naming of a pair of mountains as 'Sheba's Breasts' negates, I suggest, an entirely de-eroticised reading of the text. Read upside down, the explorers' map depicts the alignment of what appears to be a woman's breasts, navel and genital region. This anthropomorphising of the explored and effectively colonised landscape accentuates the theme of the masculine exerting power over the feminine so prevalent in late-Victorian adventure fiction.

Adventure fiction also uses femininity to represent a dangerous peril against which the masculine subject must defend himself. Dryden (2000: 3) has observed that 'these are novels that revel in the exploits of virile heroes, while reducing women to ugly "native" witches, or sultry beauties whose predatory sexuality threatens the hero's manliness'. In *She*, however, Haggard's eponymous character is a woman, entitled

'She-who-must-be-obeyed'. Ayesha is white-skinned and youthfully beautiful, despite being two millennia old, and represents a different form of empire, posing a direct threat to the masculinised British Empire represented by Horace Holly, Leo Vincey and Job. Addressing the Englishmen, she says, 'How thinkest thou that I rule this people? I have but a regiment of guards to do my bidding, therefore it is not by force. It is by terror. Mine is of the imagination', kept alive since once in a generation she will 'slay a score by torture' (Haggard, 1991: 175). By making his Englishmen dependent on an African woman for protection, without which they would be eaten by local cannibals, Haggard undercuts his earlier, simpler hierarchy of power, making a shocking challenge to masculinity.

Building on foundations of masculine identity laid by mid-Victorian realist writers such as Trollope, Haggard and Hudson's adventure fiction could present empire as the ultimate triumph of travelling English masculinity, resulting in considerable power. This powerful image of masculine heroism could be passed down to new generations of English males (as per the intensions outlined in the prefaces studied here), hiding the realities of colonial unrest indicated by numerous rebellions during the period. While within the adventure genre, Conrad's fiction marks the reemergence of concerns about the consequences of such normative gendering in terms of individual psychology and broader cultural debates. However, fiction that interrogates and deconstructs imperial masculinity is not necessarily nihilistic in nature; it does not mark the death of all aspects of nineteenth-century manliness nor its engulfment into the reputedly feminine territory of dandyism. As well as exposing the innate cracks in and internationally sustained damage to the model fin-de-siècle travelling English gentleman, these adventure narratives left an opening for new, potentially more heterogeneous, masculinities to emerge in early twentieth-century English literature and culture.

Nonetheless, twenty-first-century ideals of masculinity from birth to adulthood still echo Victorian tensions in relation to styles of dress, muscularity, the division of labour, and countless other contexts. While international adventure travel in early adulthood has become balanced between the sexes, early childhood has become less gender-neutral. While Victorian boys as well as girls wore dresses until the age of five or six, most Western children today begin to wear gender-specific styles and colours by the time they are a few months old. A higher proportion of children's fiction is now aimed at both girls and boys, contradicting the views indicated in Haggard's prefaces. The recent decision by the chain toy store 'Toys R Us' to stop labelling their products as suitable specifically for boys or for girls (Delmar-Morgan, 2013), discussions about the ethics of gender reassignment surgery on children younger than age 18 (Drescher, 2013), and a myriad of other media debates indicate a recent resurgence of arguments about the psychological effects of rigid gender boundaries. Following the rise of feminist and queer theories in the 1970s, 1980s and 1990s, this is an exciting time for masculinity studies in popular cultural and academic contexts alike.

References

Althusser, L. (1971) 'Ideology and ideological state apparatuses', in Ben Brewster (trans.) *Lenin and Philosophy and Other Essays*. London: Monthly Review Press.
Armstrong, N. (1987) *Desire and Domestic Fiction: A Political History of the Novel*. Oxford: Oxford University Press.
Carlyle, T. (1901) *On Heroes, Hero-Worship, and the Heroic in History*. London: Ginn & Co.
Conan Doyle, A. [1927] (2001) 'The adventure of the creeping man', *The Case-Books of Sherlock Holmes*. Kelly Bray, Cornwall: House of Stratus.
Conrad, J. (1994) *Heart of Darkness*. London: Penguin.
Conrad, J. (2000) *Lord Jim*. London: Penguin.
Conrad, J. (2007) *The Nigger of the Narcissus*. London: Penguin.
Conrad, J. (2008) 'Karain', *Tales of Unrest*. Teddington, Middlesex: Echo Library.
Darwin, C. (2004) *The Descent of Man, and Selection in Relation to Sex*. London: Penguin.
Dawson, G. (1994) *Soldier Heroes: British Adventure, Empire and the Imagining of Masculinities*. London: Routledge.
Delmar-Morgan, A. (2013) 'Toys R Us to stop marketing its toys by gender in wake of sexism claims', *The Independent*, 4 September.
Drescher, J. (2013) 'Sunday dialogue: Our notions of gender', *New York Times*, 29 June.
Dickens, C. (2003) *Oliver Twist*. London: Penguin.
Dixon, R. (1995) *Writing the Colonial Adventure: Race, Gender and Nation in Anglo-Australian Popular Fiction, 1875–1914*. Cambridge: Cambridge University Press.
Dryden, L. (2000) *Joseph Conrad and the Imperial Romance*. Basingstoke: Macmillan.
Dryen, L., Arata, S. and Massie, E. (eds) (2009) *Robert Louis Stevenson and Joseph Conrad: Writers of Transition*. Lubbock, Texas: Texas Tech University Press.
Gagnier, R. (1986) *Idylls of the Marketplace: Oscar Wilde and the Victorian Public*. Stanford: Stanford University Press.
Gramsci, A. (2005) *Prison Notebooks*. London: Lawrence and Wishart.
Green, M. (1979) *Dreams of Adventure, Deeds of Empire*. New York: Basic Books.
Haggard, H. R. (1887) 'About fiction', *Contemporary Review* 51(February): 172–80.
Haggard, H. R. (1995) *Allan Quatermain*. London: Penguin.
Haggard, H. R. (1994) *King Solomon's Mines*. London: Penguin.
Haggard, H. R. (1991) *She*. Oxford: Oxford University Press.
Hudson, W. H. (1968) *Green Mansions*. London: Minster Classics.

Katz, W. R. (1987) *Rider Haggard and the Fiction of Empire: A Critical Study of British Imperial Fiction*. Cambridge: Cambridge University Press.

Kestner, J. A. (2010) *Masculinities in British Adventure Fiction, 1880–1915*. Farnham, Surrey: Ashgate.

Kiely, R. (1964) *Robert Louis Stevenson and the Fiction of Adventure*. Cambridge: Harvard University Press.

Kipling, R. (1994) *The Man Who Would be King*. Ware, Hertfordshire: Wordsworth.

McLaren, A. (1997) *The Trials of Masculinity: Policing Sexual Boundaries, 1870–1930*. Chicago and London: Chicago University Press.

Melville, H. (1993) 'September 1851', in L. Horth (ed.) *Correspondence Vol. IV: The Writings of Herman Melville*. Evanston, Illinois: Northwestern University Press.

Melville, H. (1995) *Moby Dick*. Oxford: Oxford University Press.

Millman, L. (1974) 'Rider Haggard and the Male Novel; What is Pericles?; Beckett Gags.' PhD thesis, Rutgers University.

Monsmon, G. (2000) 'Of diamonds and deities: social anthropology', in H. Rider Haggard's *'King Solomon's Mines', English Literature in Transition, 1880–1920*. Vol. 43(3): 280–97.

Nelson, C. (1991) *Boys Will Be Girls: The Feminine Ethic and British Children's Fiction, 1857–1917*. London: Rutgers University Press.

Pick, D. L. (1996) *Faces of Degeneration: A European Disorder, c. 1848 – c. 1918*. Cambridge: Cambridge University Press.

Pocock, T. (1988) *Rider Haggard and the Lost Empire*. London: Weidenfeld and Nicolson.

Roberts, A. M. (ed.) (1993) *Conrad and Gender*. Amsterdam: Rodopi.

Said, E. S. (1994) *Culture and Imperialism*. London: Vintage.

Schultz, E. and Springer, H. (eds) (2006) *Melville and Women*. Kent: Kent State University Press.

Showalter, E. (1990) *Sexual Anarchy: Gender and Culture at the Fin de Siècle*. London: Bloomsbury.

Schultz, E. and Haskell, S. (eds) (2006) *Melville and Women*. Kent, Ohio: Kent State University Press.

Stevenson, R. L. (1884) 'A humble remonstrance', *Longman's Magazine*, December, 139–47.

Stevenson, R. L. (2009) *Treasure Island*. London: Penguin.

Tosh, J. (2005) *Manliness and Masculinities in Nineteenth-Century Britain: Essays on Gender, Family and Empire*. Harlow: Pearson.

Trollope, A. (2004) *He Knew He Was Right*. London: Penguin.

Wee, C. J. W. L. (1994) 'Christian manliness and national identity: The problematic construction of a racially "pure" nation', in D. Hall (ed.) *Muscular Christianity*. Cambridge: Cambridge University Press.

Wollstonecraft, M. (1987) *A Short Residence in Sweden*. London: Penguin.

Zweig, P. (1974) *The Adventurer: The Fate of Adventure in the Western World*. Princeton: Princeton University Press.

3
Heroes and Villains: Travel, Risk and Masculinity

Kristin Lozanski

> 'I think the main reason why I got into it ... was for the *adventure*, for the fun.' (Mark, 36, Australia, my emphasis)
>
> 'That traveller will say, "Okay, I'm going", because it's like, um, you know, like new place to see and nobody see it before, so it's like a small *adventure*.' (Daniel, 22, Israel, my emphasis)

Although their practices may ultimately be similar, 'travellers' often distinguish themselves from 'tourists' (Hottola, 2007). In part, this distinction is accomplished through references to risk and, as reflected in the quotes above, adventure. In this chapter, I argue that this ideal of adventure – an important signifier of 'travel' – is articulated through risk, and that risk, in turn, refracts and reproduces the norms of hegemonic masculinity. I focus on risk in particular as it is infused with masculine norms. Through encounters with 'Others', travellers emphasise their participation in risk. Travellers' narratives not only draw upon risk as a tenet associated with masculinity, but in so doing reiterate travel as a project that is available without caveat only to men enacting hegemonic forms of masculinity. Idealised travel – as opposed to tourism – is predicated upon the successful negotiation of risk and, as such, gendered privilege because this success contrasts the narrative of women's vulnerability and need for protection.

Methodology

In 2005, I conducted ethnographic fieldwork in regions of India that are popular with travellers. My research took place primarily in the states of Rajasthan, Goa, and Kerala, along with a few other destinations including Hampi and Agra. I interviewed 14 men and 15 women representing four general geographical areas: Canada/United States, Australia/New

Zealand, Western Europe and Israel. These geographic clusters reflect world areas in which much of the population possesses the affluence to travel, as well as competence in English, which enables travel and also enabled these interviews. While the majority were white, eight participants were not white, including five participants with Indian ancestry. These travellers varied in their previous travel experience. They ranged in age from 19 to mid-40s and were at various stages in their education and/or careers. Some were travelling after completing their education or military service or were between careers, while others fit their travel into a relatively continuous life at home.

Though these individuals were highly diverse, they participated in the shared subculture of travel (Sorensen, 2003/2007). The people with whom I spoke engaed in activities that were largely – and ironically, given their desire to conduct their travels independently – directed by travel guides such as the *Lonely Planet*. They collected in spaces including budget hotels, internet cafes and restaurants that Hottola (2005: 2) identifies as 'metaspaces': 'places of recovery; the behavioural and physical tourist "bubbles" where the locus of control is with the tourists rather than with their so-called hosts.' Within these spaces, there was a general camaraderie amongst travellers who sought to navigate foreign spaces and foreign cultures through an 'us versus them' framework. In this way, other travellers – including myself – became almost instant allies. Although there is a hierarchy of status associated with travel (Sorenson, 2003), common threads of identity persisted among travellers I interviewed (Lozanski, 2011). This shared identity facilitated my data collection through conversations and discussions that moved easily into an explanation of my research (Lozanski and Beres, 2007). All participants were aware of my role as a traveller and researcher; all interviews were completed with informed consent and recorded with permission.

Masculinity ⇆ travel

In this chapter, I conceptualise masculinity through Connell's (Connell, 1995; Connell and Messerschimdt, 2005) theory of hegemonic masculinity alongside Whitehead's (2005) analysis of masculinity through Hero/Villain/(Non-man) roles (see below). In setting out hegemonic masculinity, Connell identifies it as 'pattern of practice ... [that] embodie[s] the currently most honored way of being a man, it require[s] all other men to position themselves in relation to it, and it ideologically legitimate[s] the global subordination of women to men' (2005: 832). While Connell's framework seeks to understand men's differential access to masculine privilege, Whitehead formulates a more universal construction of masculinity to explain violence between men. Whitehead's analysis applies to independent travel insomuch as he emphasises a homosocial enactment of identity (Kimel, 2007), as well as episodic, and thus ongoing, constructions of identity. In Whitehead's Hero/Villain/(Non-man) paradigm, hegemonic forms of masculinity are associated with both the hero and the villain, as both demonstrate a 'core of transcendental courage in the face of danger' (2005: 413). However, given the impermanence of confrontations and danger, the requirement to perform masculinity is antagonistic, public and ongoing. In contrast to the masculinised figures of the hero and the villain who engage in physical or other forms of combat, the non-man is the individual who refuses to enter, or is excluded from, the public performance of this confrontational – and respected – form of masculinity.

Connell's and Whitehead's analytics can be transposed into the field of travel and tourism studies insomuch as many 'travellers' (that is, heroes; see also Noy, 2007) understand themselves in contrast to 'tourists' (that is, the non-man). They accomplish this distinction through their engagements with villains, a role played by both the local Other and occasionally by other travellers.[1] In the context of travel to exotic locales, travellers – both men and women – must be self-reliant, brave and ambitious, reflecting Whitehead's (2005) role of 'hero' against a backdrop of a villainised Orient. This antagonistic relationship is mediated by desire: travellers desire intimate encounters with the Other (Conran, 2006), an encounter across difference that enables claims to self-challenge and subsequent transformation. At the same time, the Other remains articulated (implicitly and explicitly; immediately and historically) through Orientalist discourse of deceit, savagery and hedonism (Said, 1979), which situates travellers within the Occidental supranational bloc (Balibar, 1991) characterised by ('gentlemanly') qualities of honesty, civility and self-control.

In this (post/neo)colonial context, those travellers who are (ostensibly) the most frugal, the most willing to travel off the beaten path, and the most culturally savvy are attributed higher status (Sorensen, 2003). In the same way that hegemonic masculinity represents 'the most honored way of being a man' (Connell & Messerschmidt, 2005: 832), travel represents the most honoured way of being a tourist, at least according to those who subscribe to the distinction from tourism. The performativity of these norms of travel and surveillance by other travellers echoes the 'homosocial enactment' central to Kimmel's (2007) analysis of homophobia. Homophobia, according to Kimmel, does not represent men's fear of homosexuals, but rather the fear that 'other men will unmask us, emasculate us, reveal to us and to the world that we do not measure, that we are not real men' (2007: 79). Thus, masculinity is performed homosocially: it is performed to convince other men of one's masculinity. Similarly, road status is enacted by travellers for judgment by other travellers, with the ever-present threat of being revealed as inadequate. In this way, both men and travellers are perpetually threatened with the possibility of having their failings revealed.

The norms of road status – the cultural norms by which backpackers define themselves against other tourists and other travellers – map onto prevailing Western norms of hegemonic masculinity as indicators of willingness to take on and endure hardship stoically, take risks and be brave, and remain in control. These Western norms are brought into sharp relief against contradictory perceptions of both the feminisation of Asian men (Malam, 2004; Shek, 2006) and the hypersexualization of Indian men (Hottola, 2013). There is clear overlap between these norms of travel and Brannon's dated (1976), though resilient, definitions of Western manhood: 'Be a Sturdy Oak,' 'No Sissy Stuff,' 'Give 'em Hell' and 'Be a Big Wheel' (cited in Kimmel, 2012: 203–4). Travel and masculinity are thus also entangled through the discourse of risk that provides the scaffolding for both travel and masculinity. Backpacker narratives are reliant upon 'participation in adventurous and risky activities' (Noy 2007: 48; Wilson and Little, 2008; Falconer, 2011) in order to obtain both cultural capital and a manly identity. While not necessarily predicated on physicality, as in Noy's (2007) trekking research, successful travellers can seek and manage risk, typically through the assumption of masculine attributes (Elsrud, 2005; Falconer, 2011) and recount this engagement with risk in their constructions of self (Desforges, 2000).

Not only do the norms of risk and masculinity overlap, but the capacity to engage risk is itself highly gendered. While women can enact norms of masculinity through risk to a degree, these norms cannot be fully taken up by women: unlike the racialised/nationed male privilege that carries across borders, the racialised/nationed privilege of women is qualified by gender marginalisation (see Hottola, 2013).

Travel as risk

As noted above, travellers often distinguished their practices from those of tourists through narratives of cultural interactions, which included overt references to risk. According to Sorensen's (2003: 856) definition of road status, non-touristic travel is predicated upon not only 'paying "local prices" [and] getting the best deal,' but also 'traveling off the beaten track, long-term travel, diseases, dangerous experiences, and more.' These latter necessities – possibilities of illness and danger, along with extended exposure to the cultural Other outside of one's cultural domain – overlap significantly with the Lupton and Tulloch's (2002: 116) definition of risk as involving 'danger, uncertainty, threat and hazard.' In this way, travellers do not seek not to mitigate risk as they would in the everyday experience of modernity (Beck, 1992), but rather participate in 'voluntary risk-taking…for the sake of facing and conquering fear, displaying courage, seeking excitement and thrills and achieving self-actualisation and a sense of personal agency' (Lupton and Tulloch, 2002: 116). Participation in risk enables travellers to enact the role of hero – an agentic identity manifesting ambition, courage and self-sufficiency (all associated with hegemonic masculinity) and thus consistent with the idealised liberal self of modernity. Thus, the 'master scripts' (Noy, 2003) of travel and risk are an important component of narrating a transformed identity (Desforges, 2000) and, in this way, engagements with risk become a source of cultural capital.

Participants' constructions of travel emphasised their understanding and acceptance of risk, insomuch as it was tangled with normative definitions of travel:

> Being able to get away from the tourists…tourist spots…[When] I travelled through Rajasthan and the North last time, I always managed, I was able to talk to the locals rather than go through the tourist track of the desert…and managed to find sort of this, um, place where the tourists weren't actually allowed to go. So, yes, that was…really good…There's definitely like a danger with it. Um, I've had a friend of mine who's done that, who's had quite bad experiences going off the beaten track in Egypt, being beaten. (Nigel, 33, Australia)

In contrast to Nigel and many of the other participants, a couple of participants indicated that they felt safer while travelling in India than at home. Leah (24, US), for instance, said she did not feel she was taking more risks by being in India, but rather that she took fewer risks in India because she '[didn't] want her body to have to be shipped back'. Hardeep (35, UK) stated that he felt safer in India because he was less likely to get beaten up by drunks than he would be in England on a Saturday night. While both of these individuals identify risk as a non-factor, or at least one that can be mediated through behaviour, that risk inheres in India is evident in both of their narratives. The risk of death cited by Leah is outside of the realm of risk understood within her home country of the United States. The perceived risk of violence in his home country may have been higher for Hardeep, a British man of Sikh ancestry. Hardeep, a traveller who could often pass as Indian while still holding the colonial privilege of a British passport, emphasises the other forms of risk inherent in travel: '[The people at home] don't realise all the risks you're taking in terms of, you don't know where you're going sometimes, you don't know, and that's the fun part of it, you're in a new city, and you gotta find your way 'round, find how to get back out again, and they don't understand that at all in any way.'

These conceptualisations of the unknown – both in terms of geographies and cultural practices that are unfamiliar and not fully understood – as constituting risk were similarly articulated by Sara (23, Israel) and Julia (28, Australia), as well as by Conrad (21, Canada): 'It is super-risky. Because it's a new place, I guess, and you don't know where you're going and different people and ya, [I] feel riskier.'

Shared in the narratives of Conrad, Hardeep and Leah are the intrinsic risks associated with being in a strange place, surrounded by strangers-cum-(potential)-villains. While other travellers identified material forms of possible risk such as food (Shannel, 28, US and Karl, 21, Sweden); medical treatment (Sara, 23, Israel); feral animals (Katrin, 28, Germany); and traffic (Maggie, 29, Canada and Ian, 29, Australia) – being in India in and of itself is understood through risk.

While there is little scholarship that specifically correlates risk with masculinity, Lupton and Tulloch (2002) indicate that risk is differentially tied to gendered subject positions, enabling men to enact masculinity and women to supersede femininity. Empirical work suggests that the contemporary mode of hegemonic masculinity, specifically as it is constructed in the so-called developed world, encourages men to participate in activities deemed to be higher-risk (Miller, 2008) and to subject their bodies to the risk of physical harm (Ryle, 2012).

In contrast to tourists (non-men), travellers articulate risk through the enactment of agency in their ongoing confrontations with the unknown. Travellers are engaged in ongoing battles with the Other, especially when the (potential) exchange of money is involved:

> Every time they come to me and say a price, I know I cannot trust the price. I need to bargain and then when I ask 'Is there a bus going from here to there?' Sometimes they say there is a bus; sometimes there isn't a bus; sometimes they want to sell you the train ticket or to take you by rickshaw, so they say there is no local bus going there, and afterwards you discover that there is, they just didn't tell you...So I just learned not to trust anything they say. (Rina, 24, Israel)

Within these encounters, Indians enact the role of the villain, a role that provides the necessary foil to the travellers' role of the hero, which is characterised by more than agentic subjectivity: assertiveness and/or aggression along with the successful management of their Otherness. These hero-villain encounters speak not only to the perceived or constructed hostility between travellers and locals, but also to the Otherness that, in the eyes of many of these travellers, characterises Indian practices, including bartering and aggressive sales. It is through the construction of (Indians as) the Other that travellers are able to actively engage in narratives of risk. While bartering does not in and of itself constitute a risk, India is constituted as a risk-scape in which even banal activities – including bartering – are narrated as inherently risky. In this way, risk does not exist as a barrier to, but a facilitator of, successful travel.

Gender, travel, risk

While narratives of risk are imperative to distinguish travel from tourism, the opportunity to engage this risk – and travel more generally – is highly gendered. Both men and women travellers with whom I spoke commented on the different experiences of men and women, given the risk associated with their respective bodies. In this way, travellers articulated a normative differentiation of experience based upon the marked vulnerability of women and the implicit impenetrability of men. Both men and women stated that women could not engage in the same activities as men, with the implication that only men could fully participate in independent travel practices:

> There's definitely a lot more freedom for me as a male, um, I mean, I can walk the streets at night without too much hassle at all, *at all*. [...] I feel really safe, just no problem at all, [... but] like just walking behind you for a while, just noticing what was going on with the looks and the comments, so that's when I feel guilty, that's when I feel awful that Western woman just can't have the freedom that I enjoy.' (Ian, 29, Australia; his emphasis)
>
> According to Sara, a 23–year–old from Israel, 'Me being a woman doesn't help [laughs]. [...] There are many things I can't do because, like travelling alone in a night bus or, [...] after dark going alone in the city.' (Sara, 23, Israel)

These comments, along with those of other participants, reflect previous findings that 97% of female travellers to India had been sexually harassed by Indian men (Hottola, 2013). Given these experiences, while interactions with locals were critical as a means to distinguish travel from tourism (Lozanski, 2010; Hottola 2007), those interactions also emphasised the ostensible powerlessness of women travellers given the perceived liability of the female body (Marcus, 1992), a liability exacerbated by travellers' narratives about hypersexualised heterosexual Indian men: 'I went into a restaurant by myself and [...] just wanted to eat and, and I was expecting like, yeah, there were going to be men there, and they were going to stare [...] but it was so, I mean, it was just so shameless and blatant.' (Susan, 26, Canada)

This understanding – alongside that of Simon, who categorised all Indian men as 'raging perverts' – reflects broader Western assumptions about what it means to be a (vulnerable/sexualised) woman, what it means to be a (predatory) Other, and what it means to be a (Western/civilised) man. Notably, the latter was never articulated as a potential source of sexualised threat or implicated in harassment, despite the virtual certainty that Western women who travel, like those at home, are at greatest risk not by stranger assault, but from those men they know (see Nichols, 2006: 6; Romito, 2008; DeKeseredy, 2011). Recalling dominant binaries of good versus evil, accounts of 'raging perverts', and other such generalisations also work to conflate villains and Indians as, in an Orientalist discourse (Said, 1979), the acts of some inappropriate individuals against female travellers are generalised onto Indian men as a whole. Indeed, the recent reporting of violent sexual assaults against both Indian and foreign women – which took place several years after this data was collected – beginning in December 2012 with the rape, and ultimately death, of a student in Delhi were correlated with a 25% drop in tourist arrivals in the three months immediately following the crime (Pradhan, 2013). However, the number of foreign female tourists assaulted violently is approximately a dozen since December 2012, out of more than 6.8 million tourist arrivals in India in 2013 (Ministry of Tourism, 2014; see also Hottola, 2013). Thus, while violent sexual assault certainly exists in India (again, in a timeframe beyond the data presented here) as well as in other jurisdictions (consider, for instance, the 1000 murdered or missing Aboriginal women in Canada [Canadian Press, 2014]), there is a persistent Orientalist premise in framing women travellers' experiences primarily through the lens of violent sexual assault.

With one exception (see Simon below), male travellers did not understand their experiences to be affected by their gender. Their lack of reflection on their masculinity is suggestive of the resilience of the male privilege originally conferred in their home countries. These men did not realise that their gender was, in fact, central to their ability to move throughout India. This subject positioning situates men's travel practices as the default experience for independent travel. As a result, female travellers were far more likely to acknowledge and question the gendered nature of independent travel than their male counterparts who, it seemed, rarely questioned the freedom, autonomy and independence afforded by their gendered privilege.

Women were complicit in the construction of travel as a possibility only fully available to men, thereby reiterating the patriarchal discourse that holds their bodies as sites of vulnerability (Marcus, 1992). Rina, 24, from Israel, said

> In India it's a bit scary because um, the Indian men, they look at you as a sex object and [...] we send like sexual vibes, it's like the way we dress... It just looks like a hooker to them, so I heard that a lot of girls that were harassed, that were touched.

In this comment, Rina draws upon the discourse of victim blame – in strong tension with her self-description as 'a girl,' signifying innocence – to make sense of the presumed inability of heterosexual Indian men to control themselves. Invoking the same explanatory rationale, Shannel shared anecdotes of female travellers who wore bikinis and were subsequently harassed, while Maggie expressed frustration with the lack of cultural awareness of the norms of modesty reflected in the relatively revealing dress of some female travellers. Given the ways in which these women identify the source of women's harassment as emanating from 'sexual vibes' and attire, it is evident that some, if not many, Western women travellers are unable to fully reconcile their liberated selves with their travelling selves.

In this way, travel operates through men's and women's assumption that only men can truly travel. While Indian men's bodies are understood as tools of violence, women's bodies are understood to be sites upon which varying degrees of violence are inflicted. These distinct experiences for men and women also reiterated dominant Western constructions of gender in which women, as the presumed embodiment of femininity, are never fully whole subjects: their possibility of 'self' – one informed by liberal notions of individualism and freedom – can never be fully enacted because of the omnipresent risk of violation. In this, travel – as opposed to the insular and presumably protected (im) mobility of tourism – is defined through an idealised masculine body, one that is invested with not only stature and racialisation, but also citizenship, currency and cultural capital.

Given the rarity and precarity of this idealised masculinity, most men were, and are, vulnerable in their travel to Other places. Simon, who was trying to get 'off the beaten track' for the first time, indicated that he felt safer than women, but 'not entirely safe' because of his own encounter:

> I had some young kids have a go at me one night in Jaipur... They were laughing and joking and pulling at my arms and then one of them pulled my bottle of water from my bag and started waving it around in the air [...] So I sweep my arm out and grabbed a hold of the bottle of water off him, and I just said, 'Hey, no!' and he was a bit shocked because I did it really sharply...and, some of the older men started getting a bit bolshie at me, started shouting. I just turned and walked, walked fast, you know?

In spite of the exceptional circumstances he encountered, I want to suggest that *all travellers* to India (or any Other places that are assumed to be less safe than their home country) are subject to potential violence, a threat implied to be universal with Simon's conclusion, 'you know.' Indeed, it is this very threat of violence set against the relative safety and security enjoyed by white, heterosexual men in the so-called Global North that helps to define India as Other. The risk and Otherness of danger and violence is desired only as potential and not as actualisation:

> Ethnological travel is contingent upon unruliness, as potential violence; meanwhile, within any form of tourism, including ethnological travel, there is little (if any) tolerance for physical manifestation of violence enacted upon the bodies of travellers and tourists. While travellers to Other places desire Otherness, which is typified through unpredictability, chaos, and disorder, they simultaneously assume a subject position outside of any actualised violence associated Otherness. (Lozanski, in press)

Unlike Simon, other men provided scripts that strongly challenged their personal physical vulnerability. The absence of vulnerability in the narratives of all but one of the men in this study recalls Whitehead's (2005) 'Hero' in that these men presented a 'core of transcendental courage in the face of danger' (p. 413), danger that inheres in India, a volatile place in which Orientalist discourses give meaning to travel. It is not individual Indian men who typically take on the role of the villain in specific encounters, but rather the abstracted *possibility* of villains within the villainous essence of India itself that satisfies the role of the hero's foil.

Protecting women/managing risk

Through the tacit claims that they were not vulnerable, the Western men travelling in India implicitly participated in the emasculation of Indian men. Women understood themselves to be and were indeed at risk of unwanted advances, touching and possibly – though unlikely – violent gang rapes, given the highly publicized sexual assaults of Indian women in Delhi (December 2012) and Mumbai (July and August 2013) and of tourists in Orchha (March 2013) and New Delhi (January 2014). In contrast, men did not present themselves as susceptible to Indian men's aggression beyond the commodified activities noted above. For male travellers, the villains identified were the 'unruly' Indians (Bhattacharyya, 1997) associated with commodified encounters – posing economic risk – rather than Indian men who posed a physical risk to women travellers. Thus, Indian men were villains vis-à-vis female travellers, but non-men vis-à-vis male travellers.

Beyond revoking Indian men's capacity to threaten them, male travellers occasionally participated in more explicit efforts to emasculate Indian men as possible aggressors (Lozanski, 2007). These actions operated through the establishment hierarchy of masculinity relative to Western women travellers. For instance, Vasu (37, US) identified efforts to protect women travellers merely through physical closeness:

> I was travelling with two women from Austria, and we were at the beach a little north ... [and] these women were in bikinis, and the guys would come down, and I realised there was this whole ritual about [how] the proximity of me to these women changed how the men reacted.... If I was far away, then the guys would be up close to the women; if I was near, or when I came back, they would all step back. (Vasu, 37, US)

In this anecdote, Vasu communicates a proprietary relationship, albeit a benevolent one, with the women. He takes up the role of the hero, recognising Indian men as villains, but only vis–à–vis female travellers. This role serves to reiterate his dominant (although racialised) status over both Indian men and travelling women.

One traveller related a more egregious display of heroic masculinity, observed in Diu, Gujarat:

> The women in the group were wearing bikinis, and while they were in the water, a group of middle-aged Indian men began approaching them and taunting them with comments like 'Nice tits.' At some

> point, one of the men in the travelers' group grabbed the Indian ringleader and dragged him out of the water. The traveler turned the Indian man around to face everyone on the beach and then pulled his pants down, saying, 'Everyone look at his penis and laugh. Now you know what it feels like to be objectified.' (Lozanski, 2007: 305)

The traveller who exposed the Indian man on the beach emasculated the man who taunted the women. This act clearly defined the Indian as a non-man through violence, not as a confrontation, but as a form of humiliating violence meant to indicate the Indian's failure of masculinity.

The penis is a metonym of masculinity. However, it only stands in for masculinity when it is under control (Stephens, 2008). As a marker of dominant masculinity, it is most likely to be invisible (Stephens, 2007). While 'the penis is the site and source of male power and must be used to define masculinity in opposition to femininity' (Brubaker and Johnson, 2008: 138), in this instance the Indian man is feminised as his penis is objectified in much the same way as women's bodies are.

In the above scenarios, male travellers from Western countries reiterate their hegemonic masculinity over the Indian men through protection of/control over Western women, protection that was not challenged by any of the women with whom I spoke. While India stands culturally as a place of risk that must be managed, Indians within it (men in particular) are successfully managed through cultural savvy, privilege and, occasionally, physicality. Indian men who threaten Western women represent demonstrations of 'exclusive violence' in which individual Indian men are subject to sexual humiliation, a form of violence that 'excludes the victim from the category of "man" as unworthy of belonging there' (Whitehead, 2005: 417).

Conclusion

Travel is dependent upon performances of masculinity in which Western travellers – both men and women – situate their practices as distinct from the pejoratively characterised 'tourist.' Not only do tenets of hegemonic masculinity underpin travel, but travel reiterates the dialectical construction of an inferior Other upon which masculinity depends. In the realm of travel, femininity equates to vulnerability, and tourists constitute non-men because of their respective failures to confront and negotiate risk and, in so doing, demonstrate the ideals of heroic masculinity-as-travel.

Indian men also constitute non-men, at least as potential physical threats. While their exclusion from the realm of competitive masculinity is perhaps counterintuitive given the hero's need for a villain, it is symptomatic of the broader feminisation that is inscribed upon the masculinities of Asian men (Malam, 2004; Shek, 2006). Moreover, Indian men are situated as antagonists of Western women, a position that reinforces the dominant status of Western male travellers (Lozanski, 2007).

In contrast to the presence of these non-men, idealised forms of travel are dependent upon active engagements with risk. Significantly, these engagements with risk are almost always recounted as successfully managed through the deployment of masculine qualities of courage, fortitude and competence. Less obviously risky encounters are situated against the risk that inheres in any commodified interaction in India in which the traveller is virtually always at risk of being deceived or swindled. Thus, the hero(ic traveller) maintains the narrative of a villain(ous India), a worthy opponent through which travel can be distinguished from tourism, and through which idealised tenets of masculinity can be continually renewed.

Note

1. The contrast between travellers and tourists, although disputed in the literature (see Bhattacharyya, 1997; Uriely, Yonay, and Simchai, 2002; Kontogeorgopoulos, 2003) is maintained by many people who travel and is achieved through narrative constructions of difference between the motivations, justifications, and practices of travellers and tourists (Munt, 1994: 116; Lozanski, 2011).

References

Azarya, V. (2004) 'Globalization and international tourism in developing countries: Marginality as a commercial commodity', *Current Sociology*, 52(6): 949–67.

Balibar, É. (1991) 'Racism and nationalism' in É. Balibar and I. Wallerstein (eds) *Race, Nation, Class: Ambiguous Identities*. New York: Verso, pp. 37–67.

Beck, U. (1992) *Risk Society, Towards a New Modernity*. London: Sage.

Bhattacharyya, D. P. (1997) 'Mediating India: An analysis of a guidebook', *Annals of Tourism Research*, 24(2): 371–89.

Brubaker, S. J. and Johnson, A. J. (2008) 'Pack a more powerful punch and lay the pipe: Erectile enhancement discourse as a body project for masculinity', *Journal of Gender Studies*, 17(2): 131–46.

Canadian Press. (1 May 2014) 'RCMP not denying report of 1,000 murdered or missing Aboriginal women,' *CBCnews.ca*. Retrieved 1 May 2014 from http://www.cbc.ca/news/ politics/rcmp-not-denying-report-of-1-000-murdered-or-missing-aboriginal-women-1.2628372.

Connell, R. W. (1995) *Masculinities*. Cambridge: Polity Press.

Connell, R. W. and Messerschmidt, J. W. (2005) 'Hegemonic masculinity: Rethinking the concept', *Gender & Society*, 19(6): 829–59.

Conran, M. (2006) 'Beyond authenticity: Exploring intimacy in the touristic encounter in Thailand', *Tourism Geographies*, 8(3): 274–85.

DeKeseredy, W. S. (2011) *Violence Against Women: Myths, Facts, Controversies*. North York: University of Toronto Press.

Desforges, L. (2000) 'Travelling the world: Identity and travel biography', *Annals of Tourism Research*, 27(4): 926–45.

Elsrud, T. (2005) 'Recapturing the adventuress: Narratives on identity and gendered positioning in backpacking', *Tourism Review International*, 9(2): 123–37.

Falconer, E. (2011) 'Risk, excitement and emotional conflict in women's travel narratives', *Recreation and Society in Africa, Asia, and Latin America*, 1(2): 65–89.

Hottola, P. (2005) 'The metaspatialities of control management in tourism: Backpacking in India', *Tourism Geographies*, 7(1): 1–22.

Hottola, P. (2007) 'The social psychological interface of tourism and independent travel', in I. Ateljevic and K. Hannam (eds) *Backpacker Tourism: Concepts and Profiles*. Bristol: Channel View Publications, pp. 26–37.

Hottola, P. (2013). 'Real-and-imagined women: Goddess America meets the world' in O. Moufakkir and Y. Reisinger (eds) *The Host Gaze in Global Tourism*. Cambridge, MA: CABI, pp. 219–231.

Kimmel, M. (2007) 'Masculinity as homophobia: Fear, shame, and silence in the construction of gender identity' in N. Cook (ed.) *Gender Relations in Global Perspective: Essential Readings*. Toronto: Canadian Scholars' Press, pp. 73–82.

Kimmel, M. (2012) *Manhood in America: A Cultural History*. New York: Oxford University Press.

Kontogeorgopoulos, N. (2003) 'Keeping up with the Joneses', *Tourist Studies*, 3(2). 171–203.

Lozanski, K. (2007) 'Violence in independent travel to India: Unpacking patriarchy and neo–colonialism', *Tourist Studies*, 7(3): 295–315.

Lozanski, K. (2010) 'Defining real India: Representations of authenticity in independent travel', *Social Identities*, 16(6): 741–62.

Lozanski, K. (2011) 'Independent travel: Colonialism, liberalism and the self', *Critical Sociology*, 37(4): 465–82.

Lozanski, K. (In press) 'Desire for danger, aversion to harm: Violence in travel to "Other" places', in H. Andrews (ed.) *Tourism and Violence*. London: Ashgate.

Lozanski, K. and Beres, M. A. (2007) 'Temporary transience and qualitative research: Methodological lessons from fieldwork with independent travellers and seasonal workers', *International Journal of Qualitative Research*, 6(2): 106–24.

Lupton, D. and Tulloch, J. (2002) 'Life would be pretty dull without risk: Voluntary risk-taking and its pleasures', *Health, Risk & Society*, 4(2): 113–24.

Marcus, S. (1992) 'Fighting bodies, fighting words: A theory and politics of rape prevention' in C. Mui and J. Murphy (eds) *Gender Struggles: Practical Approaches to Contemporary Feminism*. Lanham, MD: Rohman and Littlefield Publishers, pp. 385–403.

Malam, L. (2004) 'Performing masculinity on the Thai beach scene', *Tourism Geographies*, 6(4): 455–71.

Miller, K. E. (2008) 'Wired: Energy drinks, jock identity, masculine norms, and risk taking', *Journal of American College Health*, 56(5): 481–90.

Ministry of Tourism, Government of India. (9 January 2014) 'Achievements of Ministry of Tourism during the year 2013.' Retrieved 1 May 2014 from http://pib.nic.in/newsite/ PrintRelease.aspx?relid=102378.
Munt, I. (1994) 'The "Other" postmodern tourism: Culture, travel and the new middle classes', *Theory, Culture & Society*, 11: 101–23.
Nichols, B. (2006) 'Violence against women: The extent of the problem' in P. K. Lundberg-Love and S. L. Marmion (eds) *Intimate Violence against Women: When Spouses, Partners or Lovers Attack*. Westport, CT: Praeger, pp. 1–8.
Noy, C. (2003) 'The write of passage: Reflections on writing a dissertation in narrative methodology', *Qualitative Social Research*, 4(2): 1–20.
Noy, C. (2007) 'Travelling for masculinity: The construction of bodies/spaces in Israeli backpackers' narratives' in A. Pritchard, N. Morgan, I. Ateljevic and C. Harris (eds) *Tourism and Gender: Embodiment, Sensuality, and Experience*. MA: CABI, pp. 47–72.
Pradhan, B. (1 April 2013) 'Women tourists avoid India following sexual assaults, study says,' *Bloomberg.com*. Retrieved 1 May 2014 from http://www.bloomberg.com/news/ 2013-04-01/women-tourists-avoid-india-following-sexual-assaults-study-says.html.
Romito, P. (2008) *Deafening Silence: Hidden Violence Against Women and Children*. Bristol: The Policy Press.
Ryle, R. (2012) *Questioning Gender: A Sociological Exploration*. Thousand Oaks, CA: Sage.
Said, E. (1979) *Orientalism*. New York: Vintage Books.
Shek, Y. L. (2006) 'Asian American masculinity: A review of the literature', *Journal of Men's Studies*, 14(3): 379–91.
Sorenson, A. (2003) 'Backpacker ethnography', *Annals of Tourism Research*, 30(4): 847–67.
Stephens, E. (2007) 'The spectacularized penis: Contemporary representations of the phallic male body', *Men and Masculinities*, 10(1): 85–98.
Uriely, N., Yuval, Y. and Simchai, D. (2002) 'Backpacking experiences: A type and form analysis', *Annals of Tourism Research*, 29(2): 520–38.
Whitehead, A. (2005) 'Man to man violence: How masculinity may work as a dynamic risk factor', *The Howard Journal of Criminal Justice*, 44(4): 411–22.
Wilson, E. and Little, D. E. (2008) 'The solo female travel experience: Exploring the geography of women's fear', *Current Issues in Tourism*, 11(2): 167–86.

4

'Just Blokes Doing Blokes' Stuff': Risk, Gender and the Collective Performance of Masculinity during the Eastern European Stag Tour Weekend

Thomas Thurnell-Read

The emergence in recent decades of the stag tour as a premarital ritual undertaken by large numbers of British men sheds new, and at times vivid, light on an array of connections linking notions of masculinity and travel. Starting with cities such as Amsterdam and Dublin in the 1990s, the stag tour 'phenomenon' soon spread east, with destinations in newly developing post-Soviet states of Central and Eastern Europe such as Prague, Riga, and Budapest achieving a degree of notoriety for drunk and disorderly British men, fuelled by copious amounts of what was, for them, outrageously cheap beer and vodka, causing trouble. The streets of these cities were, according to the media, literally overrun with gangs of dishevelled and disrespectful male tourists, all eager to seek out fun, excitement, and titillation at the expense of put-upon locals, yet, equally, such antics were and still are rarely greeted with more than a prosaic shrug and the dismissive assertion that, after all, 'boys will be boys'.

Such is the nature of the intimate links between travel and masculinity that while the images coming out of Eastern Europe of half-dressed, vomiting British men were met with a large degree of contempt, such concerns were always tempered with a notably matter-of-fact resignation. Indeed, even stories of stays in foreign police cells, altercations with violent nightclub bouncers, and tragic accidents rarely led to a sustained questioning of why, indeed, it should seem so normal for such men to seek out adventures abroad. Indeed, such risks evidently further entrenched the link between this form of travel and its ability to confirm masculine status on participants. Stag tourism is therefore an inescapably gendered phenomenon. The stag tour is held up by the media in particular through programmes such as *Stag Weekends: The Dirty Secrets* (BBC, 2010) and *Boozed Up Brits Abroad* (Bravo TV, 2008), as an example of a bad masculinity, which is brash, loutish and out of control. Such can be held to represent the worst excesses of a dominant masculinity that idealises the pursuit of beer, girls, and 'good times' with friends whilst also, in many ways, being accepted as 'normal' or 'typical' and expected.

The notions of risk and release, fun and excitement, friendship and sociability, are revealed through the specific constructions of masculinity in the actions and practices of stag tourism. Drawing on ethnographic fieldwork in Krakow, Poland, this chapter explores how the stag tour is established and sustained as a distinctly masculine tourist practice. The notion of risk and adventure is integral to this construction in that it fosters a particularly desired sense of playful release that typifies the stag experience. Further, the chapter identifies the ways in which specific perceptions and performances of masculinity and femininity interrelate within the stag tour context, producing a complex dynamic of gendered interactions between stag tourists, local women and local men. Finally, in the concluding discussion, it is suggested that the masculinity displayed by stag tourists makes use of an easy slippage between a long-held serious, tough, adventurous manliness associated with travel and a more ironic, playful, leisured masculinity.

The great escape: for men only

The implicitly gendered nature of international tourism has been addressed within the literature on several, but seemingly too few, occasions (Enloe, 1989; Kinnaird and Hall, 1994; Swain, 1995; Sinclair, 1997; Pritchard et al., 2007). Enloe (1989), for example, provided an early seminal analysis by outlining how the very idea of tourism is often founded on normative notions that associate men and masculinity with adventure and exploration and women with safety and domesticity. Further still, Pritchard and Morgan (2000) argue that the tourist gaze, as first coined by John Urry (1990), is invariably that of the privileged and heterosexual male, while Edensor and Kothari (1994: 185) have demonstrated how many tourist sites are deeply masculinised and privilege masculine forms of engagement with tourism.

The value here is to acknowledge that gender pervades all levels of the tourism industry and of tourist practice. As such, Kinnaird and Hall (1994: 2) have called for a framework 'based on the recognition that tourism processes are gendered in their construction, presentation and consumption, and the form of this gendering is configured in different and diverse ways which are both temporally and spatially specific'. We may, therefore, go as far as to say that tourism 'is built of human relations, and thus impacts and is impacted by global and local gender relations' (Swain, 1995: 247). This position firstly recognises that, in being made up of gendered human relations, tourism is itself inescapably gendered, and that, secondly, the highly gendered nature of tourism is both shaped by and has the power to shape gender relationships and roles at all levels.

It would appear, then, that in order to understand a phenomenon such as stag tourism, which is so readily interpreted as being a typical and overtly 'masculine' activity, we must consider the various ways the stag tour experience relies upon notions of gender and upon heavily gendered interactions within the tourist setting. Stag tourism is a characteristically masculine pursuit but is also dependent on the gendered and sexualised relationship with the women who work in the pubs, bars, clubs and strip clubs that play host to them. Most notably, studies of sex tourism have highlighted the sexualisation of women by male tourists, which is based upon strikingly unequal power relations (Cohen, 1982; Enloe, 1989; Sanchez Taylor, 2001). In a study of the sexualised promotion of the Thai resort of Phuket, Hobbs et al. (2011) observe that industry advertising, media depictions and, increasingly, the Internet all draw heavily on gender stereotypes that allow the resort to be 'portrayed as a patriarchal world where a man believes he can live his fantasy of being the perfect hegemonic male' and where 'men are encouraged to pursue the traits of masculine hegemony' (Hobbs et al., 2011: 99). Similarly, in her research on cross-border sex tourism between the United States and Mexico, Katsulis (2010; and this volume) suggests that the confluence of gendered, classed and racialised subjectivities present in destinations such as Tijuana allow American men to 'live like kings' and 'experience a form of masculinised sexuality that might be inaccessible to them in their home culture' (Katsulis, 2010: 222). As such, the invariably male tourist is economically in a position of power over the female host, and further, this economic power is bound up with certain gendered, and frequently racialised, notions of superiority.

What emerges out of such literature is a picture of how tourist spaces and practices, for many men, offer the opportunity to work at and towards a particular performance of masculinity. Linda Malam (2004) notes, in her work on beach resorts in Thailand, that tourism spaces offer intriguing settings in which the gendered identity of both tourists and the workers who serve them may be worked upon and reconfigured. Likewise, a trend in studies of masculinity more generally has been to stress the situational and contextual nature of men's masculine performativity. Specifically, as a now substantial body of literature has identified, it is the field of leisure and the nighttime economy that has received notable attention as a site where masculinity is frequently enacted (Campbell, 2000; Blackshaw, 2003; Nayak, 2003). As this chapter identifies, the stag tour provides a context in which tourist experiences are demanded and provided according to a heavily masculinised 'script' that visibly enacts the supposed association between masculinity, adventure and dominance.

Methods and context

This chapter will explore the links between the social construction of masculinity and tourism by discussing findings from an ethnographic study conducted in the Polish city of Krakow between 2007 and 2008. Participant observation was conducted with eight separate stag tour groups from various areas of Britain and ranging in size from eight to nineteen members and in age between their early 20s and mid-30s. Access to these groups was initially secured with the assistance of a local tour company that provide mixed packages of accommodation, activities and guided tours to stag tour groups. Further contact was made with groups in the field. In all cases, the researcher made his role as a sociologist researching stag tourism in Poland clear and sought informed consent from participants. The author accompanied groups to bars, nightclubs and restaurants and, during the daytime, also spent time with groups as they partook in activities such as go-carting and sightseeing in the city's Old Town. Time was also spent observing the city's nightlife as a nonparticipant role, and formal and informal interviews with tour company workers and local residents were conducted. Additionally, diverse textual and visual content was collected, including media reports from Polish and British print and broadcast media, promotional materials from stag tour company websites and Internet forums and review websites. During research, the author drank alcohol with tour participants, though, for a number of reasons, most notably concern for recall when taking ethnographic field notes, sought to manage intake to prevent too great a level of intoxication (for in-depth methodological reflections, see Thurnell-Read, 2011a).

Gendering the stag tour

Examining a few of the websites belonging to the numerous stag tour companies offering packages of accommodation, activities and tour guide services to predominantly British customers, it becomes clear that the stag tour is geared towards a very narrow cache of typically 'male' pursuits. While this, in itself, comes as no surprise, given such companies are catering solely to male customers, it is revealing in the range of stereotypical 'masculine' interests and traits that are deployed to construct the stag tour. Popular activities offered by stag tour companies include motorsports such as go-cart racing, quad bike driving and even tank driving, firearms-based activities such as paintball games and pistol and rifle shooting, and food and drink events such as vodka tastings, brewery tours and medieval-style banquets of grilled meat washed down with jugs of beer. Such events are evidently informed by notions of masculinity as revolving around weaponry and warfare, competition, technical mastery of machines and indulgent consumption.

A further reflection of the gendering of the stag tour industry was found in the general pattern of marketing by such companies that focuses on alcohol, bar- and club-hopping and the overt use of images of Eastern European women as part of the branding of destination cities. A common, and in some cases overwhelming, feature of stag tour company sites, therefore, is the representation of Eastern European women as sexually appealing and available. The destinations chosen by stag tourists are themselves gendered in that there is an evident perception of Eastern European cities such as Krakow as being 'male' destinations (Thurnell-Read, 2011b). As such, when asked about destination choices, a characteristic response would emphasise the apparently self-explanatory, and deeply masculinised, nature of cities such as Krakow. One stag tourist, for example, matter-of-factly asserted, 'The beer is cheap and strong, and the women are hot. What more can a bloke ask for?!'

Amongst companies offering both stag and 'hen' tours, for male and female customers respectively, it was common for websites to split between stag and hen areas, the assumed intention being both to direct potential customers to the relevant content but also to build a sense of gender-segregated exclusivity. Thus, even in the planning stages, the stag tour begins its symbolic construction as an exclusively male space. Interestingly, the stag tourist noted above, when asked about his fiancée's corresponding hen party, merely commented, 'Oh, they're *just* going out in Leeds.' Here, the British city, Leeds, is held as a mundane and 'safe' *feminine* destination in contrast to the more adventurous, far-flung, *masculine* destination represented by the Eastern European city of Krakow. When stag tourists spoke of hen parties that did make foreign cities their destination, it was, indeed, far more common for more traditional beach-orientated holiday resorts to be favoured. This is supported by reviewing tour company websites that revealed a noticeably East-West divide, with stag destinations all predominantly in Central or Eastern Europe, and popular hen destinations more focused on the Western European destinations lining the Spanish Mediterranean coast and islands such as Mallorca (Andrews, this volume). Frohlick (2010) has noted the tendency for women's bodies to be the subject of a political and moral discourse concerning safety and risk while, we must note, the bodies of male travellers rarely are. This, it seems, is reflected in the tendency for 'safer' destinations to be seen as more appropriate for female hen parties while the 'danger' implicit in the stag tour weekend is left largely unchallenged.

Krakow as a stag tour destination is heavily sexualised; the women who come into contact with stag tourists through their work in the city's bars, restaurants and nightclubs, or merely by being present in the public spaces the stag tour groups also temporarily inhabit, are frequently subjects of a sexualised gaze. Indeed, many have noted this use of the female body as a commodity in the sale of tourist experiences to men. Sinclair (1997: 5), for example, observes that 'gendered and sexualised modes of behaviour and appearance are often demanded and supplied as part of tourism transactions'. Observations throughout fieldwork further support the assertion that the expectations of male tourists are frequently based on the assumption that the women of the countries they visit are 'more available and submissive than the women of their own countries' (Enloe, 1989: 36). With stag tourism, although this dynamic is most obvious in the commodification of female bodies in strip clubs, and in some instances, escort agencies and brothels, with the majority of stag tourists, their more mundane interactions with women in the tourist setting – that nonetheless replicate a strict female-male server-served binary – were evidently informed by the sense that gendered, unequal power relations were an implicit part of the stag tour phenomenon.

Risk and adventure

The myth of the noble explorer is deeply ingrained in the Western consciousness and, while a growing number of female travellers appear to be deploying such discourses of exploration and adventure in their travel narratives, it remains a typical male trope (Laing and Crouch, 2011). Recent research has identified some of the ways in which male tourists are seen to readily incorporate risky and dangerous behaviour into their experiences of travel and tourism (also see Lozanski, this volume). Thus, adventurous and physically demanding travel experiences are frequently central to the travel narratives of young male travellers (for example, see Noy, 2007). Likewise, research on young Danish tourists in coastal resorts in Bulgaria found that men reported higher levels of alcohol, tobacco and illicit drug use on holiday (Tutenges, 2012) and men were also found to be more likely to visit strip clubs and sex shows (Hesse and Tutenges, 2011).

The notion that a stag tour to a distant country is suitable for men, whereas the female equivalent, the hen party, is far more commonly restricted to domestic cities in the UK within easier reach of the participants' hometowns, therefore reflects a persistent gendering of travel along dichotomous lines. Home and away, risk and safety, are played against each other. The stag tour weekend relies heavily on a narrative of risk and adventure. While, for some, the heavy drinking might mean daylight hours spent recuperating in hotel rooms, the evening and nights saw stag groups exploring the city's streets and bars. Throughout fieldwork, stag groups would be keen to 'explore' the city, with many group members speaking of finding bars and clubs 'where the locals go to drink'. Interestingly, a further tendency was noted for many stag tourists to draw on analogies with warfare and sport to describe the trials and tribulations of the weekend. As such, a sizable stag group literally filling the dance floor of a nightclub have 'invaded the place', grazed knees from falling over during a drunken sprint across the Krakow Old Town market square become 'war wounds', and a stag is encouraged by his friends not to 'give in and be defeated' by a particularly agonising hangover felt the next morning to which 'Surrender is not an option!' Such 'cross-fertilisation' between different fields of masculinity has been noted by both Spracklen (2013: 85) and Noy (2007) and reminds us that travel, war and competitive sports have a long history of supporting hegemonic ideals of masculinity as tough, independent and adventurous.

This desire to explore the city, coupled with the disinhibiting nature of the stag tour, gives rise to a range of potentially risky or dangerous situations. In recent years, both Polish and British media have reported numerous accidents. In June 2012, for example, the death of Graeme Hartis was widely reported in the British media after the former soldier dived from his hotel balcony into a Slovakian lake during his best friend's stag tour (Poonan, 2012). The British Foreign and Commonwealth Office

(2005) issued a report in August 2005 featuring guidelines in an attempt to highlight the potential for harm and damage on overseas stag tours. In spite of this, ongoing news coverage continues to raise concerns about the strains placed on British consular services by stag tour groups in Eastern European cities (Sherman, 2006; Channel 4 News, 20 September 2007). Interestingly, Tutenges (2012) observes that the negative reputation of such destinations cultivated in the media in fact fosters interest among young male tourists drawn by the allure of a holiday environment where drunkenness, disinhibition and spontaneity are encouraged and accepted. Indeed, such negative portrayals of Eastern European cities ultimately prove counterproductive when they so readily feed into narratives of masculine travel that valorise risk, danger and adventure.

The notion of risk was evidently seen to be an integral element of the stag tour and a theme readily acknowledged by stag tour participants who held the excitement of 'causing mayhem' and 'getting messy' (similarly, see Treadwell in this volume). However, many participants also spoke favourably of engaging the services of a stag tour company that provided a tour guide as a means of ensuring that such dangers were to some extent mitigated. Indeed, stag tour guides worked tirelessly to ensure that stag tourists did not come to any harm by, for example, falling foul of local nightclub security staff, in their inebriated disorientation becoming detached from the group or, through their own drunken misadventure, injuring or otherwise harming themselves. However, stag tourists were, in many cases, not aware of how orchestrated the tour was. One attributes this to the astuteness of stag tour companies and their staffs, who recognise that stag tourists desire a sense of adventure but would, in the vast majority of cases, never wish to spend a night in a foreign prison cell or hospital bed. Thus, tour companies worked to build up rapport with bar and club owners and would strategically manage the time and duration of visits by stag groups on their guided bar crawls so as to take in many drinking venues yet never outstay the group's welcome in one particular location. The dangerous elements of the stag tour are, therefore, often finely balanced. Risk and adventure are knowingly cultivated, yet stage managed by tour companies and tour guides, and willingly consumed by stag tour participants.

Performing masculinity on the stag tour

Within the physical setting of the Krakow Old Town centre, stag tourists deploy their bodies in specifically heterosexual and implicitly masculine ways. The male body of the stag tourist expresses its heterosexuality through performing an unruly masculinity that is drunk, unrestrained and frequently seen as 'out of control' (Thurnell-Read, 2011c). This sense of performance pervades the stag tour and, particularly, places a demand on the individual members of stag tour groups to act out a particular collectively defined performance of masculinity. This section will, therefore, explore the various ways that the collective performance of masculinity during the stag tour is constructed through placing various pressures on group members to align their gendered behaviour with that of the group. Indeed, many research participants spoke of being concerned prior to the trip about 'fitting in'.

Central to Connell's (1987; 1995; Connell and Messerschmitt, 2005) conception of hegemonic masculinity is the notion of complicity where, both at a structural and an interpersonal level, the majority of men who might not achieve the hegemonic ideal of masculinity in any coherent form do, nonetheless, comply with and (directly and indirectly) condone the dominant practices and actions. Stag tourism, then, provides an interesting case for exploring how such complicity is put into action in practice. The emphasis on sociability and the importance placed on mutual engagement in shared activities therefore ought to be seen as complicity in action. The sense of being 'up for a laugh' and of willingness to participate in the shared experiences endorsed by the group is both a means of complying with the situationally defined expression of masculinity (for a more detailed analysis of the nature of homosocial bonding and friendship during the stag tour, see Thurnell-Read, 2012). However, it becomes apparent that, although the stag tour might appear to entail a sense of spontaneity and 'natural' masculine pursuits, this does not, in fact, come without significant efforts made by group members and others in specific 'supporting roles'.

Prominent in this sense is the 'best man', chosen by the groom to do much of the planning of the stag tour weekend: selecting the destination, gathering deposits from participants and, most noticeably, orchestrating the events and activities during the weekend itself. More specifically, the best man might organise and distribute fancy dress costumes or matching team shirts, as well as hold the 'kitty' used to buy rounds of drinks in bars and clubs. The success of the best man, it often seemed, relied upon his ability to strategically manage the group and encourage a sense of joviality and playfulness. To this end, it was common for the best man to have spent considerable time before the weekend itself in devising often elaborate drinking games, the playing of which would unavoidably involve rapid consumption of alcohol by all group members. Likewise, the best man would invariably take the lead in initiating various challenges and tasks that the stag had to then carry out whilst being supplied with copious amounts of alcohol in the form of 'forfeits'.

A closer reading of the drinking games devised by stag tour groups adds further light to the gendered nature of the stag tour and, in particular, how a particular collective masculinity is constructed around an adventurous, boisterous masculinity. For example, one stag was for the entirety of the weekend ordered to drink a shot of vodka, regardless of time or location, whenever they mentioned either the bride or the wedding itself. In another group, each member of the group was encouraged to tip 'three fingers', a portion of each of their drinks, into a glass from which the stag would then drink. In both cases, we can clearly see the implication of the game in terms of gender; in the former, the stag is punished for failure to distance himself from the perceived feminising effect of the looming marriage, and, in the latter, the stag is bound to the male group through his act of drinking a portion of each participant's drink.

A further important actor in the stag tour weekend was, for those groups who chose to book an organised package from a tour company, the tour guide provided to oversee their time in the city. Inhabiting a similar facilitating role to the best man, the tour guide would also be involved in ensuring the smooth running of the weekend and, like the best man, with aligning the actual festivities with expectations. Tour guides would, for example, encourage quieter members of the group to drink a shot of vodka in order to 'get into the spirit' of the weekend. In one group, for example, the tour guide danced with the groom and several other members of the group on an otherwise empty dance floor. While this act is in itself unremarkable, the guide later explained in an interview that she would often dance with members of the group as she felt they expected to receive some interaction from women during the night out. It was therefore, in her words, her job to 'flirt a bit with them and to let them flirt a bit with me'.

One common theme raised by participants was the perceived threat of local men who were seen to express hostility to the presence of stag tourists in the city. A member of Phil's group, for example, responded to the perceived threat of local men in the club, by saying, 'Don't like the look of that big fucker, looks fucking nuts', and 'Since we got here, there's been a few places we think we're going to get our heads kicked in for sure, think they want to protect their women or something.' Such talk was common amongst stag groups as they made their way through the city, evidently aware of the potential for confrontation with local men. Stag tourists would talk of some local men as being threatening because of their cropped skinhead haircuts, muscular physiques and bodily comportment.

Groups would speak of Polish men, labelling their bodies in such a way as to locate them as being symbolic of a traditional masculinity based on strength and toughness. In contrast, the bodies of British stag tourists would be spoken of by the group in a way to draw lines of difference between them and local men; for example, they might point out that British men, while not being strong and muscular, are more 'fun' and up for having a laugh. Interestingly, not having to perform a tough and aggressive, 'hard' masculinity was used by British men to draw a line of distinction between them and some of the Polish men they would at times come into contact with. Such talk serves to normalise the masculinity of British stag tourists as acceptable whilst relegating the perceived tough pose of some Polish men to a position of subordination as masculine Other. In this light, the interaction between stag tourists and others in the tourist setting illustrates how gender embodiment is constructed through the interface between various competing gendered bodily subjects.

Discussion and conclusion

This chapter has sought to show how the stag tour made by groups of British men to Eastern European cities such as Krakow is informed by dominant codes of masculinity based on notions of risk and adventure and constituted in relation to the gendered interaction that takes place within the group and between the group and others such as tour guides or groups of local men. Within stag tour groups, there is a high level of complicity that, it is argued, is analogous to a wider sense of complicity with hegemonic masculinity as identified in the work of Connell (1987; 1995; Connell and Messerschmidt, 2005). Further still, in various ways, this performance of masculinity is shown to be relational and, as such, enacted in reference to the gender of others such as, notably, local women and men and other groups of stag tourists. This concurs with the assertion made by Frohlick (2013) concerning the importance of observing the complexity of gendered interactions in tourist settings where the respective femininities and masculinities of female and male tourists and local women and men – both working in and not working in the tourist industry – interrelate as part of a wider power dynamic built upon gender, class, ethnicity and sexuality.

However, a closer reading of these discourses relating to masculinity and travel reveals some notable tensions. One aspect worth further discussion is how the depiction of masculinity oscillates between the serious and the mundane, between the adventurous risk- taking explorer and the infantilised and self-deprecating bloke 'up for a laugh' but never to be taken too seriously. While much of the performance of masculinity within the stag tour setting is overt and certainly, at times, spectacular, it is noteworthy that much of the observed 'doing' of masculinity was tinged with irony. Examples of this include knowing references to gun shooting as a 'manly' activity or reference by the group to themselves as 'just a bunch of pissed-up blokes' or, as the title of this chapter suggests, 'just blokes doing blokes' stuff'. Such comments, particularly the latter, can be read as attempts to underplay the actions of the group in their conformity to dominate masculine stereotypes. Many authors have noted the importance of irony in allowing men to enact highly gendered behaviour whilst distancing themselves from reproach or accusations of sexism or insensitivity (Whelehan, 2000; Benwell, 2003, 2004; Korobov, 2005, 2009). The influential concepts of fun, frivolity and playfulness that pervade the tourist setting are particularly well suited to the ironic assertion of masculinity.

Many participants appeared to feel that they can 'play the game' and retain masculine spaces for homosocial bonding while not allowing those times to encroach on their relationships in the work and family sphere. Indeed, the much-used phrase, 'What happens on tour stays on tour', deployed with striking frequency by stag tour participants, appears to indicate, on several levels, the wish to compartmentalise the stag tour weekend as its own arena, behaviour within which is exclusive of how participants might act at 'home'. Both Tony Jefferson's (2002) notion of 'context-specific hegemonic strategies' and Tony Coles' (2009) work on 'fields' of masculinity shed light on this. Clearly the stag tourists largely know and are aware of their enactment of hegemonic masculinity, yet it is the physical and social context that makes this so readily possible. Indeed, on numerous occasions, individuals, clearly aware of the negative reputation of stag tourists, expressed how their behaviour during the stag tour bore very little similarity to their behaviour at home and in other situations.

Thus, it seems that the tourist setting and, indeed, the very notion of tourism as based on adventure and risk-taking, provides a setting conducive to the enactment of dominant codes of masculinity. Stag tourism is experienced within a specific social space and time, identified, through an ongoing process, as a spatial-temporal location for the enactment of a certain form of playful masculinity that, at times, transgresses various aspects of social propriety. In this sense, it has been noted that in sexualised environments, such as strip clubs and gentlemen's clubs, men can act out behaviour for which they might be more readily chastised or that could be less easily achieved in more prosaic settings (Donlon and Agrusa, 2003). The tourist setting of stag tourism provides the men who partake in stag tours with a space and time for engaging in behaviour explicitly coded as masculine.

Much of the bombast of the stag tour is notable for its ostentation, and certainly draws attention and much scorn from the media, academics and the general public alike, but also, we must conclude, offers participants a readily available, and in many senses, reassuring view of masculinity. The stag tour is socially constructed to offer a space for the unproblematic assertion, and at times ironic enactment, of normative masculinity and, thus, provides a degree of certainty. In Blackshaw's (2003) study, the nighttime economy offered his participant group of men a refuge from the wider world, where workplace and home life roles were rapidly changing. In this, it is more aligned with conceptions of a 'crisis' that sees, not a wholesale erosion of masculine identity, but rather a shift to assertions of masculinity that are more compartmentalised, working within the flexibility that is evident in the identity constructions of many men.

The stag tour weekend offers an insight into wider notions of how masculinities are negotiated through embodied, relational performances within tourist spaces and through tourist practice. Both the stag tourism industry and the stag tourists themselves invest heavily in constructing an image of Eastern European cities such as Krakow based on gendered and sexualised assumptions, not to mention national cultural stereotypes. Further, throughout the stag tour weekend, various actors – certainly not just the stag tourists themselves – make active efforts to achieve and sustain a sense of collective masculinity through the gendered performance of tourist practice. Ultimately, what the men who participate in Eastern European stag tours are buying into is the idea that a time and place where masculinity can be performed and acted out, to a large extent unproblematically, can be packaged as tourist experience.

References

Benwell, B. (2003) *Masculinity and Men's Lifestyle Magazines*. Oxford: Blackwell.

Benwell, B. (2004) 'Ironic discourse: Evasive masculinity in men's lifestyle magazines', *Men and Masculinity*, 73(1): 3–21.

Blackshaw, T. (2003) *Leisure Life: Myth, Masculinity and Modernity*. London: Routledge.

Campbell, H. (2000) 'The glass phallus: Pub(lic) masculinity and drinking in rural New Zealand', *Rural Sociology*, 65(4): 562–81.

Channel 4 News (2007) 'Stag parties plaguing Eastern Europe', 20 September 2007.

Cohen, E. (1982) 'Thai girls and Farang men: The edge of ambiguity', *Annals of Tourism Research*, 9(3): 403–28.

Coles, T. (2010) 'Negotiating the field of masculinity: The production and reproduction of multiple dominant masculinities', *Men and Masculinities*, 12(1): 30–44.

Connell, R. (1987) *Gender and Power: Society, the Person and Sexual Politics*. Cambridge: Polity Press/Blackwell.

Connell, R. W. (1995) *Masculinities*. Cambridge: Polity.

Connell, R. and Messerschmidt, J. W. (2005) 'Hegemonic masculinity: Rethinking the concept', *Gender and Society*, 19(6): 829–59.

Donlon, J. and Agrusa, J. (2003) 'Attractions of the naughty-gentlemen's clubs as a tourism resource: The French quarter example', in T. Bauer and B. McKercher (eds.) *Sex and Tourism: Journeys of Romance, Love, and Lust*. New York: Haworth Hospitality Press, pp. 119–36.

Edensor, T. and Kothari, U. (1994) 'The masculinisation of Stirling's heritage' in V. Kinnaird and D. Hall (eds) *Tourism: A Gender Analysis*. Chichester: Wiley, pp. 164–87.

Enloe, C. (1989) *Bananas, Beaches and Bases: Making Feminist Sense of International Politics*. London: Pandora.

Foreign and Commonwealth Office (2005) 'British Hens and Stags: In trouble Abroad', 30 August 2005.

Frohlick, S. (2010) 'The sex of tourism? Bodies under suspicion in paradise' in J. Scott and T. Selwyn (eds) *Thinking Through Tourism*. Oxford: Berg, pp. 51–70.

Frohlick, S. (2013) 'Intimate tourism markets: Money, gender, and the complexity of erotic exchange in a Costa Rican Caribbean town', *Anthropological Quarterly*, 86(1): 133–62.

Hesse, M. and Tutenges, S. (2011) 'Young tourists visiting strip clubs and paying for sex', *Tourism Management*, 32(4): 869–74.

Hobbs, J., Na Pattalung, P. and Chandler, R. (2011) 'Advertising Phuket's nightlife on the Internet: A case study of double binds and hegemonic masculinity in sex tourism', *Journal of Social Issues in Southeast Asia*, 26(1): 80–104.

Jefferson, T. (2002) 'Subordinating hegemonic masculinity', *Theoretical Criminology*, 6(1): 63–88.

Katsulis, Y. (2010) '"Living like a King": Conspicuous consumption, virtual communities, and the social construction of paid sexual encounters by U.S. sex tourists', *Men and Masculinities*, 13(2): 210–30.

Kinnaird, V. and Hall, D. (1994) *Tourism: A Gender Analysis*. Chichester: Wiley.

Korobov, N. (2005). 'Ironizing masculinity: How adolescent boys negotiate heteronormative dilemmas in conversational interaction', *Journal of Men's Studies*, 13(2): 225–46.

Korobov, N. (2009) 'Expanding hegemonic masculinity: The use of irony in young men's stories about romantic experiences', *American Journal of Men's Health*, 3(4): 286–99.

Laing, J. and Crouch, G. (2011) 'Frontier tourism: Retracing mythical journeys', *Annals of Tourism Research*, 38(4): 1516–34.

Malam, L. (2004) 'Performing masculinity on the Thai Beach scene', *Tourism Geographies*, 6(4): 455–71.

Nayak, A. (2003) 'Last of the 'real Geordies': White masculinities and the subcultural response to deindustrialisation', *Environment and Planning D: Society and Space*, 21(1): 7–25.

Noy, C. (2007). 'Travelling for masculinity: The construction of bodies/spaces in Israeli backpackers' narratives' in A. Pritchard, N. Morgan, I. Ateljevic and C. Harris (eds.) *Tourism and Gender: Embodiment, Sensuality and Experience*. Wallingford: CABI, pp. 47–72.

Poonan, N. (2012) 'Stag holiday soldier drowns in lake', *The Sun*, 28 June 2012, http://www.thesun.co.uk/sol/homepage/news/scottishnews/4398711/Stag-holiday-soldier-drowns-in-lake.html.

Pritchard, A. and Morgan, N. J. (2000) 'Privileging the male gaze: Gendered tourism landscapes', *Annals of Tourism Research*, 27(4): 884–905.

Pritchard, A., Morgan, N., Ateljevic, I. and Harris, C. (eds.) (2007) *Tourism and Gender: Embodiment, Sensuality and Experience: Embodiment, Sensuality and Experience*. Wallingford: CAB International.

Sanchez Taylor, J. (2001) 'Dollars are a girl's best friend? Female tourists' sexual behaviour in the Caribbean', *Sociology*, 35(3): 749–64.

Sherman, J. (2006) 'Stag night revellers 'should pay for embassy crisis help'', *The Times Online*, 20 April 2006.

Sinclair, T. (1997) *Gender, Work and Tourism*. London: Routledge.

Spracklen, K. (2013) *Leisure, Sports and Society*. Basingstoke: Palgrave Macmillan

Swain, M. B. (1995) 'Gender in tourism', *Annals of Tourism Research*, 22(2): 247–66.

Thurnell-Read, T. (2012) 'What happens on tour: The premarital stag tour, homosocial bonding and male friendship', *Men and Masculinities*, 15(3): 249–70.

Thurnell-Read, T. (2011a) '"Common-sense" research: Senses, emotions and embodiment in researching stag tourism in Eastern Europe', *Methodological Innovations Online*, 6(3): 39–49.

Thurnell-Read, T. (2011b) 'Tourism place and space: British stag tourism in Poland', *Annals of Tourism Research*, 39(2): 801–19.

Thurnell-Read, T. (2011c) '"Off the leash and out of control": Masculinities and embodiment in Eastern European stag tourism', *Sociology*, 45(6): 977–91.

Tutenges, S. (2012) 'Nightlife tourism: A mixed methods study of young tourists at an international nightlife resort', *Tourist Studies*, 12(2): 131–50.

Urry, J. (1990/2002) *The Tourist Gaze: Leisure and Travel in Contemporary Societies*. London: Sage.

Whelehan, I. (2000) *Overloaded: Popular Culture and the Future of Feminism*. London: Women's Press.

5
Masculinity and the Gay Games: A Consideration of Hegemonic and Queer Debates

Nigel Jarvis

Travel to formal organised lesbian, gay, bisexual and transgendered (LGBT) sport networks and related events is a recent occurrence. This largely conceptual chapter focuses specifically on how alternative and lesser-known sport tourism events, such as the Gay Games, provide an opportunity to investigate masculinity. While much literature and research on sport and the Gay Games also relates to women's participation, this chapter focuses on the implications of gay men taking part in this athletic event. The Gay Games potentially represent a significant transgressive and alternative space in the world of sport because they involve high levels of tourist and international mobility. The two theoretical frameworks that are used in this chapter to help understand masculinity debates related to the Gay Games are hegemony and queer theory. Hegemony theory is one of the most popular and fruitful strands of the neo-Marxist approach to the sociological study of sport, while the young roots and contested nature of queer theory offers a more contemporary consideration.

Methods

In addition to a review of the literature associated with masculinity and sport tourism, a qualitative approach was used to investigate the Gay Games case study. An in-depth interview was undertaken with Marc Naismith – the current vice-president for external affairs of the Federation of Gay Games (FGG), the international body that oversees the event – in Paris on 9 May 2012. Marc helped provide valuable insights on masculinity issues associated with the games. He granted permission to tape the interview and agreed that his name could be used in the publication. Additionally, through the author's experience at three past Gay Games (Sydney, Chicago and Cologne) as a participant athlete, the chapter illustrates how gay men travelling to participate can possibly undermine traditional and conservative notions of hegemonic sporting masculinity in numerous ways. Casual conversations were undertaken with a number of gay male athletes taking part in previous games, with the researcher acknowledging his interest in exploring issues related to masculinity, sport and sexuality. Issues related to the queering of the games are also discussed, as this helps to illuminate further hegemonic debates. The researcher's keen interest in playing sport, including tennis, badminton, volleyball and softball, within both mainstream and gay club environments, helps to strengthen the critical observations made.

Sport tourism, gay men and masculinity within the Gay Games

LGBT travel represents a significant emerging and increasingly visible niche within contemporary tourism across the world. This segment represents another form of increased mobility (Urry and Sheller, 2004), not only travelling for an event like the Gay Games, but also within a tourism context. While some more recent literature has explored gay men's holiday choices and preferences (Casey, 2009; Hughes and Deutsche, 2010) and the politics associated with gay and lesbian tourism festivals like Pride and Mardi Gras (Hughes, 2006), scant research has specifically investigated issues around masculinity and gay men within a tourism context. Much academic literature exists on masculinity and sexuality in sport, but less is known about these issues in a sport tourism setting. The large number of LGBT sport events that take place all over the world provide the impetus for international gay tourism. Thus the Gay Games provide an excellent opportunity to investigate gay men's masculinities within these alternative sport tourism events.

Higham and Hinch (2009) note much research on sport tourism remains largely concentrated on high profile, mainstream mega-events and professional sports. They further argue a need for more critical explorations of less known events. Messner (1992) noted since the outset of the gay liberation movement in the early 1970s, organised sport has become an integral part of developing lesbian and gay communities. Gay athletes began an appropriation of spaces (see Williams, 1961), openly playing sport in public parks and venues, which were traditionally dominated by mainstream sporting groups and an ideology that favours and promotes heterosexuality (also see Treadwell, this volume). This created tensions and conflicts as gay men and lesbians were not originally welcomed into these places dominated by a hegemonic masculinity (Connell, 1995). The contestation and clash around space and meaning and the aspiration and desire to acquire or provide space for new types and different forms of cultural expression, such as gay sporting events like the Gay Games, is a theme that has only emerged in the sociological and cultural analysis of modern sport (Waitt, 2003; Symons, 2010; Davidson, 2012) and tourism over the past 15 years (Visser, 2008; Casey, 2009).

For those who may not be familiar with the history, growth and structure of the Gay Games, Dr Tom Waddell originally conceived of this significant hallmark sport tourism event in 1980. The mission of the FGG (2012a) is 'to promote equality through the organisation of the premiere international LGBT and gay-friendly sports and cultural event known as the Gay Games. They are built upon the principles of participation, inclusion and personal best.' Participants with diverse backgrounds are encouraged to take part, regardless of age, gender, ethnicity, disability, sporting ability, HIV status and geographic origin. Less known is that straight athletes can also participate in the event, a phenomenon known as 'inverse integration' (Elling et al., 2001).

The games have now taken place on three different continents and experienced considerable growth in the number of athletes competing and countries and types of sports represented. The original games in San Francisco had 1,300 athletes, whereas Cologne had some 10,000 competing in 2010 (FGG, 2012b). Obviously, this implies there are significant numbers of not only athletes, but officials, friends and family who will travel to take part in or watch the event, which demonstrates that the Gay Games involve high levels of international tourist mobility. Naismith stresses the games are a privately organised event that relies on individual registrations to finance and stage each quadrennial gathering. While some criticism of the games comes from cultural advocates who want to promote gay arts displays and activities within the event space, Naismith contends that despite this political opposition, the event has to be first and foremost an attractive sporting experience and thus commercially appealing to a large number of athletes and spectators.

The majority of the sports at the 2010 Gay Games in Cologne are also offered at the Olympics. These included basketball, cycling, ice hockey, softball, triathlon, football, tennis, athletics, figure skating, sailing, volleyball and weightlifting. Non-traditional sport events included billiards, bowling, chess, physique (bodybuilding) and dancing. Further, the games were the first to offer golf and women's wrestling as medal events, even before they became a sanctioned event at the Olympics.

The Gay Games constitute a powerful reaction to the homophobic discrimination and oppression in sport and have provided a safe space for participants within all categories of gender and sexuality, as well as skill level. According to Symons (2010), the global growth and development of gay sport networks, competitions and events like the Gay Games have transformed the way gay athletes experience and understand sport. Maguire (1999) believes sport helps to unite the 'global village', and in a way, the Gay Games, with its inherent mobility, can be seen bringing together an imagined global gay village. The Gay Games have also lead to similar sporting spaces like the EuroGames, held in various European cities and competing global events like the Outgames. While this paints a positive picture, there exist underlying debates surrounding masculinity and gay men that have accompanied the development of this sporting endeavour.

Sport (tourism), sexuality and hegemonic masculinity

The issue of sexuality and sport has long attracted considerable attention from academics (see Connell, 1990; Pronger, 1990, 2000; Messner, 1992; Robertson, 2003; Pringle, 2005; Caudwell, 2006; Jarvis, 2006; Anderson, 2009). Sexuality and sport falls under the wider analytical area of gender issues because, in both representational forms and in lived practices, sport is one of the cultural spheres that most explicitly generates, reproduces and publicly displays gender identities and differences (Connell, 1995; 2008; Theberge, 2000) and justifies the existing hierarchical gender order.

Historically, the most common theoretical framework for understanding the hierarchy of masculinities is Connell's concept of hegemonic masculinity (1995, 2008). While Connell's hegemony concept has come under challenge from several directions in recent years (see Beasley, 2008; and Anderson's 2009 and 2010 work on 'inclusive masculinity'), it still provides a useful tool to theorise gendered power relations among men and to understand the effectiveness of masculinities in the legitimation of the gender order within sport (see Allain, 2008; Howson, 2008; Jones and McCarthy, 2010). Indeed, Robertson (2003) believes sport represents an arena where masculinities can become unstable, evoking contradictory emotions for individual men. Still others call for research on more complex renderings of masculinity, that is, for diversity among men (King, 2008). Caudwell (2011) comments that social power relations infuse and suffuse (sport) spaces, and the human body helps construct hegemonic but also counter-hegemonic identities and subjectivities. Thus, this paper sees the merits of using Connell's work to frame the discussion about whether contemporary forms and experiences of sport tourism, such as the Gay Games, reproduce or challenge traditional hegemonic masculinity. What contradictions emerge in and around hegemonic and other forms of masculinity at the games? Do the Gay Games provide a sport tourism space that challenges hegemonic masculinity and traditional notions of sporting institutions?

The Gay Games and hegemonic masculinity

The games provide an inclusive space for male and female athletes, of all skill levels, to experience a sense of belonging to a wider gay community. As Elling et al. (2003) stated, the growth of gay sport has stimulated participation among both lesbians and gay men. When speaking to gay male athletes at past games, the majority identified that they participate to meet new friends, to play in a relaxed and comfortable environment, and to be a part of the wider gay community. Others said they wanted to take part for fitness and exercise reasons, to compete and to win medals.

Few male athletes have told me over the years that they participate for political reasons or to challenge traditional conservative sporting institutions. While a few men said they took part because they 'wanted to show there was more to gay life than gay bars', none explicitly commented they wanted to make a radical critique of sport. This did not support Waitt's (2003) exploration on why some gay men chose to travel to the Sydney Games. His research found some took part because they wanted to challenge gay stereotypes – that is, the lack of gay male participation in sports – but also to make fun of hegemonic sporting masculinities by exhibiting camp sport behaviour. Instead, the participatory reasons among the men I spoke to seem to reflect what Messner (1992) and Pronger (2000) identified as a value system and vision based on feminist and gay liberationist ideals of equality and universal participation. Through taking part in the tennis and badminton competitions myself, and observing others in a wide range of disciplines, I find most gay male athletes generally exhibit many conventional forms of traditional masculinity found within mainstream sport (for example using typical argot, displaying aggression, swearing, competing to win). Further, based on my conversations with the athletes, the men seemed satisfied with the nature of their current involvement in sport and were not politically motivated to transform and challenge existing sporting practices or structures.

The Gay Games have conflicting ideals – on the one hand, they are set up to provide a space away from oppressive heteronormativity and masculinity in mainstream sport settings, and on the other hand, they tend to be based around competition and traditional sports and universal standards. This demonstrates that they largely mirror existing mainstream sporting institutions, which are dominated by a hegemonic masculine narrative. Part of this relates to competition and winning. Critics (Pronger, 2000; Waitt, 2003) advocate the Gay Games' need to stop modelling themselves on the elite Olympics and perpetuating a competitive culture, which is part of the hegemonic masculine discourse. Gay Games' Naismith acknowledged the awarding of medals has always been a contentious issue. He counters the medals are desired by most athletes as they 'represent a symbol of their success...it is the value that participants attach to it which becomes important'.

One fundamental question about the self-organisation of the Gay Games, which largely take place in their own subcultural settings within a host city, is whether it increases the social and individual isolation of gay men (and lesbians) and, thus, is counterproductive in terms of the integration process and challenging existing traditional sporting masculinity. Hargreaves (2000: 171) observed gay sport takes place in insular 'ghettoised' spaces, also noted in the wider literature about some gay tourist destinations (see Pritchard et al., 2000; Hughes and Deutsch, 2010), and that gay sports liberation is partial and conditional – it has only come with separation and not with integration. How can a major gay event represent a clear challenge to sport as an institution when it largely occurs in spaces outside of dominant settings? To some extent, this is not true. While local gay sport clubs tend to play in more isolated settings, the Gay Games, because it is a significant hallmark sport tourism event, requires many visible spaces to accommodate the travelling athletes, officials, supporting friends and family and other spectators. Indeed, the visibility of the games is important in helping challenge stereotypes and discrimination toward LGBT people in general and athletes specifically. This is especially true for the 2014 Gay Games, awarded to Cleveland, Ohio. Naismith said 'Cleveland [a traditional blue-collar city] organisers' aim for the games is to help promote LGBT issues in the heartland of mid-west America, perceived as a more conservative part of the nation'.

The opening and closing ceremonies of the past three games I attended took place in large popular sporting venues that held the 30–40,000 athletes and spectators. Aussie Stadium in Sydney (home for Australian rugby), Soldier Field (the US's National Football League's Chicago Bears), Wrigley Field (the US's Major League Baseball's Chicago Cubs) and Rhein Energie Stadion (Bundesliga's Cologne football club) all hosted these events. On one level, these well-known stadiums were the only spaces large enough to accommodate all athletes and spectators. However, there appears to be an ironic challenge to conservative sporting institutions when the Gay Games take place in sites well known for their association with professional hyper-masculinity as typified by rugby, American football, baseball and soccer respectively (see Pronger, 1990; Messner, 1992; Allain, 2008; Anderson, 2010).

While there was some significant coverage of the games' ceremonies and participants by the local media in Sydney, Chicago and Cologne, few people in the wider population outside of the host city would likely know the games were taking place, especially on a more regional or national scale. Besides the more visible opening and closing festivities, most of the sport at the Gay Games occurs in small- or medium-sized venues such as schools, community leisure centres and public parks and courts. It is likely many people casually passing by would not be aware that LGBT athletes were playing sport. However, I talked to an older local woman in Cologne who came to the tennis courts and wanted to see 'what it was all about, as I heard the Gay Games were taking place in my city'. Many local people on the public transit in Chicago and Cologne would see my participation badge and athletic gear and would engage in conversation about sport and the games, so there was some awareness of the event in the host cities. Despite these instances of local knowledge and engagement, alternative sport tourism events, like the Gay Games, may not necessarily fully confront the dominant structure of sport to a great degree because they largely occur in spaces outside of the mainstream institutions. This means the event is organised by the little-known Federation of Gay Games, as opposed to an institution such as the International Olympic Committee.

A continuing and future struggle for the Gay Games movement is how to make them more visible and how to translate this visibility into increased social awareness and attitudinal changes in the wider population. Naismith says the FGG needs to work at making the games better known, but it is a challenge. He believes the best way forward to promote the games is by having representation in local, regional, national and international sport groups and at non-gay specific sport conferences. Naismith states it is up to the individual LGBT athletes and clubs that participate in sport in local mainstream settings to make other people and teams know that lesbians and gay men take part in sport in the first place but also compete in global networks and events like the Gay Games. Until this happens as an everyday occurrence, actual gay sports liberation is partial and conditional, as it results from separation and not integration (Hargreaves, 2000).

Whilst the Gay Games represent a radical break from past conceptions of the role of sport in society (Messner, 1992; Elling et al., 2003; Symons, 2010), they do not appear to represent a major challenge to sport as an institution. This alternative sport tourism event is not overtly challenging existing traditional notions of sporting masculinity. The Gay Games can provide a more nuanced opportunity for gender play, disruption or even resistance to hegemonic traditional conservative masculinity. It is possible that the Gay Games can at times challenge the traditional notion of hegemonic conservative masculinity, but at the same time, reinforce some of the same institutions or practices. To some extent, the athletes travel to the games in various destinations across the globe to assert their cultural power (Pritchard et al., 2000). Yet, to automatically assume that all gay male games athletes contest, modify or challenge heterosexual masculinity – or for that matter, that they all enact the same masculine roles – does not take into consideration a differentiated understanding beyond monolithic concepts of gender within sport. It does not reflect the reality that gay men are as diverse as all other groups.

While an initial and cautious conclusion has been made regarding the Gay Games and hegemonic masculinity, King (2008) advocated for more complex understandings of masculinity, while Caudwell (2011) called for research exploring counter-hegemonic identities. Further, Pringle (2005) argues hegemony theory needs to engage with other paradigms. Therefore queer theory is also deemed to be a relevant framework in order to help understand gay men's masculinities in the context of the Gay Games.

The queering of the Gay Games

At the North American Society for the Sociology of Sport conference in 2003 in Montréal, a gay academic argued that the Gay Games should challenge conservative notions of sport that are dominant in mainstream settings. She stated how events at the games should be organised and be more 'queer', boldly commenting, 'Wouldn't it be great if women and men ran the 100 meters together in high heel shoes and dresses!' Some attending academics questioned why the Gay Games were competitive in nature and were modelled on the Olympics' practice of awarding medals to athletes. Needless to say, these issues raised much debate and little agreement among conference participants. While the queering of the games and other gay sport events has been relatively quiet since, the concept again reared its head at the International Sociology of Sport Association World Congress in Havana in 2011. A speaker renewed the call for current and future gay sport events to be 'queered' and to disrupt traditional forms of sport. Thus the phenomenon of the Gay Games prompts a reconsideration of these 'queer' debates and contestations.

Gay men and lesbians gained a number of limited legal and political rights in the 1980s in some developed countries. Further, they became more accepted by wider society, advocated partly via a political strategy associated with how similar and 'normal' they were compared to heterosexual people. Duggan (2003) identified this trend as evidence of the new homonormativity, a politics that did not contest dominant heteronormative assumptions and institutions but upheld and sustained them. This upset some academics such as Pronger (2000), who wanted to stress how gay people were not similar to straight people and urged a celebration of sexual divergence. Queer theory thus built itself on social constructionism to further dismantle sexual identities and categories and took on the form of a more radical politics of difference (Irvine, 1998). The use of the term 'queer' has caused debate and controversy within the gay community and among some academics (see Caudwell, 2006), but it has been useful in terms of rethinking identity. Queer theory thus does not offer just a view on sexuality or gender, but it suggests all people can potentially reinvent their identity through themselves.

Queer theory has some relevance for sport tourism events. Pronger (2000) asserted the main goal of gay sport is the inclusion of people marginalised due to their sexual identities, as opposed to the disruption or transformation of the foundations of sport as a bodily practice. He further argued gay sports events, like the Gay Games, have not pursued a radical strategy of liberation that would see gay engagement with sport as an opportunity to transform sport's cultural conservatism. Many gay sports groups have gone out of their way to stress inclusiveness, regardless of ability, age, gender, race and sexual orientation. Indeed, this is part of the mission of the Gay Games, which has stressed friendship over the sexuality and sexual desire of the event. In other words, the games have been criticised for not promoting that many athletes will meet others for sexual gratification during the event. Thus, this could be seen as the erasure of lesbian and gay difference in the pursuit of sporting legitimation, because the games' organisers are trying to primarily promote the sport above potential sexual experiences.

One can see that gay sport and related tourism events thus help to prove the 'normality' of their athletes. Pronger (2000) declared gay sport has missed an opportunity to refigure the construction of sport as a conservative culture of desire, which could in turn contribute to the critique and transformation of oppressions that are perpetuated by conservative political perspectives more generally. Visser (2008: 1345) contends 'current debates largely approach the homonormalisation of leisure space from an unhelpful gay/queer-disempowered perspective which is inadequate to explain the development of a range of spaces in which gay/queer people seek leisure'. He further states 'homonormalised spaces are far more than heteronormativity infiltrating the gay world through a range of consumption-led processes and events (such as the Gay Games), or gay male capitulation to such normative hegemonies'. Davidson (2006) and Sykes (2006) are also critical of the Gay Games because they do not radically challenge traditional notions of sport, and indeed they are mirrored on conservative and heteronormative models of competitive events like the Olympics.

Naismith acknowledges these valid criticisms; however, he says the Gay Games must be commercially attractive and viable – otherwise the event simply would not take place. He asserts the vast majority of athletes would not register for the games if they provided different or 'queer' types of sport. However, he does believe the games are challenging to a certain degree and have 'pushed the boundaries'. They offer men's synchronised swimming, which is not provided by the Olympics. The games provide numerous events for same-sex couples in dancing and figure skating. While Naismith hopes sport played at the Gay Games may resonate with mainstream sporting institutions, he realises the games cannot at the same time reject the competitive and conservative aspects of mainstream sports in which gay athletes want to participate. He believes some 'queering' of events occurs within the sport at the games, and I would agree, based on a few instances that I have seen.

In Sydney and Cologne, I witnessed some male volleyball and tennis players wearing short skirts and wigs, although this did not seem to undermine their skill level or competitive desire to win. The 'Pink Flamingo' race is a traditional, although non-medal, event that takes place in the pool at each games. Participants wear pink flamingo hats whilst swimming and sharing strokes, and teams appear in drag before the event. Davidson (2006) acknowledges this camp extravaganza as an expression of gay pride, one that may challenge traditional notions of sport. Wellard (2006) and Jarvis (2006) see the potential of 'queer' when applied to sport but note a wide gulf between the theoretical claims of academic queer theory and the real or lived practices of those who take part in sport events. I have spoken to many gay male athletes at recent games and asked them, via casual conversations, if they were interested in running the 100 meters in drag and high heels, and they all stated clearly that they would hate it if the games moved in that direction. Naismith notes many newer and younger athletes registering for their first Gay Games never state a demand for 'queer' sports and want to play their sport to internationally recognised rules and conventions. As Hughes (2006) noted, many lesbian and gay festivals have lost their radical political roots and are more about celebration, commercialisation and tourist attraction. This seems to have much resonance when applied to the Gay Games.

While the Gay Games tries to promote inclusion and diversity, and allows men and women to compete against each other in some events, the binaries associated with gender (man/woman) causes much conflict. The FGG has a detailed policy on gender participation and ideally wants athletes to simply self-identify as their legal gender. However, Naismith has stated that the games have a number of transgendered and intersex athletes, and some women participants are concerned that those female athletes who were formally men have an unfair advantage. The FGG is aware of this complex issue and does take some steps in ensuring parity among competitors by asking for legal proof of athletes' gender. These issues associated with transgenderism and intersex, however, only represent a few cases for each games held. The FGG does not want to mirror the Olympics' policy of gender and biological sex testing.

While queer advocates (Pronger, 1990; Sykes, 2006) would welcome all athletes to be able to compete against each other regardless of their biological sex identity, the FGG again realises the practicality of offering a sport event that appeals to the vast majority of participants. Naismith believes in the queer potential of the games, as 'much gay or queer stuff takes place off of the field or court but that is where it should remain.' As seen, the Gay Games provide some opportunities for queer resistance, but this can lead to increased opposition, not only by other gay athletes, but by the wider public – instead of bringing greater acceptance of the LGBT community. Overall, despite some obvious queer instances of disruption, gay men's participation at the Gay Games largely conforms to traditional sporting structures and practices that help perpetuate Connell's (1995) notion of hegemonic masculinity.

Conclusion

There is little doubt that the considerable growth of gay sport tourism events like the Gay Games over the past few decades signifies steady progress for mobile sexual minorities in the arena of physical activity. On a very basic level, gay men taking part in sport events in general can be viewed as confronting traditional notions of sport. However, it is also clear that an alternative space like the Gay Games has many conflicts and contestations that need to be reconciled. These were acknowledged by not only Naismith, but among academics (see Pronger, 2000; Sykes, 2006; Symons, 2010; Davidson, 2012) and by some participant or potential athletes who may have been be turned off by some aspects of the games and thus did not travel to the event and take part.

The previous sections on hegemonic and queer debates demonstrate that both frameworks are useful in interpreting the actions and voices of gay male athletes and the organisers of the games. The mobility of the international sport tourists taking part at the destinations gives these spaces meaning. Thus, overall, the Gay Games do not overtly challenge hegemonic masculinity and traditional sporting institutions as a whole, but they provide some opportunities to subtly contest and disrupt these conservative ideological notions through some of their queer embodied practices.

Either way, gay male athletes, as well as lesbians, bisexual and transgendered people, have challenged pervasive views about sport. The increasing number of gay sport clubs and the popularity of global sport tourism events, such as the Gay Games, clearly illustrates that sport does not necessarily overcome social differences but sometimes provides a forum for their reinforcement. Particularly for men, gay sport events provide an opportunity to relax away from wider societal pressures and allow them to assert their cultural power as they travel around the world. Sporting success and/or participation can significantly contribute to a general process of destigmatisation, particularly for a group in society that is usually considered to be inferior or subordinate. While there are obvious positive aspects attached to the games, this chapter has highlighted the debate related to the contentious issue of masculinity. Any sociocultural phenomenon like the Gay Games is open to many readings, but this chapter has presented a selective account of the seminal masculinity debates. The FGG, as well as the key informant for this chapter, Marc Naismith, are required to play to multiple constituencies, seeking financial support from corporations and governments, but also from the LGBT sport and wider community by emphasising its economic desirability and political roots and, as much as possible, its nonconfrontational position that underpins some of the fundamental principles attached to the games.

Some debate centres on the future format of the Gay Games, which attempt to steer a non-defiant path as much as possible. Critics such as Pronger (2000) and Sykes (2006) state the games have lost their political edge and have forgotten their original reason for being and instead are now just a highly commercialised event, based on conspicuous lifestyles. Further, in a time of decreasing cultural homophobia (Anderson, 2009; 2010) is there a continued need for the Gay Games? Many people learning about them for the first time wonder about this, and Naismith says has been asked this question many times. He states, even in an ideal world without homophobia, there is still a fundamental need for the Gay Games because sport is a social activity, it is part of people's social identity and practice, and thus it is a place where LGBT athletes can meet one another. He adds, 'It is space where lesbians and gay men can meet others and share experiences, and the Gay Games provide an opportunity other than online dating sites and gay bars...Gay people need a variety of venues to share interests, and the games provide that.' For the participants, the Gay Games are a public forum for the celebration and display of gay culture with particular emphasis on conspicuous lifestyle statements. Thus, the games are still politically relevant, and although they are only subtly challenging traditional notions of hegemonic sporting masculinity, they remain a significant sport tourism event in the lives of many LGBT persons across the globe.

Finally, this chapter demonstrates how contemporary mobility and diverse engagements in alternative sport events and spaces, like the Gay Games, are such that the intersection of the two has become an important area for research. The games show how sport tourism mobility can be a force for social change. The purpose of the chapter was mainly conceptual, proposing a theoretical framework that could be adopted to guide future studies in the area. It has attempted to critically explore wider manifestations of sport-related tourism and mobility and to provide a theoretically informed insight into the study of tourism and sport in relation to hegemonic masculinity and queer debates.

References

Allain, K. A. (2008) 'Real fast and tough: The construction of Canadian hockey masculinity', *Sociology of Sport Journal,* 25(4): 462–81.

Anderson, E. (2009) *Inclusive Masculinity: The Changing Nature of Masculinities.* New York: Routledge.

Anderson, E. (2010) 'Inclusive masculinity theory and the gendered politics of men's rugby', *Journal of Gender Studies,* 19(3): 249–61.
Beasley, C. (2008) 'Rethinking hegemonic masculinity in a globalizing world', *Men and Masculinities,* 11(1): 86–103.
Casey, M. E. (2009) 'Tourist gay(ze) or transnational sex: Australian gay men's holiday desires', *Leisure Studies,* 28(2): 157–72.
Caudwell, J. (ed.) (2006) *Sport, Sexualities and Queer/Theory*. London: Routledge.
Caudwell, J. (2011) 'Does your boyfriend know you're here? The spatiality of homophobia in men's football culture in the UK', *Leisure Studies,* 30(2): 123–38.
Connell, R. W. (1990) 'An iron man: The body and some contradictions of hegemonic masculinity', in M. Messner and D. Sabo (eds) *Sport, Men and the Gender Order*. Champaign, IL: Human Kinetics, pp. 83–114.
Connell, R. W. (1995) *Masculinities*. Berkeley: University of California Press.
Connell, R. (2008) 'Masculinity construction and sports in boys' education: A framework for thinking about the issue', *Sport, Education and Society,* 13(2): 131–45.
Davidson, J. (2006) 'The necessity of queer shame for gay pride: The Gay Games and cultural events', in J. Caudwell (ed.) *Sport, Sexualities and Queer/Theory*. London: Routledge, pp. 90–105.
Davidson, J. (2012) 'Racism against the abnormal? The twentieth-century Gay Games, biopower and the emergence of homonational sport', *Leisure Studies,* 16(4): 1–22.
Duggan, L. (2003) *The Twilight of Equality? Neoliberalism, Cultural Politics and the Attack on American Democracy*. Boston: Beacon Press.
Elling, A., Knoppers, A. and de Knop, P. (2001) 'The social integrative meaning of sport: A critical and comparative analysis of policy and practice in the Netherlands', *Sociology of Sport Journal*, 18(4): 414–34.
Elling, A., Knoppers, A. and de Knop, P. (2003) 'Gay/Lesbian sport clubs and events: Places of homo-social bonding and cultural resistance?', *International Review for the Sociology of Sport,* 38: 441–56.
FGG (2012a) *Mission, vision and values,* Online. Available at: http://www.gaygames.net/index.php?id=56 (accessed 15 May 2012).
FGG (2012b) *History of the FGG and the Gay Games,* Online. Available at: http://www.gaygames.net/index.php?id=28 (accessed 15 May 2012).
Hargreaves, J. A. (2000) *Heroines of Sport: The Politics of Difference and Identity*. London: Routledge.
Higham, J. and Hinch, T. (2009) *Sport and Tourism: Globalization, Mobility and Identity*. Oxford: Butterworth-Heinemann.
Howson, R. (2008) 'Hegemonic masculinity in the theory of hegemony, a brief response to Christine Beasley's rethinking hegemonic masculinity in a globalizing world', *Men and Masculinities,* 11(1): 109–13.
Hughes, H. (2006) 'Gay and lesbian festivals: Tourism in the change from politics to party', in D. Picard and M. Robinson (eds) *Festivals, Tourism and Social Change*. Clevedon: Channel View Publications, pp. 238–54.
Hughes, H. and Deutsche, R. (2010) 'Holidays of older gay men: Age or sexual orientation as decisive factors?', *Tourism Management,* 31(4): 454–63.
Irvine, J. (1998) 'A place in the rainbow: Theorizing lesbian and gay culture', in P. Nardi and B. Schneider (eds) *Social Perspectives in Lesbian and Gay Studies: A Reader*, London: Routledge, pp. 573–88.

Jarvis N. (2006) 'Ten men out: Gay sporting masculinities in softball', in J. Caudwell (ed.) *Sport, Sexualities and Queer/Theory*. London: Routledge, pp. 62–75.
Jones, L. and McCarthy, M. (2010) 'Mapping the landscape of gay men's football', *Leisure Studies*, 29(2): 161–73.
King, S. (2008) 'What's queer about (queer) sport sociology now? A review essay', *Sociology of Sport Journal*, 25: 419–42.
Maguire, J. (1999) *Global Sport*. Oxford: Blackwell Publishers.
Messner, M. (1992) *Power at Play: Sports and the Problem of Masculinity*. Boston: Beacon Press.
Pringle, R. (2005) 'Masculinities, sport and power', *Journal of Sport & Social Issues*, 29(3): 256–78.
Pritchard, A., Morgan, N., Sedgley, D., Khan, E. and Jenkins, A. (2000) 'Sexuality and holiday choices: Conversations with gay and lesbian tourists', *Leisure Studies*, 19: 267–82.
Pronger, B. (1990) *The Arena of Masculinity*. Toronto: University of Toronto Press.
Pronger, B. (2000) 'Homosexuality and sport: Who's winning?', in J. M. McKay, M. Messner and D. Sabo (eds) *Masculinities, Gender Relations and Sport*. London: Sage, pp. 222–45.
Robertson, S. (2003) 'If I let a goal in, I'll get beat up: Contradictions in masculinity, sport and health', *Health Education Research*, 18: 706–16.
Sykes, H. (2006) 'Queering theories of sexuality in sport studies', in J. Caudwell (ed.) *Sport, Sexualities and Queer/Theory*. London: Routledge, pp.13–32.
Symons, C. (2010) *The Gay Games: A History*. London: Routledge.
Theberge, N. (2000) 'Gender and sport', in J. Coakley and E. Dunning (eds) *Handbook of Sports Studies*. London: Sage, pp. 322–33.
Urry, J. and Sheller, M. (2004) *Tourism Mobilities: Places to Stay, Places in Play*. London: Routledge.
Visser, G. (2008) 'The homonormalisation of white heterosexual leisure space in Bloemfontein, South Africa', *Geoforum*, 39: 1344–58.
Waitt, G. (2003) 'Gay games: Performing "community" out from the closet of the locker room', *Social & Cultural Geography*, 4(2): 167–83.
Wellard, I. (2006) 'Exploring the limits of queer and sport: Gay men playing tennis', in J. Caudwell (ed.) *Sport, Sexualities and Queer/Theory*. London: Routledge, pp. 76–89.
Williams, R. (1961) *The Long Revolution*. Harmondsworth: Penguin.

Part II

Masculinities, Tourism and Identity

6

Working-Class Men's Masculinities on the Spanish *Costas*: Watching ITV's *Benidorm*

Mark Casey

Popular images of working-class tourism on the Spanish *costas* (coasts), as presented in ITV1's successful comedy *Benidorm* and in the wider media, have represented the typical British tourist through images of sunburnt white bodies. Such bodies are often visibly marked as heterosexual, working class and gender appropriate in behaviour, dress and style. This imagery is underpinned through hegemonic notions of the masculine or feminine working-class body at play. The recent work of Andrews (2005, 2011a/b) has been some of the first to theorise the mass working-class tourist and their gendered identities in mass tourist resorts in Spain. Imagery used in *Benidorm* and the experiences of Andrews (2010) during her research, reflect that tourism is an embodied leisure activity that is built and sustained through human relations, which are affected by both global and local gender relations. Such imagery reminds the viewer or tourist that there are expected ways of 'doing gender' on holiday, which are informed by notions of how one 'does' masculinity or femininity back home (also see chapters by Costa and by Gilli and Ruspini, this volume). Although tourism may offer some tourists a liminoid state away from home, where daily realities can be forgotten (see Shaw and Williams, 2004), the centrality of gender to contemporary understandings of self and identity has the consequence that gender roles, expectations and performance are sustained by or even intensified within tourist sites and spaces. For example, both women and men face very different societal pressures of how they should prepare their bodies for partial nudity on the beach and just how much of their bodies they are able to display whilst on holiday (Obrador Pons, 2007; Small, 2007).

The daily embodied experience of masculinity for men (and women), and the worth and value given to it for men's identities, has the consequence that understandings of masculinity have to engage with everyday quotidian spaces 'at home'. However this chapter argues that understandings of masculinities must also theorise how everyday 'foreign spaces and sites' that are experienced whilst on holiday inform and are informed by men and the multiple ways in which they do masculinity. Multiple tourist sites require a plurality of performances of masculinity, bringing various masculinities into contact, and at times into conflict (see Ward, this volume). Gender is critical to the construction and experience of tourist sites, attractions, landmarks, spaces and places. Tourism itself has been created, sustained and consumed within societies that are gendered. Because of this, it follows that 'tourism processes are gendered in their construction, presentation and consumption' (Pritchard and Morgan, 2000a: 116). As Wearing and Wearing remind us, tourism involves gendered tourists, gendered hosts, gendered tourism marketing and gendered tourism objects that each reveal power differences between women and men, and privilege the (heterosexual) male gaze above others (1996: 231).

One of the key thinkers on tourism, John Urry (1990) has been critiqued for the privileging of the male gaze within his writing (Pritchard and Morgan, 2000b). In particular Urry's 'tourist gaze', although initially very influential in thinking, can be understood as problematic in a number of assumptions it makes that appear to be underpinned through a centrality of (heterosexual) male values and men's relationships with the everyday and with tourism. The centrality given to the world of paid employment within the tourist gaze centres the realities of middle-class men as defining which tourism practices and sites should be valued. Through the importance given to paid employment for men's identities and masculinity (Connell, 2002), the positioning of work as 'the other' to tourism centres male values and norms within this binary relationship. As Urry (2002: 2) suggests, 'tourism is a leisure activity that presupposes its opposite, namely regulated and organised work'. He goes on to claim that 'it is one manifestation of how work and leisure are organised as separate and regulated spheres of social practice in "modern" societies' (p. 2). In contrast, the world of unpaid domestic chores or childcare that is often the domain of women (Silver and Goldsheider, 1994) is not positioned as 'the other' to tourism and touristic practice (see Costa, this volume). Even with new technologies allowing paid work to creep into holiday time (BBCNews.co.uk, 22.08.11), the world of paid employment is still likely to stay at a geographical distance at least. However, the reality of childcare, cooking and cleaning will follow most women with children across the globe on their travels.

Urry's 'gaze' has also witnessed critique for not examining embodied experiences of tourism, and how these are likely to be different for men and women (Pritchard et al., 2007). As both Edwards (2006) and Ahmed (2007) have suggested, the study of masculinities in tourism studies has not only been absent, but tourism studies have taken a white, middle-class, heterosexual, hegemonic position. This position has assumed what identities, lifestyles and ways of doing tourism require explanation or investigation by researchers, and in turn, which ones require little or no explanation at all.

This chapter will first discuss the resort of Benidorm before outlining the television show itself and the key characters that can be found within it. The chapter then draws on a number of these characters to examine how multiple masculinities are presented within the show that allow the viewer to witness the diverse ways that working-class men are portrayed as 'doing masculinity' when on holiday by the popular media. The chapter will conclude by suggesting that more work needs to be undertaken on the intersection of social class, masculinity and tourism.

Benidorm

Benidorm and the wider coastal areas of Spain attract millions of tourists in search of low-cost tourism. The arrival of mass tourism onto the Spanish coasts during the 1950s brought overseas travel to both the working- and middle-class masses in the UK, and was to transform the lives of Spanish citizens and the wider culture of Spain. In 1951, there were around 1.2 million overseas visitors to Spain exploring a relatively 'undeveloped' country that was slowly recovering from its civil war, one in which the Franco regime emerged successful (Turner and Ash, 1975). During the 1950s, in what was then the 'Costa Azul', the small fishing village of Benidorm sat within its large sweeping bay, surrounded by farmland and the homes of fishermen (Tremlett, 2006). The then-mayor of Benidorm, Pedro Zaragoza, was to become the father of modern tourism in Spain. It was his drive to develop Benidorm into a tourist destination (underpinned through the personal support of General Franco himself) that transformed this small fishing village into one of the largest tourist resorts in the world (Tremlett, 2006).

The latent resources of Benidorm, from its coves, beaches and its 'old town', tied together with the very low-cost of living for foreign visitors and three hundred days of sunshine, created a perfect setting for the new tourist (Busby and Meethan, 2008). The vision of Zaragoza was central in the creation of the Benidorm of today. Not only did the 'Costa Azul' become the 'Costa Blanca' to appeal more to foreign tourists, Zaragoza gave support for the easy planning of tourist accommodation. Developers were quick to build new properties on the beach's edge, before moving inland and turning orange groves into sites of swimming pools, bars and palm trees. Benidorm has a relatively small resident population of just 50,000 inhabitants, yet by the turn of this century, the town had more than half as many hotels as New York. Benidorm and its immediate environment receive three times more tourists than Cuba, and half as many as visit the whole of Greece (Modrego et al., 2000). The buildings that crowd the skyline of Benidorm are present in the opening credits of *Benidorm*, a familiar sight to millions of tourists. The imagery reminds the viewer that Benidorm is similar in size to a city. As I have suggested elsewhere (Casey, 2013), the crowded skyline of Benidorm presents an image of an unsophisticated high-rise city. Indeed throughout the television show the hotel's rooms are presented as cramped, dated, dirty, hot, 'foreign' and uncomfortable. As O'Reilly (2000: 19) observes, the media represents such mass tourist resorts as portraying 'pollution, decay, [and] overcrowding'.

Benidorm continues to be popular but with working-class, low-income and low-spending tourists, who seek out 'others like them'. For example, a one-week holiday in Benidorm can be purchased in the UK for as little as £121 per person (Travelsupermarket.com, 2013), which is relatively accessible even for those on the UK's minimum wage of £6.31 per hour (BBCNews.co.uk, 15.04.13). However, as Sheerin (2011) notes, Benidorm has become something of a 'nightmare resort' due to its large geographical spread. The looming, decaying figures of some of the earlier high-rise hotels, tied in with overcrowded beaches, drunken tourists and litter presents the resort as rather downmarket (Ritchie, 1993; Robinson, 1996). The resort's latent resources of the 1950s, from its beautiful cove, sandy beaches, unspoilt waters and village atmosphere can no longer be utilised to secure higher-spending middle-class tourists in search of the 'authentic Spain'. It is questionable if Benidorm's recent attempts to regenerate its seafront and attract more upmarket hotels have been successful. For Tremlett (2006), the Benidorm of the early twenty-first century has echoes of the British working-class resort of Blackpool (see Webb, 2005), a common claim made by one of the key characters on *Benidorm*, Mick Garvey: 'It's Blackpool with sun'.

In a resort like Benidorm, it is difficult to avoid 'others like you',[1] with its crowded streets and beaches allowing working-class British tourists to meet other working-class British tourists, creating a sense of belonging and British ownership both in the resort and in the show. These claims have been echoed in work by Andrews (2005, 2010, 2011a/b) on the Mallorcan resort of Magaluf. Due to Magaluf's appeal to working-class British tourists, the British press and middle-class travellers alike ridicule it, earning the nicknames 'Megaruff' and 'Shagaluf'.[2] Similar complaints have been made against Benidorm, which has become 'a huge joke', a place the British press go to sneer at the working-class at play (Tremlett, 2006: 109).

Laughing with *Benidorm*?

The TV show, *Benidorm*, first aired on British television in 2007. The show follows a tradition of other fictitious British television programmes and films set in and around the Spanish *costas* (see Casey, 2010a). The television show, created by Derren Litten, secured high ratings, being aired at prime time on Friday evenings on British terrestrial channel *ITV1*. Its success was mirrored in the increase in running time of each episode from 30 minutes in its first season to an hour in seasons two to six. For Beeton (2005) and Kim et al. (2009), television series that run for several seasons, such as *Benidorm*, allow viewers to develop strong attachments to characters and locations; they often want to visit filming locations themselves. Both the inside and outside filming of *Benidorm* is undertaken on location within a real hotel (given the fictional name 'the Solanas Hotel').

The setting and filming of the show in Spain may have less to do with a celebration of the 'exotic' or the difference Spain may offer to the viewer, and more to do with its familiarity. Both the show and resort offer familiar elements of low-cost package holidays, from arrival on the low-cost airline easyJet, high-rise/low-cost hotels, crowded beaches, all-inclusive food and drink that is 'not too foreign', nightly British-themed entertainment, and of course, the Spanish sun. As I have argued elsewhere (2010b), the advent of low-cost air travel in Europe has made the distant and exotic increasingly familiar.

Tourism and the draw of (once) exotic locations have had an important place in British television history. Earlier travel programmes such as *Wickers World* (1959–90), *Wish You Were Here?* (1974–2003) and *Holiday* (1969–2007) have been theorised as prioritising a scopic approach that has privileged sightseeing over other elements of how people 'do' holidays. More recent reality television programmes such as *Ibiza Uncovered* (2004–7), *Geordie Shore: Magaluf Madness* (2011), *Benidorm ER* (2011–),

Holiday ER (2012–) and *Sun, Sex and Suspicious Parents* (2011–), along with the fictional *Benidorm*, have shifted the focus onto the mass package holiday and the working-class tourist body at play. Both positive and negative embodied experiences of working-class package tourism are presented for the viewer to enjoy, from characters getting a tan, binge drinking, having sex, fighting, getting food poisoning or breaking a bone (also see parallels with Ward and Treadwell, this volume). Such television may remind the viewer that 'the world is grasped through the body' (Crouch, 2002: 217), but for Skeggs (2009) reality television can present working-class bodies as deficient, lacking, inadequate and requiring improvement. The middle-class viewer can tune into such programmes whilst gazing upon a distant and somewhat problematic 'other', who is safely removed from their own quotidian realities.

Central to the show is the Garvey family, who appear to fit the stereotype of the British urban underclass (Murray, 1996; Jones, 2011). This characterisation draws on imagery in which the Garveys rely on state benefits, have little intention of working and have a 15-year-old daughter who has had a baby by an unknown father (see Willis, 1977; Jones, 2011; Sparrow, 2011). In series one, the father of the family, Mick, funds the family holiday to Benidorm from illegally gained state benefits. As the series develops, Mick is presented as a big-hearted husband and father who often utilises bravado and comedy in his attempts to mask his inability to hold down secure employment. Mick is out of time and place in an economy that is witnessing the loss of low-paid manual jobs, with a continued shift to service sector employment. Within the show, he willingly undertakes the role of the primary entertainer for his children (Chantelle and Michael), whilst his wife, Janice, tries to manage her unruly mother (Madge) and her offensive views.

Schänzel and Smith (2011) have shown how fathers often take on the primary responsibility of children's entertainer in holiday settings, and also facilitate the mother's activities and interests. Mick's actions within the series vary: benefit fraud, long-term unemployment, defending and being supportive of his daughter's unexpected pregnancy, running a successful business and then losing the business and all its inherited wealth. His take on the world is reminiscent of Willis's working-class lads in his study, *Learning to Labour* (1977). Mick, as with Willis' lads, is aware that he is on the periphery of society, with bravado and humour empowering him in his management of his daily struggles. As with Mac an Ghaill's (2000) more recent study, there is now very little 'shop floor' work left in the UK, in which Mick's humour and skills could have a place. Instead, he finds temporary use, value and belonging in the working-class setting of the Solanas Hotel.

Initially, the middle-class couple, 'Martin and Kate', serve as a stark contrast to the Garvey family. Through their desire to access 'authentic Spain' away from Benidorm, they are marked out as 'the outsiders'. Their discomfort about having to stay at the Solanas Hotel presents them out as the 'uninvited other', their middle-class values spoiling the fun of the working-class tourists. Although middle-class, Martin, like Mick, is presented as an example of failing manhood in the early twenty-first century (Hayward and Mac an Ghaill, 2003). Both characters fail to live up to hegemonic notions of masculinity, with key women in their lives reminding them of their failures *as men*. Throughout each season, Mick has to manage Madge's comments that he has 'failed as a man, husband and father' through his inability to provide financially for his family. For Martin, his failure to provide a middle-class holiday in Spain for Kate, as well as his inability to make her pregnant due to his very low sperm count, are utilised to present Martin as a weak man. The ability for men to father children, and thus indicate to other men that they have a 'normal' sperm count and are not homosexual (through their sexual encounters with women), serves as a key indicator of perceived masculinity and virility (Daniels, 2006; Schänzel and Smith, 2011). For Martin, his low sperm count suggests that he is not a 'real man'. His character is presented as subordinate to his wife (season one, episode three), the waiter Mateo – with whom his wife has a sexual encounter (season one, episode three), his girlfriend, Brandy (season three) and his mother (season three). Through Martin's inability to make claims to markers of manliness and hegemonic masculinity, he is presented as subordinate to female demands. As Flood (2002) asserts, it is deterministic to assume that the hegemonic form of masculinity always works in a manner that 'guarantees male dominance over females'.

Geoff Maltby is a regular visitor to Benidorm and the Solanas Hotel. Like both Mick and Martin, Geoff is presented as an example of a failing man – an unmarried virgin, undertaking his holidays with his retired mum, Noreen, with whom he lives (despite being in his mid-30s). Key to Geoff's failure in his manhood is his lack of clear employment 'back home', and initial inability to secure a girlfriend or lose his virginity. As Martin (2010) has shown, for many heterosexual men, the loss of their virginity and sexual experience with women is often utilised to make claims to being a 'real man' and dominant notions of heterosexual masculinity. In season three, Geoff attempts to lose his virginity, which is presented as the character wishing to claim a sexually successful heterosexual male identity. However, his efforts at Internet dating are presented as desperate and fail when he meets a transvestite named Lesley. His discomfort at meeting Lesley is in stark contrast to Lesley's own comfort as a transvestite living in Benidorm's 'old town'. After this experience, Geoff's mother and other holidaymakers at the Solanas Hotel assume he's gay. When they throw a 'coming out party' for Geoff (one in which his mother and other characters dress up as The Village People and perform 'YMCA'), he is horrified at the public assumption that he is gay. The actions of his mother and other tourists remind the viewer how tourists not only gaze at tourist sites, but can and do 'gaze' upon each other's actions and listen into other's conversations and holiday experiences across the pool. In seasons four and five, Geoff's character is no longer present on the show, although Noreen is. Geoff's absence is attributed to his becoming a father, which presents him as moving on in his independence as a man and economically (there is little need to holiday with his mother in a low-cost resort).

Although it is possible to theorise heterosexual sexual activity and its visibility through fatherhood as an indicator of successful male identities (Mallon, 2004), limits exist as to who can make claims to such notions of masculinity. For Taylor (2010), sexual virility and activity is often theorised as belonging to the young – older generations are presented as either non-sexual or sexually threatening beings. The increasing use of and access to Viagra, however, is challenging assumptions about the male body, its sexual abilities and ideas of maleness and masculinity in later life (see Daniels, 2006; Li, 2006). On *Benidorm*, Donald and Jacqueline are a white working-class couple in their mid-50s from Northern England. Neither Donald's nor Jacqueline's bloated and pale bodies fit the image of a sexually attractive couple, although both are highly sexually active. Unlike some of the other characters, Donald is presented as a man who is secure in his masculinity and sexuality; fleeting homosexual encounters from his past or present do not trouble his masculinity in his eyes or those of his wife. In fact, by season five, Donald is presented as extremely virile. In his discussions with other characters, he admits he may be the father of dozens of children whom he has never met. His claims and confidence in his virility are at odds with those of Martin. Donald, as with the attractive Spanish bar tender Mateo (see below), is one of the most sexually successful characters in the show. As work on men and travel has shown, the ability to secure sexual experiences when travelling (with tourists or local populations), is presented as key to a good holiday and to many men's identities (see Waitt and Markwell, and Katsulis, this volume). Indeed those gay men in my research who could not secure sexual experiences whilst travelling were understood by other gay men as having an 'inferior' gay male identity and to have failed to experience a successful holiday (Casey, 2009).

Gavin and Troy, a mixed-class couple (Gavin is middle class in his tastes and values and his dislike for Benidorm, while Troy is clearly identifiable as working class, partly through his love for the resort), are presented in stark contrast to Donald and Jacqueline. Both characters challenge earlier media portrayals of gay men as exclusively affluent, handsome, urban and promiscuous. Their limited incomes, imperfect bodies and the love and flaws present in their relationship make Gavin and Troy refreshingly real. In seasons one and two, their arrival at the Solanas Hotel brings them into a dysfunctional heterosexualised space. The hotel and its poolside, although potentially homophobic, are settings in which Gavin and Troy are but one form of the diversity to be found in Benidorm. The characters' ability and desire to leave the confines of the all-inclusive Solanas Hotel and to explore the queer and stylish streets of Benidorm's 'old town' present them as confident and adventurous, wishing to access a more authentic Spain away from the tourist crowds (Curtin, 2010). However, Gavin's continued concern with his weight in each series, particularly after he is called a 'fat puff' by another holidaymaker, presents him as somewhat feminine in his traits and concerns (Davis et al., 2005; Alexiou, 2012). Gavin's concern is in striking contrast to that of the multiple overweight male characters and extras who are present on the show and confident in their shorts, with their large, sunburnt stomachs spilling out. His concern is not presented as something 'a real man' would consider. The heterosexual male working-class body on holiday is presented as a site of excess alcohol and food consumption (see Andrews, 2010), unconcerned with attaining a feminised slim, toned and hairless torso as epitomised by the contemporary male metrosexual (Gill et al., 2005; Simpson, 2011). Although neither Gavin nor Troy fit hegemonic ideals of masculinity, it is through their diverse portrayals in the show – as camp, funny, bitchy, brave, real, caring, independent and so on – that they remind the viewer that masculinities are multiple, social-historic constructions that are performed through diverse and distinct male bodies (Connell, 2002: 29).

Mateo is the only Spanish character regularly on *Benidorm*. His character initially supported the stereotype of Spanish men – handsome, hypersexual and a little bit lazy (see Cleminson, 2004). Mateo is presented as successful not only through his role as a waiter, but as a matador, a flamenco dancer and a lifeguard, which encompass representations of manliness, virility and successful heterosexual masculinity.

As work on Spanish cinema and the Spanish male body has shown (Fouz-Hernandez and Martinez-Exposito, 2007), Spanish men have been presented as virile, hypersexual and attractive to the foreign female gaze. Their looks and bodies are presented as at once threatening and sexually alluring. Although initially Mateo appears to confirm his masculinity and maleness through his successful sexual activity with female tourists and his perfect body, his masculinity and sexuality become a source of questioning for the other characters present. In season one, Mateo is found having sex with Troy (above). This indiscretion is presented as a one-off event, although other characters use the encounter to label him a 'Spanish poof'. As the work of Malam (2008) has shown, male tourists will often perceive the qualities and traits they expect (male) bar staff to possess. If local male bar staff disrupt codes of behaviour (as perceived by other male tourists), conflict can occur. Mateo's and Troy's sexual encounter reminds the viewer that 'real (heterosexual) men' would not allow themselves to engage in sexual activity with other men (Hockey et al., 2007).

Mateo's manliness is further cast into doubt when the infertile Martin beats him at arm wrestling, which develops over Martin's wife, Kate. Their arm wrestling is not only presented as a personal duel, but as a duel between Spain and the UK. Through Martin's win, the UK is presented as superior in its masculinised identity to that of a losing and effeminised Spain (Plain, 2006). In seasons four and five, Mateo is increasingly portrayed as desperate in his sexual quests and threatening through his potential to spread sexually transmitted infections. As each season develops, Mateo is not portrayed as the type of man who possesses traits associated with successful manhood – security, comfort, protection, career development, partnership or fatherhood (Schänzel and Smith, 2011).

Conclusion

Although a fictitious comedy, *Benidorm* and its diverse male characters represent to the viewer the distinct and multiple ways men's identities and masculinity are performed on holiday. In utilising popular images of mass tourism on the Spanish coasts, the show presents known or imagined Benidorm landscape and bodies. As with other similar mass (working-class) tourist resorts in Spain, the male bodies in Benidorm are either 'in place' and belong (such as Mick or Mateo) or they are potentially 'out of place' (such as Martin, Gavin and Troy). Initially, the male working-class heterosexual body is presented as having the most valid claims to a visibility and presence in the high-rise-filled streets of Benidorm. Mick Garvey's inability to work and to provide economic support for his family, along with his general childlike humour, are not presented as problematic or that dissimilar to other tourists present. Although such male traits – once characteristic on the factory floor (Willis, 1977) – are increasingly out of time and place, in Benidorm, they have worth and value, albeit for only a week or two. The presence of middle-class Martin, which is juxtaposed against working-class Mick, at first reflects the idea that socioeconomic factors are not the only ways in which men can fail *as men* in contemporary British society. Martin's low sperm count, and therefore his weakened masculinity, represent the importance of fertility and sexual prowess for male identities. Such limited sexual success is shared by Geoff, whose desire to prove himself to be heterosexual reflects the continued positioning of homosexual identities as a distinct 'other' to heterosexuality. As with the other male characters, Gavin and Troy highlight the diverse ways that men do masculinity, with their gay male identities intersecting with the classed and (hetero)sexualised spaces of the Solanas Hotel. Through his age and his obese body, Donald is physically representative of the stereotypical male working-class British tourist abroad (as portrayed in the programme's opening credits). However, his diverse sexual success and tastes, along with his strong, happy and successful relationship with his wife, allow viewers to see Donald as a man who is secure and comfortable in his masculinity. Finally, there is Mateo, who through his handsome looks and sexual desires is stereotypical of hegemonic notions of (Spanish) masculinity as held by the British. But as with the other male characters, he is flawed in his abilities. Mateo cannot maintain an honest long-term relationship or commit to his career as a waiter. His failing masculinity intersects with his identity as a 'shifty' foreign other, who is sexually dangerous to both women and men. As the diverse characters in the show suggest, men's identities – and the masculinities they make claims to whilst on holiday – must be theorised alongside the other complex and shifting identities they may possess, and the spaces and places where they are performed.

Notes

1. This paper would suggest that the positioning of the Spanish body in the TV show *Benidorm*, in particular the character of Mateo, suggests that it is difficult to meet non-British individuals in the resort. If a Spanish person becomes visible within *Benidorm*, it is not through their claims to a visibility

in a Spanish town as a resident, but through their role as an employee in the tourist industry. In fact, the Spanish body is presented as the Other within Benidorm, out of place in a resort designed and maintained for the British tourist.

2. The term 'Shagaluf' is drawn from the word 'shag', a British colloquialism for sexual intercourse, which invokes uninhibited and gratuitous heterosexual activity in the resort.

References

Ahmed, R. (2007) 'Rethinking masculinity', *The Review of Communication*, 7(2): 203–6.

Alexiou, J. (2012) 'Gay body self-hatred explained', *Out Magazine*, August.

Andrews, H. (2005) 'Feeling at home: Embodying Britishness in Spanish charter tourist resort', *Tourist Studies* 5(3): 247–66.

Andrews, H. (2010) 'Contours of a nation: Being British in Mallorca', in J. Scott and T. Selwyn, (eds) *Thinking Through Tourism* (ASA Monograph). Oxford: Berg Books, pp. 27–50.

Andrews, H. (2011a) 'Porkin' pig goes to Magaluf', *Journal of Material Culture* 16(2): 151–70.

Andrews, H. (2011b) *The British on Holiday: Charter Tourism, Identity and Consumption*. Clevedon: Channel View.

BBCNews.co.uk (2011) *Working on Holiday: Your Views*. Available at: http://www.bbc.co.uk/news/world-14618606 (accessed 22 September 2011).

BBCNews.co.uk (2013) *Minimum Wage to Increase to £6.31*. Available at: http://www.bbc.co.uk/news/business-22153007 (accessed 15 May 2013).

Beeton, S. (2005) *Film-induced Tourism*. Clevedon: Channel View Publications.

Busby, G. and Meethan, K. (2008) 'Cultural capital in Cornwall: Heritage and the visitor', in P. Payton (ed.) *Cornish Studies Sixteen*. University of Exeter Press: Exeter, pp. 146–66.

Casey, M. (2009) 'Tourist gay(ze) or transnational sex: Australian gay men's holiday desires', *Leisure Studies* 28(2): 157–73.

Casey, M. (2010a) *Benidorm*. Presented at the 4th International Selicup Conference, Palma de Mallorca. 22–24 October 2010.

Casey, M. (2010b) 'Low-cost air travel: Welcome aboard?' *Tourist Studies*, 10(2): 175–91.

Casey, M. (2013) 'The working class on holiday: British comedy in Benidorm and classed tourism', in *Journal of Tourism Consumption and Practice*, 5(1): 1–17.

Cleminson, R. M. (2004) 'The significance of the fairy for the cultural archaeology of same-sex male desire in Spain, 1850–1930', *Sexualities*, 7(4): 412–30.

Connell, R. W. (2002) 'Masculinities and globalisation', in H. Worth, A. Paris and L. Allen (eds) *The Life of Brian: Masculinities, Sexualities and Health in New Zealand*. Dunedin: University of Otago Press, pp. 27–42

Crouch, D. (2002) 'Surrounded by place: Embodies encounters', in S. Coleman and M. Crang (eds) *Tourism Between Place and Performance*. Berghahn Books; Oxford, pp. 207–18.

Curtin, S. (2010) 'What makes for memorable wildlife encounters? Revelations from "serious" wildlife tourists', *Journal of Ecotourism*, 9: 149–68.

Daniels, C. (2006) *Exposing Men: The Science and Politics of Male Reproduction*. New York: Oxford University Press.

Davis, C., Karvinen, K., and McCreary, D. R. (2005) 'Personality correlates of a drive for muscularity in young men', *Personality and Individual Differences*, 39: 349–59.

Edwards, T. (2006) *Cultures of Masculinity*. New York: Routledge.

Flood, M. (2002) 'Between men and masculinity: An assessment of the term masculinity in recent scholarship on men', in S. Pearce and V. Muller (eds) *Manning the Next Millennium: Studies in Masculinities*. Perth: Black Swan, pp. 203–13.

Fouz-Hernández, S. and Martínez-Expósito, A. (2007) *Live Flesh: The Male Body in Contemporary Spanish Cinema*. London: I B Tauris.

Gill, R., Henwood, K. and McLean, C. (2005) 'Body projects and the regulation of normative masculinity', *Body & Society*, 11(1): 37–62.

Hockey, J. L., Meah, A. and Robinson, V. (2007) *Mundane Heterosexualities*. Basingstoke: Palgrave MacMillan.

Jones, O. (2011) *Chavs*. London: Verso.

Kim, S., Long, P. and Robinson, M. (2009) 'Small screen, big tourism: The role of popular Korean television dramas in South Korean tourism', *Tourism Geographies* 11(3): 308–33.

Li, J. J. (2006) *Laughing Gas, Viagra to Lipitor: The Human Stories behind the Drugs We Use*. Oxford: Oxford University Press.

Mac an Ghaill, M. (2000) 'New cultures of training: Emerging male (hetero)sexual identities', *British Education Research Journal*, 25 (4): 427–43.

Malam, L. (2008) 'Bodies, beaches and bars: Negotiating heterosexual masculinity in Southern Thailand's tourism industry', *Gender, Place and Culture*, 15(6): 581–94.

Mallon, G. P. (2004) *Gay Men Choosing Parenthood*. New York: Colombia University Press.

Martin, P. (2010) 'These days virginity is just a feeling: Heterosexuality and change in young urban Vietnamese men', in *Culture, Health and Sexuality*, 12(1).

Modrego, F., Domenech, V., Llorens, V., Torner, J. M., Abellan, M. and Manuel, J. R. (2000) 'Locating a large theme park addressed to the tourist market: The case of Benidorm', in *Planning Practice and Research*, 15(4): 385–95.

Murray, C. (1996) *Charles Murray and the Underclass: The Developing Debate*. London: IEA Health and Welfare Unit.

Obrador Pons, P. (2007) 'A haptic geography of the beach: Naked bodies, vision and touch', *Social & Cultural Geography*, 8(1): 123–41.

O'Reilly, K. (2000) *The British on the Costa del Sol*. London: Routledge.

Plain, G. (2006) *John Mills and British Cinema: Masculinity, Identity and Nation*. Edinburgh: Edinburgh University Press.

Pritchard, A. and Morgan, N. J. (2000a) 'Constructing tourism landscapes: Gender, sexuality and space', *Tourism Geographies*, 2(2): 115–39.

Pritchard, A. and Morgan, N. J. (2000b) 'Privileging the male gaze: Gendered tourism landscapes', *Annals of Tourism Research*, 27(2): 884–905.

Pritchard, A., Morgan, N., Ateljevic, I. and Harris, C. (eds) (2007) *Tourism and Gender: Embodiment, Sensuality and Experience*. Wallingford: CABI.

Ritchie, H. (1993) *Here We Go: A Summer on the Costa del Sol*. London: Penguin Books.

Robinson, M. (1996) 'Sustainable tourism', in M. Barke, J. Towner and M. T. Newton, (eds) *Tourism in Spain: Critical Issues*. Wallingford: Cab International, pp. 401–25.

Schänzel, H. A. and Smith, K. A. (2011) 'The absence of fatherhood: Achieving true gender scholarship in family tourism research', *Annals of Leisure Research*, 14(2–3): 129–40.

Shaw, G. and Williams, A. (2004) *Tourism and Tourism Spaces*. London: Sage.

Sheerin, B. (2011) *My Life: A Coach Trip Adventure*. London: Michael O'Mara Books.

Silver, H. and Goldsheider, F. (1994) 'Flexible work and housework: Work and family constraints on women's domestic labour', *Social Forces*, 72(4): 1103–19.

Simpson, M. (2011) *Metrosexy: A 21st Century Love Story*. Amazon.

Skeggs, B. (2009) 'The moral economy of person production: The class relations of self-performance on "reality" television', *The Sociological Review*, 57(4): 626–44.

Small, J. (2007) 'The emergence of the body in holiday accounts of women and girls', in A. Pritchard, N. Morgan, I. Ateljevic and C. Harris (eds) *Tourism and Gender: Embodiment, Sensuality and Experience*. Wallingford: CABI, pp. 73–91.

Sparrow, A. (2011) *England Riots: Cameron and Miliband Speeches and Reaction*, in The Guardin.co.uk, Available at: http://www.guardian.co.uk/politics/blog/2011/aug/15/england-riots-cameron-miliband-speeches (accessed 15 August 2011).

Taylor, J. (2010) 'Queer temporalities and the significance of the "music scene" participation in the social identities of middle aged queers', *Sociology*, 44(5): 893–907.

Taylor, J. S. (2001) 'Dollars are a girl's best friend? Female tourist's sexual behaviour in the Caribbean', *Sociology*, 35(3): 749–64.

Travelsupermarket.com (2013) *Holidays in Spain. Available at:* http://www.travelsupermarket.com/c/holidays/spain/Benidorm/39/ (accessed 23 July 2013).

Tremlett, G. (2006) *Ghosts of Spain: Travels Through a Country's Hidden Past*. London: Faber and Faber Ltd.

Turner, L. and Ash, J. (1975) *The Golden Hordes: International Tourism and the Pleasure Periphery*. London: Constable.

Urry, J. (2002 [1990]) *The Tourist Gaze. Leisure and Travel in Contemporary Societies*. London: Sage.

Wearing, B. and Wearing, S. (1996) 'Refocusing the tourist experience: The flaneur and the choraster', *Leisure Studies*, 15: 229–44.

Webb, D. (2005) 'Bakhtin at the seaside: Utopia, modernity and the carnivalesque', *Theory, Culture and Society*, 22(3): 121–38.

Willis, P. (1977) *Learning to Labour: How Working Class Kids Get Working Class Jobs*. Farnborough: Saxon House.

7

'You Get a Reputation If You're from the Valleys': The Stigmatisation of Place in Young Working-Class Men's Lives

Michael R. M. Ward

Since the 1990s, there has been a large amount of research conducted on both sides of the Atlantic that has looked at the negative implications of living in communities that are 'racially', economically and social marginalised (see Shields, 1991; Campbell, 1993; Walker et al., 1998; Reay, 2000; Williams and Collins, 2001; Watkins and Jacoby, 2007; Wacquant, 2008; Keene and Padilla, 2010; Rhodes, 2012). A lot of this work has examined the consequences that stem from living close to unhealthy environments (see also Bullard, 1999; Phillimore and Moffatt, 1999; Bush et al., 2001; Acevedo-Garcia et al., 2003; Entwisle, 2007), through a lack of access to key community services (Gordon et al., 2004; Watkins and Jacoby, 2007) and through residing in areas suffering high levels of crime, deprivation and unemployment (Walker et al., 1998; Sampson et al., 2002). However, as Wacquant (2007, 2008) has pointed out, due to the deindustrialisation process that has taken place in Western countries over the past four decades, certain communities and neighbourhoods have become more 'socially marginalised' than others. Deindustrialisation has therefore resulted in residents not only becoming physically confined to decaying environments, but these environments also act as symbolic and powerful signifiers of class and 'racial' identities (see also Reay, 2000; Skeggs, 2004; MacDonald et al., 2005; Rhodes, 2012). These characteristics, as Goffman (1963: 3) argues, become 'deeply discrediting', stigmatising those who are attributed with them and reducing them from 'a whole and usual person to a tainted, discounted one'. External perceptions of a specific place, therefore, come to define a whole group of people.

However, what happens when people leave a particular stigmatised place? Some studies have looked at how the identities and reputations of a particular place can accompany individuals when they relocate, or try to relocate, to a new destination (see Hayden, 2000; Galster, 2007; Keene and Padillia, 2010). But there has been very little work within tourism studies and on men and masculinities that has looked at how the reputations of a particular place or community accompany inhabitants when away from that geographical area. Studies have shown how particular characteristics of tourist destinations attract certain labels (Kneafsey, 2000; Andrews, 2005, 2006; Thurnell-Reed, 2011); how reputations about certain holiday resorts emerge (Malam, 2004; Mordue, 2005); and how the spatial order of tourist sites are negotiated (MacCannell, 1976; Edensor, 1998, 2001), but there is little research on how young men experience 'tourist encounters' (Crouch et al., 2001) when they temporarily leave a particular stigmatised place for a tourist destination. In this chapter, I argue that for a group of young working-class men from a community in the South Wales Valleys (UK), their tourist encounters are often tempered by the historic classed and gender codes of the region (accent, perceived behaviours and external stereotypes) that accompany them when away from their community.

This chapter begins by extending Goffman's arguments surrounding stigma, and I link this to young working-class men and social exclusion, before moving on to look at the socioeconomic context of the South Wales Valleys. Drawing on ethnographic material and in-depth interviews, in the next two sections, I closely examine these young men's narratives, investigating what happens when young working-class men leave their community temporarily, and also talking with them about their fantasies of more greater mobility that allows these young men to explore ideas of escape from their mundane everyday lives and from where they live.

Stigma, social exclusion and young working-class masculinities

Stigma is one of the consequences that accompany being defined as different to social standards of normalcy. Goffman (1963) identified three separate types of stigma: abominations of the body such as physical deformities; blemishes of individual character such as mental illness; and the 'tribal stigma' of 'race', nation, religion, class and social milieu. Acting as spatial representations of structural inequalities, marginalised communities are examples that reinforce segregation through popular and political discourses, resulting in the pathologisation of those who reside there (see Campbell, 1993; McDowell, 2003; Wacquant, 2008; Jones, 2011; McDowell, 2012). Those who live in such 'vilified' places are therefore not only marred by the stigma of 'race' and class, but also through what Wacquant (2007: 67) terms the 'blemish of place' (see also Shields, 1991). Here stigmatising traits are attributed to those who inhabit a specific community, and a place comes to define a people (Hayden, 2000). Therefore as a result of such powerful representations, people become labelled when they come from a particular setting of origin (Curtis, 2004; Rhodes, 2012) and their stigmatisation belongs to specific representations of that place. One explanation for this discrediting of place and communities is said to result from the cultural dispositions of those who live there (Wacquant, 2008). It is also attributed to the multiple other stigmas associated with both people and the places they live in, such as poor living conditions, high levels of crime and unemployment, families who are dependent on state benefits and also industrial waste, technological stigma (e.g., nuclear power plants) and air pollution (Bush et al., 2001).

As Shields (1991: 22) has argued, place can have multiple meanings: 'the same place, at one and the same time, can be made to symbolise a whole variety of social statuses, personal conditions and social attitudes'. Shields suggests such place images can come to form a 'place-myth' (1991: 61). This then allows for a specific space, a geographic area, to become an imagined space, socially and culturally symbolised with specific activities and behaviours. But these place-myths emerge not only through symbolism but also social, economic and culture conditions. This place-myth can then accompany individuals who relocate to other areas via images produced through TV, print media, films and advertising.[1] Myths held about a place can encourage certain behaviours and have an impact on tourist encounters (Crouch et al., 2001).

While negative attributes are assigned to those who may live in particular communities, as Wacquant (2008) has pointed out, this 'Othering' has multiple meanings within different contexts and with different individuals. McDowell (2007, 2012) has argued that young working-class men who live in stigmatised places are most often associated with fears of disorder, disrespect and delinquency. Their class backgrounds, their accents and their (often) aggressive performances of masculinity are seen as 'redundant' (McDowell, 2003) in a deindustrialised society (see Willis, 1977; Mac an Ghaill, 1994; Winlow, 2001; Nayak, 2006; Kenway et al., 2006). These more traditional performances of masculinity are particularly disadvantageous to working-class young men in terms of educational success and access to higher education. They are also less likely to move into professional occupations or to find employment in lower paid service sector work because they lack the social and cultural attributes valued by employers in such fields. As Goffman (1963: 9) argues, this results in individuals being 'disqualified from full social acceptance'. Current political and media discourses further support this representation; young working-class men are constructed as lazy, unwilling to work, 'feckless', violent and excessively heterosexual (McDowell, 2012; also see Andrews and Treadwell, this volume). As a result of these powerful representations, such men are deemed to demonstrate a moral, cultural, physical and social threat to an otherwise 'respectable' late modernity. A current example of this 'moral panic' (Cohen, 1972) is symbolised in the UK through the derogatory figure of the 'chav' (see Nayak, 2006, 2009; Jones, 2011 for a further discussion). This chapter examines the process of spatial stigma young working-class men experience temporarily during tourist encounters. I now turn briefly to describe the deprived community and outline the methods used in this study.

Context and methods

In understanding the identities of these young men and why reputation is important, we must first examine the social context. Developing at the end of the nineteenth century to feed the growth in iron manufacturing, the South Wales Valleys were once a major contributor to the British coal industry (Williams, 1985) and one of the largest industrial centres in the country, employing up to a quarter of a million men (Grant, 1991; Rees and Stroud, 2004). A strong division of labour accompanied these communities, where distance from anything seen as 'feminine' was essential for a strong masculine identity, which would enable the communities to survive (Walkerdine, 2010). Men earned respect for working arduously and for 'doing a hard job well and being known for it' (Willis, 1977: 52). Kenway and Kraack (2004) suggest that these roles were often seen as heroic, with punishing physical labour involving different degrees of manual skill and bodily toughness, creating a tough, stoic masculinity. Male camaraderie, which was established through physicality and close working conditions underground, was also developed through jokes, storytelling, sexist language and banter at the work site. This was further supported through institutions such as miners' institutes, chapels, pubs, working men's clubs and sports. Rugby union (and to a lesser extent boxing and football) in particular still hold powerful positions in the culture of the locale: influencing those who play it, those who watch it, those who reject it and those who are deemed unfit for it (Holland and Scourfield, 1998; Howe, 2001; Harris, 2007).

After the Second World War, despite the nationalisation of the industry in 1947, coal mining in the region continued to weaken, and large numbers of collieries were closed. However, during the 1980s and onwards into the 1990s, due to economic restructuring policies of the Conservative government led by Margaret Thatcher, the region underwent rapid deindustrialisation (Williams, 1985; Smith, 1999; Day, 2002) and struggled to reinvent itself in the 'new modernity' (Beck, 1999). This acute collapse, coupled with the decline of the manufacturing industry, led to a drastic increase in economic inactivity (see Brewer, 1999; Fevre, 1999). The area is now characterised by what Adamson (2008: 21) terms a 'triangle of poverty' with low levels of educational attainment and high levels of unemployment, health inequalities and poor housing across the region (see also Winlow, 2001 and Nayak 2003, 2006, for research undertaken in other communities with similar post-industrial issues).

This chapter draws on findings from an Economic and Social Research Council-funded ethnographic study that was conducted over a two-and-a-half year period and looked at the diversity of a group of white, working-class young men within the former industrial town of Cwm Dyffryn[2] situated in the South Wales Valleys. The overall aim was to investigate how masculinities were formed, articulated and negotiated by one school-year group at the end of their compulsory schooling, and then to subsequently follow them through their different post-16 educational pathways. This was conducted in the same school sixth form (ages 16–18) and within other educational institutions that some of the young men opted for after their General Certificate of Secondary Education (GCSEs) qualifications. Nayak (2003: 148) has argued that 'young people's gender identities cannot be adequately comprehended within the microcosm of the school institution alone'. This research was therefore undertaken across multiple other arenas in order to gain what Geertz (1973: 6) called a rich, 'thick description' of their lives, which led to a more meaningful and intricate understanding of how they understood and represented their world.

I personally knew the research area, as I grew up there, so I was able to form close relationships with many of the participants in my study through a shared biographical history. At the time the research was conducted, I was also a Further Education lecturer, so I was familiar with the education system in the area. The head teacher of the high school where much of the research was conducted granted access. The fieldwork included participant observation supported by extensive field notes, focus group interviews, ethnographic conversations and more formally recorded one-on-one interviews over the research period. These interviews were fully transcribed, and along with the detailed field notes, coded using a CAQDAS package for key themes. At school and (as the research period progressed) college fieldwork included observing and actively participating in different lessons, 'hanging around' in the sixth form common room and various canteens during break and dinner times, playing football and Scrabble and attending school events such as prize nights, parents' evenings, school trips and sporting occasions. Outside their different educational institutions (once the young men invited me into these other areas of their lives), I was also able to spend time in a variety of other settings. These included sitting in cars in car parks and driving around the town, attending nights out in pubs and night clubs in the town centre and in the larger cities of South Wales, going to live music events, the cinema, eighteenth birthday parties, cafes and shops, playing computer games and attending university open days and places of work (such as sports centres and supermarkets). I also used the social networking site Facebook as a means of keeping in touch and becoming involved in organised events. In this chapter, I concentrate on the meanings and interpretations that these young men discussed with me in relation to the stigma that they felt came with being from a certain deprived area when temporarily away from it on holiday. During interviews, the topics of reputation, mobility and escape emerged organically and are discussed in the following sections.

Reputation of place

During interviews with the young men, it became clear that they felt the area they lived in had an undesirable reputation. This was something that they commented on when I asked them to describe where they lived:

TOMO: It's a shit hole.[3]

SEAN: It's pretty rough...kind of a rough place to live. There's nothing here no more. My dad said Cwm Dyffryn used to be known as the Vegas of the Valleys with loads of clubs and stuff. It's just shit now!

ALAN: I can't wait to leave for uni anyway and get the fuck out of here and never come back.

During a group interview, some young men further described their hatred for living in the area but also described some of the problems they felt were occurring in their community that might have an impact on the negative reputation of the area:

BRUCE: I hate it around here. It's a dump, like, proper shit, no jobs, nothing to do, all drugs and stuff.
JACK: Things have got worse, too, innit? Kids like thirteen walking round with babies, innit?
CLUMP: Like eleven-year-old walking 'round, drinking cans [of alcohol] in their hands and kids smoking at the age of six.
JACK: Breaking into schools and old people homes – it's just wrong!

Besides the frequent negative references that the young men themselves made to their community through describing the lack of employment opportunities, drug use and their opinions on underage pregnancies, drinking and smoking (see also Ward 2012, 2013), they also reflected on the stereotypes they felt people living outside the area held towards them because they were from a particular place:

MW: So being from this area then, do you think there is a reputation?
GREG: Definitely! I went down to Pembrokeshire[4] on holiday, and we saw a woman who asked me and my friend where we were from and we said Cwm Dyffryn, and she stepped back, she literally stood up and walked back. It's like, 'You're from the Valleys!' I mean I know Cwm Dyffryn is rough, but I been to rougher places. But you can tell they don't like us. They think we are all ruffians.

Greg, on his holiday in Pembrokeshire, felt that coming from a community in the South Wales Valleys was discrediting. Here he states that a woman who he had met on holiday, who lived outside the area he was from, was stereotyping him as 'rough' because she had preconceived ideas about the type of place that Greg lived and was mapping this onto the type of person he would then be. For her, he embodied the roughness and undesirability of Cwm Dyffryn. This idea of 'roughness' is linked to outsider perceptions around inhabitants of this working-class community as undesirable, tough and something to be avoided. It is recognised through the accent that places Greg as being from a specific area, and he is guilty by association with it.

Cresco and Clive also discussed how the stigma of being from a deprived working-class community was transported with them in the following discussion about a foreign holiday. Here Cresco and Clive express anger at the way a father assumed they were troublemakers because of their accents and behaviour:

CRESCO: Well, I've been to Spain and um, we pulled [kissed] two girls there, and their father came out and went, 'Stop talking to them taffy[5] boys.'
MW: Ah, right.
CRESCO: And I went to him 'Why'd you call us taffy boys?' 'Because you're from the Valleys. I know that because you've got an accent.' I was like, 'What!' 'And the way you act as well.'
MW: Really?
CRESCO: Yeah, I was like, 'The way I act? What do you mean by that?' He goes, 'Playing loud music, constantly drinking.' I'm like, 'Yeah, I'm on holidays.' [*laughs*]
MW: So you think it's a bad reputation then?
CRESCO: Do you reckon we have?
CLIVE: I think it depends on the person and who's judging you.
CRESCO: Yeah, but this was the girls' father.
CLIVE: Yeah I know, harsh though, wasn't it!

It is clear here that the father of the two girls that Cresco and Clive had kissed or 'pulled' was extremely wary of these young men. To him they symbolised not only a sexual threat, but a threat of contamination where their loud music and 'constant' alcohol consumption is positioned as representing the threatening nature of the South Wales Valleys. This was present in their South Wales accents, but also in the showy performance of their threatening masculinities that were deemed to be dangerous (Connell, 1995). Again, like the woman in Greg's experience above, the father already had preconceived ideas about where these young men were from and what they were like as a consequence of coming from a stigmatised locale (Shields, 1991; Campbell, 1993; Hayden, 2000; Keene and Padilla, 2010). Cresco and Clive's performance of a macho masculinity (Mac an Ghaill, 1994) displayed through drinking, playing loud music and pursuing heterosexual conquests, is a further problem for the father of the unnamed girls: In his eyes, this sort of behaviour is not fully acceptable (even on holiday!) and something he should protect his daughters from. Others also reflected on how through the media, the area they were from was associated with negative stereotypes that aided the creation of a discredited self (Goffman, 1963):

> TOMO: Yeah, you get a reputation if you're from the Valleys like... the TV and that always focus on it... it's seen as bad, innit? [...] Seen as scrubbers and unemployed, dole bums, druggies and stuff.

For Tomo, these derogatory reputations of place are identified by the symbolic dangers of unemployment (as signified by dependence upon social welfare or the 'dole'), drugs and dirtiness. These unsavoury and unpleasant remarks are instilled through jokes, insults, disparaging comments and stories about the degeneracy and fecklessness of the place he lives.

As the discussions above have shown, Cwm Dyffryn, and the South Wales Valleys more generally, are highly stigmatised places. The effects of this stigma have been shown to be felt both directly and indirectly. Through being subjected to name-calling, stereotypical opinions and negative reactions to their accents and performances of masculinity, the young men experienced a direct consequence of their 'spatial stigma' (Keena and Padilla, 2010). Much of this can be attributed to deindustrialisation since the early 1980s and the subsequent social and economic inequalities that have developed as a direct consequence of this process (Rees and Stroud, 2004; Adamson, 2008; Walkerdine, 2010). Indirectly, these views and opinions of their community as stigmatised and something to be avoided also came through media and political representations of the area supported by wider representations of young working-class men. This meant that the young men felt they further suffered from a spoiled identity (Goffman, 1963). However, can the idea of travel and exotic holiday destinations also be an arena through which to challenge stigma and act out escape fantasies from this?

Challenging stigma – temporary and permanent mobilities

The concept of travel and holidays as an opportunity to escape from mundane everyday life has been well-documented within tourism studies (MacCannell, 1976; Malam, 2004; Andrews, 2006; Thurnell-Reed, 2011). However, there has been little attention paid to young working-class men's perceptions of holiday destinations, adventures and the fantasy of travel as a means to escape a deprived, stigmatised locale. In this final section of the chapter, I look more closely at how this group of young working-class men saw travel and holiday destinations as a form of independence and adventure, but also as an arena through which to act out ideas of fantasy and escape from the problematic position that they occupied:

MW: So, would you consider moving outside of Wales or the Valleys?
ALAN: Ah, yeah, definitely, it's too small a world to be stuck in the Valleys.
CARR: I want to move, aye, hate it around here. Love to move to America, be awesome!
SCOTT: I'll probably move. I don't like this town that much It annoys me.
MW: What annoys you about it?
SCOTT: Generally everything's falling apart. People shout at you and swear at you for no real apparent reason. Seriously, I forgot how many times I've been walking home and someone has shouted something at me from a car. I remember standing on this corner on fireworks night, and this guy just leaned out of his car and shouted, 'Get a haircut you freak!' at me and drove off!

Here Alan, Carr and Scott all reflect on how they want to leave the area and move elsewhere in slightly different ways. For Alan and Carr, the community they live in is viewed as small and uninteresting, with bigger and better places (e.g., America) to be discovered elsewhere. Sometimes, as in Scott's case, the desire to escape was due to not quite fitting into his hometown. He had grown his hair long, and in an area where traditional displays of masculinity were the default reference point, this association with femininity brought with it abuse from others as he walked about the town. If he could leave, Scott felt he could escape some of the abuse he got within his own community for not quite fitting it. He illustrates here that for him, living in the community was not a pleasant experience, and the idea of travel and mobility was one potential way to create an identity and a new self elsewhere.

Living in the area and expressing desires of independence and escape were also discussed when talking about their best memories from school:

MW: So what's [*sic*] the best memories that you're going to take away with you from school then?

Brad: I'd say skiing.[6] I'm going to go to the sixth form till Christmas just so I can go skiing again. It was brilliant, never normally get a chance to do that, like, I think the school got some money from somewhere or something. It was epic to get to ski and see all that snow and stuff. Pretty shit to come back here to this dump!'

For Brad, his best memory from school is not about the academic opportunities that his education could have given him; it is about the time he went skiing, which gave him the chance to get away from Wales and try something that he would normally never have done. This opportunity to escape was tempered by the disappointment he felt at returning home, to what he describes as a 'dump' after the holiday had finished. Through the school's subsidised trips, Brad was free to embody a different identity, and this access to travel helped him escaped his stigmatised position. Others also talked about using travel, and especially exotic destinations, as a way to live out their fantasies of escape and sexual conquests:

DAVIES: We're gonna set up a beach bar in Barbados.
HUGHESY: Ibiza ...
DAVIES: Narrr! Barbados, mun more fun in it. Loads of women out there and just drink rum all day. [*laughs*]
BUNK: Australia or New Zealand for me, somewhere like that ...
WAYNE: Spain.
BAKERS: Yeah, go to Magaluf, just get steaming, like, loads of women there as well!

Escape here for these young men is not just about getting out of Cwm Dyffryn and South Wales, but also a way for them to validate their masculinity through imagined sexual conquests. Davies and Bakers emphasis heterosexual motives along with drinking large amounts of alcohol, which will support their other activities away from home, where they assume these pleasures can be easily consumed. Exotic destinations such as Barbados, Australia and New Zealand are also a long way from the deprived community they live in and truly represent a different world to the one they inhabit. A fantasy is being acted out here not just in terms of sexuality, but also in terms of mobility and distance from their community.

Conclusion

In the study of identity, identity boundaries and how experiences of place shape reputation, a number of authors have explored how certain deprived communities and the people who live in them are increasingly likely to become stigmatised by both political and media discourses (Waquant, 2007; Keene and Padilla, 2010; Jones, 2011; Rhodes, 2012). This chapter expands this work by showing that for a group of working-class young men, the deindustrialised place they inhabit and the stigma that is associated with it is transported with them to different places when they leave it on holiday. The young men in this study describe how they feel they have a reputation because they are from a deprived community and are associated with negative stereotypes. The stigmatisation seems to have an impact on how they experience being away from their community and how they refer to it. I argue, then, that these young men suffer from a spoiled, discredited identity (Goffman, 1963) due to their working-class accents and traditional performances of masculinity, which are viewed by others as threatening.

The place they come from is spoiled, but so, too, individuals who are known to be from this place become spoiled by association. As they travel, other tourists see them as a threat of contamination, almost as if these young men can pass on the undesirable qualities of the South Wales Valleys to them. What is also apparent is that even though these young men feel stigmatised when they leave the locale temporarily on holiday, the idea of travel to exotic holiday destinations also brings with it fantasies of escape from the discredited position they find themselves in. The longing these young men expressed for fleeing to destinations such as Barbados and Ibiza are not only physically a long way from their homes, but also areas to experience different leisure pleasures and places to act out sexual fantasies. Tellingly, the aspirations to travel are characterised by a desire for the space to enable an enactment of an unproblematic masculinity, whereas the actual experiences reported by the young men show just how readily their 'home' identity might disrupt their pursuits of a 'new self' through the mobility of tourism. The experiences of the young men as tourists suggests a need to consider how the different processes that contribute to the construction of spatial stigma and masculine identities occur. The multiple disadvantages these young men (and others like them) experience by coming from deprived communities did not arise organically. They are the result of historic social and political practices that continue to create exclusion and difference. However, while travel would seem to have the power to reconfigure class and gender identities, access to these opportunities for those most likely to benefit from it are restricted.

Notes

1. See, for example, recent media representations of communities and their inhabitants such as the ITV programme *Benidorm* and the MTV reality television shows *Geordie Shore* and *The Valleys*.
2. The name has been changed to maintain the anonymity of participants, but chosen to reflect its history and geography.
3. All names of participants have been changed.
4. Pembrokeshire is a county on the coast of West Wales, around a hundred miles away from Cwm Dyffryn.
5. Taffy (or Taffi), is a colloquial name for people (mainly men) from South Wales that is taken from the river Taff which runs through the area. The term is also used pejoratively and deemed ganti-Welsh.
6. Cwm Dyffryn High School, despite being situated in a highly deprived community, ran an extensive programme of subsidized school trips every year with skiing and foreign language holidays to Europe.

References

Acevedo-Garcia, D., Lochner, K. A., Osypuk, T. L. and Subramanian, S. V. (2003) 'Future directions in segregation and health research: A multilevel approach', *American Journal of Public Health*, 93(2): 215–21.

Adamson, D. (2008) 'Still living on the edge?' *Contemporary Wales*, 21: 47–66.

Andrews, H. (2005) 'Feeling at home: Embodying Britishness in a Spanish charter tourist resort', *Tourist Studies*, 5(3): 247–66.

Andrews, H. (2006) 'Consuming pleasures: Package tourists in Mallorca', in K. Meethan, A. Anderson and S. Miles (eds) *Tourism, Consumption and Representation: Narratives of Place and Self*. Wallingford: CABI, pp. 217–35.

Beck, U. (1999) *World Risk Society* Cambridge: Polity Press.

Brewer, T. (1999) 'Heritage tourism: A mirror for Wales', in D. Dunkerley and A. Thompson (eds) *Wales Today*. Cardiff: University of Wales Press, pp. 149–63.

Bullard, R. (1999) 'Dismantling environmental racism in the USA', *Local Environment*, 4(1): 5–19.

Bush, J., Moffatt, S. and Dunn, C. E. (2001) 'Even the birds round here cough: Stigma, air pollution and health in Teesside', *Health and Place*, 7(1): 47–56.

Campbell, B. (1993) *Goliath, Britain's Dangerous Places*. London: Methuen.

Cohen, S. (1972) *Folk Devils and Moral Panics: The Creation of the Mods and Rockers*. London: MacGibbon and Lee.

Connell, R. W. (1995) *Masculinities*. Cambridge: Polity.

Crouch, D., Aronsson, L. and Wahlström, L. (2001) 'Tourist encounters', *Tourist Studies*, 1(1): 253–70.

Curtis, S. (2004) *Health and Inequality: Geographical Perspectives*. London: Sage.

Day, G. (2002) *Making Sense of Wales*. Cardiff: University of Wales Press.

Edensor, T. (1998) *Tourists at the Taj: Performance and Meaning at a Symbolic Site*. London: Routledge.

Edensor, T. (2001) 'Performing tourism, staging tourism, (re)producing tourist space and practice', *Tourist Studies*, 1(1): 59–81.

Entwisle, B. (2007) 'Putting people into place', *Demography*, 44(4): 687–703.

Fevre, R. (1999) 'The Welsh economy', in D. Dunkerley and A. Thompson (eds) *Wales Today*. Cardiff: University of Wales, pp. 61–74.
Galster, G. (2007) 'Neighbourhood social mix as a goal of housing policy', *European Journal of Housing Policy*, 7(1): 19–43.
Geertz, C. (1973) *The Interpretation of Cultures: Selected Essays*. New York: Basic Books.
Goffman, E. (1963) *Stigma: Notes on the Management of a Spoiled Identity*. New York: Simon and Shuster.
Gordon, D., Kay, A., Kelly, M., Mandy, S., Senior, M., and Shaw, M. (2004) *Targeting Poor Health: Review of Rural and Urban Factors Affecting the Costs of Health Services and Other Implementation Issues*. Cardiff: National Assembly for Wales.
Grant, R. (1991) *Cynon Valley in the Age of Iron*. Stroud: Alan Sutton Publishing.
Harris, J. (2007) 'Cool Cymru, rugby union and an imagined community', *International Journal of Sociology*, 27(3/4): 151–62.
Hayden, K. (2000) 'Stigma and place: Space, community, and the politics of reputation', in N. Denzin (ed.) *Studies in Symbolic Interaction, Volume 23*, Emerald Group Publishing Limited, pp. 219–39.
Holland, S. and Scourfield, J. (1998) 'Ei gwrol ryfelwyr. Reflections on body, gender, class and nation in Welsh rugby', in J. Richardson and A. Shaw (eds) *The Body and Qualitative Research,* Aldershot, Overbuy, pp. 56–71.
Howe, P. D. (2001) 'Women's rugby and the nexus between embodiment, professionalism and sexuality: An ethnographic account', *Football Studies,* 4(2): 77–91.
Jones, O. (2011) *Chavs, The Demonization of the Working Class*. London: Verso.
Keene, D. E. and Padilla, M. B. (2010) 'Race, class and the stigma of place: Moving to "opportunity" in Eastern Iowa', *Health and Place,* 16(6): 1216–23.
Kenway, J. and Kraack, A. (2004) 'Reordering work and destabilizing masculinity', in N. Dolby, G. Dimitriadis, and P. Willis (eds) *Learning to Labor in New Times*. London: RoutledgeFalmer, pp. 95–109.
Kenway, J., Kraak, A. and Hickey-Moody, A. (2006) *Masculinity Beyond the Metropolis*. Basingstoke: Palgrave.
Kneafsey, M. R. (2000) 'Tourism, place identities and social relations in the European rural periphery', *European Urban and Regional Studies* 7(1): 35–50.
Mac an Ghaill, M. (1994) *The Making of Men*. Buckingham: Open University Press.
MacCannell, D. (1976) *The Tourist*. London: Macmillan.
MacDonald, R., Shildrick, T., Webster, C. and Simpson, D. (2005) 'Growing up in poor neighbourhoods: The significance of class and place in the extended transitions of 'socially excluded' young adults', *Sociology,* 39(5): 873–91.
Malam, L. (2004) 'Performing masculinity on the Thai beach scene', *Tourism Geographies*, 6(4): 455–71.
McDowell, L. (2003) *Redundant Masculinities?: Employment, Change and White Working Class Youth*. Oxford: Blackwell.
McDowell, L. (2007) 'Respect, deference, respectability and place: What is the problem with/for working class boys?' *Geoforum,* 38: 276–86.
McDowell, L. (2012) 'Post-crisis, post-Ford and post-gender? Youth identities in an era of austerity', *Journal of Youth Studies*, 15(5): 573–90.
Mordue, T. (2005) 'Tourism, performance and social exclusion in "Olde York"', *Annals of Tourism Research*, 32(1): 179–98.
Nayak, A. (2003) '"Boyz to men": Masculinities, schooling and labour transitions in deindustrial times', *Educational Review*, 55(2): 147–59.

Nayak, A. (2006) 'Displaced masculinities: Chavs, youth and class in the post-industrial city', *Sociology*, 40(5): 813–31.

Nayak, A. (2009) 'Beyond the pale: Chavs, youth and social class', in K. Sveinsson (ed.) *Who Cares about the White Working Class?* London: Runnymede, pp. 28–35.

Phillimore, P. and Moffatt S. (1999) 'Narratives of insecurity in Teesside: Environmental politics and health risks', in J. Vail, J. Wheelock and M. Hill, M (eds) *Insecure Times: Living with Insecurity in Contemporary Society.* London: Routledge, pp. 137–53.

Reay, D. (2000) 'Children's urban landscapes: Configurations of class and place', in S. Munt (ed.) *Cultural Studies and the Working Class, Subject to Change*, London: Cassell, pp. 151–66.

Rees, G. and Stroud, D. (2004) *Regenerating the Coalfields, the South Wales Experience*. Tredegar: The Bevan Foundation.

Rhodes, J. (2012) 'Stigmatization, space, and boundaries in deindustrial Burnley', *Ethnic and Racial Studies,* 35(4): 684–703.

Sampson, R., Morenoff, J. and Gannon-Rowley, T. (2002) 'Assessing "neighbourhood effects": Social processes and new directions in research', *Annual Review of Sociology,* 28: 443–78.

Shields, R. (1991) *Places on the Margin: Alternative Geographies of Modernity*. London: Routledge.

Skeggs, B. (2004) *Class, Self, Culture*. London: Routledge.

Smith, D. (1999) *Wales, A Question for History*. Bridgend: Poetry Wales Press Ltd.

Thurnell-Read, T. (2011) 'Off the leash and out of control: Masculinities and embodiment in Eastern European stag tourism', *Sociology,* 45(6): 877–911.

Wacquant, L. (2007) 'Territorial stigmatization in the age of advanced marginality', *Thesis Eleven*, 91(1): 66–77.

Wacquant, L. (2008) *Urban Outcasts: A Comparative Sociology of Advanced Marginality*, Cambridge: Polity.

Walker, G., Simmons, P., Irwin, A. and Wynne, B. (1998) *Public Perceptions of Risks Associated with Major Accident Hazards*. Research report prepared for the Health and Safety Executive.

Walkerdine, V. (2010) 'Communal beingness and affect: An exploration of trauma in an ex-industrial community', *Body & Society*, 16(1): 91–116.

Ward, M. R. M. (2012) 'The Emos: The re-traditonalization of white, working-class masculinities through the "alternative scene"'. *Cardiff School of Social Sciences Working Paper Series*, Cardiff University.

Ward, M. R. M. (2013) 'The Performance of Working-Class Masculinities in the South Wales Valleys.' Unpublished PhD thesis, Cardiff University.

Watkins, F. and Jacoby, A. (2007) 'Is the rural idyll bad for your health? Stigma and exclusion in the English countryside', *Health & Place*, 13(4): 851–64.

Williams, D. R. and Collins C. (2001) 'Racial residential segregation: A fundamental cause of racial disparities in health', *Public Health Reports*, 116: 404–16.

Williams, G. A. (1985) *When Was Wales?* Harmondsworth: Penguin.

Willis, P. (1977) *Learning to Labour, How Working Class Kids Get Working Class Jobs*. Farnborough: Saxon House.

Winlow, S. (2001) *Badfellas: Crime, Tradition and New Masculinities*. Oxford: Berg.

8

'I Don't Want to Think I Am a Prostitute': Embodied Geographies of Men, Masculinities and Clubbing in Seminyak, Bali, Indonesia

Gordon Waitt and Kevin Markwell

Insights into the relations between men, masculinities, travel and the gay tourism industry are typically ethnocentric, privileging Western perspectives (Waitt and Markwell, 2006). Our chapter addresses this ethnocentric bias by focusing on the narratives of men who live their lives as Indonesians – particularly those who have migrated to Bali having learnt of the commercial gay venues in the district of Seminyak. Bali is still not an internationally-recognised 'gay destination' akin to Mykonos in Greece or Sitges in Spain, yet the district of Seminyak, over the past decade or so, has become increasingly popular as a tourist destination for predominantly mature gay men who live their lives as gay in Europe, North America and Australia. Furthermore, many Indonesian men consider Seminyak a place of 'sexual freedom'. In this chapter, we explore the encounters of men who live their lives as Indonesian with those who become Western tourists in and through commercial night-time economies sustained by the gay tourism industry. These nightclubs are pitched by the tourism industry as the centre of queer/gay life in Seminyak, Bali. Our aim is to provide what an embodied geographical perspective can offer to better understand the relationships between travel, men, masculinities and sexualities. We conceive nightclubs as always spatial, multiple, fluid, embodied, performative, in-the-making, unstable and endlessly differentiated. We understand nightclubs as spatially contingent upon the temporal flows of emotion and affect at the intersection of the social constructions of sexuality with age, gender, ethnicity and class. Nightclub bodies and spaces allow for the circulation of emotions and affect such as love, comfort, disgust, anger and care. Through our focus on who feels they belong or do not belong in the nightclubs in Seminyak, our objective is to extend insights into travel, men, masculinities and sexualities.

We begin by reviewing a range of literature on nightclubs. Next, we discuss our conceptual framework, which draws on the work of post-structuralist, feminist geographers. We then outline the context of the fieldwork and methods. Our discussion explores the embodied geographies of two participants, Ali and Adi [pseudonyms] to provide insights to how encounters between sexed, gendered, aged, classed and racialised bodies play out in nightclubs. Ali points to the importance of the embodied geographies of a transnational love when encountering mature white bodies in nightclubs, while Ali illustrates the challenges and pain of negotiating embodied geographies of romantic love in and through nightclub spaces. Both underscore the emotional labour involved in blurring the boundary between host and guest in the nightclub, so as not to be configured as a sex worker. We conclude by thinking about embodied geographies as one approach to furthering scholarship on men, masculinities and travel.

Nightclubs: travel, men and masculinities

Nightclubs are widely understood as a normative context for dancing, consuming alcohol or drugs and pursuing sexual partners (Laumann et al., 2004; Hughes, 2006). The commercial opportunities presented by nightclubs are just one example of how the market is used to establish, maintain, express and transform sexual subjectivities and group identifications (Chauncey, 1994; Bernstein, 2001; Hughes, 2002; Chatterton and Hollands, 2003). Gendered, aged and sexualised nightclub spaces are often integral to tourism destination marketing pitches, because they are settings that encourage sexualised encounters among participants through a mix of lighting, music, alcohol, talk, bodily gestures and poise. Nightclubs are therefore often positioned as a key requirement of the so-called pink economies (Binnie, 1995; Brown, 2000), popularly dubbed the 'gay scene,' and 'gay tourism' (Hughes, 2006).

Scholars approach nightclubs from a number of disciplinary perspectives. Underpinning sociocultural examinations are Bourdieu's notion of cultural capital (Thornton, 1994; MacRae, 2004), Turner's rites of passage (Northcote, 2006) and Connell's (1995) concept of hegemonic masculinity. The latter strand of literature explores the set of practices that sustain the conventional ideals around gendered, masculine and heterosexual norms. For example, Thurnell-Read (2012) illustrated how hegemonic masculinity is reproduced by British stag weekenders in Poland through visits to nightclubs, strip-clubs and sporting activities based on the intersection of gender power relations within a capitalist economic system. Grazian's (2007) work illustrated the social status gained by college men in Philadelphia, USA, who arguably (re)produce a hegemonic version of masculinity. This is cultivated and privileged by the ways college men prepare and use their bodies when attending particular nightclubs. Furthermore, he pointed to the importance of sexual and sexist joking as a crucial element of sustaining gender power relations through the subordination of those men and women who are seen to contradict (hegemonic) masculine ideals (also see Thurnell-Read, this volume). Crucially, ideal or hegemonic versions of masculinity in nightclubs are not solely based on gender power relations, but intersect with age, sexuality and ethnicity and social class.

Malbon's (1999) ethnographic study of London's nightclubs sensitively highlights how clubbing creates choreographies of belonging, through the ways people habitually prepare and use their bodies (through dancing, clothing, accessories, haircuts, tattoos, gestures, ways of walking, flirting and innuendo) to create particular forms of sociality. Nightclubs offer varying opportunities for people to enjoy themselves together that may be empowering by either challenging or reproducing conventional gender and sexual norms. Valentine and Skelton (2003) illustrate how pleasures may be derived from dressing up to go out, and from being looked at by others, whereas Taylor (2010) highlights the pleasures of extended engagement in clubbing in the gay scene for some people aged 40+ years – derived from queering temporal schemes that operate under the assumption of a life course. This has often resulted in utopian visions of commercial nightclubs as creating a supportive community space of 'true' self-expression and belonging (see Prior, 2008).

Yet, at the same time, Valentine and Skelton's (2003) work highlights the paradoxical qualities of the choreographies of belonging. Discussing the role of nightclubs in crystallising lesbian and gay communities and subjectivities, these authors (2003: 357) coined the term 'forced transitions' to highlight how a person may orient their mannerism, dress, talk or haircut to 'fit' the choreographies of nightclubs, while remaining uncertain of their own sexuality. Who is felt to belong, or not, contributes in important ways to forging subjectivities along preexisting sets of ideas and conditions. Likewise, the work of Browne (2007), Casey (2007), Hughes and Deutsch (2010) and Thurnell-Read (this volume), illustrate how the choreographies of belonging may operate to marginalise those in nightclubs who do not conform to particular dominant understandings of sexual attractiveness along lines of emerging homonormative discourses of ethnicity, gender, sexuality, body shape or age.

Malbon (1999) also considered the emotional bonds and affective ties triggered by the materiality of nightclubs in mediating choreographies of belonging – the lights, music, drugs and alcohol. Similarly, the work of Jayne, Valentine and Holloway (2010) on drinking in the English nighttime economy underlines the importance of alcohol in mediating people's bodies to become communal. For Jayne et al. (2010), the affective and emotional bonds forged by alcohol may, paradoxically, enhance a sense of either insecurity or togetherness. The affective properties of sweat, alcohol, tobacco and other drugs and music may (dis)connect clubbers' bodies to/from each other and the nightclub space. For example, Caluya (2008) discussed, in the context of the gay clubbing scene of Sydney, Australia, how ideas of race operate to configure social and spatial borders by both categorising men and territorialising space. At the same time, he points to how emotions and bodily affects play a crucial role in 'border crossings' between self and Other, and non-ethnic and ethnic, within these spaces.

This scholarship is helpful to us in providing a conceptual framework to conceive nightclubs as shared-space. Firstly, we foreground bodies. This requires thinking relationally. Thus, attention is given not only to the way gender is choreographed in relation to ideal, or hegemonic, representations, but also to the emotional bonds and affective ties between different bodies when they are place-sharing. How masculinity and sexuality are felt in a particular context may help rupture or reconfigure ideas about gender. Secondly, we examine the importance of the personal. Clubbing involves social practices framed by cultural, economic and political structures, but the embodied practice and lived experience is also extremely personal and varies among individuals. Masculinity, sexuality and other dimensions of our selves may be thought of as embodied histories that may be triggered through particular encounters. Thirdly, we consider the paradoxical attributes of nightclubs. At the same time, nightclubs may be a space of both belonging and alienation that requires paying attention to how bodies become gendered, sexed and racialised within unfolding power relationships. Finally, the role of the past in forging the particularities of place is analysed. Each nightclub is embedded within a specific constellation of trajectories. To conceive of nightclubs as the outcome of ongoing reciprocal relationships between bodies sharing space allows us to pay attention to how the emotional and affective qualities of personal geographies are entangled in the historical weight of wider cultural, economic and social structures including those of heteropatriarchy and heteronormativity.

Context and method

The work of Boellstorff (2005) underscores how difficult it is to overemphasise the historical weight of heteropatriarchy and heteronormativity implicit in the imaginaries of Indonesian citizenship. He points to how in legal, educational and Islamic institutions, a very rigid social distinction is maintained between the categories of 'man' and 'woman'. While a national bill to criminalise homosexuality was defeated in 2003, the law recognises the existence of only heterosexual men or women. Boellstorff (2005) argues that since the 1970s, some Indonesian men, at least in Jakarta, started calling themselves 'gay'. In doing so, these men articulated subjectivities that lay outside an Indonesian archaeology of knowledge. Indeed, Boellstorff (2005) also suggests some Indonesians may think of gay as a Western version of the better-known Bahasa Indonesian term *waria* (a male-female transgender person). Furthermore, Boellstorff (2005) points out that throughout popular Indonesian culture, the dominant image of gay men is as selfish – for not conforming to the quintessential social norms and duties of Indonesian heterosexual family life.

One contemporary example of how state-sanctioned heteropatriarchy plays out is how, in 2012, organisations representing Islam successfully lobbied law enforcement agencies to prevent the pop singer, Lady Gaga, from performing in Jakarta. Lady Gaga is known for LGBT (lesbian, gay, bisexual and transgender) rights advocacy in her music and through her Born This Way Foundation (Loinaz, 2012). It is perhaps not surprising, then, that same-sex relations are generally treated with secrecy in Indonesian society, and that many same-sex-attracted men will engage in heterosexual relationships and, ultimately, marriage. Across various ethnic and religious social groups in Bali, Java, Lombok and Sumbawa, heterosexual masculinity is configured as natural, moral and essential to individual, family and national life.

Important to our project is the configuration and reconfiguration of Bali as an international gay tourist destination and the consequent place-sharing arising from the arrival of international visitors. Atkins (2012) discusses the economic and cultural imperatives that underpin the growth of gay tourism in Bali-as-paradise. Gay tourism guidebooks and travel websites often warn of the social norms of heterosexuality within Indonesia as a whole and, specifically, in Bali. However, while homosexuality is not considered a legitimate expression of sexual expression by many Hindu Balinese, they are perhaps more accepting of sexual diversity, generally, than Muslim Indonesians.

Today, the district of Seminyak, is positioned as the hub of an 'open gay lifestyle' in Indonesia (see http://www.utopia-asia.com/indobali.htm). Seminyak is a coastal resort district and part of the Kuta-Legian tourist corridor, approximately 10 km south of the capital, Denpasar. Online posts also operate to make sexuality visible to travellers and help territorialise Seminyak as queer space, by identifying certain bars as 'gay', and including affective images of bartenders, 'drag shows' and 'go-go boys'. In Indonesia, Seminyak's main street, Jalan Dhyana Pura, is synonymous with gay tourism.

Clustered together on Jalan Dhyana Pura, some of the nightclubs and bars are explicitly racialised and sexualised by the emergent gay tourism industry. For example, being aware of the privileged position held by some Asian men as more desirable for the white, middle-class Western gay tourism market, nightclubs frequently portray Indonesian men in short shorts as youthful, boyish, toned, muscular, hairless, clean-shaven and dark-eyed. For many people walking along Jalan Dhyana Pura at night, those clubs operating outside the social norms of heterosexuality are instantly recognisable by their rainbow flags, branding, music and entertainment. The nightclubs are a spectacle in themselves, with the dance music, shows and lighting becoming a magnet for Indonesian men as well as for international tourists, regardless of sexuality. Large groups of Indonesian men often congregate outside the nightclubs, or sit watching the comings and goings from the wall across the street or from their motorbikes.

In Bali, the nightclubs of Seminyak are only one of many locations where men can meet men for sex. Other sites include cruising areas (*tempat ngeber*), bars, cafes and via Internet sites. These locations are distinguishable by the relative levels of discretion they provide and by their affordability. The *tempat ngeber* are often listed in travel guidebooks and online and offer possibilities for men to hang out together for free. In contrast, people who share the nightclubs require economic, social, symbolic and Western cultural capital. To participate in nightclub spaces requires learning (at least some) English, and also knowledge about the music, dress, dance styles and aesthetics of Western gay clubbing culture as well as ready access to financial resources needed to purchase drinks and other services. Consequently, the nightclubs of Seminyak are paradoxical spaces, given that it is not often clear who constitutes the host and guest. Power relationships are incredibly fluid, given that Indonesian men could be constructed as either hosts – as someone who is a citizen of the host nation and knowledgeable about Indonesia – or as guests of the Western visitors. Equally, visitors from overseas could readily be positioned as guests, but also hosts as part of the dominant Western gay culture that prevails in the bars and nightclubs.

The methodology used in this research combined semi-structured interviews alongside participant observation conducted during fieldwork in Bali during 2009. To interpret our data, we used a form of analysis termed by Gubrium and Holstein (2009) as 'narrative ethnography'. Our own narratives, and those we were told, were interpreted not as 'factual-truthful accounts' nor for linguistic content, but as phenomenological-interactions constitutive of embodied sociocultural worlds. These analytical techniques offered tools for understanding more about the intersections between embodied knowledge, travel, men, masculinities and sexualities.

Participants were invited to participate in a project exploring their experiences of tourism and tourists. The ten men who took part were recruited through online social-network sites like Gaydar – where people can register to access the profiles of men to facilitate connections that may result in friendship, casual sex or longer-term relationships. Participants normally requested to meet us at our hotel and conduct the interview outside in the hotel grounds where nobody could overhear our conversation. Having explained the research aim, we asked for permission to record each interview. The ten participants were a socioeconomically diverse group, ranging in education levels, employment and religious beliefs. The participants ranged in age from their mid-twenties to mid-thirties. Only two were born in Bali. The most recent migrant had lived in Seminyak for two years, the longest thirteen years. All participants lived alone. None spoke of 'partners'. Five spoke of social relationships that configured 'boyfriend(s)' with Western tourists.

The semi-structured interview was divided into three sections to enable participants to tell a life narrative: 'important things about you', 'sexuality in Indonesia' and 'tourism and tourists'. Interviews lasted from between around forty minutes to two hours. As all participants spoke a good level of conversational English, interviews were conducted in English. Each recorded interview was later transcribed verbatim.

Conducting this research, we remained mindful of how we negotiated our positionality with participants. On the one hand, those taking part readily identified us as older, affluent, white, foreigners; part of the dominant host Western culture that frames the Seminyak gay tourism industry. As older, white and recently-arrived visitors from Australia, our participants were eager to meet us, constituting us as guests who were eager to learn about Indonesia and their lives. On the one hand, we played the role of host, welcoming the men to the hotel where we were staying and providing drinks and food as a means of generating a space to talk about their social and sexual encounters. Cognisant of our paradoxical positioning as both guest and host, participant observation of the nightclub spaces also provided an opportunity to understand how various discursive, emotional and affective relationships unfold along Jalan Dhyana Pura.

Ali – cultivating the cosmopolitan gay subject

When we met Ali in Seminyak he was 30 years old, a Hindu and a hotel receptionist. Ali explained that he grew up in the north of Bali, studied tourism at a college in Denpasar and had worked in restaurants. He was fifteen when he began experiencing sexual arousal and urges relating to men, but he knew nothing about being gay until attending college in Denpasar and encountering holidaymakers.

Importantly, for this project, Ali spoke of how the sensations and emotions of his encounters with the bodies of white men that help sustain his sexuality only became possible after his migration to the tourism spaces of Bali:

ALI: I don't know why – I have only had sex with white guys, so I know from practice, I know that I have feelings, I like him.
KEVIN: So, are you not attracted to other Indonesian guys?
ALI: Never. I can see that they are cute, but I never had wanted to have sex with them.
KEVIN: Okay, so why do you think that is?
ALI: I don't know. [*laughing*] ... Maybe, I always think that Western boy more, I don't know, open minded – and relaxed and I don't know, I can talk to them better and more attracted to Westerner.

Ali illustrates how bodily sensations and subjectivities are co-constituted. Ali is either too shy or unable to explain why his sexual urgings are for the bodies of white men. The important point here is that irrespective of how Ali may understand Indonesian men as 'cute', the set of affects that he has learnt to identify with being sexually aroused spurs him to seek sex with the bodies of white men. The nightclubs of Seminyak figured strongly in Ali's narrative:

KEVIN: Do you go out very much to clubs?
ALI: Every weekend.
KEVIN: Which are you favourite clubs?
ALI: Mixwell and Qbar. I go where the places are this year. I think just like everybody.

He spoke about how socialising in the nightclubs of Seminyak facilitated sex with visitors, and of building an international network of 'good friends', rather than 'boyfriends', around these sexual relationships with holidaymakers. By cultivating relationships with tourists that involved sex, as being 'good friends,' he underscored that there was no hurt, guilt or deceit with having multiple partners. His bodily emotions and subjectivities are trained and co-constituted through practices cultivated within the nighttime economy.

Ali spoke of the importance of designer clothes, looks, gestures, alcohol and perhaps most importantly, English-language skills. How Ali dresses, talks and gestures enabled his body to be sexualised in the context of the nightclub. When recognised as the cosmopolitan gay subject by holidaymakers, Ali gains proximity to the bodies of white men.

Ali felt uneasy about going to the nightclub alone. He spoke about the tensions he felt among some Indonesian men:

ALI: If I go to bars, I always go with a friend at least. I don't like to go there alone, to the gay bars... It is from other [Indonesian] boys there; they can be very rude. Rude, because I am not their friend. They just look at me – with a very bad look.
KEVIN: So if you went to [the popular local nightclub] Mixwell by yourself – you would feel a bit threatened by some other local guys that might see you as a competitor and want you to go away – because you might get a guy and they don't. So there is a bit of competition then?
ALI: Yeah – that is the thing.

Ali is not alone in pursuing same-sex intimacies with the bodies of affluent white men. The nightclubs are places of surveillance and they can be highly territorialised, with friendship networks regularly gathering almost on a nightly basis in the same locations. A glance may help reconfigure spatial boundaries between friendship networks.

Touch is crucial to negotiating subjectivities in the nightclub. Ali spoke about being touched by tourists on eroticised bodily zones such as the bottom or crotch without his consent. Ali is angered by this unauthorised affection. Such unsanctioned sexualised experiences may translate into bounded sets of understandings that legitimise the sexualised subjectivity of a sex-worker. In Ali's words, he is positioned as 'a money boy': 'Sometime[s] people do it like, grab me, or touching me and it's not nice. Yeah, and they think they can do it to anybody. I don't like at all. I don't want, ah, people to look at me the same way as a money boy.'

Ali illustrates the importance of sight and touch in terms of how bodies are performed and negotiated in relationship to each other (Johnston, 2012). Many men who are tourists struggle to successfully identify who is a sex worker and who is not. For some overseas visitors, all Indonesian men in the club are potential sexual partners. For Ali, touch without consent is confirmation that he has failed to live up to his own ideals and understanding of his self as the cosmopolitan gay. Furthermore, consented touch works alongside indirect exchanges of money and gifts for companionship, to further blur the boundaries between casual sexual relationships and purchased sex, according to Ali:

> Well, I have sex with a guy because I want, I like to. Some think I want money, and that make[s] my very angry because, some do think I am a money boy. But then, I cannot really blame them because there are so many money boys at the clubs – I have sex with them because I want to – but if you give me money – then I am not going to say no, because who doesn't want money? [*laughter*].

Ali's narrative echoes those of Dhales' (2002) study of Indonesian 'beach boys' who provide romantic companionship to white women tourists, and Collins' (2007) work on the sexual labour of male companions and their relationships to white men tourists. Sexualised subjectivities and space hinge on resisting the commodification of hospitality by blurring the boundary between host and guest, sex-worker and boyfriend.

Adi – cultivating the romantic gay subject

Adi was born in Jakarta in 1985 and migrated to Bali in 2008. Until then, he lived with his family, of whom he said: 'I love a lot'. He claimed a gay Muslim identity, by reconstructing Holy Scripture in a gay affirmative manner and participating in social spaces of groups that cultivate his sexuality in positive terms. At senior high school, he explored same-sex intimacies by 'hooking-up' in Jakarta with Indonesian men through the Internet. A turning point in his narrative came following the death of his father and older brother. He left school to find employment to support his mother and his two younger sisters. Adi narrated a life of family arguments, depression, drug and alcohol abuse and a series of low-paid jobs as a machine operator, waiter, security worker and masseur. Adi described losing his self-worth after accepting payment for sex mostly with married Chinese men under the commodified circumstances of a massage parlour in Jakarta:

> But it's massage plus plus [meaning massage and some form of sexual activity] ... I don't have a choice, so I take that job. I'm working as a massage boy for three months, yeah, at least three months ... I get money, I get a room to stay, but I lose my dignity, you know what I mean ... you become so ... nothing. I mean people pay you because you give sex, and meanwhile, I know I can get money with my brain, not only with my dick or my butt. So I [am] saving and saving and then I move to Bali.

Explaining his decision to migrate to Seminyak, Adi invoked notions of romantic love, in terms of intimacy with a Western tourist as a life partner: 'I have to be honest, I have some ridiculous expectation that I will meet *a bule* [a slang term meaning foreigner], I will meet a tourist Western, that he like me, he will help me, blah blah blah. At that moment I think about that, I must confess. That's one of my reasons why I moved to Bali.'

Finding romantic love was crucial to Adi's decision to move to Bali. His dream was to find a Western tourist to fall in love with as a life partner who would also provide financial security. This dream was underpinned by a set of ideas that invoked love in its romantic aspects. Furthermore he reproduced narrow, idealised and limited versions of Western men:

> With people from the Western culture, you can expect more ... when you have a relationship with Indonesian people you don't get new things, same person, same culture, same background, same sex [*giggles*] ... you don't get a new thing, that's why mostly boys in Bali like Western ... also they have expectation they get much pleasure, fun and money from [them].

For Adi, Western men become more rewarding life partners because of assumed differences from Indonesian men. Silenced in Adi's narrative is how Western men also bring hurt, pain and frustration from negotiating intracultural differences not only in ideas about appropriate relationships between the people in coupled relationships (such as language, gifts and gender roles), but also relationships with family members, neighbours and various social groups (see Bystydzienski, 2011).

Given the information Adi had about the nightclubs of Seminyak, the clubs initially appeared a logical starting point for Adi for 'going out' and 'meeting up' with a Western tourist in order to find romantic love. Reflecting on his experience, Adi recalled:

> I expected to meet a boyfriend and to build a good relationship, and I dreaming that he will see that I will have something, not only sex, but it's not easy. I then spend three months: go to bar every night but not working... I'm just go to the bar and sit down, get a drink and that's it... but the other men living here more longer... they are very clever to get attention from the Westerner... if I meet ten new people, and I spend a day with them, nine people of them will not believe that I'm gay because of how way I look, how do I talk... that's why I not get Western[er] in three months even though I go the bar every night... because I'm not attractive enough for them... so I have to work because I need money... and I got a job as a receptionist.

In the nightclub, Adi is seemingly not recognised as sexually desirable by many Western tourists. He suggests that there is much competition between Indonesian men to attract the attention of Westerners. He illustrates how bodies and spaces are co-constituted through practice. Adi suggests that he initially lacked knowledge of the choreographies that sustain the sexualised cosmopolitan gay subject in the context of the nightclub, including what to wear and what to say. Furthermore, informed by discourses of romantic love, Adi's body – when going out to the nightclubs of Seminyak – is enmeshed in particular practices that enable a particular sexualised subjectivity to flourish, not just around sex, but also caring, kissing, sharing and bonding. Consequently, experienced through a romantic loving body, he spoke of his disappointment and hurt when some Western tourists he met in the nightclubs failed to live up to this ideal:

> I'm like you, when I meet Westerns in a bar, and he buy me a drink, and we dance, and we then go to room and have sex, and after they give me money, I think used, I feel so bad... mostly I turn back the money... I ask: 'Why you give me money?'... 'I'm not paying for

> sex, just to say thank you'...'Don't pay with money. I can be more comfortable. If you want to thank me, there are other, nicer way to say thank you. Take me to dinner, or just to walk on the beach. It's more nice'...I don't want to think I am a prostitute.

Adi aligned romantic love with his subjectivity. He draws on dominant discourses of sex in Indonesia to set his moral compass to marginalise sex-workers as immoral, and stigmatise lust or open relationships. Therefore, for Adi, the embodiment of romantic love resulted in offers of payment by Western tourists for the pleasures of sex leaving him feeling 'bad' and 'used'. Instead, for Adi, the preferred way of showing gratitude requires doing romantic love, and includes purchasing dinner or walking together along the beach. These intimacies point to sensory practices and emotional intensities that help blur the categories of sex-worker/boyfriend and host/guest, and stabilise a sexualised subjectivity and the desires that he associates with romantic love.

Adi explains how many Western tourists understand payment as negating any further social relationship or obligations:

> I've been with a lot of Western people, and mostly they treat Indonesian people not as a human [but] just like a sex machine. I have sex with you, I pay you, we have fun, then the other [next] day, we meet again, I don't know you, you don't know me...It is shit, yeah...I don't know what the reason is...maybe he doesn't want me to disturb with him and his new local boy, but it's not right.

Adi's comments alert us to how being paid for sex may reassert boundaries between national and foreigner, self and other. Adi points to how antagonism is felt between bodies in the nightclub by subverting his notions of romantic love. Inequalities are felt in the nightclub whenever Adi encountered a tourist who has failed to live up to his ideal. Having secured full-time employment, he abandoned his quest for finding financial security through romantic love in the nightclubs of Seminyak. Instead, the affective longing for romantic love and intimate partnership orientated him towards gay Internet dating sites: 'I forget for a while to get boyfriend from club. I choose other option, which is Internet...so I go to the 'net and I do a good job [*laughs*]...I meet one, two, three Western people in Internet...it's more nice, more comforting.'

In comparison to the nightclubs, Adi evokes the emotion of comfort to describe the discursive spaces on the Internet. Unlike his experience of nightclubs, Adi described how, through the website Gaydar, his embodied knowledge of online rapport reproduces normative romantic notions of intimacy and love.

Conclusion

Our chapter explores the embodied geographical knowledge of nightclubs that are pitched as the 'heart of gay life' in Seminyak to explore the relationship between travel, men, masculinity and sexuality. In doing so, our chapter helps address the lack of scholarship on the gay tourism industry that includes Indonesian men's experiences. We argue that an embodied geographical knowledge of nightclubs draws attention to the sexualised politics and emotional economy at work in these spaces. We focussed on the lived experiences of two Indonesian men. Their narratives defy understandings that sexuality and hospitality in the tourism industry are simply commodities exchanged between nationals/hosts and foreigners/guests (see Bishop et al., 1998). The embodied geographical knowledge of the nightclub suggests the importance of the affective push and emotion relationships triggered by sensual body-to-body encounters in helping understand the power geometries that shape sexualised subjectivities in tourism spaces. Alongside social, cultural and economic structures that marginalise gay Indonesian subjects, the affective capacities and emotional relationships between bodies are an integral component to blurring and reconfiguring nightclub spaces and subjectivities. We encourage others to consider embodied geographies to extend scholarship on travel, men, masculinities and places.

References

Atkins, G. L. (2012) *Imagining Gay Paradise: Bali, Bangkok and Cyber-Singapore*. Hong Kong: Hong Kong University Press.

Bernstein, E. (2001) 'The meaning of the purchase: Desire, demand, and the commerce of sex', *Ethnography*, 2(3): 389–420.

Binnie, J. (1995) 'Trading places: Consumption, sexuality and the production of queer space', in D. Bell and G. Valentine (eds) *Mapping Desire: Geographies of Sexualities*. Routledge: London, pp. 182–99.

Bishop, R. and Robinson, L. S. (1998) *Night Market: Sexual Cultures and the Thai Economic Miracle*. New York: Routledge.

Boellstorff, T. (2005) *The Gay Archipelago: Sexuality and Nation in Indonesia*. Princeton and Oxford: Princeton University Press.

Brown, M. (2000) *Closet Space: Geographies of Metaphor from the Body to the Globe*. London: Routledge.

Browne, K. (2007) 'Drag queens and drab dykes: Deploying and deploring femininities', in K. Browne, J. Lim and G. Brown (eds) *Geographies of Sexualities*. Hampshire: Ashgate, pp. 113–24.

Bystydzienski, J. M. (2011) *Intercultural Couples: Crossing Boundaries, Negotiating Difference*. New York: New York University Press.
Caluya, G. (2008) '"The rice steamer": Race, desire and affect in Sydney's gay scene', *Australian Geographer*, 39(3): 283–92.
Casey, M. (2007) 'The queer unwanted and their undesirable "otherness"', in K. Browne, J. Lim and G. Brown (eds) *Geographies of Sexualities*. Hampshire: Ashgate, pp. 125–36.
Chatterton, P. and Hollands, R. (2003) *Urban Nightscapes: Youth Cultures, Pleasure Spaces and Corporate Power*. London: Routledge.
Chauncey, G. (1994) *Gay New York: Gender, Urban Culture and the Making of the Gay Male World, 1890–1940*. London: Flamingo.
Collins, D. (2007) 'When sex work isn't work: Hospitality, gay life, and the production of desiring labour', *Tourist Studies*, 7(2): 115–39.
Connell, R. (1995) *Masculinities*. Cambridge: Polity Press.
Grazian, D. (2007) 'The girl hunt: Urban nightlife and the performance of masculinity as collective activity', *Symbolic Interaction* 30(2): 221–43.
Gubrium, J. and Holstein, J. (2009) *Analysing Narrative Reality*. London: Sage.
Hughes, H. (2006) *Pink Tourism: Holidays of Gay Men and Lesbians*. Wallingford: CAB.
Hughes, H. (2002) 'Gay men's holiday destination choice: A case of risk and avoidance', *NS* 35(4): 540–54.
Hughes, H. and Deutsch, R. (2010) 'Holidays of older gay men: Age or sexual orientation as decisive factors?' *Tourism Management*, 31: 454–63.
Jayne, M., Valentine, G. and Holloway, S. (2010) 'Emotional, embodied and affective geographies of alcohol, drinking and drunkenness', *Transactions of the Institute of British Geographers, International Journal of Tourism Research*, 4(4): 299–312.
Johnston, L. (2012) 'Site of excess: The spatial politics of touch for drag queens in Aotearoa, New Zealand', *Emotion, Space and Society*, 5(1): 1–12.
Laumann, E. O., Ellingson, S., Mahay, J., Paik, A. and Youm, Y (eds) (2004) *The Sexual Organization of the City*. Chicago: University of Chicago Press.
Loinaz, A. L. (2012) 'Lady Gaga banned in Indonesia following Islamic protests'. Available at: http://au.eonline.com/news/316301/lady-gaga-banned-in-indonesia-following-islamic-protests (accessed 16 September 2012).
Northcote, J. (2006) 'Nightclubbing and the search for identity: Transitions from childhood to adulthood in an urban milieu', *Youth Studies*, 15: 1–16.
MacRae, R. (2004) 'Notions of "us and them": Markers of stratification in clubbing lifestyles', *Journal of Youth Studies*, 7(1): 55–71.
Malbon, B. (1999) *Clubbing Dancing, Ecstasy and Vitality*. London: Routledge.
Prior, J. (2008) 'Planning for sex in the city: Urban governance, planning and the placement of sex industrial premises in inner Sydney', *Australian Geographer*, 39(3): 339–52.
Taylor, J. (2010) 'Queer temporalities and the significance of 'music scene' participation in the social identities of middle-aged queers', *Sociology*, 44: 893–907.
Thornton S. (1994) 'Moral panics, the media and British rave culture' in T. Ross and A. Rose (eds) *Microphone Fields: Youth Music and Youth Culture*. London: Routledge, pp. 177–92.
Thurnell-Read, T. (2012) 'What happens on tour: The premarital stag tour, homosocial bonding and male friendship, *Men and Masculinities*, 15(3): 249–70.

Valentine, G. and Skelton, T. (2003) 'Finding oneself, losing oneself: The lesbian and gay "scene" as a paradoxical space', *International Journal of Urban and Regional Research,* 27: 849–66.

Waitt, G. and Markwell, K. (2006) *Gay Tourism: Culture and Context.* Haworth Press, New York.

9

Ephemeral Masculinities? Tracking Men, Partners and Fathers in the Geography of Family Holidays

Rosalina Costa

Whereas holidays commonly represent leisure time away from work (for adults) or school (for children), family holidays seem to evoke a more complicated picture. Significantly, the notion of a holiday as a time 'without the watch' (Daly, 1996), where there is no need to manage family schedules linking house–school–work, is particularly heuristic. Additionally, the physical and/or psychological distance from the world of paid work helps to foster aspirations of an increased chance of absolute and unconditional enjoyment for parents and children being together. Parents recognise that the short time they spend during the year with children may somehow be filled by the investment in a holiday time together (Gillis, 2000). Even though children have other opportunities for holidays (e.g., with the grandparents or in summer camps), the possibility of enjoying quality time together is perceived as a temporary opportunity, almost ephemeral, to be the family that the constraints of the daily life do not allow.

Often desired and planned by adults for different places outside the familiar environment and 'away from home', holidays mark a 'distinct time' in the annual calendar of families. Time and place are, thus, fundamental coordinates to understand not only what family holidays actually are but also what they are to be: different and socially constructed experiences of dreams and anticipation. This is especially true for both middle-class and working-class families, since both daily face the lack and the pressure of time. For these, holidays are moreover idealised as a time and space away from unpaid household work and the subsequent gender division (Sinclair, 1997; Coltrane, 1998; Carr, 2011). However, holidays with small children always involve, in some degree, domestic or care work. How do individuals take into holidays the structure of gender roles that support quotidian routines? In the absence of paid work, which rules do men and women follow to allocate and complete seemingly daily tasks? Because individuals are now less defined by the traditional gender roles (Wharton, 2005; Lindsey, 2011), it is interesting to explore the ways they deal with such challenges.

In answering these questions, family leisure research has been traditionally dominated by the individual experiences of women's/mothers' point of view. Only more recently did men/fathers come to be recognised as important actors with a specific voice in the leisure experiences and practices of the family (Harrington, 2006; Shaw 2008; Schänzel and Smith, 2011). This chapter intends to add further discussion on this particular topic, bringing together the sociology of the family contributions into ritual literature (Van Gennep, 1909; Durkheim, 1912; Turner, 1967, 1969; Segalen, 1998) and travel and tourism studies (Graburn, 1989, 2001; Urry, 1990; Holden, 2005). Based on this plural perspective, the analysis of family holidays is conceptualised through the lens of *family practices* (Morgan, 1996, 1999, 2011) and, in particular, *family rituals* (Bossard and Boll, 1950; Wolin and Bennett, 1984; Imber-Black and Roberts, 1993; Etzioni and Bloom, 2004).

The discussion presented in the following pages is anchored in data derived from a broader sociological study into contemporary family rituals (Costa, 2011).[1] The core argument is that in heterosexual couples, the family–centred holiday presents an opportunity for the everyday gendered division of domestic labour and children care work to be negotiated and temporarily restructured. Through its focus on the role men play in family holidays, this chapter makes an important contribution to debates about masculinity, travel and familial relationships within heterosexual couples. While one needs to be attentive to the complex realities of modern family life and intrafamily relations towards travel and tourism, studying the traditional nuclear family remains an exciting and fruitful work to the extent that one cannot neglect the fact that this model is (still) behind many experiences and representations of families on holidays (Marshment, 1997; Schänzel, Yeoman and Backer, 2012).

The following section summarises the existing debates around family holidays and introduces the main concepts of family practices and family rituals. The methodology is presented in the section 'Following the tracks', with a specific focus on the construction of the sample, the data collection methods and procedures of analysis. The discussion of data is then structured through a three part and inductively constructed sequence around leaving home, being on vacation and returning back home. Finally, the chapter ends by pointing out some concluding remarks and the ways in which talking about ephemeral masculinities allows one to rethink the engendered experience of men on holidays.

Family holidays: crossed gazes on relaxation, fun and magic

The image of the family holiday is built upon the sense that it is a time for parents to be with their children, in a different environment from the rest of the year, and specifically devoted to escape, relaxation, fun or even magic. On the one hand, holidays are the time of the year when parents can focus on their parental role, unhindered by the competing demands of careers and paid work. On the other hand, they constitute an added opportunity for children to live their status as children (Prout, 2005). The photographs that underpin 'the family gaze' (Haldrup and Larsen, 2003) perfectly match this golden image of 'familyness', closeness and intimacy. In fact, family holidays are so child-centred (Cross, 2004; Obrador, 2012) that when describing them, individuals often use the comparison to another time – prior to having children – in which the vacations were more self-determined. At the heart of the difference between the 'before' and 'after' is the fact that contrary to what happened before having children, now everything from the type of food to mealtimes, must be carefully planned and scheduled (Costa, 2011).

Consistent with the idea that holidays are highly child-centred, the media and the tourism industry constantly publicise images of happy families in an environment of 'ideal' family holidays (Gillis, 1996; Marshment, 1997). In addition, the suspension of reality (Bryman, 2004) experienced in such an environment and the blurring of the boundaries between the adults and the children's worlds in it seems to be completely opposite of the educational mobilisation that characterises the everyday practices of nuclear families (Montandon and Perrenoud, 2001). However, the holiday desires of children and parents are sometimes at odds with each other with the contradictory expectations of adults seeking rest and children seeking fun. In fact, a more detailed analysis of those images unveils the ways in which tourist marketing seems to try to balance two things – fun and play for kids, rest and relaxation for parents. Indeed, marketing often involves the provision of services specifically designed for children (e.g., different playing areas for each age group, special or adapted pools for young children, kids' clubs, special menus and even babysitting facilities), which allow parents to relax while others look after their children (Obrador, 2012).[2]

Undoubtedly, family holidays are a fruitful window into the understanding of contemporary family dynamics, as illustrated by recent research output involving partners (Hilbrecht et al., 2008; Schänzel, 2010), the parent-child (Jeanes, 2010), or nuclear and extended family relationships (Gram, 2005). However, academic research on tourism has mainly focused on the individual lonely, detached and desocialised tourist and overlooked different group dynamics claiming for domesticity and sociality, namely families and children (Carr, 2011; Obrador, 2012; Schänzel, Yeoman and Backer, 2012). As Obrador puts it, 'families form the consumer base of many tourist resorts and attractions and yet tourist research has rarely taken notice of children's and families' holiday experiences' (2012: 402). This article helps to correct this gap. Specifically, it is concerned with masculinities and gender relations in family holidays. This is especially relevant as there is a lack of research into fathers and fathering on holiday in tourism studies (Schänzel and Smith, 2011), even though increasing research into the topic is claiming a place of its own right in the contiguous area of the family and leisure studies (Dermott, 2008; Jackson, 2012; Oechsle, Müller and Hess, 2012).

To understand the experiences of men on family holidays, the analysis that follows makes use of the heuristic power enhanced by the concept of family practices. In this endeavour, I closely follow the work of David Morgan (1996; 1999; 2011), according to whom *family practices* are presented as a powerful theoretical and conceptual tool able to capture the flow, fluidity and meaning of contemporary families. Morgan suggests that this task might be done by understanding families not for what they are, but for what they do. Instead of taking as a starting point 'finished' categories such as family, marriage or parenthood, Morgan advocates that one looks at the many family-related activities undertaken by partners, parents, children and relatives, examining at the same time meanings, expectations and obligations associated with them. The family is therefore seen not as a fixed entity but as the changing product of the complex daily interactions of family members. Through the focus upon the dynamics of 'personal life', the traditional study of the family unit can be extended and enriched (Smart, 2007).

Specifically, this chapter will analyse the particular kinds of family practices fitting into a larger category – family rituals – that take place during family holidays. According to Wolin and Bennett (1984), one can distinguish between family celebrations, family traditions and patterned family interactions. Family rituals thus comprise days or occasions during the year or the life of the individuals, as well as moments in the daily routine. In short, family rituals can be defined as any prescribed practice arising from family interaction, targeting a specific purpose and holding a symbolic or special meaning (Bossard and Boll, 1950; Wolin and Bennett, 1984; Imber-Black and Roberts, 1993; Fiese, 2006).

In this chapter, I narrow the analysis to the study of family holidays perceived as family rituals. The chapter draws heavily on the work of the anthropologist Victor Turner whose work took the classifications of Van Gennep (1909) in *Rites of Passage* and deepens the distinction between 'preliminary', 'liminal' and 'post-liminal' rituals (Turner, 1967, 1969). His analysis focuses more specifically on the liminal phase, in which the individual shows the peculiarity of escaping from classical sociological classifications, since he/she is located between two phases. This characteristic of being in an intermediate position, neither wholly one thing nor another and putting the individual in a situation of some invisibility from a social perspective became known as being 'betwixt and between' (Turner, 1964). Various studies of tourism have incorporated this contribution from anthropology arguing that the tourist experience can be understood as a bounded liminal experience, aiming for either fun or escape, far from the structure, roles and routines of everyday life (Graburn, 1989, 2001; Urry, 1990; Holden, 2005). Regarding family holidays, such conceptualisation means that one can envisage them as an occasion where families go out of regular time and locations into 'special protected time and space' (Imber-Black and Roberts, 1993: 210). Having this conceptual background, the next section presents and details the methodological decisions behind this study and the empirical approach to family holidays.

Following the tracks: an empirical look at family holidays

This chapter draws on the qualitative analysis of the accounts of family holidays provided by individuals interviewed in the context of a larger sociological inquiry aiming an in-depth understanding of family rituals (Costa, 2011). A total of 30 middle-class individuals, 15 men and 15 women, with a mean age of 38 years old, living in an urban medium-sized city (Évora) in the south of Portugal (southern Europe) in diverse family contexts and with at least one small child between the age of 3 and 14 years old, were interviewed for this study. The study focused on middle-class individuals because they daily face the lack and the pressure of time while being best equipped financially to 'invest' in family holidays (Carr, 2011). Informants were identified from their educational and professional capitals. Explicitly, they were selected empirically from a minimum level of education, which included the completion of secondary education, and occupations within the first focus groups of the Portuguese National Classification of Occupations (IEFP, 2001).[3] A purposive and snowball procedure allowed for the construction of a 'theoretical sampling' (Glaser and Strauss, 1967). Specifically, I followed a process of sampling through multiple cases, the 'sample by homogenisation' (Pires, 1997). All interviewees were unknown to the researcher and their participation was voluntary and without payment.

Original data were collected in 2009 through episodic interviews (Flick, 1997, 1998), a particular kind of a semi-structured interview aiming at the detailed description of a concrete experience and related meaning through the form of a narrative. The interviews were carried out individually and conducted in several contexts: in different spaces of the everyday life (in the house of the respondents or in their workplaces), as well as in distinctive periods during the day, week and year. The average length of interviews was of an hour and 40 minutes. Insofar as the advantages of carrying out joint interviews are well-documented, in this study, the option for individual interviews is justified since the aim was to generate a wealth of detail from a few cases, not to contrast and compare both members of the couple's accounts (Bjornholt and Farstad, 2012).

All interviews were digitally recorded and later subject to a verbatim transcription before being coded and interpreted through a thematic and structural content analysis (Bardin, 1977) using the qualitative data analysis software NVivo. The result is a reconstruction of the original accounts, now intertwined with the researcher's critical discussion of the data.[4]

Deriving from the particular sample studied, the family holidays discussed hereafter refer to holidays involving a heterosexual nuclear family (parents and children) at a specific location outside the family residence. As observed, this comprised a very limited period of typically either one or two weeks devoted to getaways, relaxation and fun. In all the cases, this always involved, in some degree, some days for holidays at the beach, during the summer.

The next section reports data analysis and discussion and is anchored upon an analytical strategy that implies 'following' men within the dimensions of partnering and fathering along the diverse geographies of family holidays. When presenting and discussing the empirical findings on family practices and meanings, a special focus on men and their correlated dynamics to gender issues is undertaken. This strategy is based upon a two part argument. Firstly, it accepts the limits of physically following these men through firsthand observations. Hence, from the moment they leave home, during all of the period they are away, and again when they return home, their practices will be reconstructed across time and space through the accounts collected during the interviews. Secondly, owing to the added value of using the timeline with the three key moments of 'leaving' home, 'being' on vacation, and 'returning' back home were not aprioristic defined as dimensions of analysis. Instead, they have emerged as meaningful categories in the broader context of the qualitative content analysis and proved to be particularly heuristic.

Although the focus throughout this chapter is on men, it is important to stress that their practices come to us from a dual source: either by their own accounts or, alternatively, through the narratives that their female partners elaborate on their behaviour. Hence, using the gender dynamics as a 'codification of knowledge' (Aitchison, 2001), further understanding is gained as information coming from the two sides of the story is overlaid on such a complex puzzle of family interaction. By the end, this chapter suggests that studying men using the lens of travel and tourism offers both the common reader and the researcher a fruitful opportunity to demonstrate the wider contribution of this topic into the social understanding of 'new masculinities' and contemporary family lifestyle.

Leaving home: the invisible luggage

For interviewees, holidays are mainly a time away from the tight time pressures and gendered division of labour felt at home throughout the rest of the year (also see Gilli and Ruspini, this volume). The time of leaving home for vacation is a striking example. Maria dos Anjos makes a paradigmatic description that crosses the accounts of both men and women with regard to the time that formally marks the beginning of the holiday season away from home:

> We leave when everything is ready. After that we leave. We put our stuff in the car and drive away, never with the stress of having to arrive at a certain time, or 'at that time, I have to be in such a place' No! [...] I prepare the suitcases and...put the bags on the doorstep. Then he [husband] puts them into the car, and manages the spaces inside the car. Then we leave. Then sometimes along the way he asks

> me, 'Did you bring this?' 'Did you bring that?' (Maria dos Anjos, 41, married, mother of a girl of 13 and a boy of 11 years old)

Similarly, when asked to talk about family holidays, one of the husbands, Eduardo, is particularly enthusiastic about the holidays as a moment of family getaways. He is the one responsible for putting in motion the longstanding yet exciting process of going on vacation:

> Let's say that I'm in command of the logistics [*laughs*]. So, where are we going this year? This year we go ... we think ... 'What if this year, we went over there?' Let's say the decision is joint, but after that the logistics, that is mine [...] It is a task that we start earlier, because we have to book it, and it is a hard work ... Then everything will be ready ... 'Two weeks to go.' 'One week left.' 'Let's go!' Well, it's really a great excitement! (Eduardo, 37, married, father of a girl of 10 and a boy of 7 years old)

These two excerpts clearly show that whilst the anticipation around the holidays is experienced in the plural – as a family project – the ways in which women and men talk about the actual preparation for such trips away is perceived as strongly gendered. Maria dos Anjos speaks of the moment when 'we leave', or when 'we put our stuff ...', and Eduardo discusses how before the summer, they (as a family) always think, 'Where are we going this year?' or, 'What if this year, we went over there?' However, a careful and attentive analysis of the discourses shows that when talking about the preparation for holidays, suddenly the use of the plural 'we' is replaced by the singular 'I' or 'he/she' [the partner]. Hence, Maria dos Anjos then clarifies that 'I prepare the suitcases' and '[I] put the bags ...' while he [husband] 'puts them into the car', and '[he] manages the spaces inside the car.' Eduardo also clarifies that despite being a family project and the result of a joint decision, he is in command of the holiday's logistics: '[...] the decision is joint, but after that, the logistics, that is mine.'

Pedro, a 35-year-old father of sons aged 7 and 4, describes the work behind the moment of leaving home for holidays in a similarly gendered way: 'In terms of preparation [of the suitcases], 75% are hers [his wife] and 25% is mine [laughs]. Then, the storage is mine. The car has a good trunk [...] *and we want to take it all*' (Pedro, 35, married, father of two boys, 7 and 4 years old).

Preparation for the holidays thus gives visibility to the persistence of the division of labour by gender in the heterosexual family (Aitchison,

2001). The members of the couple are unanimous in recognising that women are more often responsible for preparing the bags, and for selecting and storage of toys and clothes for children and adults, while men frequently take on storage of bags in the trunk of their car and driving to the destination. This fact allows for understanding why sometimes men question or blame women for any failure, as well as for any eventual overloading of bags on vacation.

Furthermore, these results come to reinforce the idea of the permeability of the boundaries between the private/domestic and public/holiday spaces as far as gender is concerned (Aitchison, 2001). In fact, holidays do not establish an absolute and utter opposition between these two spheres; rather, they point out their porosities and bridges. In these seemingly innocuous accounts, the interviewees thus give a clear indication of how gender distinctions on which rests the division of tasks in the daily life will accompany them in their holidays.

On holiday: the secret of gender (seemingly) on standby

Notwithstanding some recent changes causing a greater participation of men in domestic life, Portuguese families still experience a very unequal household division of labour. While Portuguese women participate strongly in the paid work economy, especially on a full-time basis, they also undertake the majority of household chores – both in number of tasks and time spent on them – such as cooking, washing, and cleaning (Aboim, 2010; Wall, Aboim and Cunha, 2010). This is also true in the studied sample, where data shows that during the year, a pre-defined allocation of roles in the couple enables the tasks' accomplishment, often with a greater weight for women than for men. What about vacation time? How is the structure of gender roles transported from the daily life to this special time, (apparently) away from routine? These questions are particularly important, since holidays are greatly attached to paid work rhythms and, thus, very pervasive in an egalitarian discourse of the holidays as a right of both members of the couple.

Empirical data suggests that gender differences during the vacation period seem to be smoothed over or even disappear from the individual discourses. Men and women thus tend to speak about holidays as different from daily life: genderless, as if there were no gender divisions structuring the tasks that unavoidably form everyday existence. Several strategies were utilised by couples to avoid confrontation concerning the unequal distribution of tasks, which otherwise may become visible on holidays. The first strategy is a role reversal, by which men who usually do little housework during the year allow themselves to participate more or more actively during the vacation period. This strategy, reported by the women and assumed by men, is possible since holidays are perceived as a 'special time' (Imber-Black and Roberts, 1993). This ephemeral setting in the annual calendar of families seems to allow routine tasks such as cooking or bathing the children to be undertaken by those who do not usually do them: men. However, these exceptions only come to confirm, as we shall see later, the rule of gender distinctions on which the household's routines across the year are based.

This is the case for José, who admits cooking more on holidays than during the rest of the year: 'Yes, of course I cook more at that time [on holidays] than in the rest of the year! [...] It's because of time. Mainly because there's more time to do it...' (José, 38, married, father of a girl of 3 years old). José admits that on holidays, he cooks more than during the year, although generally they go out to eat at night. However, especially at lunch, after the beach, he is the one who most often deals with the meal. The greater availability of time (and the consequent lower pressure to be effective within a short period) enables José to participate without the risk of not being efficient. That would, indeed, constitute a problem during the year, but not on vacation time.

When talking of the time when she was married, Dora also admits that her ex-husband did things differently on vacation than he did during the year. The social representation around the holidays and the understanding that the right to holidays is for the two members of the couple forces an injunction to share, or at least in a bigger share than the usual: 'The two of us are on vacation; it is meant for the two to be on vacation! Not that I get to do the things and you do not. [...] We were both seated; things could be undone; we were watching a movie; we both watched the movie, then we get up, and we both do things. Or vice-versa' (Dora, 33, divorced, mother of a girl of 14 years old).

Let us now return to Maria dos Anjos, already quoted, who during the rest of the year almost singlehandedly performs all domestic chores while also working as a nurse. During that period when she goes to the beach, the few tasks that she has to undertake, such as washing the dishes for breakfast or lunch, are completed by her husband. Usually he is away from such tasks during the year, it is as if he recognises that she also needs a vacation from that kind of work:

> Why? I do not know... maybe because he [husband] also wants to give me a little more rest during those periods, so the task of washing the dishes is assumed by him on vacation. He immediately assumes

> this responsibility, as soon as he is on vacation. For me, it's great! It's great! Sometimes I say, 'Let me. I'll do it', but he goes 'No, no, on holidays the task of washing the dishes is mine.' Because during the year, his availability is also lower. He makes 12-hour shifts, and is less available. Then, on vacation, this task... I don't argue! It's his; it is his task to wash the dishes!' [*laughs*] (Maria dos Anjos, 41, married, mother of a girl of 13 and a boy of 11 years old)

However, in most of the situations, the question of 'who does what' is not resolved through a greater sharing of tasks. Instead, the buying of services allows in some cases the feeling of 'rest' that women, in particular, experience on vacation. The investment, savings and effort channelled for a fully paid holiday in which there is no need to do anything is the second strategy by which these couples achieve the suspension of gender differences in the distribution of tasks, as far as holidays are concerned.

Joana complains that during the year, her husband spends most of his time in front of the computer. The division of labour at home is in the proportion of '90% for me and 10% for him', she adds, yet, on vacation, 'it is completely different'. According to Joana, it is the absence of paid work for both that encourages equality when accomplishing the tasks that need to be done, even on holidays:

> It is incredible, but since there [on holidays], we no longer have to be on the computer to work; we are more free... everything is much more divided. As far as cooking goes, because I'm on vacation, I'm not worried about cooking. We buy precooked food. No! 'Wait a moment!' Food is only for the baby and, and... baby food can be purchased, and it is very good! He likes the same. I'm not worried with that, if the goal is to rest, it is to rest! Then we share things differently, very different. We should transpose it to here [the daily life]... but then... 'Oh, I have to finish an article.' 'Oh, and now someone called me and so on.' 'I have to finish something because I'm going to attend a congress somewhere.' (Joana, 35, *de facto* union, mother of two boys, ages 6 and 1)

The equal division of tasks Joana perceives is not at the expense of greater participation by her husband; rather that equality is 'purchased' through acquiring goods, namely precooked food, including baby food. Nonetheless, Joana recognises that being on holiday also allows her partner to get closer to and spend more time doing things with their children.

Regarding fatherhood, holidays also allow greater involvement with the children, apparent in the planning, anticipation, and decision-making processes behind the choice of destinations and the daily activities there. More than during the year, parents, and fathers in particular, tend to listen more attentively to their child(ren), so they can meet their desires, take them where they want to go, and thus make them happy (Oechsle, Müller and Hess, 2012). A clear example of this is António, who admits totally giving in to the interests of his two children when organising family holidays and leisure travel in general:

> We have a map of Portugal, and then we put a pushpin in all the places where we've been, with the date and all that. [...] And usually we even go in January to the BTL[5] in Lisbon to collect brochures and information on places, activities to do with kids. [...] This Saturday, we go down [to the Algarve[6]] for a fortnight's holiday [...] at their [his children] wish, this year we will go also to Gibraltar [...]. We also agreed that at least one day we must go to Seville. This is because the last time we didn't have the time to visit the Gold Tower[7] and some other palace, which was already closed by the time we were there. Anyway, we have to go one day to Seville! The older one [his oldest son] also wants to go a day to Huelva because he wants to watch some game there. (António, 41, married, father of 2 boys, aged 13 and 10)

Interviewed just a few days before going on holiday to the Algarve, the detailed description that António gives in this excerpt clearly illustrates the anticipation of this period as a time when he can be 100% dedicated to his children (Daly, 1996; Harrington, 2006).

Regarding intimacy, the holidays also seem to allow for a greater proximity and intimacy between partners, who are often set apart due to the multiple demands and requests of daily routines. Eduardo has been married for 12 years, and he believes that holidays are important to create new opportunities for encounters between both members of the couple:

> The routines, the work, the food, the house, the school, the children... I mean... it stresses, and... it creates some friction... sometimes you get fed up with it! There are frictions... And these moments [family holidays] are also important for us to see that... so that we can realise that... so that we can evaluate things. Taking advantage of being together... Taking advantage of the moment. Let's say: to find us again. (Eduardo, 37, married, father of a girl of 10 and a boy of 7)

Besides the moment of leaving home, gender is visible throughout the holiday period. The insistence on women not having to do 'anything' is based on gendered notions of what a woman wants/needs from the time spent away from home. The same happens with men's insistence on doing chores, or helping with the kids. The men I interviewed have a different and unusual time on holidays, regarding domestic work, and care and intimacy with their partners and children. This seems to be triggered by the apparent absence of gender forces that structure the yearly routines, which confirms the experience of the holiday as a bounded liminal timespan experience (Graburn, 1989; 2001). Particularly, the holiday period can be perceived as a liminal one (Turner, 1964), allowing for the rise of behaviours that otherwise would not be 'normal', namely an extreme participation and engagement by men with the tasks and social such as cooking and childcare.

Additionally, this behaviour also meets the expectations of their female partners and children, who anticipate a more present and closer partner and father. In the cases analysed here, deriving from a specific sample where both partners work full-time yet with a clear inequality as far as the participation in housework is concerned, the ephemeral masculinities seem to be a way for men to approach the mainstream cultural images of 'fatherhood in late modernity' (Oechle, Müller and Hess, 2012). However, as we saw, this is ephemeral and only occurs in the temporary and bounded limit of a week or a fortnight's experience.

Back home: where are the men?

While the scenario of leaving home for holidays is always reported by the men with reference to themselves and their female partners, returning home is most frequently reported as a women's matter. Indeed, men rarely reported the homecoming and the related tasks. Perhaps naively, this leads one to ask: 'Where are the men when these families return home after holidays?'

To address this question, I now introduce Filipa. She is 39 years old, married to a military officer, and the mother of three children – two boys, ages 9 and 7, and a girl, age 3. Filipa is generally responsible for almost all the housework during the year. Usually the family goes to the Algarve to have some beach time for the kids. In fact, according to her, holidays are for enjoying time spent with the children. When travelling, Filipa carries just the right quantity of clothes, so there is no need to wash or iron during vacation: 'Clothes... we do not carry much! Each one goes with only a small bag [*laughs*] [...] After we return from vacation, *then the mother* does the laundry! [*laughs*]. However, only when we come back.' This excerpt confirms how the issue of the laundry, and the underlying gender rules are not really changed on vacation, only 'postponed'. When arriving home, Filipa rapidly embodies the gender division of work from which she was (seemingly) kept away. The end of the liminal experience of holidays thus brings up the effectiveness of the unthinking routines, which again meets with their persistent gender divisions (Kaufmann, 1997).

Summing up, men tend to emphasise the experience of family holidays as a time of either the family or their own, allowing them to perform full-time the role of partner and father. Still, women insist on incorporating in their discourses descriptions of family holidays' images of a time away from work. This happens because work means different things for men and women. While the absence of paid employment seems to be a sufficient argument to allow these men to be on holiday, women seem to need a further guarantee that no kind of work will interpose on their rest, including unpaid domestic chores. Hence, knowing that Portuguese women are the ones who, within the relationship, tend to face the double shift on a daily basis, both longing for and actually being on holiday is to guarantee absolutely no work for themselves. This somehow forces men to engage more in tasks, while women run from them. Seemingly contradictory, this maximises the advantage that men draw from the holidays: gladly, they can benefit more than their female partners from the longed-for breaking up of their daily lives, and, moreover, engage in new masculinities (Wall et al., 2010). Subtly, this is the power of the gender, which even covertly, by the egalitarian right of holidays, prints their determinations with absolute strength and efficacy.

Conclusion

This chapter has addressed the issue of travel, family and relationships. Bringing together contributions from family, leisure and tourism studies and bound by the sociological perspective, it discussed the ephemeral character of some masculinities arising in the special setting of heterosexual nuclear family-centred holidays of working couples with children and their related mundane routines.

Specifically, as holidays with small children always involve in some degree domestic or care work, this text addressed the ways individuals take the structure of gender roles that support everyday routines (e.g., cooking, laundry and child care) into holidays. In so doing, I focused on the gender dynamics and followed the men along the diverse geographies of family holidays: from the exact moment they left home, during the period away, and when returning home. In that specific journey, I was guided by their own and their female partners' words.

The qualitative in-depth analysis that I followed supports the conclusion that gender is a pervasive and structuring variable that indeed 'travels' with families on holidays. The moment of departure and leaving the house is symbolic of how gender works within the family. Metaphorically, it travels with the couple, just like the luggage they carry. Therein, the physical displacement that many individuals experience when desiring and planning holidays away from home, unveils much more a continuum of daily life than a break from it.

In the context of a particular and small-scale sample, limited to nuclear and heterosexual families, the data revealed a strong discursive injunction to equality encouraged by the absence of paid work on holidays. As a result, the question of 'who does what' is partially resolved by reverting to a greater sharing of tasks or even role reversal by using the paid market economy (e.g., buying precooked food), or by putting gender on standby, thus delaying in time the confrontation – for women – with these tasks when they return home. Through the acquisition of goods or services such as eating out, or by taking a sufficient amount of clothes so that there is no need to wash or iron, women, in particular, experience a feeling of rest on holidays. At the same time, this leaves men freer to experience new or different dimensions in their roles as both partner and father during this period. However, I suggest that this temporary experimentation with alternative gender relations within the family, most evident in the alternative arrangements made for domestic tasks during the holiday, is soon abandoned on returning home.

In conclusion, making use of the ritual literature, I argue that holidays can be seen as a bounded liminal time experience. Empirical findings confirm that family-centred holidays set for a special and protected time and space *in*, *with* and *for* the family. Away from both paid and unpaid work, family holidays seemingly allow for the suspension of gender differences, yet end up reinforcing them, while encouraging a traditional male bread winner role in order for the husband to provide the desired family holiday experience once a year. In heterosexual couples that face a daily gendered division of housework and child care, being in the specific scenario of family-centred holidays allows for the emergence of ephemeral masculinities pervasive to partnering and fathering that fade away when they return home.

Focusing on masculinities on holidays, this study definitely enlarges the contemporary visibility of the significance of the family in tourism and challenges the binary and tight oppositions underpinning the Western understandings of tourism as desocialised and set apart from the sphere of the home and the everyday life of the individuals.

Notes

1. The major research mentioned above was developed at ICS – UL, University of Lisbon (Portugal), with a grant from FCT, the Portuguese Foundation for Science and Technology (Ref. SFRH/BD/38679/2007), and supervised by Ana Nunes de Almeida, whom I thank deeply.
2. See, for instance, the British website Child friendly at http://www.child-friendly.co.uk/. Therein, family holidays are advertised specially for the regions of Caribbean, Cuba, Croatia, Cyprus, Egypt, Greece, Italy, Portugal, Spain, Turkey, Canary Islands, Balearics and Dubai (accessed on 7 January 2013).
3. Particularly the ones included in Group 2 – Intellectual and Scientific Occupations; Group 3 – Technicians and Associate Professionals; Group 4 – Clerical Support Workers, and Group 5 – Services, Protection and Security Workers.
4. All excerpts are identified by a pseudonym given by the researcher, followed by the age of the informant, and a brief indication of the parental status and sibling context.
5. 'Bolsa de Turismo de Lisboa'. It is an international tourism fair that takes place annually in Lisbon, Portugal.
6. Algarve is the main sun and beach Portuguese tourism region, located in the south of the country.
7. Orig., Torre del Oro, Seville, Spain.

References

Aboim, S. (2010) 'Gender cultures and the division of labour in contemporary Europe: A cross-national perspective', *The Sociological Review,* 58(2): 171–96.

Aitchison, C. (2001) 'Gender and leisure research: The 'codification of knowledge', *Leisure Sciences,* 23(1): 1–20.

Almeida, A. N. and Vieira, M. M. (2006) *A Escola em Portugal – Novos Olhares, Outros Cenários*. Lisboa: ICS.

Bardin, L. (1977) *L'Analyse de Contenu*. Paris: Presses Universitaires de France.

Bjornholt, M. and Farstad, G. R. (2012) 'Am I rambling?: On the advantages of interviewing couples together', *Qualitative Research,* September 2012.

Bossard, J. H. S. and Boll, E. (1950) *Ritual in Family Living – A Contemporary Study*. Philadelphia: University of Pennsylvania Press.

Bryman, A. (2004) *The Disneyization of Society*. London: Sage.

Carr, N. (2011) *Children's and Families' Holiday Experiences*. London: Routledge.

Coltrane, S. (1998) *Gender and Families*. London: Pine Forge Press.

Costa, R. (2011) *Pequenos e Grandes Dias: Os Rituais na Construção da Família Contemporânea*. PhD Thesis in Social Sciences – specialization 'General Sociology'. University of Lisbon: Institute of Social Sciences of the University of Lisbon (ICS-UL). http://hdl.handle.net/10451/4770.

Cross, G. (2004) 'Just for kids: How holidays became child centered', in A. Etzioni and J. Bloom (eds) *We Are What We Celebrate – Understanding Holidays and Rituals*. New York: New York University Press, pp. 151–64.

Daly, K. J. (1996) *Families and Time – Keeping Pace in a Hurried Culture*. Thousand Oaks: Sage.

Dermott, E. (2008) *Intimate Fatherhood: A Sociological Analysis*. New York: Routledge.

Durkheim, É. (1912) *Les Formes Élémentaires de la Vie Religieuse. Le Système Totémique en Australie*. Paris: Les Presses Universitaires de France.

Etzioni, A. and J. Bloom (eds) (2004) *We Are What We Celebrate – Understanding Holidays and Rituals*. New York: New York University Press.

Fiese, B. H. (2006) *Family Routines and Rituals*. New Haven and London: Yale University Press.

Flick, U. (1997) 'The episodic interview: Small-scale narratives as an approach to relevant experiences', Series Paper, London. Available at: http://www2.lse.ac.uk/methodologyInstitute/pdf/QualPapers/Flick-episodic.pdf (accessed 29 October 2010).

Flick, U. (1998) *An Introduction to Qualitative Research*. London: Sage.

Gillis, J. R. (1996) *A World of their Own Making. Myth, Ritual, and the Quest for Family Values*. Cambridge: Harvard University Press.

Gillis, J. R. (2000) 'Our Virtual Families: Toward a Cultural Understanding of Modern Family Life', *The Emory Center for Myth and Ritual in American Life – Working Paper*, 2. Rutgers University/Department of History. http://www.marial.emory.edu/pdfs/Gillispaper.PDF.

Glaser, B. G. and Strauss, A. L. (1967) *The Discovery of Grounded Theory: Strategies for Qualitative Research*. Chicago: Aldine Publishing Company.

Graburn, N. (1989) 'Tourism: The sacred journey', in V. L. Smith (ed.) *Hosts and Guests: The Anthropology of Tourism*. University Of Pennsylvania Press, Philadelphia, pp. 17–23.

Graburn, N. (2001) 'Secular ritual: A general theory of tourism', in V. L. Smith and M. Brent (eds) *Hosts and Guests Revisited: Tourism Issues in the 21st Century*. New York: Cognizant Communications Corporation, pp. 42–50.

Gram, M. (2005) 'Family holidays. A qualitative analysis of family holiday experiences', *Scandinavian Journal of Hospitality and Tourism*, 5(1): 2–22.

Haldrup, M. and Larsen, J. (2003) 'The family gaze', *Tourist Studies*, 3: 23–46.

Harrington, M. (2006) 'Family leisure', in C. Rojek, S. Shaw, and A. Veal (eds), *A Handbook of Leisure Studies*. Houndsmills: Palgrave Macmillan, pp. 417–32.

Hilbrecht, M., Shaw, S. M., Johnson, L. C. and Andrey, J. (2008) '"I'm home for the kids": Contradictory implications for work-life balance of teleworking mothers', *Gender, Work and Organization*, 15(5): 454–76.

Holden, A. (2005) *Tourism Studies and the Social Sciences*. London: Routledge.

IEFP (2001) *Classificação Nacional de Profissões – versão 1994*, 2nd ed. Lisboa: IEFP.

Imber-Black, E. and Roberts, J. (1993) *Rituals for Our Times: Celebrating, Healing and Changing our Lives and Our Relationships*. New York: Harper Perennial.

Jackson, B. (2012) *Fatherhood.* New York: Routledge.

Jeanes, R. (2010) 'Seen but not heard? Examining children's voices in leisure and family research', *Leisure/Loisir*, 34(3): 243–59.

Kaufmann, J.-C. (1997) *Le Coeur à l'ouvrage – Théorie de l'action Ménagère*. Paris: Édition Nathan.

Lindsey, L. L. (2011) *Gender Roles: A Sociological Perspective*. Upper Saddle River, New Jersey: Pearson.

Marshment, M. (1997) 'Gender takes a holiday: Representation in holiday brochures', in M. Thea Sinclair (ed.), *Gender, Work and Tourism*. London: Routledge, pp. 15–32.

Montandon, C. and Perrenoud, P. (2001) *Entre Pais e Professores: Um Diálogo Impossível?* Oeiras: Celta.

Morgan, D. H. J. (1996) *Family Connections – An Introduction to Family Studies*. Cambridge: Polity Press.

Morgan, D. H. J. (1999) 'Risk and family practices: Accounting for change and fluidity in family life', in E. B. Silva and C. Smart (eds) *The New Family?* London: Sage, pp. 13–30.

Morgan, D. H. J. (2011) *Rethinking Family Practices*. Hampshire: Palgrave Macmillan.

Obrador, P. (2012) 'The place of the family in tourism research: Domesticity and thick sociality by the pool', *Annals of Tourism Research*, 39(1): 401–20.

Oechsle, M., Müller, U. and Hess, S. (eds.) (2012) *Fatherhood in Late Modernity: Cultural Images, Social Practices, Structural Frames*. Opladen: Verlag Barbara Budrich.

Pires, Á. (1997) 'Échantillonnage et recherche qualitative: Essai théorique et méthodologique', in J. Dans Poupart et al. (eds) *Enjeux Épistémologiques et Méthodologiques*. Montreal: Gaëtan Morin, pp. 113–67.

Prout, A. (2005) *The Future of Childhood*. London: Routledge.

Schänzel, H. A. (2010) *Family Time and Own Time on Holiday: Generation, Gender, and Group Dynamic Perspectives from New Zealand*, PhD thesis in Philosophy in Tourism Management, Victoria University of Wellington.

Schänzel, H. A. and Smith, K. A. (2011) 'The absence of fatherhood: Achieving true gender scholarship in family tourism research', *Annals of Leisure Research*, 14: 143–54.

Schänzel, H., Yeoman, I. and Backer, E. (eds.) (2012) *Family Tourism: Multidisciplinary Perspectives*. Bristol: Channel View Publications.

Segalen, M. (1998) *Rites et Rituels Contemporains*. Nathan: Paris.

Shaw, S. (2008) 'Family leisure and changing ideologies of parenthood', *Sociology Compass*, 2: 688–703.

Sinclair, M. T. (1997) 'Issues and theories of gender and work in tourism', in M. Thea Sinclair (ed.), *Gender, Work and Tourism*: London: Routledge, pp. 1–14.

Smart, C. (2007) *Personal Life – New Directions in Sociological Thinking*. Cambridge: Polity Press.

Turner, V. (1967) *The Forest of Symbols: Aspects of Ndembu Ritual*. Ithaca, N.Y: Cornell University Press.

Turner, V. (1969) *The Ritual Process. Structure and Anti-Structure*. New York: Aldine de Gruyter.

Turner, V. W. (1964) 'Betwixt and between: The liminal period in rites de passage', in American Ethnological Society, *Symposium on New Approaches to the Study of Religion: Proceedings*. Seattle: University of Washington Press, pp. 4–20.

Urry, J. (1990) *The Tourist Gaze*. London: Sage.
Van Gennep, A. (1909) *The Rites of Passage*. London and Henley: Routledge and Kegan Paul.
Wall, K., Aboim, S. and Cunha, V. (2010) *A Vida Familiar no Masculino. Negociando Velhas e Novas Masculinidades*. Lisboa: CITE.
Wharton, A. S. (2005) *The Sociology of Gender: An Introduction to Theory and Research*. Oxford: Blackwell Publishing.
Wolin, S. J. and Bennett, L. A. (1984) 'Family rituals', *Family Process,* 23(3): 401–20.

Part III

Sex, Sexuality, Tourism and Masculinity

10

Taiwanese Men's Wife-Finding Tours in Southeast Asian Countries and China

Chun-Yu Lin

The total number of migrants over the last two decades who have gone to Taiwan for the purpose of international marriage currently stands at more than 450,000. The greatest proportion of them are women, and most come from Mainland China (MC) and Southeast Asian countries (SEA), such as Vietnam, Indonesia and Thailand.

The phenomenon relates not only to the intensive economic interactions between SEA, MC and Taiwan, but also to the rapid growth of commercial matchmaking agencies in recent years. Many matchmaking agencies began as travel agencies that organised trips to SEA and MC. However, many of these companies have extended their businesses to offer 'wife-finding tours' for Taiwanese men wishing to travel to SEA and MC to meet women and then, if the tour is successful, marry them. On these tours, Taiwanese men usually stay for a week or ten days, during which time they can meet as many SEA and MC women as they wish until they choose one they are satisfied with. Many women in my research met their Taiwanese husband by using one of these commercial matchmaking agencies.

The international marriage matchmaking industry, through which women and Taiwanese men pair up, represents a wider phenomenon of the commodification of intimacy and the commercialisation of women's bodies. In addition, the matchmaking process subsequently enhances men's sense of masculine power, and this characterises my analytical framework. Based on document analysis and in-depth interviews with 50 women immigrants in Taiwan, I will discuss the mechanisms and processes that commercial matchmaking agencies use to bring both Taiwanese men and foreign women together. From some women's accounts, commercial matchmaking agencies arranged for them to stay in a 'brides-to-be' camp until a Taiwanese man chose them. In such camps, women not only take training courses to learn how to be suitable candidates for marriage but also work for those companies, providing domestic labour and doing other tasks within the camps themselves. This chapter will show how matchmaking companies prepare women who stay with them, and discuss how the gendered expectations and interaction between both Taiwanese men and foreign women work in the matchmaking process.

Matchmaking agencies and the research method

In recent years, advertisements from commercial matchmaking agencies that offer matchmaking services to foreign brides from many countries, such as Vietnam, Indonesia and China, could be seen on billboards in the street and in the advertisement sections of newspapers in Taiwan. The advertisements of profit-oriented matchmaking agencies usually mention a wife-finding tour:

> Our company provides a speedy matching service for a reasonable price which includes a return ticket between Taiwan and Vietnam for Taiwanese men. Our services include not only matchmaking, but also a wedding banquet, help for new couples to pass the marital interview, and help with the procedure for applying for the official certification of marriage. (*The Journalist*, 4 April 2009: 8)

Commercial matchmaking companies guarantee a 'convenient', 'speedy' and 'all-inclusive' service which is very attractive to many Taiwanese men, as it ensures they will be able to take a wife back to their home country (Wu et. al., 2008: 144). The marriage matchmaking business in Taiwan is growing rapidly, and it has been supplying services to foster international marriages since the 1990s (Chang, 2001: 2; Weng, 2007: 5). For Wang (2007), these international marriages can be understood as 'commodified cross-border marriages' (p. 707).

Current studies in Taiwan have examined the blossoming phenomenon of the international matchmaking business, seeing it as one result of the unbalanced development of global capitalism, and as the commodification of interpersonal relationships (Hsia, 2002: 3, 61; Chang, 2001: 20). Further still, some have suggested that it reflects the law of supply and demand in a market economy (Simons, 2001: 91; Wang and Chang, 2002: 93; Luehrmann, 2004: 872). There is no explicit service that exists in the matchmaking businesses; however, women are commodified during the matchmaking process, and some are treated as products. In this sense, international marriage matchmaking agencies view both marriage and brides-to-be as products that can bring significant profits to the company (Hsia, 2000: 48; Wu, 2008: 12).

This research combines three methods: participant observation, interviews and document analysis. I participated in two government-funded literacy programmes and was allowed to be a teaching assistant; through this, I gained access and made contact with marriage immigrants. I conducted in-depth interviews with 50 SEA and MC women in Taiwan. I used Mandarin or Taiwanese as appropriate to conduct the interviews, as these are the main languages in Taiwan, which I have in common with most of the interviewees. In addition, I collected related second-hand resources used by marriage matchmaking agencies as material to analyse. I found that my position shifted and that I was both insider and outsider during the fieldwork. On the one hand, as I am a woman who speaks Mandarin and Taiwanese and has had experience living abroad as they have, and also am a participant in literacy programmes, some women saw me as an insider. On the other hand, I was considered an outside, as I am single and do not have experience of what it feels like to be a wife, mother and daughter-in-law, and I was sometimes the one who checked their attendance when the literacy programme organiser was busy.

This chapter focuses on the marketisation of international marriage in these regions of the world and discusses what services commercial matchmaking agencies offer, how these companies' matchmaking processes operate, and how they bring Taiwanese men and brides-to-be together. More specifically, it explores the kinds of mechanisms that matchmaking companies use to train women to meet Taiwanese men's marital expectations. How do gendered interactions work in the matchmaking process? I will begin with a brief discussion of masculinity before looking at the factors that drive Taiwanese men to travel outside of Taiwan in order to join a wife-finding tour. This chapter will explore what wife candidates are asked to do when they stay with an agency. How does the matchmaking process serve as an opportunity for Taiwanese men to perform their masculinity?

Masculinity and gender relations in Taiwan

Scholars refer to masculinity as 'the pattern or configuration of social practices linked to the position of men in the gender order, and socially distinguished from practices linked to the position of women' (Connell, 2001: 44). It is a socially constructed result, rather than a nature-born characteristic, and it is practised within a system of gender relations (Connell, 2005: 84). Men learn how to perform in a manly way through socialisation and various institutions – for example, the family, school, sport (Treadwell and Jarvis, this volume) and the army (Connell, 2008: 240). It is common for women's roles to be culturally scripted so as to be supportive and subordinate to male partners (Bui and Morash, 2008: 192).

Masculinity is related to cultural phenomena, and the forms of masculinity are various in different historical periods and locations (Beynon, 2002: 58, 62). People 'do gender' differently in different cultures and, as Connell (2005) has noted, there is a need to recognise how men perform multiple masculinities across varying cultures (p. 58, 76). In Taiwan, adult men are expected to be masculine by taking on the responsibility for being successful in business and in their careers (Hwang, 2007: 269). The gendered criterion for choosing a spouse in Taiwan is that men have to have 'three highs': higher stature, education and income than their female partners (Hwang, 2007: 280). Men are usually expected to have a higher social position than women in Taiwan's patriarchal marriage system, and this makes it difficult for low-skilled male labourers to find a partner in the domestic marital market (Hsia, 2002: 162; Chung, 2004: 14). In addition, Confucianism has a significant influence on gender relations in Taiwanese society (Lu and Chen, 2002: 223). To produce descendants and continue the family line is viewed as filial piety (in Chinese, *bu xiao you san wu hou wei da*) (Lu et al., 2006: 52). Thus, heterosexuality is valued, and a married couple is expected and encouraged to have children in Taiwanese culture.

In recent years, these expectations relating to marriage, gender relations and familial responsibilities have to some extent conflicted with a rapid increase in women in higher education and with their professional careers. As a result, there is said to be a 'marriage squeeze' that leaves well-educated women and poorly educated men 'stranded' in the marriage market (Jones, 2004: 10, 30). Many men do not want to marry a woman who has a higher social and/or educational position, and at the same time, women do not want to 'marry down' (Jones, 2004: 20), or beneath them. Chen's research (2001) found that many Taiwanese husbands feel a sense of inferiority if they pursue Taiwanese women, because the men do not have a degree and have a low income (p. 13). Thus the traditional masculinity culture motivates Taiwanese men to turn to seek an 'ideal/traditional wife' from a neighbouring country (Chen, 2001: 52).

The trend towards international marriage between Taiwanese men and SEA and MC women is related to Taiwan's economic development on the international stage and the opening up of some policies by the government. For example, the announcement of the 'return/family visit' policy in 1987 opened a gate for the Taiwanese to visit China, and interactions as well as exchanges across the Taiwan Strait have mushroomed since then (Chao, 2002). In addition, the 'go-south' policy in 1994 opened up the market for investment in and trade with Southeast Asian countries (Hsia, 2002: 164). Since that time, more and more migrant workers from Southeast Asian countries have entered Taiwan. An image of modernisation which is influenced by globalisation and Taiwan's economic development causes many SEA and MC women to dream of marrying a Taiwanese man (Chao, 2002). The numbers of SEA and MC marriage immigrants have increased rapidly since 1995 (Hsia, 2002: 164).

According to Connell (2005), in terms of a man's power and social position, there are four masculinity types: hegemonic, subordinate, complicit and marginalised (Connell, 2005: 76–86). Most Taiwanese men who go on a tour tend to be in the category of marginalised masculinity, in comparison to the hegemonic masculinity of the dominant group (Hwang, 2007: 281; Cheng, 2011: 13), and they look for a feeling of masculine power. Many Taiwanese men are in a disadvantaged socioeconomic position (e.g., working in industry, agriculture or fishing) (Hsia, 2000: 341; Hsia, 2002: 172–5), and many of them are elderly singles, widowers or disabled (Chao, 2002: 2). I found a significant range of age differences between interviewees and their Taiwanese husbands. For example, one married couple was comprised of a 39-year-old Chinese wife and her Taiwanese husband, who was 82 years old.

In addition, the expectations of other family members, who are also influenced by traditional Chinese culture, cause many single Taiwanese men to experience the pressure of being expected to continue the family name by reproducing within marriage (Hsia, 2002: 85, 190). Some women told me that their Taiwanese husband had relatives and/or family elders who accompanied them on a wife-finding tour in order to 'help' them find a wife to bring home. Southeast Asian countries and China are seen as backward, because their social and economic development is relatively lower than in Taiwan, while SEA and MC women are perceived as more primitive and obedient than Taiwanese women (Wang and Tien, 2006: 4). Thus, the desire to fulfil the expectations of hegemonic masculinity appears to motivate many Taiwanese men to travel abroad to find a wife (Chen, 2001: 50–1).

Scholars found that many Taiwanese men have gendered stereotypes and expect to maintain their masculine position within their relationship (Wang and Tien, 2006: 11). For example, men do little housework and expect their wives to cook, do the washing and care for children because this is positioned as feminised women's work. Although more and more Taiwanese women work outside the household, and may earn money to support the family finances, women are still expected to undertake the majority of domestic chores (Wang and Tien, 2006: 10). Some Taiwanese men, if they cannot find a wife who has these traditional 'virtues', and who is submissive to a man's masculinity, will turn to neighbouring countries where they expect to find an ideal/traditional wife. Such women may be less educated, unlike their Taiwanese counterparts who seek gender equality and a career outside of the home (Wang and Tien, 2006: 11, 19).

Potential wives in 'brides-to-be' camps

In order to satisfy Taiwanese men's ideals and desire to find an ideal/traditional wife, matchmaking companies make extensive efforts to prepare women to attract men in accordance with the expected gender performance of femininity (Wang and Tien, 2006: 21). SEA and MC women's accounts show that some of them took a training course in a 'brides-to-be' camp, where they prepared themselves as appropriate candidates to be the wife of a Taiwanese man. Whilst resident in such camps, the women reported undertaking a range of tasks in preparation for their eventual selection by a Taiwanese man. As such, before women meet Taiwanese men participating in a wife-finding tour, they are asked not only to do some work on their body, such as slimming to achieve and maintain a slender body shape, but also to use their body to do work, for example, learning how to do housework.

Xiang, a Cambodian woman, told me that when she was 19 years old, a commercial matchmaking agency arranged for her to live with many Cambodia women in a 'brides-to-be' camp before she met her Taiwanese husband. Xiang described the 'brides-to-be' camp as being similar to 'many students living in crowded accommodation'. Women pay a matchmaking agency for their accommodation and meals when they stay in a camp. For example, a 22-year-old Vietnamese woman, Ding, said: 'My mother put our only house in pledge for a loan with a bank in Vietnam, in order to pay the matchmaking agency 2,000 US dollars in advance, before I met my Taiwanese husband'. Agencies collect hundreds of Vietnamese women in a building and hire a warden to be responsible for the future brides' daily lives (Chang, 2001: 102). The main requirement for women to stay in such a camp is that they meet different foreign men who have joined a package tour arranged by a matchmaking agency (Kung, 2005: 10).

Xiang complained to me that 'the agency did not give us enough food, so we always felt hungry.' I asked her why and she explained that, 'because there are many women who live together there, it is very crowded, and everyone can have only limited space – it is even difficult for women to walk or do some easy exercises – so the matchmaking agency is afraid we will become fat if we eat too much.' Bordo (2003) notes that females are trained in subjectivity and subordinated through many forms of mass media in their everyday lives, and one requirement to make a female body admirable is to be slender (pp. 19, 26, 169–70). The way in which matchmaking agencies ask brides-to-be to maintain an attractive (slim) body shape for the male gaze reflects this gender stereotype. Agencies believe that a thin female body has greater potential to attract Taiwanese male clients; this is therefore a common business strategy employed by these companies.

Besides asking future brides to maintain their weight, Xiang also told me that they were taught to make an effort to attract the attention of men as much and as soon as possible. The agency keeps reminding and encouraging brides-to-be to ensure that they make a good impression of themselves on the visiting Taiwanese men. When men came to visit, Xiang explained that, 'I made an effort to give every man a big smile', a strategy she used to draw the attention of Taiwanese men. Consequently, keeping their bodies thin and making themselves look charming are the most important skills that women are taught and encouraged to work on in the camps.

'Emotional labour' refers to women workers' bodies that are appropriated by employers, and they are required as part of their work obligations to perform their emotions in a proper way to serve clients (Wolkowitz, 2006: 76–7, 82). For example, female flight attendants are asked to 'walk softly through the cabin' and 'make eye contact with passengers' – a gendered performance to make customers enjoy their service (Ibid.). This requirement for 'body techniques' from women workers is usually associated with a stereotype of femininity (Steinberg and Figart, 1999: 10). Xiang's performance of smiling at Taiwanese men illustrates her interacting with clients in a desired way, and this makes men link Xiang to an acceptable, subservient wife image.

Some agencies run simple Mandarin or Taiwanese language-training courses for future wives and teach women how to cook Taiwanese cuisine, clean the house and use modern electronic equipment (Wu, 2008: 38; Wu et al., 2008: 147). During this socialising process, women gain knowledge about Taiwan and household skills, and they perform femininity as they learn how to play the role of a 'good' Taiwanese wife. These mechanisms are similar to how businessmen 'manufacture' and 'transform' new materials into finished, polished products. Agencies package women to satisfy how Taiwanese men imagine and characterise a potential wife as being 'beautiful, virtuous, agreeable and good at housekeeping' (Han, 2003: 166–67). Similar to Foucault (1977: 136), who investigated the training process to produce soldiers with docile bodies, agencies transform women to make them more 'Taiwanese-like' by enforcing characteristics and gender expectations that the agencies think Taiwanese men prefer. An ideal feminine performance and gendered body are constructed through the gendered regulatory process of training courses that produce the gendered bodies, docile and passive to the demands of men.

Being a powerful man: gendered interaction in the matchmaking process

The ways that matchmaking agencies make Taiwanese men feel that they have priority and advantages over women are constructed not only by the companies' advertisements, but also in the later stages of the matchmaking process. For instance, I received a Vietnamese bride introduction service document from a matchmaking company that promoted a domesticated image of Vietnamese women and women's virgin bodies in words and images. It states that 'Vietnamese are traditional women who put their husband as the top priority in their own life, they are good at helping the family business', and tells male clients that 'you can choose any one you want from hundreds of Vietnamese women with our company'. Service items that promote the benefits of women are included in matchmaking business deals, and the matchmaking agencies try to make their male clients feel the added value that the companies provide.

Mei Yum, a 40-year-old woman who is originally from Guilin, China and who left there ten years ago, provided an example. She felt that the matchmaking process is comparable to how 'a king chooses his concubines'. Similarly, Xiang and Ding told me that matchmaking agencies offer Taiwanese men hundreds of women to see and select from during their stay. 'Don't be shy, you can select some women that you feel are good first, and have dates with them, and then choose one from them, otherwise, you perhaps cannot find anyone at the end of the matchmaking process' (Kung, 2005: 13). In Kung's (2005) research, a Taiwanese man finally decides which woman he wants to marry after he has seen three hundred Vietnamese women (p. 13). Mei Yum said, 'My husband told me that I was the sixtieth Chinese women he saw that day.' Maximising Taiwanese men's choice allows them to have the illusion that they are important, and an oversupply of women is needed to create this effect. An unequal power relationship between men and women is created by the way agencies arrange the matchmaking process. In comparison with seeking to boost women's femininity, Taiwanese men's masculinity is constituted and aggrandised by the matchmaking process. As such, it is not difficult to see how men who have grown used to occupying a position of marginalised status in Taiwanese society might feel a great and seductive sense of power during the matchmaking process.

When Xiang's husband and his aunt went to Cambodia to look for a potential wife four years earlier, several Cambodian women were chosen and had dates with Xiang's husband. Xiang's husband and his aunt took some potential brides shopping in a department store and went to restaurants for dinner on a few occasions. During these dates, the aunt observed the women's behaviour and movements. Xiang's husband told her that when they went shopping, she brought clothes that had a conservative style, rather than tighter and more revealing items. In addition, Xiang did not choose expensive gifts, even though she knew that the family would have paid for them. When she went to restaurants with other Cambodian women, some only ate a little, and some of them were too shy to eat, while Xiang showed her appetite and ate as much as she desired. Like many East Asian countries, Taiwanese share food from communal dishes into individuals' rice bowls. Xiang put food into other Cambodian women's bowls, and this behaviour impressed the aunt and made Xiang's husband decide to choose her in the end.

Xiang said that she realised that the reason why her husband chose her was because she was different from many Taiwanese women. This reason is very similar to how one Taiwanese male informant in Wang and Tien's research (2006) described their foreign wife. Such women are seen as having more traditional values than Taiwanese women at home, as they 'do not have strong material demands and are willing to live a plain life, submissive and willing to serve their husband' (p. 19). The matchmaking process of dating and observation of potential wives reinforces a perception of potential brides as submissive, docile, family-oriented, good at managing the household and ready to marry. Linking to this, the normally marginalised masculinity of many of the men who partake in tours is elevated and reinforced by the performance of 'traditional' femininity which, it is important to note, is all the while underpinned by the economic disparity between the potential bride and the men who travel to judge and select her.

Bordo (2003) notes that a rule of femininity construction requires women to develop an other-oriented characteristic (pp. 169–70). The aunt and Xiang's husband were impressed that Xiang served food to other immigrant women. Xiang's behaviour was viewed as considerate of her friends, showing a feminine virtue, and her behaviour while dating made her all the more admirable. Xiang told me that she was lucky that she only stayed in a brides-to-be camp for eight days and then she was chosen by her husband.

The 'wife-finding tour' can be seen as similar to sex tourism in some ways, but they are different. O'Connell Davidson (2001) defines 'sex tourists' as 'tourists who enter into some form of sexual-economic exchange with women, men or children resident in the host destination' (p. 7). Women's accounts in my fieldwork show that there is no sexual behaviour or contact between Taiwanese men and women because the matchmaking agencies have to maintain the 'quality' of the women in their care to attract potential clients. However, the way that matchmaking agencies prepare women to be proper wife candidates and provide training courses for women follows a gendered stereotype.

Matchmaking agencies promote their business by using the strategy of constructing an image of multiple wife candidates for men to choose from. This is similar to sex-tourism business owners who assert a natural difference of 'race' and 'culture' between the West and others and seek to emphasise the supposed superiority of the male client (Sanchez Taylor, 2006: 53). Similarly, British stag tourists share an image of East European destinations, such as Poland, and the women present, as a 'third-world country' (Thurnell-Read, 2012: 811). The watching or chasing of Polish women is identified as a masculine practice by stag tour participants, where such women are positioned as sexually different from British women 'back home' (Thurnell-Read, 2012: 811). The notion of a 'different other' in the Taiwanese context refers not only to women's ethnicity, but also to the difference between the perception of women in Taiwan and in SEA or China. Xiang's husband perceived Xiang as standing out from many Taiwanese women in terms of her conservative style of dress and her other-oriented performance. The various options for choosing different women and the feeling of consuming 'difference' in the matchmaking process make Taiwanese men feel that their own masculinity, which is questioned and marginalised in their own country, is enhanced.

The matchmaking companies charge not only women, but also Taiwanese men, for their matchmaking service. In comparison to what the matchmaking companies charge the women, the money that Taiwanese men pay to join a 'wife-finding tour' is significantly higher. In Kung's (2011) research into the matchmaking business in Vietnam, the price that Taiwanese men paid to a company ranged from US $10,000 to $16,666 in total (p. 87). Although many men who go on a tour are from the working class in Taiwan society, they can afford the expense of such participation. These Taiwanese men spend a relatively small amount of money to have an all-inclusive matchmaking service and gain a significant sense of dominant power that they cannot have in their home country. It is similar to non-elite North American male tourists who spend a little money to have sex with Costa Rican women, which makes them feel a sense of having superior masculine (and racial) power and transnational mobility while on holiday (Rivers-Moore, 2012: 860). It is also similar to Sanchez Taylor's findings (2001): that some Western heterosexual tourists' privilege in relation to women is partly contingent on their social, economic and political status being higher than that of local Caribbean men (p. 761).

Although matchmaking agencies charge fees to both Taiwanese men and women, women generally do not have the opportunity to choose their marital partner. While Taiwanese men may choose their marriage partner in the matchmaking process, the services that agencies offer to Taiwanese men and women are on a basis of asymmetrical information. For example, the training courses that agencies run prepare women for men, rather than ask men if they are ready to be a husband. In addition, agencies usually oversupply women for Taiwanese men during the matchmaking process, and they have opportunities to date whichever women they want. Taiwanese men can ask questions of women in the matchmaking process, but women do not have the opportunity to ask questions in return (Kung, 2011: 108).

From most matchmakers' experiences in Vietnam, nearly 99% of Vietnamese women would not reject a Taiwanese man (Shen, 2002: 23–5). Kung (2011) found that Vietnamese women almost always accept men in the matchmaking process because there are many potential brides who hope that they can marry a foreign man (p. 106). All the immigrant wives that I interviewed accepted their Taiwanese husband and got married to them. I found that some women not only felt that having a Taiwanese husband is 'fashionable', but also, the image of

Taiwanese men is that they are richer, responsible, less lazy and less violent than local men. For example, Ding said, 'My brother in-law is a lazy Vietnamese man. He never takes on the responsibility to share the housework or earn money to support the family; my sister works really hard; she has to do everything from morning to night.' Ding described her negative feelings about Vietnamese men by saying, 'I decided to find a commercial matchmaking company after I heard many Vietnamese women's experiences about how good their Taiwanese husbands are.' However, women's 'Taiwan imagination' (Chen, 2003: 2) may not last long; some wives told me that they were disappointed when they found out that their Taiwanese husband avoided doing any housework.

From Ding's bad impressions of Vietnamese men, it seems that she naturalised and generalised Vietnamese men's behaviour and viewed this gendered stereotype of Vietnam as a result of a typical representation of a certain ethnicity. Commercial matchmaking agencies construct an image of Taiwanese men as 'perfect' and Taiwan as a desirable place to live. Local men, such as Vietnamese men, consequently become the last marital choice of Vietnamese women (Chen, 2003: 36–41). Matchmaking agencies' strategy of constructing Taiwan and Taiwanese men's positive image is similar to the discourse of some white European and North American male heterosexual tourists who exaggerate other cultures as being 'close to the state of nature'. For example, in the work of O'Connell Davison (2001), such male tourists described the Dominican Republic as 'a lawless and corrupt place' (p. 6, 14). Some Western tourist men construct a different type of masculinity by identifying themselves as 'civilised', 'enlightened', 'modern' and 'nice', in contrast to local men's 'violent', 'conservative' and 'closed-minded' nature (O'Connell Davison, 2001: 6; Rivers-Moore, 2012: 855–6). Nevertheless, no matter what women's marital expectations are, or what kind of man they are looking for, matchmaking companies manipulate men's demands and women's imagination, and together with the asymmetric opportunities to make choices in the matchmaking process, they ensure that most women are very unlikely to reject men's choices.

Conclusion

In this chapter, I have illustrated the complexity of gender issues between Taiwanese men participating in wife-finding tours and SEA and MC wife candidates, and the strategies that matchmaking agencies use to promote their businesses. During women's stays in 'brides-to-be' camps, they are taught to show their femininity by presenting a docile, submissive image to meet the expectations of the Taiwanese men they meet. These companies are, in essence, selling a service based on the provision of a context in which both masculinity and femininity are stage-managed to facilitate gender relations. The typical tour in which men seek a wife is similar in some respects to Western men's sex tourism visits to developing countries. Both male participants in both sex and wife-finding tours can spend a relatively small amount of money in exchange for services that they cannot usually have in their home country, and they can achieve an elevated but temporary feeling of heterosexuality in that particular space. The sense of being an economically privileged, powerful, masculine man is seen in both men's sex tourism and the wife seeking process. However, the absence of sexual encounters in the Taiwanese men's search makes them different from the phenomenon of Western men's sex tourism. By analysing how such agencies prepare women to be proper wife candidates in 'brides-to-be' camps and exploring the agencies' strategy of oversupplying potential wives to maximise Taiwanese men's options, this chapter finds that Taiwanese men's sense of masculinity is inflated in the matchmaking process. Such a mechanism allows male tourists to believe that they have the power to dominate women on these tours. Matchmaking companies' arrangements make potential brides gradually lose control of their own destiny in the process.

References

Bordo, S. (2003) *Unbearable Weight: Feminism, Western Culture, and the Body.* University of California Press.

Beynon, J. (2002) *Masculinities and Culture.* Maidenhead: Open University Press.

Bui, H. and Morash, M. (2008) 'Immigration, masculinity and intimate partner violence from the standpoint of domestic violence service providers and Vietnamese-origin women', *Feminist Criminology,* 3: 191–215.

Chang, S.-M. (2001) 'Marketing international marriages: Cross-border marriage business in Vietnam and Taiwan', *The Graduate School of Southeast Asian Studies.* Tamkang University.

Chao, Y.-N. (2002) 'Citizenship, nationalism, and intimacy: A case study of marriage between mainland brides and glorious citizens in Taiwan', *Review East Asia: Global, Region, Nation, Citizen' in The Annual Conference of the Cultural Studies Association.* Taipei, Taiwan.

Chen, L.-Y. (2001) 'Brides from Southeast Asia – An in-depth report with the perspective of post-colonial feminism', *The Graduate Institute of Journalism.* Taipei: Taiwan University.

Chen, P.-Y. (2003) 'Taiwan imagination and gaps: 19 Vietnamese brides' stories in Puli', *The Graduate School of Southeast Asian Studies.* Chaiyi City: National Chi Nan University.

Cheng, D.-L. (2011) 'The perception of masculinity in Taiwan', *Department of Speech Communication*. Taipei: Shih-Hsin University.

Chung, T.-F. (2004) 'A study on the lived experiences of those Taiwanese men who married foreign spouses', *The Graduate Institute of Family Education*. Chaiyi City: National Chiayi University.

Connell, R. W. (2008) 'A thousand miles from kind: Men, masculinities and modern institutions', *The Journal of Men's Studies,* 16(3): 237–52.

Connell, R. W. (2005) *Masculinities*. London: Polity Press.

Connell, R. W. (2001) 'Studying men and masculinity', *Resources for Feminist Research,* 29: 43–56.

Foucault, M. (1977) *Discipline and Punish: The Birth of Prison*. London: Penguin.

Has, J.-L. (2003) 'Maid or wife? The cross-border migration of women labour: Mainland China brides in Taiwan', *Community Development Journal Quarterly,* 101: 163–75.

Hsia, H.-C. (2002) *The Phenomenon of Foreign Brides Under the Internalization of Capitalism*. Taipei: Tonsan Publication.

Hsia, H.-C. (2000) 'Transnational marriage and internationalization of capital – The case of the "foreign bride" phenomenon in Taiwan', *Taiwan: A Radical Quarterly in Social Studies,* 39: 45–92.

Hwang, S.-L. (2007) 'Masculinity and manliness', in S. L. Hwang and A. Y. Mei-Hui (eds) *Gender Dimension and Taiwan Society*. Taipei: Sage, pp. 268–92.

Jones, G. W. (2004) 'Not "when to marry" but "whether to marry": The changing context of marriage decisions in East and Southeast Asia', in W. Gavin and K. Jones (eds) *(Un)tying the Knot: Ideal and Reality in Asian Marriage*. Singapore: National University of Singapore, pp. 3–56.

Journalist, The (2009) 'Advertisement of Vietnamese brides.' 4 April 2009, p. 8.

Kung, I. C. (2011) 'Subordination of "Vietnamese brides" and the matchmaking marriage between Taiwan and Vietnam', *Taiwan: A Radical Quarterly in Social Studies,* 82: 85–122.

Kung, I. C. (2005) 'The matchmaker agency and discipline: The birth of Vietnamese brides', *2005 Annual Conference on Southeast Asian Studies in Taiwan*. Chaiyi City: National Chi Nan University. Available at: http://www.dseas.ncnu.edu.tw/data/2005_seastw/2k5_seas_tw/%E9%BE%94%E5%AE%9C%E5%90%9B.pdf.

Lu, L. and Chen, H-H. (2002) 'An exploratory study on role adjustment and intergenerational relationships among the elderly in the changing Taiwan', *Research in Applied Psychology,* 14: 221–49.

Lu, L. and Kao, S.-F. and Chen, F.-Y. (2006) 'Psychological traditionality, modernity, filial piety and their influences on subjective well-being: A parent-child dyadic design', *Psychological Research in Chinese Societies,* 25: 243–78.

Luehrmann, S. (2004) 'Mediated marriage: Internet matchmaking in provincial Russia', *Europe-Asia Studies*, 56(6): 857–75.

O'Connell Davidson, J. (2001) 'The sex tourist, the expatriate, his ex-wife and her "other": The politics of loss, difference and desire', *Sexualities,* 4(1): 5–24.

Rivers-Moore, M. (2012) 'Almighty gringos: Masculinity and value in sex tourism', *Sexualities,* 15: 850–70.

Sanchez Taylor, J. (2006) 'Female sex tourism: A contradiction in terms?' *Feminist Review,* 83: 42–59.

Sanchez Taylor, J. (2001) 'Dollars are a girl's best friend? Female tourist sexual behavior in the Caribbean', *Sociology*, 35(3): 749–64.
Shen, H.-J. (2002) 'Stairway to heaven?: Power and resistance within the commodified Taiwanese-Vietnamese marriages', *Sociology Department,* National Tsing Hua University.
Simons, L. A. (2001) 'Marriage, migration and markets: International matchmaking and international feminism', *The Faculty of the Graduate School of International Studies*. Denver: University of Denver.
Steinberg, R. J. and Figart, D. M. (1999) 'Emotional labour since: The managed heart', *The Annals of the American Academic of Political and Social Science,* 561: 8–26.
Thurnell-Read, T. (2012) 'Tourism place and space: British stage tourism in Poland', *Annals of Tourism Research,* 39: 801–19.
Wang, H.-Z. and Chang, S.-M. (2002) 'The commodification of international marriages: Cross-border marriage business in Taiwan and Vietnam', *International Migration,* 40: 93–116.
Wang, H.-Z. and Tien, J.-Y. (2006) 'Masculinity and cross-border marriages: Why Taiwanese men seek Vietnamese women to marry', *Taiwan Journal of Southeast Asian Studies,* 13: 3–36.
Wang, H.-Z. (2007) 'Hidden spaces of resistance of the subordinated: Case studies from Vietnamese female migrant partners in Taiwan', *International Migration Review,* 41(3): 706–27.
Weng, B.-L. (2007) 'The ladder to the heaven or the hell-the role of matchmaking agency in the international marriage', *The Graduate School of Southeast Asian Studies*. Chaiyi City: National Chi Nan University.
Wolkowitz, C. (2006) *Bodies at Work*. London: Sage.
Wu, C.-C. (2008) 'The traditional matchmakers? Or the marriage agencies? The roles and problems of the Southeast Asian matchmakers', *Sociology Department*. Taipei: Soochow University.
Wu, C.-C., Tseng, H.-Y. and Chun, Y.-I. (2008) 'The marriage market with/out them? The power-dependence relationship between bachelors and matchmakers in South-Eastern Asia', *Journal of Border Police*, 133–78.

11

Risky Business: How Gender, Race, and Culture Influence the Culture of Risk-Taking among Sex Tourists

Yasmina Katsulis

The San Diego trolley runs from the border entry gate to Mexico throughout San Diego County. Next to cars, this is arguably the most popular way to travel to and from the border. Often, this venue attracts young, loud, drunk and bragging males on their way home from the Tijuana red-light district (*la Zona Norte*) area, especially on the weekends. One particular day, I sat behind two older men who were quietly laughing and sharing their experiences of the day. I interrupted their conversation to discuss my research, and they proceeded to show me a few pictures of their *novias* ('girlfriends') – sex workers with whom they had established regular rapport.[1]

Responding positively to my project, the customers proceeded to tell me about an online forum they participated in – one completely devoted to Tijuana sex tourists. Customers participate in order to share information, advice and personal experiences, to socialise with one another and to organise offline social activities, where they share 'war stories', beer, and sometimes, women. In an introduction to the Club Hombre (2003) website, a newcomer (or 'newbie') is greeted with the following:

> So you want to go to Tijuana, TJ. The weekend party town [for] San Diego teens, supreme gringo tourist trap, and most importantly, a paradise for the American single adult traveler. You've heard the stories of gorgeous women who can be had for the price of a couple lap dances at your local strip club. But you've also heard tales of scams & rip-offs, muggings and she-males. You want to know the truth about the risks and how to avoid becoming a victim. You also need information on where to find the hottest 'chicas' ['girls'] in TJ, which chicas provide

> the best service, and how to get the best possible deal. Well you've come to the right place.

With the Internet, what was once a relatively small network of sex tourists from Southern California has been transformed into several burgeoning online communities. Data for this paper was drawn from the most frequently visited site during 1999–2003.

Sex tourism, and the discourse that surrounds this set of activities, can be conceptualised as a form of *conspicuous consumption*, allowing sex tourists to display and derive pleasure from wealth, power and social status relative to others. Chow-White (2006: 884) suggests that the Internet 'enables sex tourists to build deeper connections between the racialisation, sexualisation and commodification of sex workers' bodies and Western masculinity'. Further, Williams and Lyons (2008: 79) argue that this online identity work enables participants 'to better articulate both normative and potentially idealised notions of masculinity and heterosexuality, regardless of whether or not they are materially practiced and thus representative of the "truth" of men's experiences'. And, as I have argued, sex tourism involves 'the cultivation, and experience, of a particular form of masculine subjectivity that relies upon (and exploits) historical differences in power and privilege. In imagining, and making meaning of these differences through conversations online, customers are able to create a complex, gendered subjectivity that is continually reimagined and reinscribed' (Katsulis, 2010: 211). Thus, for consumers, the parameters of the purchase extend far beyond the physical encounter and into the discursive.

Like sex tourism itself, the virtual world provides opportunities for men to inhabit a largely hegemonic form of masculine subjectivity that might not be readily available to them in other venues (Katsulis, 2010). Sex tourists increasingly interact with one another both on and offline (Bishop and Robinson, 1998; Davidson, 2001; Soothill and Sanders 2007; Sanders, 2008a, 2008b; Hobbs et al., 2011), and online encounters provide a welcome opportunity for research into the social complexities of the identity work that shapes their activities. The emerging literature on gender identity work online provides a framework for this analysis (Kendall, 2002; Katsulis, 2010).

Setting

Located ten miles from San Diego, California, the city of Tijuana, Mexico has been an international sex tourist destination for several decades

(Katsulis, 2009). Structural inequalities have long been pervasive along the border. Tijuana's primary point of entry is the busiest land-based border crossing in the world (USGSA, 2013), and the city is both an international tourist destination and the primary migration corridor into the United States. In 2012, over 28 million documented crossings occurred at the San Ysidro border checkpoint alone.[2] Tijuana is a postmodern city and tourist mecca that welcomes the world's consumers with open shops, restaurants and nightclubs. Tijuana caters to the young and old, and its downtown streets pulsate with music, voices and laughter. The city supplies the fulfilment of needs and desires on every level: food, beer, liquor, voyeurism, sex and recreation. Its geographic importance has stimulated substantial growth in tourism, services, and manufacturing.

The local sex industry is regulated through worker registration, mandatory health screenings, site licensing and tolerance zones. It can be characterised as a bifurcated industry with both a formal (and legal) and informal (and illegal) sector – not all sex workers or establishments comply with legal requirements. The majority of customers from the United States regularly visit Tijuana's most infamous and commercialised red-light district, la Zona Norte (literally, 'the northernmost zone').

The ethnosexual landscape

Online forums provide a sense of the larger political economy and social dynamics in which the sex industry is embedded. Baudrillard (1991: 2) suggest that 'Seduction ... never belongs to the order of nature, but that of artifice – never to the order of energy, but that of signs and rituals'. For example, studies of written materials (online message forums, LISTSERV lists and guidebooks) produced by 'hardcore' sex tourists and 'sexpatriates' have often been characterised as 'aggressively heterosexist, deeply misogynist, and profoundly racist' (Davidson, 2001: 8; see also Bishop and Robinson, 1998; Hughes, 1998/9; Chow-White, 2006), performing a kind of 'imperial masculinity through which men reinstate and live out their economic, racial, and geopolitical power' (Rivers-Moore, 2012: 851).[3,4]

The racial imaginary at play among these sex tourists goes beyond the physical, as the imagery is conflated with racialised sexual stereotypes of *mexicanas* as more 'naturally' passionate and sexual (Katsulis, 2010). Through these controlling images, the intrinsic asymmetry in the encounter is legitimated and naturalised (Nagel, 2003). Racialised stereotypes are reinforced by how sex workers perform (and market) their cultural and gendered identities (see Katsulis, 2010: 216). Moreover, the ethnosexual landscape is coloured by ideas about the Other, wherein Mexico itself is described 'as a passive whore to be fucked over' (Castillo and Tabuenca, cited in Carroll, 2006: 367).

I argue that sex tourism can be thought to operate on several core cultural logics, wherein: (1) sex with the Other (in this case, women of colour from less affluent countries) is perceived as superior (or at least desirable) to sex with those at 'home'; (2) the ethnosexual landscape is characterised by commonly held stereotypes of the Other as more 'natural', 'traditional' and 'authentic', as well as cheap, easily exploitable and replaceable; and (3) gendered sexual subjectivity is cultivated and enhanced through the praxis of these encounters. A racialised sexual hierarchy is erected, naturalised and continuously reinscribed; structural differences in power and affluence are conflated with perceived differences in race, culture and sexuality.

Demographic profile

Due to the paucity of research in this area, it is unknown how the sample in this study compares with other sex tourists. To get a sense of demographics, the study employed existing customer-generated polls, which provided an aggregate profile but did not allow for tying any particular narrative excerpt to any specific background characteristics.

Most characteristics do suggest a stereotypical profile – that of an unmarried, white middle-class male, between the ages of 26 and 39 living within 100 miles of the US–Mexican border. Most visited the red-light district about once a month, spending between $500–$7,500 US dollars a year on paid sexual activities in Tijuana. Most had travelled to at least one other country as a sex tourist on a previous occasion, most commonly Thailand and Costa Rica.

Of particular interest in this sample, however, is not so much the 'typical' (or modal) characteristics found, but the unexpected range of diversity in the sample. Class backgrounds appeared diverse – there were a number of affluent professionals participating in the forums alongside working- and middle-class men from a variety of occupational and ethnic backgrounds. With respect to income, for example, while 6% made less than $24,000 US dollars each year, three times as many participants (18%) made more than $126,000 US dollars each year. However, the vast majority (79%) could be considered middle income, earning between $41,000–$125,000 US dollars per year.

Although most studies of sex tourists describe them as white, male, and/or from more affluent settings, we must acknowledge that male privilege is not a homogenous phenomenon, and internal class and race relations between men suggest that multiple masculinities, dominant, subordinate or marginalised, are always in play (Connell and Messerschmidt, 2005). Sex tourism is itself grounded in particular locales with historically specific actors and power relations. However, our understanding of more diverse subcultures within sex tourism is just beginning to emerge (Williams, et al., 2008; Mitchell, 2011; Törnqvist, 2012), and we do not yet know how socioeconomic class and income influence the material and symbolic landscape of sex tourism; when class is discussed in the literature on sex tourism, sex tourists are often simply characterised as relatively affluent in comparison to males in destination or host cities. Future research should make a concerted effort to examine the salience of class-based differences on both a symbolic and material level in other contexts.

The social context of risk

Customers utilise the forum to share insight about how to avoid or navigate a wide range of risks associated with their activities, including sexually transmitted infections, substance use, victimisation, street violence, heartbreak, sexual addiction, exhaustion, depression and social stigma. Perceptions and responses to risk are informed by far more than medicine and science. Although medical messages of risk-reduction (e.g., use a condom with every sexual encounter) emerge, factors such as peer norms and ideas around a highly masculinised sexuality are more constant. Peer guidance about risk-taking takes the form of detailed trip reports, question and answer areas and topical discussions on particular health topics. Discursive activities create a framework through which multiple social processes around risk-taking occur, including masculine socialisation, sexualisation, desensitisation, minimisation, medicalisation and normalisation. These are not mutually exclusive categories; customers might participate in each of these simultaneously.

Risk-taking: an adventurer's guide

Risk-taking is seen as an inescapable (and for some, pleasurable) aspect of sex tourism. In this context, it is an expression of personal autonomy, masculine privilege, adventure, pleasure, indulgence and sexual liberty. The function of the forum as advice column for fellow peers is referred to explicitly within website advertisements: 'This entire site is a kind of how-to book on "How to Fuck in Dirty, Nasty, Filthy Foreign Countries that Have No Rules and Still Have Fun and Survive Without Taking Some Nasty Disease Home to Your Wife, Children, and Girlfriend"' (latinlover).

This quote illustrates the complex dichotomies created between home and the Other. The Other is portrayed as an inhabitant of a liminal zone of dirt, filth and disease, which is at the same time a no-holds-barred locale where risk-taking is tolerated and expected. Sex tourists can establish a clearly demarcated border between what is acceptable behaviour at 'home' and the loosening of these social controls on behaviour elsewhere. The consequences of risk-taking are acknowledged only in terms of its contamination of the body at home, and familial bodies (wife, children, girlfriend), which represent cleanliness and respectability, are worthy of protection; whereas the foreign Other is already polluted, precisely because of the perceived lack of social controls that act to heighten pleasure, risk and desire for tourists.

Chow-White (2006: 887) notes, 'The Internet and sex tourism websites are not only tools for information exchange and product distribution, but also structuring devices that create and reproduce myths of sex worker sexuality, race and male dominance'. For some, risk must be viewed as a potential trade-off for a pleasurable lifestyle that allows for masculine indulgence and adventure: 'We cannot live our lives in a cocoon and hide ourselves from all the risks in life, if we did we would have a very empty and meaningless life. The best we can do is to be aware of all of the risks that we are exposed to and take the precautions that are available to us to reduce the level of risk' (gringo_seguro).

The structural risks faced by the Other are rendered invisible within the framework of risk reduction as a personal choice and entitlement. Risk-taking is valorised as an *entitlement* regardless of the consequences – a liberty, a right of citizenship. Online, sex tourists often expressed open defiance and rejection of the idea that they should alter their adventure-seeking pursuits. Sex tourists socialise one another to look risk resolutely in the eye and not back down, and to embrace their shared risks. Many acknowledge that risk is part of the allure. The very idea of risk is eroticised and masculinised and heightens their sense of virility and pleasure. This is particularly true in reference to the eroticisation of 'barebacking' (having sex without a condom) as a cultural practice framed through the rugged individualist ethos often associated with hegemonic masculinity (Carroll, 2003). Unprotected sex is both pleasurable and purposeful within some social contexts (Sobo, 1995; Higgins and Hirsch, 2007; Higgins et al., 2008), and connections between barebacking, sexual closeness and autonomy have been well established in the AIDS literature in reference to men who have sex with men (Díaz, 1999; Parker, 1999; Junge, 2002; Carballo-Dieguez and Bauermeister, 2004; Shernoff, 2005).

Sex tourists may endorse the pleasure of entitlement associated with barebacking whilst still acknowledging, minimising, or denying the risks involved. Personal philosophies on barebacking include weighing risk with pleasure, freedom and quality of life – risk-taking as a form of private entitlement and choice. There is a strong 'rugged individualist' sentiment that reflects US perceptions of male privilege more generally (Kimmel, 1996; Carroll, 2003). For decades, the link between gender inequality and women's risk for sexually transmitted infections has been well established (Farmer, et al., 1996; Zierler and Krieger, 1998). Thus, the pleasures experienced in connecting male entitlement, barebacking and sexual autonomy further complicates the picture for their female partners, who might concede to sex without condoms simply because they are structurally disadvantaged. Thus, in the pursuit of sexual pleasure and autonomy, men may further diminish the sexual autonomy of their female partners (Oriel, 2005).

There are those who minimise risks associated with barebacking, and this covers several strategies: citing the scientific literature, which suggests men are less likely (than women) to become infected with HIV through vaginal sex; the belief that they are able to 'tell' if someone is infected; and the assertion that 'their' sexual encounter was unique, not something provided to other customers. Others simply do their best to accept it as part of the territory:

> There is not enough space in this web site to list the dangers of fucking a prostitute in Tijuana. If you're worried about getting hurt (like getting mugged by a bandito) dying (by getting AIDS from a prostitute) or damage to your manhood (by getting a permanent case of genital herpes) then don't fuck prostitutes in Tijuana… [I]n the zone, come on guys. Every time you go there you have a pair a dice in your hands hoping you don't crap out and catch something deadly. (taco_loco)

Risk-taking: a fool's game

Risk discourse also includes sharing strategies to mitigate their exposure to risk, the portrayal of risk-taking as a fool's game. An ethos of personal (sometimes communal) responsibility also shapes the framework around risk-taking. Many *veteranos* ('old-timers') are vocal in expressing their concern that other (presumably less experienced) customers would risk the health of their peers (fellow customers), sex workers, or their own families, simply for the sake of self-indulgence. Differences in age and experience shape their discursive influence in setting the tone around risk. This approach acknowledges barebacking as regrettable, even expected, but deeply irresponsible, particularly as it relates to the ethics of a brotherhood of equals (e.g., a version of 'bro' code that is specific to sex tourists).

In this setting, there is a basic need for detailed sexual health prevention information that goes beyond simply using a condom for sexual encounters. Discussions show that customers often rely on information provided by other customers in order to stay safe, healthy and satisfied with their activities across the border and express dissatisfaction about the inadequacy of formal sex education:

> Damn public education, one lousy slide show about VD was my entire sex education. I learned more pertinent health info from this board than everywhere else combined. Sometimes it raised more questions for me that I then researched to find out... [A] seminar on 'Staying Healthy while having Fun with Sex Workers' would be a great public service. (TJ_cruiser)

Many expressed frustration that their clinical care providers were untrained in this area and that they had no one else to turn to. The most common concern is that condoms don't protect against the common infections they are exposed to while in Tijuana:

> Lemme tell ya bro – that condom only covers about 50% of what is rubbing skin to skin. Especially when those girls pour on the KY jelly and the juice is all over the damn place. Condoms don't do shit to prevent herpes. If it weren't so undesirable, I would have invented the condom that covers shaft, balls, and around the base of the penis on top and sides... only then, assuming it doesn't break, would you be 100% safe from herpes. Just remember, although the condom helps, these are PROSTITUTES that have sex with ANYBODY who will pay them... No matter how careful you are, DON'T BE FOOLED... there ARE risks! (Big_Juan)

In this quote, the customer frames his fear of contamination in several ways: first, that skin-to-skin contact is a necessary part of the encounter, and that only a hazmat suite would prevent contamination at that level;

second, that lubricant is literally 'poured' on and is 'all over the place', absent are the kinds of controls that one assumes one might have with women at home; third, these Other women will have sex with anyone. They do not discriminate against men (like him), and they do not obey the kinds of social controls that shape behaviour of women at home. They are simply prostitute bodies, and any other parts of their identity and humanity are rendered invisible.

Partner and venue choice as risk reduction

Condom use is situated within particular settings and relationships. The use of condoms is not strictly a medical decision (e.g., to prevent infection or pregnancy): it is a part of a complex social and emotional world between specific actors, each of whom carry in their minds expectations, hopes, and desires about the potential meaning of their encounter. Even an individual who endorses condom use as the only way to prevent infection will likely have particular settings or relationships where they will prefer not to use one. A situation that is considered 'safe' is less likely to necessitate condom use as a precaution against STIs; a situation that is considered potentially dangerous is more likely to encourage condom use. However, the equation of risk is only part of the picture.

Customers report they are less likely to use condoms with both commercial and non-commercial partners if they see them regularly (Hooykaas et al., 1989; Waddell, 1996). It is worth noting that some regular customers, especially those who come to view their service providers with a more emotional, romantic, familiar, girlfriend-like status, view sex without condoms as a sign that their relationship with the sex worker has become more intimate. Although some authors have made distinctions between sex (male) and 'romance' (female) tourists (see Törnqvist, 2012 for a critique), the interest in girlfriend experiences (GFE) among sex tourists is well documented (Bauer and McKercher, 2003; Rivers-Moore, 2012). This is not to suggest that *all* sex tourists look for this quality in their sexual experiences, but that both men and women (as customers) can and do blur the line between a strictly financial transaction and a more emotionally engaged experience (though within certain boundaries).

Customers seem less likely to use a condom with particular sex workers, especially those who are known to them, or those found in a place that is known to them. A growing sense of familiarity with a particular place or person can include a perception that while 'his' sex worker always uses condoms with all of her customers (and therefore is 'safe' from infection), there is an element of trust and familiarity that allows her to decide that condom use with him is unnecessary. This misperception is reinforced by an important marketing tactic on the part of a successful sex worker – she is being paid not only for sex, but also to make each of her customers feel special, and one way to indicate the special qualities of their relationship is to allow certain lines to be crossed (kissing, her cell phone number, dinner and a movie, willingness to stay overnight and, of course, sex without condoms). Such tactics increase the likelihood of regular patronage, higher tips and enhanced customer satisfaction. Knowing this, highly experienced men post precautions against falling into this sense of familiarity: 'Bottom line lets not get too comfortable with the chicas no matter how well you know them' (el_toro, *original emphasis included*).

Similarly, customers warn each other not to get 'carried away' in their encounter with the Other: 'Animal instinct can change everything in a few seconds. A guy can be smart and have great intentions, but if you're lonely/drunk/tired and you connect with the right girl this shit can happen faster that you can think about it' (mr_cajones).

In this excerpt, authentic male sexuality is portrayed as uncontrollable, animalistic, a drive barely contained by the individual: a sexuality which is managed at 'home' due to social controls. Ironically, the customer has been contaminated with the Other, a place and a person without social controls, without rational thought, where concerns can be kicked to the curb – by immersing himself in that place and with that person, he has become the Other, at least momentarily. 'Connecting' with the 'right' girl is the exact encounter sought, but not always obtained; however, it is also precisely the one in which the customer is most vulnerable to contamination. A relationship-based perception of risk should be seen as a part of a larger cultural model of risk that parallels ideas about trust, familiarity, safety, and Othering (Morris et al., 1995; Plumridge et al., 1996; Marquet and Bajos, 2000; Outwater et al., 2000).

Partner and venue choice are intertwined. A sex worker is seen to reflect the qualities of that venue. Perceptions about safety are based on reputation, environmental cues and personal experiences. Environmental cues, such as employee dress codes, cleanliness, rich textures such as marble and glass and the presence of English-speaking employees and music appear to signal that the establishment is safe and friendly (and therefore healthy) for US tourists. Outdoor venues are often portrayed as rife with poverty, danger, risk, and disease, whereas indoor venues are more likely to be discussed in terms of safety. Additionally, women new to the trade (including minors), are perceived as less likely to have been exposed to infection, whether they work indoors or out on the street. However, there is also widespread agreement that economic desperation influences the likelihood and willingness of sex workers to have sex without a condom. Class thus informs strategies for partner selection in myriad ways, including the appearance of health, wealth, cleanliness, youth and beauty.

Conclusion

Wilkinson (2001: 126) argues that public perceptions of risk are never fixed or constant but, rather, subject to the dynamics of cultural processes in which people are liable to be actively involved in creating and holding to a range of contrasting points of view. The use of an online forum, then, is one way to observe this process while allowing for such inconsistencies. Individual perceptions of risk are socially constructed through a combination of personal experience and peer group discussions. Online discussions provide a complicated perspective on the social life of risk as something that is simultaneously *normalised, sexualised, minimised* and *avoided.*

There is a dearth of theoretical frameworks through which to understand the social and discursive context of risk within sex tourism. Where risk is endemic, regular participation in such activities leads to a far more complicated social context. Customers discuss risk and share reduction strategies for their activities because, in fact, they perceive their own actions as problematic and often desire to limit potential consequences while still engaging in those activities. Discourse about perceived vulnerability is also a part of risk management. Some accept risks willingly, while still seeking to reduce the consequences of those activities whenever possible. Some customers enjoy risk-taking activities, not in spite of risk, but precisely because they are risky. Many customers valorise the willingness to take risks, revelling in the accompanying sense of masculine bravado.

Online, sex tourists acknowledged that risk of danger was part of the attraction to going to Tijuana. This did not translate directly into risk reduction behaviour. Although most developed strategies to reduce or avoid risks associated with their activities, risk perceptions and behaviours were shaped by the larger ethnosexual landscape. Risk reduction strategies are part of a *social context of risk*, which includes not only culturally informed strategies for the prevention of infectious disease, but also the negotiation of risk, pleasure and personal entitlement.

Notes

1. I was intrigued, as early research studies about customers did not directly address social context (Morse, et al., 1992; Fajans, et al., 1995; de Graaf, et al., 1997; Worm, et al., 1997; Tabrizi, et al., 2000; Xantidis and McCabe, 2000; Parsons and Halkitis, 2002). In contrast, literature on sex tourists pays closer attention to culture and power in the social and emotional worlds of male, female, sex, and 'romance' tourists (Davidson and Taylor, 1999; Davidson, 2001; Bauer and McKercher, 2003; Taylor, 2006; Phua, 2009; Mitchell, 2011; Rivers-Moore, 2012; Törnqvist, 2012; Mendoza, 2013). The mail order bride industry and other transnational relationships have also been examined (Seabrook, 1996; Clift and Carter, 2000; Mullings, 2000; Brennan, 2004).
2. This includes pedestrians, personal vehicle and bus passengers traveling into the United States from Mexico only. The number does not exclude multiple crossings by the same individual. Comparable figures for southbound crossings are unavailable.
3. Davidson (2001: 10) describes sex tourists as often troubled by developments (in their country of origin) 'which they perceive to undermine a "natural" hierarchy that is classed, gendered and "raced." So enamored with their position within this ethnosexual landscape, some become "sexpatriates", permanently leaving their country of origin. More generally, the idea that there is a kind of overall panic or response to the perceived erosion of collective (white) male domination has been referred to as part of a larger so-called "crisis of masculinity" (Connell, 1995, 1998).
4. Campbell (2007: 273) notes the romanticisation of sexual and cultural difference DOES go both ways.

References

Bauer, T. and McKercher, B. (2003) *Sex and Tourism: Journeys of Romance, Love, and Lust*. New York: Haworth Hospitality Press.

Bishop, R. and Robinson, L. (1998) *Night Market: Sexual Cultures and the Thai Economic Miracle*. London: Routledge.

Brennan, D. (2004) *What's Love Got to Do with It? Transnational Desires and Sex Tourism in the Dominican Republic*. Durham: Duke University Press.

Carroll, A. (2006) '"Accidental allegories" meet "the performance documentary": Boystown, señorita extraviada, and the border–brothel–maquiladora paradigm', *Signs*, 31(21): 357–96.

Carroll, B. E. (2003) *Individualism. In American Masculinities: A Historical Encyclopedia*. London: Sage.

Campbell, H. (2007) 'Cultural seduction: American men, Mexican women, cross-border attraction', *Critique of Anthropology*, 27(3): 261–83.

Cantú, L. (1999) 'Border Crossings: Mexican Men and the Sexuality of Migration.' Unpublished PhD diss., University of California, Irvine.

Carballo-Dieguez, A. and Bauermeister, J. (2004) '"Barebacking": Intentional condomless anal sex in HIV-risk contexts – reasons for and against it', *Journal of Homosexuality*, 47(1): 1–16.

Chow-White, P. A. (2006) 'Race, gender and sex on the net: Semantic networks of selling and storytelling sex tourism', *Media, Culture & Society,* 28(6): 883–905.

Clift, S. and Carter, S. (2000) *Tourism and Sex: Culture, Commerce and Coercion.* London: Pinter.

Connell, R. W. (1995) *Masculinities.* Cambridge, UK: Polity Press.

Connell, R. W. (1998) 'Masculinities and globalization', *Men and Masculinities,* 1(1): 3–23.

Connell, R. W. and Messerschmidt, J. W. (2005) 'Hegemonic masculinity: Rethinking the concept', *Gender and Society,* 19(6): 829–59.

de Graaf, R., van Zessen, G., Vanwesenbeeck, I., Straver, C. J. and Visser, J. H. (1997) 'Condom use by Dutch men with commercial heterosexual contacts: Determinants and considerations' *AIDS Education and Prevention,* 9(5): 411–23.

Davidson, J. O. C. (2001) 'The sex tourist, the expatriate, his ex-wife and her 'other': The politics of loss, difference and desire', *Sexualities,* 4(1): 5–24.

Davidson, J. O. C. and Taylor, J. S. (1999) 'Fantasy islands: Exploring the demands for sex tourism', in K. Kampadoo (ed.) *Sun, Sex, and Gold: Tourism and Sex Work in the Caribbean.* Lanham, MD: Rowman and Littlefield, pp. 37–54.

Díaz, R. M. (1999) 'Trips to fantasy island: Contexts of risky sex for San Francisco gay men', *Sexualities,* 2(1): 89–112.

Fajans, P., Ford, K. and Wirawan, D. N. (1995) 'AIDS knowledge and risk behaviors among domestic clients of female sex workers in Bali, Indonesia', *Social Science and Medicine,* 41(3): 409–17.

Farmer, P., Conners, M. and Simmons, J. (eds) (1996) *Women, Poverty and AIDS: Sex, Drugs and Structural Violence.* Monroe, ME: Common Courage Press.

Higgins, J. A. and Hirsch, J. S. (2007) 'The pleasure deficit: Revisiting the "sexuality connection" in reproductive health', *Perspectives on Sexual and Reproductive Health,* 39(4): 240–7.

Higgins, J. A., Hirsch, J. S. and Trussell, J. (2008) 'Pleasure, prophylaxis and procreation: A qualitative analysis of intermittent contraceptive use and unintended pregnancy', *Perspectives on Sexual and Reproductive Health,* 40(3): 130–7.

Hobbs, J. D., Pattalung, P. N. and Chandler, R. C. (2011) 'Advertising Phuket's nightlife on the Internet: A case study of double binds and hegemonic masculinity in sex tourism', *Sojourn (Singapore),* 26(1): 80–104.

Holt, T. J. and Blevins, K. R. (2007) 'Examining sex work from the client's perspective: Assessing johns using on-line data', *Deviant Behavior,* 28: 333–54.

Hooykaas, C., van der Pligt, J., van Doornum, G. J., van der Linden, M. M. and Coutinho, R. R. (1989) 'Heterosexuals at risk for HIV: Differences between private and commercial partners in sexual behaviour and condom use' *AIDS,* 3(8): 525–32.

Hughes, D. (1998/9) 'men@exploitation.com', *Trouble & Strife,* 38: 21–7.

Junge, G. (2002) 'Bareback sex, risk, and eroticism: Anthropological themes (re-) surfacing in the post-AIDS era', in E. Lewin and W. L. Leap (eds) *Out in Theory: The Eergence of Lesbian and Gay Anthropology.* Chicago, IL: University of Illinois Press, pp. 186–221.

Katsulis, Y. (2009) *Sex Work and the City: The Social Geography of Health and Safety in Tijuana, Mexico.* Austin, TX: University of Texas Press.

Katsulis, Y. (2010) '"Living like a king": Conspicuous consumption, virtual communities, and the social construction of paid sexual encounters by US sex tourists', *Men and Masculinities,* 13(2): 210–30.

Katsulis, Y., Durfee, A., Lopez, V. and Robillard, A. (2014) 'Predictors of workplace violence among female sex workers in Tijuana, Mexico', *Violence against Women* (in press).

Katsulis, Y., Durfee, A., Lopez, V. and Robillard, A. (2010) 'Female sex workers and the social context of workplace violence in Tijuana, Mexico', *Medical Anthropology Quarterly*, 24(3): 344–62.

Kendall, L. (2002) *Hanging Out in the Virtual Pub: Masculinities and Relationships Online*. Berkeley, CA: University of California Press.

Kimmel, M. S. (1996) *Manhood in America: A Cultural History*. New York: Free Press.

Marquet, N. and Bajos, J. (2000) 'Research on HIV sexual risk: Social relations-based approach in a cross-cultural perspective', *Social Science and Medicine*, 50(11): 1533–46.

Mendoza, C. (2013) 'Beyond sex tourism: Gay tourists and male sex workers in Puerto Vallarta (Western Mexico)', *International Journal of Tourism Research*, 15(2):122–37.

Mitchell, G. (2011) 'TurboConsumers™ in paradise: Tourism, civil rights, and Brazil's gay sex industry', *American Ethnologist*, 38(4): 666–82.

Morris, M., Pramualratana, A., Podhisita, C. and Wawer, M. J. (1995) 'The relational determinants of condom use with commercial sex partners in Thailand', *AIDS*, 9(5): 507–15.

Morse, E. V., Simon, P. M., Balson, P. M. and Osofsky, H. J. (1992) 'Sexual behavior patterns of customers of male street prostitutes', *Archive of Sexual Behavior*, 21(4): 347–57.

Mullings, B. (2000) 'Fantasy tours: Exploring the global consumption of Caribbean sex tourism' in M Gottdeiner (ed.) *New Forms of Consumption: Consumers, Culture, and Commodification*. Lanham: Rowman and Littlefield, pp. 227–8.

Nagel, J. (2003) *Race, Ethnicity, and Sexuality: Intimate Intersections, Forbidden Frontiers*. New York: Oxford University Press.

Oriel, J. (2005) 'Sexual pleasure as a human right: Harmful or helpful to women in the context of HIV/AIDS?', *Women's Studies International Forum*, 28(5): 392–404.

Outwater, A., Nkya, L. and Lwihula, G. (2000) 'Patterns of partnership and condom use in two communities of female sex workers in Tanzania', *AIDS Care*, 11(4): 46–54.

Parker, R. (1997) 'Migration, sexual subcultures and HIV/AIDS in Brazil' in G. Herdt (ed.) *Sexual Cultures and Migration in the Era of AIDS*. Oxford: Clarendon Press, pp. 55–69.

Parker, R. G. (1999) *Beneath the Equator: Cultures of Desire, Male Homosexuality, and Emerging Gay Communities in Brazil*. New York: Routledge.

Parsons, J. T. and Halkitis, P. N. (2002) 'Sexual and drug-using practices of HIV-positive men who frequent public and commercial sex environments', *AIDS Care*, 14(6): 815–26.

Phua, V. C. (2009) 'The love that binds: Transnational relationships in sex work', *Sexuality & Culture*, 13(2): 91–110.

Plumridge, E. W., Chetwynd, S. J., Reed, A. and Gifford, S. F. (1996) 'Patrons of the sex industry: Perceptions of risk', *AIDS Care*, 8(4): 405–16.

Rivers-Moore, M. (2012) 'Almighty gringos: Masculinity and value in sex tourism', *Sexualities*, 15(7): 850–70.

Sanders, T. (2008a) 'Male sexual scripts: Intimacy, sexuality and pleasure in the purchase of commercial sex', *Sociology,* 42: 400–17.

Sanders, T. (2008b) *Paying for Pleasure: Men Who Buy Sex*. Devon, UK: Willan Publishing.

Seabrook, J. (1996) *Travels in the Skin Trade: Tourism and the Sex Industry*. London: Pluto Press.

Shernoff, M. (2005) *Without Condoms: Unprotected Sex, Gay Men, and Barebacking*. New York: Routledge.

Sobo, E. J. (1995) *Choosing Unsafe Sex: AIDS Risk Denial among Disadvantaged Women*. Philadelphia: University of Pennsylvania Press.

Soothill, K. and Sanders, T. (2005) 'The geographical mobility, preferences and pleasures of prolific punters: A demonstration study of the activities of prostitutes' clients', *Sociological Research On-Line,* 10(1). Available at: http://www.socresonline.org.uk/10/1/soothill.html.

Tabrizi, S. N., Skov, S. and Chandeying, V. (2000) 'Prevalence of sexually transmitted infections among clients of female commercial sex workers in Thailand', *Sexually Transmitted Disease,* 27(6): 358–62.

Taylor, J. S. (2006) 'Female sex tourism: A contradiction in terms?' *Feminist Review,* 83: 42–59.

Törnqvist, M. (2012) 'Troubling romance tourism: Sex, gender and class inside the Argentinean tango clubs', *Feminist Review,* 102: 21–40.

USGSA (2013) 'Fact Sheet' (San Ysidro Land Port of Entry). Available at: http://www.gsa.gov/portal/category/105703 (accessed 16 Oct 2013).

USDOT (2013) Research and Innovative Technology Administration, Bureau of Transportation Statistics, based on data from the Department of Homeland Security, US Customs and Border Protection, Office of Field Operations. Available at: http://transborder.bts.gov/programs/international/transborder/TBDR_BC/TBDR_BC_Index.html (accessed 16 Oct 2013).

Waddell, C. D. (1996) 'Female sex work, non-work sex and HIV in Perth', *Australian Journal of Social Issues,* 31(4): 410–25.

Wilkinson, L. (2001) *Anxiety in a Risk Society*. London: Routledge.

Williams, S., Lyons, L. and Ford, M. (2008) 'It's about bang for your buck, bro: Singaporean men's online conversations about sex in Batam, Indonesia', *Asian Studies Review,* 32(1): 77–97.

Worm, A. M., Lauritzen, E. and Jensen, J. P. (1997) 'Markers of sexually transmitted diseases in seminal fluid of male clients of female sex workers', *Genitourinary Medicine,* 73(4): 284–7.

Xantidis, L. and McCabe, M. P. (2000) 'Personality characteristics of male clients of female commercial sex workers in Australia', *Archives of Sexual Behavior,* 29(2): 165–76.

Zierler, S. and Krieger, N. (1998) 'HIV infection in women: Social inequalities as determinants of risk', *Critical Public Health,* 8(1): 13–32.

12
Recognising Homoeroticism in Male Gay Tourism: A Mexican Perspective

J. Carlos Monterrubio and Álvaro López-López

Aspects of sexuality in mainstream tourism have gained academic attention; however, in the main, discussions of sexuality and tourism have focused on the experiences of gay men (Hughes, 2002; Visser, 2003; Hughes and Deutsch, 2010; Melián-González et al., 2011; Waitt and Markwell, 2006, this volume). Gay men[1] have increasingly been identified as a profitable market by the travel and tourism industry. Their alleged economic power has led both industry and scholars to recognise gay communities as a niche market with a high propensity to take holidays (Holcomb and Luongo, 1996; Pritchard et al., 1998; Roth, 2004; Ersoy et al., 2012). For gay tourists, sexual identity plays an important role in defining travel and holiday patterns; the need to escape from everyday social restrictions attached to homosexuality and constructing identities becomes a catalyst for gay tourism.

When analysing the relationship between sexuality, tourism and identities, some issues deserve special attention. First, it must be recognised that the gay population is not homogenous (Hughes, 2004: 65). Differences in economic, social, cultural, as well as sexual conditions of this population question the assumption that the men's economic power and the construction of their identities are always present in so-called 'gay tourism'. With regard to identity as a determinant factor for gay tourism (Hughes, 1997), it should be noted that gay identity is only one of the several subjectivities that men may attach to their erotic interactions with other males when holidaying. Multiple practices, meanings and identities coexist among men and these may not necessarily be determinant factors for the tourism experiences they seek.

This chapter aims to acknowledge that homoeroticism, as recognition of the diverse erotic interactions and sexual practices and meanings among males, is a relevant issue in defining gay men's tourism experiences. It starts by questioning the alleged economic power of gay tourism and the gay concept as alien to some cultures. Then, it considers some of the main reasons for gay travel in order to show that sexuality, and various subjective implications, may influence tourism patterns. Finally, in order to understand the diverse sexual meanings and identities among men, a 'homoerotic' framework is presented. The main findings of a previous study on tourism and homoeroticism in Mexico are offered as an illustration of the complexity of such interactions and their effect on tourism.

Gay tourism as a lucrative market

Within the large diversity of sexualities and identities, it has been commonly argued that gay men and lesbians qualify as a profitable market group (Pritchard and Morgan, 1997; Russell, 2001; Stuber, 2002). This well established idea is largely associated with what seems a general consensus that gay men and lesbians have more available incomes for leisure. This is presumably due to the reduced number of economic expenses as compared to those of some heterosexuals. Although not all gay men are able to get involved in travel and tourism due to economic limitations (Casey, 2009), the perception of gay men and lesbians qualifying as a lucrative market is often related to allegedly large discretionary incomes, more discretionary leisure time and the specific interests that represent a preference to be high-spenders. Market reports, such as that of Community Marketing, an organisation that helps companies better understand and more effectively reach the lesbian, gay, bisexual and transgender (LGBT) communities, have found that gay men and lesbians have the largest amount of disposable income of any niche market. They have also suggested that the gay market is a sizeable niche since LGBT consumers, for instance, makes up 5% or more of the US consumer market (www.communitymarketinginc.com, 21 September 2012). The attention that companies such as American Airlines, American Express, Absolut Vodka, Ford, Avis and Virgin have paid to gay and lesbian themes in their marketing reveal the economic value of the LGBT communities (Hughes, 2006). While travel and tourism companies have become interested in the economic power of LGBT communities, their implications are not limited to market issues; travel must be regarded as a serious business that can challenge and transform identities and local and global cultures (Casey, 2009: 159).

The alleged economic power of gay men and lesbians has been echoed in various industries, including travel and tourism. The fact that tour operators, cruise companies, travel agents, accommodation providers, airlines and destinations have become interested in the gay and lesbian holiday market (Peñaloza, 1996; Hughes, 2006) suggests that such industries have acknowledged the profitability of the gay community. Nonetheless, this does not reflect a greater acceptance of lesbians or gay men; they may be just another market to exploit. As Guaracino (2007) demonstrates, this has been clearly evidenced in the large number of companies and destinations that have joined the gay tourism market; gay travellers have constituted a rapidly growing and highly lucrative segment of the tourism industry (Prichard et al., 1998).

The common perception that gay travel and tourism is a profitable market segment seems to emerge largely from market research findings and scholarly work. Market research indicates that gay people tend to take holidays, and that multiple holidays are also more common amongst gay consumers (Mintel, 2000). In some countries (for example Australia), gay travellers have been reported as being more likely than the total travelling population to holiday overseas (Roy Morgan Research, 2003). The finding that the number of gay men and women holding a valid passport (85% and 77% respectively) surpasses that of heterosexuals in some countries like the United States reinforces the perception that gay travel is in reality a lucrative market (www.communitymarketinginc.com, accessed 21/09/12).

From an academic point of view, these perceptions have also been supported. Academic studies have argued that gay male couples have higher average incomes and more discretionary time and money for travel than their heterosexual counterparts (Holcomb and Luongo, 1996; Roth, 2004). Recent surveys on gay tourists have also revealed that over half of respondents travel abroad, and that comfort and good food, resting and relaxation are very important reasons for travel (Ersoy et al., 2012). While gay men's alleged economic power and discretionary time may be questioned due to current conditions including civil partnership/marriage and the right to adopt, other factors including sexuality play a significant role in defining male gay tourism.

Sexuality and reasons for travel

The relationship between sexuality and travel is seen in heterosexual and non-heterosexual tourism. Concerning heterosexuality, aspects such as romance (Jeffreys, 2003), honeymoons (Lee et al., 2010), premarital stags

(Thurnell-Read, 2012), sex tourism and male identity (Krhuse-Mount Burton, 1995), rationalisations of sex tourists (Garrick, 2005), and many other holiday practices suggest that sexuality is often a reason to travel, regardless of sexual preference.

However, sexuality plays a very important role in gay travel. The work conducted by Clift and Forrest (1999) on gay men's holiday motivations is one of the first studies to focus on the significance of the dimensions of sexuality, particularly gayness, in holiday choice. In their study, they found that comfort, rest and relaxation as well as guaranteed sunshine are very important to gay men. Of course there is no reason to believe that these factors are exclusive to gay men as they are important to perhaps most holidaymakers, irrespective of sexuality. However, for approximately a third of Clift and Forrest's sample, social dimensions of a holiday, and particularly the gay character of a holiday, were also found to be important. The authors claim that men who visit a destination with a gay reputation will be motivated by opportunities to socialise with other gay men and perhaps to be sexually active with new partners.

The idea that sexuality plays an important role in gay travel decision-making has been supported by other academic studies, particularly those from wealthier countries. By investigating the travel motivations of gay and lesbian tourists in the UK, for example, Pritchard et al. (2000) conclude that sexuality does influence choices of accommodation, booking methods (see also Poria and Tailor, 2001), destinations and packages. They indicate that sexuality has a different significance for gay men, and that the need to feel safe and comfortable with like-minded people and to escape from heterosexism are key influences on choice of holiday for gay men and lesbians. The same authors suggest, too, that the need to escape from the pressures of being gay in a heterosexist world is commonly perceived as the most important motivating factor for taking a holiday.

In this vein, it can be postulated that in addition to relaxation and rest, the reasons for gay travel might be particularly associated with sexuality and identities. Because very often gay men – and other men who are erotically related to other men – live in areas dominated by heterosexism, holidays and other leisure activities have a particular importance as they offer the most significant opportunity for constructing, confirming and/or changing their sexual identity (Hughes, 1997; Cox, 2001): in other words, a chance to be themselves. Through his pioneering work on gay men's travel and tourism, Hughes (1997) notes that although the choice of a gay identity may be particularly painful due to society's reaction to homosexuality, many men see the acceptance of a gay identity as quite fundamental. The acceptance of this identity may entail following certain leisure activities, including tourism, in a spatial concentration of pubs, clubs, cafés, shops, residences and public space, which permit the validation of a gay identity through relationships with others. This has led to the assertion:

> A vast majority of leisure travel or tourism by gay men can be described as identity tourism. Travel to gay spaces makes it possible for gay men to express themselves freely and to experiment with gay identities that are circumscribed at home...The gay identity, thus, fuels gay men's need to travel and the act of travelling provides them new insights about themselves and paves the way for perhaps a new and certainly more positive conception of self. (Herrera and Scott, 2005: 260)

Thus, the evidence suggests that, within the context of sexualities, the reasons for much gay travel can be linked to particular needs: for sociability with other gay men, for access to gay spaces, to avoid homophobia, and to be more open about their sexual identity. Although the importance of sexual identity in travel and tourism is not exclusive to gay men, the (de)construction, confirmation and change of sexual identities are recognised as fundamental reasons for gay travel and tourism. The adoption of a gay identity for some men thus becomes a need for specific leisure activities, including holidaying (Casey, 2009).

Why raise these issues at all?

When recognising the alleged economic power of gay men for travel and tourism and the significance of these for the performance of sexualities, some observations should be borne in mind. First, a detailed literature review shows that several conclusions have been drawn from specific population segments in the so-called developed countries. According to Anon (cited in Visser, 2003), gay tourism statistics generally reflect the consumption behaviour of the wealthier Western segments of the gay population; therefore figures on economic power may have more to do with the economic status of certain social strata than with a characteristic inherent in the gay community.

Second, it needs to be acknowledged that while some findings come from a large sample, they are by no means representative of the *whole* gay population – if such a population can ever be accurately defined.

Although some researchers have undertaken surveys on a large scale (Browne et al., 2005: 38), their findings are limited to specific times, settings, groups, purposes and research methodologies; this fact represents a problem of obtaining representative samples. Third, considering that estimating the size of the gay market in terms of holidays is difficult, because of the hidden nature of the population (White, 2001), many of the existing surveys very often incorporate the views and experiences only of those who are open about their sexual orientation and behaviour, or who self-define as gay. The economic and sociocultural conditions and experiences of those who are not open about their sexual identity or do not self-define as gay or with any other label are thus underrepresented.

Furthermore, gay men are not homogenous; differences amongst them are not only present in terms of economic power, but are also present through social, cultural and political conditions that may influence their holiday patterns. Such conditions are exclusive to the specific environment in question and may differ significantly from one society to another. Therefore, when analysing the alleged economic power of the gay community as related to tourism and travel, and its implications for sexual identities, it should be noted that figures applicable to some gay populations may not be applicable to others. Attention therefore should be paid to the specific conditions of these including, on the one hand, the economic, social and political circumstances and, on the other, the sexual practices, meanings, identities and subjectivities that are culturally defined within the population studied. In this vein, in countries where culture plays a considerable role for (self) defining, understanding and performing sexualities, the holidaying of the gay men may not necessarily be determined only by their economic power, but also by the particular meanings attributed to their sexuality.

Homoeroticism

When analysing the literature on sexualities as related to travel and tourism, particularly on homosexuality, it is noted that the gay identity has become an overwhelmingly dominant term to refer to homosexuals, however they self-define. The 'gay' concept has been the dominant label for men who identify themselves by their interest in other men and who have sexual experiences with other men (Nardi, 2000). But very importantly, it is closely linked to the adoption of certain forms of thinking and behaving that comprise the gay identity (Hughes, 1997) and thus an ideal label for the customisation of leisure experiences. This identity does not imply only a sexual preference itself but a politically subversive lifestyle consisting of the adoption of specific behaviours, desires, economic power and attitudes, among others (Hernández, 2002).

Although very frequently adopted for the analysis of sexualities, tourism and marketing purposes, the notion of 'gay', and particularly the gay identity, it often ignores the complex continuum of sexual practices, desires and meanings. By understanding 'gay' as 'people who are sexually attracted to other people of the same biological sex' (Hughes, 2006: 1), it often ignores the complex continuum of sexual practices, desires, identities and meanings that people perform and assign to their experiences with people of the same sex. These issues will eventually define certain aspects of their lives including leisure and travel patterns.

Sexual practices and the meanings related to these are possibly innumerable, for they are culturally determined, either individually or collectively. Therefore a gay identity may not be adopted or not even be recognised by some males. As Cantú (2002) observes, the gay identity, as understood in Anglo-Saxon contexts, does not exist in other specific cultural contexts and can thus be understood as constructions alien to other cultural and social landscapes such as those in Mexico (p. 141), where the naming of a man sexually related to other men is socially constructed not only on the basis of biological sex of the sexual partner, but, among several other factors, on the roles of being 'active' or 'passive' in sexual acts (Carrier, 1995; Murray, 1995; Chant and Craske, 2002).

Because gay destinations have become contested sites across geographical spaces (Wait and Markwell, 2006: 22), tourist spaces are socially shaped. Various meanings, subjectivities and identities are constructed among males within them. Mobility provided by travel and tourism allows some men the ability to (de)construct, compare, contrast, negotiate, adopt or reject a specific gay identity.

Bearing this in mind, the concept of 'homoeroticism' provides a more inclusive framework for the understanding of sexualities, identities, desires and subjectivities among people of the same sex. Unlike the notion of 'gay', it becomes a more fruitful framework for the analysis of erotic interactions among males and the diverse social practices and meanings that emerge from such interactions. The concept of homoeroticism, as suggested by Núñez (2001), recognises, on the one hand, that individuals who are somehow sexually and emotionally related or orientated to people of the same sex may adopt a large number of erotic practices including, but not exclusive to, kisses, hugs, caresses, genital stimulations, looks, fetishist games and anal penetration. On the other hand, homoeroticism acknowledges that those men erotically related to other men may (not) self-define as gay, homosexual, or any other ethnocentric, academic or medical categorisation already established. Unlike individuals who hold a gay identity, not all men who are erotically linked to other men will adopt this as a sexual identity; therefore the adoption of any homosexual behaviour does not necessarily imply the adoption of a sexual identity that may eventually influence ones' holiday patterns. A large number of gay men will not participate actively in gay travel and tourism, as has been often reported and presumed in the case of those who are self-defined as or utilise a gay identity position.

Recognising homoeroticism in gay tourism in Mexico

Homosexuality in Mexico has for long been socioculturally defined not on the basis of the biological sex of the participants but commonly on the gender roles that males perform in the sexual act (Carrier, 1995). 'Homosexual' is commonly used to refer to an effeminate man who is understood to take the 'woman's role' in sex with another man (i.e. one who retains the passive role during sexual intercourse) (Prieur, 1996). In Mexican culture, while a man's feminine appearance is a signal that he wants to be penetrated, men who penetrate men are not frequently stigmatised in several regions in Mexico as documented by Córdova (2003, 2005) on the coast of Veracruz. In fact, value is given to men who penetrate women or other men and never take the passive role in the sexual act (Prieur, 1996). Hence, men who maintain a dominant sexual script by playing the active role in the same-sex act are less likely to identify themselves as homosexual and do not risk their heterosexual identity in doing so; their masculinity remains intact (Cantú, 2000; Chant and Craske, 2002).

It has been claimed that the gay identity is used and holds cultural meaning in Mexico since the mid-1990s (Murray, 1995). However, not all men who have erotic links with other men necessarily hold a specific (homo)sexual identity or self-define as gay. For some, the gay identity, either as a term or a distinguished and visible lifestyle, may not even exist in their minds. Many men in Mexico do get involved in several homoerotic practices (for example sexual acts) or desires without adopting gay-related lifestyles or consumption patterns including tourism. Tourism and other leisure activities are thus not always associated to homoerotic practices or to a specific lifestyle related to sexuality. As a matter of fact, because travel and sophisticated leisure activities are commonly related to gay lifestyles, many gay men (not self-defined as gay) may actually opt for avoiding gay tourism in order not to be associated with a gay identity or culture.

In Mexico, many cultural groups generally disapprove of homosexuality (Carrier, 1995). The social environment in some parts of the country still remains repressive and often violent towards public same-sex affection (Anodis, 2013). When gay men become visible, there is the possibility of violence, especially against effeminate men, who may even end up being murdered (Reding, 2000; Anodis, 2013). In order to live an undisturbed openly gay life in many parts of Mexico, including the larger cities, gay men need to hide their behaviour and identity. Tourism can then provide gay men in Mexico with the opportunity to perform their sexuality away from the social constraints of home; touristic spaces in Mexico, particularly those that have been regarded as gay destinations, represent a choice for men who seek be erotically involved with other men in an anonymous way. The existence of bars, cafes, accommodation, dance clubs, saunas/baths, cruising and male sexual services in destinations such as Acapulco, Guadalajara, Mexico City and Puerto Vallarta are frequently advertised in international gay guides such as *Spartacus* (Bedford, 2007) and thus represent an alternative for some to perform the complexity of their sexuality in Mexico, but away from their own quotidian lives.

Since the 1980s, and more predominantly in the 1990s, academic work on tourism and homoeroticism in Mexico has increased. This work, however, has often analysed tourism and homoeroticism as independent entities; the links between both phenomena have thus been ignored. Only recently has the relationship between tourism and homoeroticism gained academic attention in Mexico. A book edited by López and Van Broeck (2012), and the chapter summarising the findings of López and Van Broeck published in 2010, may be regarded as pioneering works in Mexico, for they first look at the multiple interactions among males in different tourism contexts in the country. These works bring together the experiences of tourism and homoeroticism of seven important tourism destinations in Mexico namely Tijuana, Puerto Vallarta, Veracruz, Acapulco, Guadalajara, Cancun and Mexico City. While Tijuana, Guadalajara and Mexico City are urban destinations, the rest are beach-oriented resorts.[2] The value of these investigations lies in their analysis of sex service among male tourists and male sex workers in tourism contexts where, based on 100 in-depth interviews, the complex relationships between homoeroticism and tourism are discussed. In addition to sex work and worker-client relationships, issues such as sexual self-definition and broader experiences are part of these investigations. For the purpose of this chapter, the four main findings are now presented.

The first relevant issue, as found by López and Van Broeck (2010, 2012), concerns the identity of the individuals studied. Both national and international tourists and local males having affective-sexual connections hold wide-ranging sexual identities. The variation ranges from those men who self-define as heterosexual, bisexual or gay (whose identities seem to be globalised) to those self-defining as *chichifos, chacales* and *mayates* (whose identities go beyond international language and are region-specific). Those who self-define as heterosexual (or who used the word 'man' for self-definition), claimed they had sexual practices with other men because of economic necessity; among these men were those who stated that although sexual relations with other men are not necessarily what they desired sexually, they could end up enjoying it. Sex workers reported that their tourist clients held various sexual practices and identities such as gay or even heterosexual. Some of these male tourists frequently travel with their families (wife and children) and find ways to have sex with other men; thus, this suggests the presence of gay sex within heterosexual tourism.

The second relevant finding is related to the notion of masculinity. Despite the large diversity of sexual identities, it was evident that for both tourists and sex workers, masculinity is fundamental in a sexual interaction. One's masculinity is used as a measure that determines the behaviour and bounds of each man within the sexual interaction, where masculinity becomes a power relation that influences sexual relationships between males. The notion of masculinity, as perceived by informants, is unclear and associated with having a body language and lexicon that are not 'effeminate' and having a large penis or the act of penetration during sex.

With regard to the third finding, there is a recent debate about whether tourists performing homoerotic practices tend to interact sexually more as compared with heterosexual tourists. This is based on the assertion that a gay identity, in comparison with other male identities, adds to itself sexual experience as something intrinsic and open (Hernández, 2005). Consequently, it is not surprising that several gay destinations around the globe have not only spaces for socialisation and leisure (such as restaurants, cinemas, theatres, etc.) but also specific amenities that encourage sexual interaction (such as saunas, dark rooms in dance clubs, male sex work in open and closed spaces, massage parlours, cruising areas, etc.). In this vein, Mexican cities with a large number of tourists, such as Puerto Vallarta, Guadalajara, Acapulco, Cancun, and Mexico City, have gay spaces where homoerotic interactions among males are facilitated.

Finally, in the case of sexual interactions between tourists and locals in the destinations studied, López and Van Broeck (2010, 2012) identify two types of geographic spaces. On the one hand, what they term 'traditional spaces' are identified in all these cities. These sites are close to the centre of each city and visited by men adopting traditional gay identities in search of clandestine sexual relations with other men; this commonly happens in cinemas, pubs, and public toilets, to mention but a few. These areas are often difficult to reach for tourists, but undoubtedly are visited by some. On the other hand, gay spaces are also identified. This is largely due to the fact that knowledge around gay identity is increasingly recognised by larger sectors of the Mexican society. In such gay venues, which are more visible than traditional places, sexual encounters are possible in sites such as hotels, dance clubs, saunas and dark rooms. Gay tourists are therefore more likely to have access to these spaces, particularly since they are widely advertised through conventional means such as the Internet and gay guides.

Conclusion

Globally, gay identity shapes holiday patterns significantly. While much of the existing work on sexuality and tourism derives from the context of developed countries, in Latin American contexts, and more specifically in Mexico, the term 'gay' is insufficient to encompass the complex multiplicity of identities, practices, meanings and subjectivities among males as demonstrated by the cases described here. The notion of gay as commonly interpreted and applied in Western countries, therefore fails to explain travel and tourism experiences of men who are erotically linked to other men in different cultural contexts.

In this vein, the term 'homoeroticism' becomes more suitable to refer to a diversity of practices and the large number of relationships among men who emotionally and sexually relate to other men both in tourism and non-tourism contexts. If sexuality influences holiday patterns, how these patterns are defined should then be revised by recognising that gay identity is only one of various possible homoerotic practices and identities that males can adopt. This then suggests that the travel and holiday patterns of homosexual men are not only defined by their alleged economic power but also, and perhaps more importantly, by how men self-define and how they construct their sexual subjectivities. If such constructions are socially and culturally defined, it is reasonable to assert that men's behaviours, sexual activities, emotions and identities will be diverse and dependent on geographical location and the historical background of any given community/country. As such, the mobility represented by travelling to and within tourist spaces must be seen to play a significant role in the construction of sexual identities.

The homoerotic framework offered in this chapter thus calls for more empirical testing in different cultural contexts. While it acknowledges the multiple and complex identities, desires and subjectivities among males, such a framework may guide other researchers to study the diverse social practices and meanings that emerge from male interactions in specific cultural contexts. Future findings guided by this framework may not only be different from those of the notion of 'gay' but also from those encountered in the context of Mexican cultures.

In the same vein, with regard to the diverse erotic interactions and sexual meanings among males in tourism contexts, one particular issue still needs to be explored. How applicable the recognition of homoeroticism is to other cultures, especially to those in developed countries, remains unknown. If we take into account that cultures vary from place to place, and that male subjectivities are culturally defined, such as the empirical evidence here provided, special effort must be made to identify gay identities and any other label in each specific cultural context. Tourism will by no means signify the same to all gay men; it will hold special value for the construction of some men's identities but it may remain insignificant for the sexual identities of others. Because of this, it is increasingly important to explore the lives of men who have same-sex experiences away from the Global North. This will certainly lead to a more inclusive understanding of tourism as related to homoerotic practices.

Notes

1. For the purpose of this chapter, the concept 'gay' is adopted to refer basically to multidimensional homosexual interactions among men. Although the notion of gay is somehow restraining for encompassing the multiple and culturally defined relationships among men, gay will be used to refer to such relationships in the subsequent sections. The corresponding discussion will be offered in turn.
2. For a detailed touristic description of each of the destinations, see López and Van Broeck (2012).

References

Anodis (2013) *En el Distrito Federal Continúa la Homofobia*, Available at: http://anodis.com/nota/23673.asp (accessed 18 September 2013).

Bedford, B. (2007) *Spartacus International Gay Guide*. Berlin: Bruno Gmünder.
Browne, K., Church, A. and Smallbone, K. (2005) *Do it with Pride: Lesbian, Gay, Bisexual and Trans Lives and Opinions*. Survey Report. University of Brighton.
Burton, S. (1995) 'Sex tourism and traditional Australian male identity', in M. Lanfant, J. Allcock and E. Bruner (eds) *International Tourism. Identity and Change*. London: Sage, pp. 192–204.
Cantú, L. (2000) 'Entre hombres/between men: Latino masculinities and homosexualities', in P. Mardi (ed.) *Gay Masculinities*. London: Sage Publications, pp. 224–47.
Cantú, L. (2002) 'De ambiente: Queer tourism and the shifting boundaries of Mexican male sexualities', *GLQ: A Journal of Lesbian and Gay Studies*, 8(1–2): 139–66.
Carrier, J. (1995) *De Los Otros: Intimacy and Homosexuality Among Mexican Men (Between Men-Between Women)*. New York: Columbia University Press.
Casey, M. (2009) 'Tourist gay(ze) or transnational sex: Australian gay men's holiday desires', *Leisure Studies*, 28(2): 157–72.
Chant, S. and Craske, N. (2002) *Gender in Latin America*. London: Latin American Bureau.
Clift, S. and Forrest, S. (1999) 'Gay men and tourism: Destinations and holiday motivations', *Tourism Management*, 20(5): 615–25.
Córdova, R. (2003) 'Mayates, chichifos y chacales: Trabajo sexual masculino en la ciudad de Xalapa, Veracruz', in M. Miano (ed.) *Caminos Inciertos de las Masculinidades*. Mexico: INAH/Conaculta, pp. 141–61.
Córdova, R. (2005) 'Vida en los márgenes: La experiencia corporal como anclaje identitario entre sexoservidores de la ciudad de Xalapa, Veracruz', *Cuicuilco*, 12(34): 217–38.
Cox, M. (2001) 'Gay Holidaymaking: A Study of Tourism and Sexual Culture'. PhD thesis, University of London, London.
Ersoy, G., Ozer, S. and Tuzunkan, D. (2012) 'Gay men and tourism: Gay men's tourism perspectives and expectations', *Procedia, Social and Behavioural Sciences*, 41: 394–401.
Garrick, D. (2005) 'Excuses, excuses: Rationalisations of western sex tourists in Thailand', *Current Issues in Tourism*, 8(6): 497–509.
Guaracino, J. (2007) *Gay and Lesbian Tourism: The Essential Guide for Marketing*. Oxford: Butterworth-Heinemann.
Hernández, P. (2002) 'No Nacimos ni nos Hicimos, sólo lo Decidimos. La Construcción de la Identidad Gay en el Grupo Unigay y su Relación con el Movimiento Lésbico, Gay, Bisexual y Transgenérico de la Ciudad de México'. Master thesis, Escuela Nacional de Antropología e Historia, México.
Hernández, P. (2005) 'El movimiento lésbico, gay, bisexual y transgenérico y la construcción social de la identidad gay en la Ciudad de México', in *Memorias de la II Semana Cultural de la Diversidad Sexual. México.* Instituto Nacional de Antropología e Historia, México.
Herrera, S. and Scott, D. (2005) 'We gotta get out of this place! Leisure travel among gay men living in a small city', *Tourism Review International*, 8(3): 249–62.
Holcomb, B. and Luongo, M. (1996) 'Gay tourism in the United States', *Annals of Tourism Research*, 23(3): 695–726.
Hughes, H. (1997) 'Holidays and homosexual identity', *Tourism Management*, 18 (1): 3–7.

Hughes, H. (2002) 'Gay men's holiday destination choice: A case of risk and avoidance', *International Journal of Tourism Research,* 4(4): 299–312.
Hughes, H. (2004) 'A gay tourism market: Reality or illusion, benefit or burden?', *Journal of Quality Assurance in Hospitality & Tourism,* 5(2/4): 57–74.
Hughes, H. (2006). *Pink Tourism: Holidays of Gay Men and Lesbians*. Oxfordshire: CABI Publishing.
Hughes, H. and Deutsch, R. (2010) 'Holiday of older gay men: Age or sexual orientation as decisive factors?', *Tourism Management,* 31(4): 454–63.
Jeffreys, S. (2003) 'Sex tourism: Do women do it too?' *Leisure Studies,* 22(3): 223–38.
Lee, C., Huang, H. and Chen, W. (2010) 'The determinants of honeymoon destination choice: The case of Taiwan', *Journal of Travel & Tourism* Marketing, 27(7): 676–93.
López, A. and Van Broeck, A. (2010) 'Sexual encounters between men in a tourist environment: A comparative study in seven localities in Mexico', in N. Carr and Y. Poria (eds) *Sex and the Sexual During People's Leisure and Tourism Experiences.* Newcastle: Cambridge Scholars Publishing, pp. 119–42.
López, A. and Van Broeck, A. (2012) *Cuerpos Masculinos en Venta: Turismo y Experiencias Homoeróticas en México. Una Perspectiva Multidisciplinaria.* Mexico: Instituto de Geografía, Universidad Nacional Autónoma de México.
Melián-González, A., Moreno-Gil, S. and Araña, J. (2011) 'Gay tourism in a sun and beach destination', *Tourism* Management, 32(5): 1027–7.
Mintel. (2000) *The Gay Holiday Market.* London: Mintel International Group.
Murray, S. (1995) 'Homosexual categorization in cross-cultural perspective', in S. Murray (ed.) *Latin American Male Homosexualities.* Albuquerque: University of New Mexico Press, pp. 3–32.
Nardi, P. (2000) 'Anything for a sis, Mary: An introduction', in P. Mardi (ed.) *Gay Masculinities.* London: SAGE, pp. 1–11.
Núñez, G. (2001) 'Reconociendo los Placeres, Desconstruyendo las Identidades. Antropología, Patriarcado y Homoerotismos en México', *Desacatos. Revista de Antropología Social,* 6: 15–34.
Peñaloza, L. (1996) 'We're here, we're queer and we're going shopping! A critical perspective on the accommodations of gays and lesbians in the US market place', in D. Wardlow (ed.) *Gays, Lesbians, and Consumer Behavior: Theory, Practice and Research Issues in Marketing.* New York: Haworth Press, pp. 9–41.
Poria, Y. and Tailor, A. (2001) 'I am not afraid to be gay when I'm on the net: Minimising social risk for lesbian and gay consumers when using the internet', *Journal of Travel and Tourism Marketing,* 11(2/3): 127–42.
Prieur, A. (1996) 'Domination and desire: Male homosexuality and the construction of masculinity in Mexico', in M. Melhuus and K. Stoles (eds) *Machos, Mistresses and Madonnas.* London: Verso, pp. 83–107.
Pritchard, A. and Morgan, N. (1997) 'The gay consumer: A meaningful market segment?', *Journal of Targeting, Measurement and Analysis for* Marketing, 6 (1): 9–20.
Pritchard, A., Morgan, N., Sedgley, D. and Jenkins, A. (1998) 'Gay tourism destinations: Identity, sponsorship and degaying', in C. Aitchinson and J. Fiona (eds.) *Gender, Space and Identity: Leisure, Culture and Commerce.* Eastbourne: Leisure Studies Association, pp. 33–46.

Pritchard, A., Morgan, N., Sedgley, D., Khan, E. and Jenkins, A. (2000) 'Sexuality and holiday choices: Conversations with gay and lesbian tourist', *Leisure* Studies, 19(4): 267–82.

Reding, A. (2000) *Mexico: Update on Treatment of Homosexuals. Question and Answer Series*. New York: Ins Resource Information Centre.

Roth, T. (2004) 'Opportunities in the gay and lesbian tourism marketplace', *e-Review of Tourism Research*, 2(4): 78–82.

Roy Morgan Research (2003) *Key Trends in Gay Leisure Travel*. Melbourne: Roy Morgan Research Pty Ltd.

Russell, P. (2001) 'The world gay travel market', *Travel & Tourism Analyst*, 2: 37–58.

Stuber, M. (2002) 'Tourism marketing aimed at gay men and lesbians: A business perspective', in S. Clift, M. Luongo and C. Callister (eds) *Gay Tourism: Culture, Identity and Sex*. London: Continuum, pp. 88–124.

Thurnell-Read, T. (2012) 'Tourism place and space: British stag tourism in Poland', *Annals of Tourism Research*, 39(2): 801–19.

Visser, G. (2003) 'Gay men, tourism and urban space: Reflections on Africa's gay capital', *Tourism Geographies*, 5(2): 168–89.

Waitt, G. and Markwell, K. (2006) *Gay Tourism: Culture and Context*. London: The Haworth Hospitality Press.

White, A. (2001) 'In the Pink', *Leisure Management*, 21(3): 42–4.

Part IV
Embodying Masculine Travel

13

The Lads Just Playing Away: An Ethnography with England's Hooligan Fringe during the 2006 World Cup

James Treadwell

While much academic literature on the topic of English football hooliganism suggests that unruly travelling fans are inappropriately socialised (Williams et al., 1984; Dunning et al., 1987) other academics, such as Ruggiero (2000: 10) have suggested in contrast that 'in fact, theirs is a form of excessive socialisation, which takes the form of an excessive attachment to their locality, city, or country…their violent behaviour may signify a perfectly acquired, if excessive national identity'. However, these accounts of English football supporters are problematic, not least because they do not consider how the fans behave in comparison with similar travelling English holidaymakers and say little about how or why they enact specific forms of masculine identity. This chapter seeks to rectify that by drawing on research undertaken with English male supporters of the national team over several nights during the football World Cup hosted in Germany in 2006. Data was gathered in both the Red Light District of Amsterdam prior to the tournament and then in Frankfurt, Germany, for two nights before England's first game. Amsterdam provided a location that many English supporters decided to travel through en route to the tournament, in part because it allowed relatively easy direct rail links to the host cities in Germany.

It is fair to suggest that the participants, drawn from several domestic English football hooligan gangs, could be described as holding 'a position of some priority' amongst 'those male fans who have regarded the appropriate support of the team as centring on drinking, singing and fighting and who have, since the mid-1980s, adopted very expensive "casual" styles of dress' (see King, 1997: 331–2). In the last decade, the official bodies of the UK national government and the Football Association for England and Wales (the FA) have been driven by a desire to clamp down on unruly fandom, while some academic commentators expressed concern about the potential marginalisation and alienation of some traditional working class fan groups (Crabbe, 2004; Crabbe and Brown, 2004). 'England' supporters, when following the national team, have often homogenously been negatively characterised as aggressive, drunken, nationalistic and xenophobic, and racist, ignoring the fact that in reality the problematic supporters were a minority (Crabbe, 2004; Stott and Pearson, 2007).

While the disorderly conduct of some English football supporters is not fictitious or imagined, problematic aspects of tourist behaviour such as drinking, fighting, risky sex and drug consumption are not confined to travelling English football fans. Such is also present in the work of Briggs (2013) in his ethnographic work with young British people on holiday in Ibiza. Less extreme variants of this problematic British travel are also seen in work on stag tour weekends taken by British men in Eastern Europe (Thurnell-Read, this volume), Casey's writing on the working class masculinity of tourists represented in the television series *Benidorm* (Casey, this volume), and Karen O'Reilly's work on 'The British on the Costa Del Sol' (2001), which also gives some useful contextual discussion of the 'Brits Abroad Problem' where drunken and offensive behaviour are seen as central features of particular forms of travel. This suggests some commonalities between some travelling England football supporters and some English male tourists more generally and, arguably, this offers an insight into national mores of masculinity and the construction of contemporary identities.

It is for a minority group of travelling England football supporters ('the lads' featured here) that the travel experience is both understood and consumed on the premise that it represents a distinct physical, social and symbolic place and time within which masculine behaviour is enacted. Much like these other forms of risky travel involving males, the 'England away' is both performative and embodied. The male body plays a central role through the consumption of alcohol (and often illicit drugs) and through corporeal practices involving riskiness, danger and the seeking out of violent encounters. As this chapter will discuss, this enacted transgression is regarded positively by such football fans and serves as both a release from social restraint and an ideologically barrier separating them from 'other' mainstream travelling supporters and a wider sense of respectable, though undesirable, middle class masculinity.

While some holiday products are specifically marketed to consumers as an opportunity for unrestrained hedonism, many such tourism experiences are in fact, as Briggs has recently argued in the context of Ibiza, characterised by standardised excessive consumption (Briggs, 2013). As a consequence, individuals are not freed, but increasingly constrained (Nayak 2003, 2006; McAuley, 2006; Winlow and Hall, 2006; Briggs, 2013).

This process has coincided with an epochal shift from industrial to consumer capitalism, which has seen a reworking of youthful hedonism as a contemporary variant of consumerism that has come to hold dominant sway in postindustrial Britain (and other nations). As such, various studies have sought to locate leisure activities built around the 'youthful' pursuit of pleasure, often linked to music and alcohol consumption, within a wider context of cultural change (e.g., see Tomsen, 1997; Hayward, 2002; Winlow and Hall, 2006; Nayak, 2006; Briggs, 2013). Such heavily marketed party lifestyles, often involving excessive consumption of designer drinks, designer clothes and designer drugs, are positioned as an edgy youthfulness which is in turn sold to many people now well into their thirties and beyond as authentic, liberating and unique, even if they are in reality none of these things (Winlow and Hall, 2006; Briggs 2013).

Violent Brits abroad

Perhaps a problem with sociological work on 'tourism', such as John Urry's *The Tourist Gaze* (2002), is the historical tendency to overstate the importance of cultured gazing over and above the importance of sensory pleasures and embodied practices such as drinking alcohol to excess, having casual sex with new partners, or fighting. For particular groups of tourists, such practices can be a common and often desired part of their holiday experience. For example, comparing a sample of some 3000 British, German and Spanish holidaymakers aged 16–35 years accessed in the departure areas of the airports in Ibiza and Majorca (Spain), Hughes et al. (2009) suggested that predictors of violence were being male, young, British, and being a frequent drinker of alcohol and user of cannabis or cocaine during the holiday period. This suggests that in and of itself, being British might be regarded as a significant predictor of violent behaviour. Thus, a particularly British (or English) form of brawling drunken hedonism seems rather static across different contexts, whether football features or not. It also suggests a problematic variant of English masculinity that is exported abroad on holiday, which is sometimes evident at and around football spectatorship, but that may also be apparent in other contexts as well.

In the 1980s, English football hooliganism abroad became a major political embarrassment for the government, and this period coincided with a growing body of the academic literature on the topic. In particular, the subject came to be almost wholly associated with a small collective of sociologists based at the University of Leicester, (commonly known as the 'Leicester school' – see, Dunning et al., 1988; Dunning et al., 1991; Dunning, 1994), who provided an analysis of football crowd violence couched in terms of the explanatory power of the violent propensities of certain types of people who attend football matches both domestic and international.

These authors, in various permutations, suggested that the propensity to physical violence created through particular forms of socialisation became increasingly marginalised as growing sections of the working class were incorporated into mainstream society and its mainstream values. However, amongst pockets of what they termed the 'rough working class' there existed a latent simmering violence that sometimes boiled over in the context of football and territorial local rivalries. Hence, for the Leicester academics, it was the disproportionate presence of these individuals among football supporters, such as English 'hooligans abroad' (Williams, et al., 1984) that positioned violence as predominantly the preserve of the rough, undersocialised English male (Dunning et al., 1991).

However, more generally, this period also witnessed the decline of industrialism and the rise of a service-based consumer society. The British economy, in particular, saw the rapid decline of heavy industries followed by a then equally rapid transition to a bourgeoning neocapitalist service-based marketplace built on consumption and social administration. While this postindustrial economy established itself as the principal form of economic life, the transition also served to demand ever more eviscerated and domesticated forms of masculinity as tied to the new dominant organising principles to which traditional working class masculinities bore increasingly less relevance. Yet at the same time, much of the new leisure that is on offer in such places and spaces is built, for example, on youthful and hedonistic consumption cultures that are simultaneously valorised and criticised (Hayward, 2002; Winlow and Hall, 2006). Various sociological analyses have contextualised all too well these social shifts and, while it is beyond the scope of this chapter to revisit them all in the fullest of detail, it is worth noting that there is a growing body of empirical work that contextualises well these wider socioeconomic changes in relation to young men (e.g., Nayak, 2006; Winlow and Hall, 2006). This wider understanding of socioeconomic transition is important in understanding the behaviour of football hooligans. Indeed, as the UK government have themselves suggested:

> English football disorder cannot be removed from its wider social context. In many ways it is a manifestation of a wider social problem of alienated young males demonstrating their frustration in an anti-social and violent way. It occurs in high streets up and down the country every weekend. Mediterranean holiday resorts are equally at risk. (Home Office 2001: 15)

Yet not all tourist groups are subject to similar legislative restrictions and controls, and it has been around travelling football support that restrictive practices enacted through legislature have come to the fore. For example, concerns have been raised about the Football (Disorder) Act 2000 and its impact upon civil liberties and human rights, as bans were permitted based only on suspicion of involvement in disorderly conduct and allowed for the imposing of travel bans on supporters who have not been convicted of any involvement in disorder. Prior to the 2006 World Cup, English and Welsh courts imposed some 3,500 banning orders on fans to prevent these 'known hooligans'[1] travelling abroad during the period of the finals (see James and Pearson, 2006). While not all English male holidaymakers are regarded as dangerous and potentially problematic, travelling football supporters have often been presented *en masse* as a homogenous problematic and latently violent group meaning that the nuances and differences amongst travelling football supporters is frequently overlooked (Crabbe 2003, 2004; Stott and Pearson, 2007). Even if there is ample evidence that suggests that violence and disorder are not restricted to such groups (Hughes et al., 2009; Briggs, 2013) this is often forgotten, so the violent English football fan becomes the norm, rather than the exception. Arguably, however, understanding this highly masculine 'exceptionalism' may be at the core of understanding how these men craft and create a separate and distinct masculine identity, as the data presented below demonstrates.

Casual sex, casual drugs and casual violence – English football lads abroad

Darren[2] tells me he cannot believe he has not been subject to a football-related travel ban by the UK authorities. His current banning order for football violence expired some nine months ago and, while he has again been involved in football-related disorder with his gang in the UK, he has not yet been banned again. He is therefore in possession of his own passport, and for that reason is drinking Amstel Lager and smoking cannabis (a behaviour officially prohibited in the bar we are in) with me[3] in a bar on the Canal in central Amsterdam near the edge of the Red Light District. From here, we will travel onwards to the site of England's first qualifying game in Frankfurt by Intercity-Express train.

Darren is happy to identify himself as a member of a football hooligan gang and is quick to offer stories of his previous exploits to other groups, including a number of young men sitting at a nearby table wearing England football shirts. We learn that, like us, they are en route to Germany for the tournament. Three of the group are university students, and the fourth is a manager in a retail store. Using cheap rail tickets, but without tickets for matches, they are taking the opportunity to sample the cannabis Amsterdam has on offer and stroll round the Red Light District. However, unlike Darren and his friends, they have only been 'window shopping'[4] [looking at the prostitutes, but not paying for sexual services]. This topic is what first unites the groups in conversation as Ray, one of Darren's friends, regales the student group with his account of the events of the previous evening:

> I was fucking this whore last night, lads. My balls are still sore, the way she was going at my foreskin, too. I properly smashed her. She was fucking good, really into it. The night before, me and Daz, we fucked these two slags, were snorting coke off their tits and arses, then fucking them, a couple of blueys [Viagra] and smashed them both in. It was a good night, but tonight I can't be fucked. Tonight I am out for a drink and that, but you never know what might happen later, eh, boys? (From field notes[5])

There are clear differences between the social class and backgrounds of the two groups, borne out in the behaviour and attitudes that are demonstrated in this holiday setting. Such distinctions are apparent for instance in attitudes to race and violence. The students are visibly uncomfortable (as am I, but trying not to show it) with some of the group's more extreme racism. It is clear that different attitudes are at play when, for example, Darren tells Paul, the most vocal of the students, that the Red Light District is 'dangerous', especially 'if you are English in a small group': 'There will be groups of nigger robbers and fucking Ajax lads looking to target small groups, so stick with us. If we will all stick together, it will be sound. We will all have a few drinks; we are all English lads, aren't we?' (Darren, from field notes)

Later in the evening, after the students have left, having declined the offers to join us (at least in part, I suspect, due to a growing discomfort with the racism and talk of violence emanating from the lads), Darren is becoming more vocally racist and aggressive by the minute. We are in another bar now; this time it is exclusively being used by English fans who are flouting the prohibition on drinking and smoking cannabis together. Many of them seem more overtly hostile, and I note that the atmosphere is changing as the effect of the alcohol seems more obvious than that of the cannabis. There is an air of aggression. Darren begins to sing, initially on his own, but then with others around him: 'If it wasn't for the English, you'd be Krauts. If it wasn't for the English, you'd be Krauts. Oh, if it wasn't for the English.'

The bar owner stares; it has little effect. Already Darren is onto another topic: he tells me how he was hoping some Dutch lads might 'fancy a pop' at the English and instigate a fight, but recognises that the possibilities are looking ever less likely. However, he is still intent on violent confrontation, and when I ask, 'What's the plan?' he responds with a smile: 'We will move on in a bit, go for a wander, see what happens.' He winks at me.

> I know the only chance we will get of a kickoff is with the fucking Turks or Somali scum round here, the nasty greasy Islamic pricks...I fucking hate them, the knife-carrying, dirty, fucking blade merchants. We need to stick together and smash the dirty Muslim bastards. You will get them 'round here, and we might get something with them.
>
> I work hard, and when I get away, I want to snort some Charlie [cocaine], smoke a few spliffs, have a good session [drink a lot], and get into a tear up [fight]. That is what it's all about. It's like, out here, you don't give a fuck, do you. No Mrs, no kids, no work – it's just a bit of fun and freedom, like back to your youth for a few days while you escape. Most lads have commitments at home; they are not like this back home, well, like, maybe weekends, but mostly it's just the odd one or two a season. Out here it's just like, freedom. (Darren, field notes from Amsterdam)

This quote of course highlights the perceived 'liberating' nature of the away trip for such men. Such themes of freedom from normality and routine are commonly talked of during my time amongst the group, and serve to demonstrate how, for some travelling England supporters like

Darren, this holiday is in itself perceived as part of an adventurous and exciting time where the use of illicit substances and seeking out risk-laden violent experiences are simply an alternative form of recreation (Ayres and Treadwell, 2012). Yet their own self-recognition of such behaviour and justifications for it are inexorably intertwined with how these behaviours are for them the normal gendered practices of back home:

> You work hard for the coin to do this, don't you, and I do I know. Most of what I earn now, I am sensible with, but now and again, I like to treat myself, so I will go and blow a grand or two. That's what it is: most of the University types don't get it, because it's a working class man's mentality, isn't it? It's just the same as it is when single lads are away in Ayia Nappa or Ibiza. It's a thing that comes from your background, your upbringing – it's almost fucking bred into you. We don't think about it, lads like me: it's just what we do on holiday. (Steve)

> A lot of people don't get it: it's not just about the football, it's about the freedom. The more you try and control us, the more you tell us what to do, the more we will say fuck you, just do our own thing. We are not the majority. We are the few – we are the ones the authorities hate, but we are the proper lads who have football in our veins. We will fight over it because it's just what we have always done on holiday, wherever, if it's football or not, on a holiday, what lads like is a few drinks, have a fight, get a fuck. (Craig)

Frankfurt – 'it's all kicking off'

We are gathered in the red-light district in Frankfurt, which has become a gathering spot for increasingly rowdy groups of travelling England fans. While the UK government, the English Football Association and the German authorities have been keen to encourage fans to behave in a culturally sensitive way, there has already been trouble. The crowd of England fans, easily identifiable by the predominance of Stone Island, Hackett and OneTrueSaxon t-shirts,[6] are drinking; many are visibly intoxicated, and the atmosphere is one of tension and hostility. Groups of German Hells Angels bikers who control the bars as part of a formal arrangement with bar owners, somewhat to the concern of the authorities who regard them with nearly as much suspicion as the potentially violent English fans, are just about managing to keep the drunken English fans in order. However, increasing numbers of boiler suited and riot helmet- carrying police are arriving by the minute.

Marty, a previously convicted member of a notorious football firm from Northern England who is currently awaiting a court date in the UK, looks at the massed ranks of German police, and gives an insight into how, for him, the tourist experience is inexorably connected with issues of male space and mobility:

> This is what it's all about! ... The whole thing of take your family to an England match, wife and kids and that ... What the fuck is that about? I am not being funny, but England away, it's not for fucking families; it's not supposed to be, never has been. As far as I am concerned, you do that and then get offended by the way the lads are behaving, you are a cunt, maybe you shouldn't have brought your Mrs and kids, eh?

Like Darren, Marty is travelling with an all-white, all-male group of 'lads' from his club. I meet and introduce myself as a researcher to several of them and spend some time in their company. Archie, another of the group, has his say on the appeal to fans by Sven Goran Erikson, the then- England team manager, 'not to sing the song about ten German bombers,' an appeal that had been widely circulating in the English press. 'Why the fuck should he tell us not to sing that song?' Archie asks. 'Fucking cunt. He should stay out of it like they did in the war, the fucking bell end. If I want to sing about shooting down German planes, then I fucking will. It's called free speech, and I get that choice because we fucking won the war.'

The lads surrounding him are in fits of laughter. Reece, who has previously been convicted of football-related offences but is not currently subject to a banning order, has been following England abroad for several years. Marty, one of his older travelling companions approaches us and puts his arm around Reece in an almost paternal way, and tells me,

> I was with him on our first away ... and we were having it with their old bill [i.e., the police], streets full of tear gas, rubber bullets whizzing past us, and he comes running up and says, all grinning and proud as punch like, 'I have just thrown my first bottle.' I patted him on the back, passed him another one, and says, 'Fucking get that one at them,' and all. I was, like, 'I am proud of you, lad.' I was, as well – properly proud of him. (Marty)

Later that day, I am talking to Archie when the familiar face of Darren and his friends comes into view. He is smiling, and they seem even more animated than when I last saw them. As Darren gets close to me, he calls my name and comes over. We exchange greetings, and I introduce him to Archie. Darren smiles at me and shows me his hand: his knuckles are bleeding, and he has clearly hit somebody. He laughs as he tells me they have just 'had it' [had a fight] with some German football fans in a back street, and he has 'smashed' [physically assaulted] one of them. He laughs: 'Now it feels like I am on holiday. Now, I can relax.'

This theme of relaxation through violence (or relaxation through risk) is a frequent and recurring one in the accounts I heard. Situationally and psychologically, it seems to act to separate the participants off from the others who they perceive as 'less authentic' supporters. For the lads, there is an evident collective perception that other fans are feminised or not real men due to their unwillingness to look for such challenges and confront them head on. This, for them, is the essence of their holiday experience. Having a good time requires such hedonistic release, which in turn will be used to separate them and make them distinct from others back home, or following the national team in a conformist manner:

> You don't wanna live a boring life. It's just looking for laughs, be it going 'round the red light, or fucking finding yourself in a row with ten lads in a quiet little side street. That is what being England away is for me, a bit of a laugh and an adventure. You always know you will come back with stories and that, and the lads who didn't go will regret it. (Reece, field notes, Frankfurt)

While there has been a drive emanating from the UK Home Office, the English Football Association and some supporters' groups to 'remarket' the image of England supporters in order to create a more socially inclusive supporter base (Crabbe 2003, 2004; Crabbe and Brown 2004), the likes of Reece clearly still operate well outside of such initiatives. For many of the self-defined 'lads', this is simply a process driven by a 'commodification' of 'their sport' (see Giulianotti, 2005), which is underpinned by a desire to drive them away from it and instead appease middle-class fan groups who they see as less worthy and less deserving due to their lack of commitment. The significance of this distinction is clearly evident in the following quotes:

> This cunt says to me, 'Why are you singing that "No surrender"? What has that got to do with football?' We all turned to him and say, 'How long have you been following England?' He says, 'I have been for five years.' I was like, 'So, you have not been following England

for years like us then, to shitholes. Like Landsdowne Road in '95? So fuck off, you cunt, before I spark you out.' I could not fucking believe it. I fucking bet he follows Arsenal and is some middle-class accountant type cunt. (Steve)

I have had Turks chase me with knives. I seen a kid stabbed to death in a phone box in Amsterdam – nothing to do with football, but still bad to see, like... I seen people get properly kicked in... had some proper good kickings myself, too... had guns pulled on us, been tear gassed loads of times, had rubber bullets ping 'round my head; but it's a laugh, isn't it? It's just part and parcel of following the team. That's why I always turn up, not like these new fucking middle-class fucking dickheads: would they go to some of the grotty Eastern European shitholes I have been in on cold, dark nights? Would they – fuck. (Darren)

Conclusions

For men like those documented here, football fandom, tourism and holidays are inexorably connected with intoxication, xenophobia, aggression and violence and serve as important cultural and emotive resources for scripting a particular, and for them, powerful masculine identity. This identity separates them from the 'uncommitted' others and from the banal routines of life at home. Their masculinity, and indeed their nationality, is affirmed through just such tourist practices. In their worlds, being 'on the away' implies violence, drug use and casual sexual encounters that are all valorised as expressions of an authentic, unrestrained and unapologetic masculinity where 'anything goes'. Theirs is an aggressive masculinity that now jars with many modern sensibilities about what manliness should be (Winlow, 2001; Nayak, 2003, 2006) and with the 'official' attempts to reframe what football supporters should be and how they should behave.

The material presented here suggests that these men engage directly in violent behaviours in the context of football spectatorship and other less acknowledged behaviours (the buying of sex, the use of illicit narcotics, excessive drinking) in part because it occurs in a perceived liberated space and setting that encourages these men to escape perceived normal societal constraints. Because such risky behaviours are regarded as inherently liberating, they conform to deeply held notions of masculinity and of what being an 'English' male is about, and they appear to help these men retain an increasingly obsolete sense of identity. Their English masculinity is both unwanted and discouraged by government and football authorities and often by the dominant socioeconomic, political and cultural context in which they exist and are commonly marginalised from (Crabbe, 2004). Yet, for them, it is what 'being a man' and being English abroad is all about. At least in part, this might tell us of the state of manhood for some men in the contemporary moment. If one simply looks a little deeper behind the articulated logic of celebration and freedom, and the superficial notions of hedonistic excess, what is at least partially revealed is often instrumentalism and violence that characterises a sort of pyrrhic celebration that locks them into a subordinate position (Willis, 1977). That is not to suggest that the males here do not enjoy those behaviours, but rather that, for all their articulation of liberation, the reality is freedoms to consume that are often quite shallow, and incredibly conformist masculinity that is free to drink, to fuck, to fight and to offend (see Briggs, 2013).

In recent years, academic sociology has afforded a great deal of attention to the participation of young people[7] in cultures of heavy alcohol use and recreational hedonism, both while on foreign holidays but also 'back at home' as part of the burgeoning sociological study of the night-time economy (e.g., see Hall and Winlow, 2006; Thurnell-Read, 2011; Briggs, 2013). While the practices of English tourists involving heavy drinking, drug use and other risky 'deviant behaviours' have recently become the source of popular media and academic attention, there are wider questions to be asked about the realities of such articulated nonconformity and hedonism. To what extent does such tourist behaviour really liberate individual participants, or, alternatively, is it simply a means of conforming to emergent contemporary social mores expectations of how young working class men might assert and preserve a specific iteration of masculinity in the face of wider socioeconomic transitions (Winlow and Hall, 2006; Briggs, 2013).

While the experience of the holiday for many people might connote the idea of nonconformity and freedom, the reality for the lads is that of a rather narrow formula of excessive drinking, taking drugs, and engaging in violence. Indeed, this can be viewed as neither liberating nor nonconformist, but narrowly conformist to quite specific excessive consumerist and hedonistic masculine norms. Even while, in the lads' world, anything and everything goes – down one's throat, up one's nose, on the end of the penis or the end of the fist – such excesses are, by and large, a predictable and, for them, expected part of the experience of tourism. For some men, it is the very ugly face of 'the beautiful game' that is the core aspect of a good 'English' holiday.

Notes

1. This term should not be accepted uncritically, however, given the rather loose definitional elements of football banning order legislation. (For an overview, see Hopkins, 2013).
2. All names used are pseudonyms to offer participants some degree of anonymity. Direct participants gave informed consent to be involved in the first study for which this research was gathered (an examination of the links between football hooliganism and criminal behaviour) and were accessed through gatekeepers and contacts already known to the author. They were drawn from several football hooligan gangs, and ages and ethnicity varied, although most were male, aged between mid-twenties and early forties. For this research, the author spent time with three specific groups, and in all around 20 core individuals were drawn from different football hooligan firms in central and northern England. Some of these individuals were known to the authorities for involvement in football violence, though none were actively prohibited from attending the World Cup formally due to football banning orders, although several were on bail and due before the criminal courts at the time of study. The research received ethical clearance from the author's then-host university, Birmingham City University.
3. As researcher I recognise the somewhat contentious nature of such behaviour, even if the activity is permissible in the specific context of the research. Of course, drinking and smoking affect the ability to take notes and recall, along with the ability to avoid personal danger if a situation arose. However, here I am also reminded of Craig Ancrum's (2012) view of ethnographic engagement and the call for 'authentic encounters' in criminal cultures and appeals for greater intellectual honesty about the specific behaviours, specifically his observations that a researcher 'has no choice but to marginalise the formalised world of ethics committees' and formalised guidelines and proceed under the somatic guidance of his or her ethical code' (Ancrum, 2012: 113). Indeed, in doing so, the result can be better understanding and better insight, and simply put, the richness of data and understanding must be held up against any accusations that the researcher has 'gone native'.
4. Quotes from members of the group taken verbatim from recorded field notes are identified here by italics.
5. During the fieldwork stage, notes and quotes were kept in a mobile telephone, either in the form of recording, or through a process of recording extensive quotes in the note function of the mobile phone. Such new technology has made the keeping of field notes easier than days previous where ethnographers describe the carrying of note pads and pencils (see Armstrong, 1993), and clearly has a less direct and obvious impact on the setting and individuals being studied.
6. These are brands now associated with and preferred by football casuals (King, 1997) and other groups such as the Far-Right English Defense League.
7. Though, as this demonstrates the term 'youthful' is in itself problematic, as the sample I encountered ranged from teenage years into the fifty something years. The same is true of some drinking violence, which cannot simply be linked to youthful groups see (Tomsen 1997, Winlow, 2001).

References

Ancrum, C. (2012) 'Stalking the margins of legality: Ethnography, participant observation and the post-modern underworld', in S. Winlow and R. Atkinson (eds) *New Directions in Crime and Deviancy*. Oxon: Routledge, pp. 113–126.

Ayres, T. and Treadwell, J. (2012) 'Bars, drugs and football thugs: Alcohol, cocaine use and violence in the night time economy among English football firms', *Criminology and Criminal Justice,* 12(1): 83–100.

Armstrong, C. (1993) 'Like that Desmond Morris?', in D. Hobbs and T. May (eds) *Interpreting the Field: Accounts of Ethnography*. Oxford: Oxford University Press, pp. 3–44.

Briggs, D. (2013) *Deviance and Risk on Holiday: An Ethnography of British Tourists in Ibiza*. Basingstoke: Palgrave Macmillan.

Crabbe, T. (2003) 'The public gets what the public wants: England football fans, truth claims and mediated realities', *Leisure, Sport and Tourism*, 38(4): 413–25.

Crabbe, T. (2004) 'Englandfans – A new club for a New England? Social inclusion, authenticity and the performance of Englishness at "home" and "away"', *Leisure Studies* 23(1): 63–78.

Crabbe, T. and Brown, A. (2004) 'You're not welcome anymore: The football crowd, class and social exclusion', in S. Wagg (ed.) *British Football and Social Exclusion*. London: Routledge, pp. 26–46.

Dunning, E., Murphy, P. and Waddington, I. (1991) 'Anthropological versus sociological approaches to the study of soccer hooliganism: Some critical notes', *Sociological Review*, 39(3) 459–78.

Dunning, E. (1994) 'The social roots of football hooliganism: A reply to the critics of the Leicester school', in N. Bonney, R. Giulianotti, and M. Hepworth (eds) *Football Violence and Social Identity*. London: Routledge.

Dunning, E., Williams, J. and Murphy, P. (1987) *The Social Roots of Football Hooliganism*. London: Routledge.

Giulianotti, R. (2005) 'Sport spectators and the social consequences of commodification: Critical perspectives from Scottish football', *Journal of Sport and Social Issues*, 29(4): 386–410.

Hall, S. (1997) 'Visceral cultures and criminal practices', *Theoretical Criminology* 1(4): 453–78.

Hayward, K. (2002) 'The vilification and pleasures of youthful transgression', in J. Muncie, G. Hughes and E. MacLaughlin (eds) *Youth Justice*. London: Sage, pp. 80–94.

Hopkins, M. (2013) 'Ten seasons of the football banning order: Police officer narratives on the operation of banning orders and the impact on the behaviour of "risk supporters"', *Policing and Society,* pre-published online 9 April 2013.

Hughes, K., Bellis, M., Whelan, G., Calafat, A., Juan, M. and Blay, N. (2009) 'Alcohol, drugs, sex and violence: Health risks and consequences in young British holidaymakers to the Balearics', *Adicciones*, 21: 265–78.

James, M. and Pearson, G. (2006) 'Football banning orders: Analysing their use in court', *The Journal of Criminal Law*, 70(6): 509–30.

King, A. (1997) 'The postmodernity of football hooliganism', *British Journal of Sociology*, 48(4): 576–93.

Nayak A. (2003) 'Last of the "real Geordies"? White masculinities and the subcultural response to deindustrialisation', *Environment and Planning D: Society and Space*, 21(1): 7–25.
Nayak A. (2006) 'Displaced masculinities: Chavs, youth and class in the post-industrial city', *Sociology*, 40(5), 813–31.
Ruggerio, V. (2000) *Crime and Markets: Essays in Anti Criminology*. Oxford: Oxford University Press.
Stott, C. and Pearson, J. (2007) *Football Hooliganism, Policing and the War on the English Disease*. London: Pennant Books.
Tomsen, S. (1997) 'A top night: Social protest, masculinity and the culture of drinking violence', *British Journal of Criminology*, 37(1): 90–102.
Thurnell-Read, T. (2011) '"Off the leash and out of control": Masculinities and embodiment in Eastern European stag tourism', *Sociology*, 45(6): 977–911.
Urry, J. (1990/2002) *The Tourist Gaze*. London: Sage.
Williams, J., Dunning, E. and Murphy, P. (1984) *Hooligans Abroad*. London: Routledge and Keegan Paul.
Willis, P. (1977) *Learning to Labour: How Working Class Kids Get Working Class Jobs*. Farnborough: Saxon House.
Winlow, S. and Hall, S. (2006) *Violent Night: Urban Leisure and Contemporary Culture*. Oxford: Berg.
Winlow, S. (2001) *Badfellas: Crime, Tradition and New Masculinities*. Oxford: Berg.

14
What is Old and What is New? Representations of Masculinity in Travel Brochures

Monica Gilli and Elisabetta Ruspini

Introduction

One aim of this chapter is to understand how men and masculinities are portrayed in Italian travel brochures. A second aim is to understand if new and changing masculinities are an area of interest for the tourism industry. Previous research (Richter, 1995) demonstrates that, despite the diversity of the market, the portrayal of the industry-produced marketing materials, aimed at potential tourists, privileges the gaze of hegemonic masculinity over others. However, according to Pritchard and Morgan (2000), further analyses of gendered tourism marketing are needed: today, in many countries, gender identities and relationships are formed and maintained in an environment of greater choice of how people can live their lives than has been possible in past generations. Following Pritchard et al. (2007), research on tourism-based representation of gender needs to further explore the ways in which masculinities are recreated in the everyday practices of men. Research should aim to generate greater and more nuanced insights into men's lives and the complexities of contemporary masculinities (Thurnell-Read, 2011).

The relationship between masculinity and tourism marketing materials will be addressed through a document analysis on a selected set of Italian travel brochures promoting holiday villages/resorts;[1] the brochures chosen for analysis in this article are mainly aimed at an Italian audience. The following major elements support this choice. Starting from the 1960s, tour operators made large investments in the development of holiday villages: in Italy, the first holiday village was opened in Tuscany in 1951 by Club Mediterranée, followed by a second

(Cefalù, Sicily) in 1957. Second, holiday villages welcome all types of visitors and are family-friendly destinations: many offer services and leisure activities designed for both parents and children. The holiday village thus seems a microcosm to test the privileged position of heterosexuality and traditional masculinity. Third, Italy is a context where masculinity is apparently univocal and one-directional: the passionate, fiery, dark Mediterranean male, the vigorous playboys populating Italian beaches (an image heavily exploited in the marketing of some tourist resorts). This one-dimensional, hyper-simplified model has constituted the distinctive hallmark – curse and delight – of many generations of men (dell'Agnese and Ruspini, 2007).

The term 'hegemonic masculinity' is one key concept for our analysis: the current, dominant form of masculinity within a society. It is normative, and it regulates the behaviour of men in terms of socially accepted manhood (Connell and Messerschmidt, 2005). The hegemonic definition of masculinity is constructed in relation to various subordinated masculinities as well as in relation to women. Connell (1995) suggests that this can be understood as the collection of practices that subordinate women to men in the maintenance of patriarchy. According to Kimmel and Levine (1992) men organise the conceptions of themselves as masculine by their willingness to take risks (see Katsulis, this volume), by their ability to experience pain or discomfort without submitting to it, by their drive to accumulate constantly (i.e., money, power, sexual partners, experiences) and by their resolute avoidance of any behaviour or feeling that might be constructed as 'feminine'. Hegemonic masculinity, indeed, exists in contrast with that which is feminine (Kimmel and Aronson, 2010): it is a complete renunciation of everything feminine. As Benecke argues (1997), in rituals such as sports, sex and work, men constantly invent and renew their masculine identities as they learn to repress and reject all 'feminised' behaviour. Such rituals become crucial 'proving grounds' where masculinity is tested and asserted. However, hegemonic masculinity is incompatible with, and unattainable for, most men and yet it is ultimate goal, and the standard by which most men measure their own sense of masculinity. Therefore, the current structure of gender identification for males is fragile and exists as an unrealistic framework in which men can define their self-identity.[2] What emerges is the question of how a dominant hegemonic masculinity, based around heterosexuality, rationality, anti-femininity, self-control and competitiveness, relates to other masculinities and femininities which, in any number of ways, contradict or fail to fit in with it (Thurnell-Read, 2011).

Today gender identities and relationships are formed and maintained in a world that no longer has fixed models of life and universal certainties risk (see for example Lyotard, 1979; Beck, 1992; Touraine, 2005). The transition from modernity to contemporary[3] modernity has been demarcated by radical transitions in the last few decades, including globalisation, sectorial deindustrialisation and the destandardisation and increasing precariousness of labour, along with rising education levels and recurrent economic and political crises. These have been accompanied by a restructuring of intergenerational relations and the transformation of gender identities and family models. Social change has favoured a drawing closer of male and female life courses both from the structural point of view (an increase in women's employment and schooling, delayed entry into adult life, a shared and lesser inclination for marriage and procreation, and the assumption by women of responsibilities that previously belonged exclusively to men, etc.), and in the way in which life courses are desired, planned, constructed, and redefined by individuals themselves (Oppenheim Mason and Jensen, 1995; Beck-Gernsheim, 2002; Janssens, 2002; Hantrais, 2004; Oinonen, 2008; Lamanna and Riedmann, 2009; Ruspini, 2013).

Within this dynamic context, the desire to discover (or rediscover) the terms and values of one's specific masculinity seems to be increasing (see for example Anderson, 2009; Ruspini, Hearn, Pease and Pringle, 2011) and this is true also for Italy (among others, Zajczyk and Ruspini, 2008; Ruspini, 2009). The number of men willing to question the stereotyped model of masculinity is growing. More men are taking a role in activism against violence against women: a small but increasing number of men are working to make change, both in their personal lives and through wider public advocacy (see for example Hearn 1998, 1999; Ruspini et al., 2011). The frequency with which men are challenging the gender segmentation in paid work is also growing, with more working in formerly overwhelmingly female-held occupations – counselling, nursing, and elementary teaching (Donaldson, 1993). Men in non-traditional jobs are challenging the traditional and socially approved relationship between labour outside the home, production and masculinity (Perra and Ruspini, 2013). If we broaden the focus on the desegmentation of paid work to include unpaid work, more interesting things occur. Increasingly, husbands, partners and fathers are involved in family life, and fathers assume greater responsibilities after the birth of their children (see for example Pleck, 1985; Doucet, 2006; Dermott, 2008; Featherstone, 2009).

Another tension emerges, that between the traditional virility model and the need to look after one's image and health. While on the one hand, the model of male subjectivity historically constructed is based on the repression of the body, on the other hand, biological imperatives and the changed conditions and styles of life demand a different attitude to one's body, which needs an increasing quantity of care. Today the ideological sphere of reference of masculinity has widened to include a greater diversity of physical styles, with beautification as another component of masculinity (Miller, 2003). These changes challenge the conditionings imposed by the static, one-dimensional and hegemonic model of masculinity and lead to explicit breaks with 'traditional' forms of masculinity.

Starting from these premises, we focus on the way contemporary touristic images express and inscribe a number of conceptions of masculine identity. Tourism is a gendered experience (Kinnaird and Hall, 1994; Rao, 1995; Pritchard and Morgan, 2000; Swain and Momsen, 2001). As tourism itself is a product of gendered societies, it follows that its processes are gendered in their construction, presentation, and consumption. The holiday experience may also contribute to the formation and to the changes in gender identities. People (especially if young) consider their travelling experiences to be an integral part of their identity and as an essential way to enrich their lives. An extensive research project[4] on young men aged 25–39 by Discovery Network EMEA/UK suggests that traditional travel and holiday gender splits may be disappearing.

Shall we then expect that the dynamic and growing tourism industry – which, according to the World Travel and Tourism Council, has the highest potential for growth of any industry – is changing (and will change), in order to accommodate male identity changes and the search for alternatives coming from (male and female) tourists, tired of encountering the rigid and hyper-simplified reproduction of gender identities and roles? If yes, how and how much?

The relationship between masculinity and tourism

The relationship between men and tourism still needs to be fully explored. The social science literature on masculinity and tourism is principally focussed on the link between heterosexuality and the hegemonic masculinity culture. Some specific issues are the Grand Tour, sex tourism, tourism and transgression.

The Grand Tour usually refers to the travel of Englishmen to Italy, where they acquired the taste deemed essential to the gentleman and the paintings and other objects of virtue to display in their country estates (Cohen, 1996, 2001). On the one hand, the Grand Tour was a cultural and educational practice for élite males in eighteenth-century England, and it played a key role in the elaboration of discourses on language and national identity (Cohen, 1996, 2001). These Grand Tourists broadened their knowledge by visiting art collections and touring ancient remains. On the other hand, the Grand Tour was a convenient vehicle for well-heeled young men to broaden their sexual knowledge: the sexual choices ranged widely, with readily available street prostitutes and whorehouses, as well as paid mistresses and married women who were prepared to have sexual liaisons (Black, 1985). Following Cheek (2003), the practice of sending young men on a Grand Tour to complete their educations regularised the notion that normative adult male sexuality depended on experiencing and then abandoning sexual relationships with foreign women.

Concerning the literature on the relationship between masculinity and sex tourism, research from around the world has shown that men who purchase sex are not a homogenous group; they represent all ages, nationalities, races and social classes (Hughes 2004; Monto, 2005; Jõe-Cannon, 2006; Ricardo and Barker, 2008). There is also a body of research indicating that men may seek sex workers to experience certain models of gender relations and submissive femininities and to reinforce their own feelings of masculinity, power and control in relationships (O'Connell Davidson, 2001; Månsson, 2004; Yokoto, 2006). Following Rivers-Moore (2012), men who are not exceptionally privileged at home are able to step closer to the ideal of hegemonic masculinity while on holiday: sex tourism is a masculinising practice carried out by men who are, by and large, unable to secure hegemonic masculinity at home. In addition, a growing body of research on the specific versions of manhood that have emerged among mine workers, truck drivers and other groups of men who migrate or are highly mobile in their work, links that manhood to sex workers (see for example the research by Campbell, 2001 on South African mine workers).

As far as the connection between masculinity, tourism and transgression is concerned, we mention the studies by Sönmez et al. (2006) on the American 'spring break', by Thurnell-Read (2011) on premarital stag party tours, and by Treadwell (this volume) on football hooliganism. The quantitative research of Sönmez et al. (2006) on attitudes about binge drinking and casual sex as part of American college students' spring break – a vacation from school or college during the spring term, lasting about a week – argues that high-risk behaviours were traced back to situational disinhibition in settings encouraging sexual and emotional transience and to liminality (a sense of inbetweenness involving a temporary loss of social bearings). Further, high-risk behaviours in these settings seemed to be strongly associated with situational factors (the resort and what it entails). The study by Thurnell-Read (2011) argues that the behaviour of stag tour groups offers considerable insight into masculinity and that the meanings attributed to such behaviour reveal complex construction of contemporary British masculinities. Here the male body plays a central role through the consumption of alcohol, its effects upon the body and the use of bodies by stag tourists to foster an ethos of playfulness and enact a transgressive release from social restraint. This release from normative pressures concerning the male body is, however, only temporary, and therefore acts to support the ritualistic reinscription of a wider hegemonic masculinity.

With this chapter, we wish to introduce a perspective that tries to discuss the relationship between masculinity, tourism and individual and social change, not only from the sexual point of view (see, for example, the study by Edelheim, 2007). More specifically, we would like to understand whether the tourism industry is paying attention to male identities distant from the hegemonic masculinities detailed by Connell (1987, 1995) and Kimmel (1995, 1996). This has been partially accomplished by the growing body of research on gay tourism. Gay tourism is increasingly seen as a powerful market segment (Pritchard et al., 1998; Russell, 2001). Moreover, gay masculinity may break down many gender stereotypes, allowing for an unbiased, more complete understanding of masculinity and gender. However, these perceptions are in contrast with the scant information about this type of tourism (Hughes, 2002). There are still a limited number of studies related to male homosexuality and holidays and most of the available information refers to English-speaking countries (see, for example, Forrest, 1999; Hughes, 2002; Casey, 2009; Melián-González et al., 2010).

We will try to answer different questions: Which male models and bodies are included in tourism promotion materials? Are these transformations in contemporary masculinities reflected in the brochures? Does the man who comes at us through the tourism promotion material seem to reinforce the social order or to challenge it? Do gender stereotypes and the myth of hegemonic masculinity (Connell, 1987, 1995; Donaldson, 1993, Kimmel, 1995, 1996) permeate this aspect of the tourist supply side? Do the language and imagery of promotion still privilege the male, heterosexual gaze above all others?

The idea behind our analysis is that, as Wernick argues (1987: 293), a constant reassembly of masculinity through the marketplace would encourage men to question and reform self-identities that had previously appeared immutable.

Methodology

Italy is an interesting context for analysis of the relationship between masculinity and tourism for two reasons. The first is that the image of Italy is that of a country in which the inhabitants are 'romantic' and 'passionate'. The model of the strong and virile man, with Mediterranean looks, is almost a stereotype and has helped nurture visitors' imaginations since the times of the Grand Tour (dell'Agnese and Ruspini, 2007). At the same time, however, Italy is a country with a strong Catholic influence, and the Catholic Church has an important role in family policies, continuing to favour the model of a traditional family and, associated with that, an equally traditional idea of masculinity and femininity.

The document analysis was carried out on the travel brochures of the two most important Italian tour operators in terms of sales. They are the market leaders in holiday village breaks: Alpitour and Eden Viaggi (Gervasio, 2011: 19). Four Alpitour brochures and five Eden Viaggi brochures were analysed in the period June-August 2012. The selection took account of various target segments, from the middle-low to the middle-upper segment. Generally, the middle or middle-low segment is associated with destinations that are close to Italy (countries on the Mediterranean, such as Egypt, Turkey and Croatia), while the middle-upper segment corresponds to more distant and exotic destinations (such as Maldives, Central Africa and the Caribbean). The analysis thus involved the texts and especially the images (N = 448: Eden Viaggi, n= 323, Alpitour, n= 125) contained in nine travel brochures. Generally, the pictures of the brochures seek to minimise human presence. A photo of a crowded swimming pool or a beach has a negative impact on the purchaser, who moves on to the next page looking for pictures that better portray the dream of a holiday in an 'exclusive and uncontaminated' place – hence, a certain lack of people on display. We have excluded from our analysis group shots, in which it was hard to distinguish the numerous social situations, and have concentrated on the scenes in which the interaction among the social actors clearly defines their roles and the division of labour among them. Two masculinity models were analysed: tour operators' staff members (such as travel assistants and tourist entertainers) and male tourists. Our analysis specifically focussed on gender preferences in determining the recipient of the text, male influence on values and/or socially desirable situations, and male connotation of space and holiday activities.

How (changing) men are portrayed in travel brochures

Let us start with some general considerations. First of all, the social panorama offered by the Alpitour pictures is more diversified and less idealised: the photographs include people of all ages (while people appearing to be aged over 60 are practically nonexistent for Eden Viaggi), and of all shapes and sizes (while in Eden Viaggi there are only slim women and muscle-laden men). In some group shots (Alpitour brochures), it is also possible to see people in wheelchairs. Therefore, one could say that Alpitour has photographed 'real customers' while Eden Viaggi exclusively uses professional models. The second point, common to both tour operators, is that there is no association between the cost of the holiday and the representation of men: in general, even if the two tour operators make offers for a mid- to high-segment of the tourist market, they do not seem to use any particular element of social distinction.

We now move to the analysis of the pictures portraying tour operators' staff members such as tourist assistants and tourist entertainers (these pictures correspond to a little more than 18% for both tour operators). Numerous photographs show men and women together, wearing company t-shirts in relaxed and active poses. Alpitour is well known for investing significantly in organised entertainment and accompanies tourists every step of the way during their holiday. In general, entertainers (both male and female) are always young; this is no surprise given that being a holiday entertainer is a typical entry level duty for any tour operator, as it presupposes high social and sporting skills and recreational ability.

Eden Viaggi very commonly uses individual images of male rather than female tourist assistants/entertainers (30 men and 4 women), while the difference is less pronounced for Alpitour (4 men and 2 women). However, it must be said that of the 30 male pictures, 26 men are chefs, a highly regarded, specifically male profession, which immediately indicates the quality of a holiday resort (Ruspini, 2005: 29). The 4 female colleagues are portrayed dressed as assistants, following that representation of women in terms of availability and being of assistance that has a long tradition in tourism advertising (see for example Richter, 2005). Men in the Alpitour brochures are windsurfing in 2 images, lifeguarding in one, and again, playing the role of chef in one. Women are more passive: one is attending a food-laden table, the other is looking out over the sea. For both tour operators, pictures also show women in sports lessons (pool exercises, gymnastics, etc.) with 9 women and 5 men, but the percentage is higher if we look at leisure activities with children (the so-called kids clubs) with 18 women and just one man.

In our opinion, the images of workers/staff members show some openness toward new models of femininity. In fact, the representations of women's staff members are much more similar to their male colleagues in terms of clothing, attitude and activity. This helps in creating a certain uniformity between the genders.

Comparing the two tour operators, there is greater role flexibility on the part of Alpitour than in the more traditional Eden Viaggi brochures. Also, Alpitour's insistence on collective images, with mixed groups of staff members jumping up and down and laughing, seems to us to help promote an equality of representation between the sexes.

Let us now move on to the representations of tourists (77.7% of Alpitour images are devoted to tourists, whereas the corresponding percentage for Eden Viaggi is 69.5). Here, the image of women in travel brochures (N = 80) is more frequent than that of men (N = 24). Once again, Eden Viaggi seems more traditional in its gender representations, but only in terms of relative frequencies (Women are featured nearly 5 times more than men; in Alpitour's brochures, the ratio is not even double). In terms of the activities, the division of labour is similarly traditional, with men dedicated to outdoor sport: in 24 cases of male activity the most passive (one case) involves relaxing on an air-bed in the pool, and 2 cases of contemplating the countryside (but from a rock which brings to mind extreme outdoor sports). In all the other cases the depicted activities are quite demanding outdoor sports (windsurfing, kitesurfing, sailing, trekking and climbing) or competitive sports (tennis, golf). There is no picture of a man by himself in his room or elsewhere in the hotel. As we will see, on the other hand, there are a lot of these if he is part of a couple.

Of the 80 female tourists portrayed in the brochures, just 7 are playing a sport, and only one of the sports could be called extreme (abseiling down from a waterfall). For the remainder, it is swimming and snorkelling/diving, a non-competitive activity that implies a contemplation of the sea. In 7 cases, the woman is reading in her room; in 32 cases she is relaxing (sleeping, sunbathing, seeing to her physical wellbeing). If she is walking, she does so in the hotel or by the sea, but she is never dressed for trekking; if she looks out at the countryside (4 cases), she does so from her room or from the hotel, but never in outdoor situations. In no case does the male tourist receive a service from a staff member by himself, whereas this is frequent for women: out of 23 cases, 22 are female tourists being massaged (6 times by a man, 16 times by a woman). The last picture portrays a tourist talking at the reception desk. The man is massaged only if his partner is also being massaged (in this case the massage can also be given by a woman). This reinforces the idea of heterosexuality.

The couples (N = 93) are portrayed in a fairly standard form: he is always dark-haired, young, slim and muscular (except in one case); she is often more fair-haired as a contrast. Elderly couples can be found only in the brochures of Alpitour. When the man is in a couple, his presence almost always has a romantic side: out of 93 couples portrayed, most envisage a series of 'romantic' activities: walking hand-in-hand on the beach, relaxing in the pool, looking at the sunset from the hotel, a candlelit dinner. Only a handful of images offer different activities, usually outdoor sport, with the man taking the more active or controlling role; his attitude is often protective. While 7 times a woman is shown with a female friend, in various activities, the only image of a male couple relates to fitness and is cropped so as to presuppose other participants. Heterosexuality seems to be a given. As for the man, there are numerous pictures of him surrounded by friends (of both genders) largely involved in sporting activities: in 31 cases, the man dominates the group, whereas the woman dominates the image in just 5.[5]

On 14 occasions, the couple is shown with children. In 11 cases, it is only the woman who is with the children, in 5 cases the man. Fatherhood seems to be envisaged more as an occasional activity than as an opportunity. The children are almost always very young and rarely adolescent.

Conclusions

The aim of this chapter was to understand how men and masculinities are portrayed in Italian travel brochures and whether the Italian tourism industry is paying attention to changing male identities.

We have analysed two groups of men: tour operators' staff members and male tourists. Generally, in our view, there is more openness towards new models of masculinity in the group of staff members than in that of tourists. Gender differences in roles are few: both women and men are required to have a certain physical and sporting robustness, although the scenes of staff members working with tourists mainly show women running things. Finally, we noted greater role flexibility and a more equal gender representation on the part of Alpitour compared to the more traditional Eden Viaggi.

The pictures portraying tourists offer stereotyped and hyper-simplified models of masculinity. The man is engaged, by himself or in a group, in physical sporting activities, and is totally absent from intellectual and contemplative activities (Caldas-Coulthard, 2005). In the activities away from the holiday resort, the man is the 'explorer' who can master new situations, even alone. Every possible ambiguity is excluded, whether of potential gay interest, or of an erotic non-marital situation. In the couple, the man is romantic and protective towards his partner. His body is shown as healthy, young and muscular.

To sum up, the brochures do not seem to reflect any of the transformations that are occurring in contemporary (Italian) masculinity, and gender stereotypes appear very strong. Here and there some innovative elements do exist, but they seem random and unplanned and do not reflect any clear marketing policy. Where present, such elements offer new profiles of femininity, in the direction of greater dynamism and activism, rather than of masculinity. However, one could say that a prudently neutral and politically correct panorama of gender is given. The two tour operators seem more interested in adopting a defensive posture, protecting themselves from any criticism, rather than in taking an opportunity for development in a period of intense social change. Of course, replications of this study should be made in relation to other destinations and markets in order to confirm the trends and suggestions made here.

Acknowledgements

Monica Gilli gratefully acknowledges the financial support from 'Regione Lombardia', with the European Social Fund (ESF), for her research activity.

Notes

1. Elisabetta Ruspini wrote the 'Introduction' and 'The Relationship Between Masculinity and Tourism' sections; Monica Gilli wrote the 'Methodology', 'How (Changing) Men are Portrayed in Travel Brochures' and 'Conclusions' sections.All websites cited in the chapter have been consulted in the period June–August 2012.
2. http://web.grinnell.edu/courses/lib/s01/lib397-01/ReStructuring_Masculinities/documents/gaymasc.pdf.
3. The term 'contemporary modernity' is used here to create a synthesis of the notions of 'second', 'late', 'high', 'new', or 'post' modernity (see for example Lyotard, 1979; Giddens 1990, 1991; Beck, 1992; Bauman, 1992; Eisenstadt, 2005; Taylor-Gooby, 2005; Beck and Grande, 2010).
4. Discovery Network EMEA/UK conducted the research project in 2007. The aim of the project was to understand cultures, lifestyles, hopes, fears, motivations and aspirations of European young men aged 25–39. More than 12,000

men in 15 European countries were studied and researchers consulted some of the world's leading experts on male behaviours. The main issues covered were delaying milestones; work pressures; play; relationships; fatherhood; friendship; image; knowledge; balancing; leisure activities; home; technology; politics; health; travel; money; success; and the blurring of roles. Available at http://ctamecap.com/Campaigns/180.

5. In the group scenes, we have defined as the protagonist the person doing the action, or, when there is no particular action, the person who has been photographed in close up.

References

Anderson, E. (2009) *Inclusive Masculinity: The Changing Nature of Masculinities.* London: Routledge.

Bauman, Z. (1992) *Intimations of Postmodernity.* London: Routledge.

Beck, U. (1992) *Risk Society: Towards a New Modernity.* London: Sage Publications.

Beck-Gernsheim, E. (2002) *Reinventing the Family. In Search of New Lifestyles.* Cambridge: Polity Press.

Beck, U. and Grande, E. (2010) 'Varieties of second modernity: The cosmopolitan turn in social and political theory and research', *British Journal of Sociology,* 61(3): 409–43.

Benecke, T. (1997) *Proving Manhood: Reflections on Men and Sexism.* Berkeley: University of California Press.

Black J. (1985) *The British and the Grand Tour.* London: Routledge.

Byrne Swain, M. and Henshall Momsen, J. (2001) *Gender/Tourism/Fun(?)* New York: Cognizant Communication Corp.

Caldas-Coulthard, C. R. (2005) 'Rappresentazioni di genere nel discorso sul turismo. Il caso del Brasile [Gender Representations in the Discourse of Tourism: The Brazilian Case]', in E. dell'Agnese and E. Ruspini (eds) *Turismo al Maschile, Turismo al Femminile. L'esperienza del Viaggio, il Mercato del Lavoro, il Turismo Sessuale [Male Tourism, Female Tourism].* Padova: Cedam, pp. 211–30.

Campbell, C. (2001) 'Going underground and going after women: Masculinity and HIV transmission amongst black workers on the gold mines', in R. Morrell (ed.) *Changing Men in Southern Africa.* Pietermaritzburg: University of Natal Press, pp. 275–86.

Casey M. (2009) 'Tourist gay(ze) or transnational sex: Australian gay men's holiday desires', *Leisure Studies,* 28(2): 157–72.

Cheek, P. (2003) *Sexual Antipodes: Enlightenment, Globalization, and the Placing of Sex.* Stanford: Stanford University Press.

Cohen, M. (1996) *Fashioning Masculinity: National Identity and Language in the Eighteenth Century.* London: Routledge.

Cohen, M. (2001) 'The Grand Tour. Language, national identity and masculinity', *Changing English: Studies in Culture and Education,* 8(2): 129–41.

Connell, R.W. (1987) *Gender and Power.* Cambridge: Polity Press.

Connell, R.W. (1995) *Masculinities.* London: Polity Press.

Connell, R.W. and Messerschmidt, J. W. (2005) 'Hegemonic masculinity: Rethinking the concept', *Gender and Society,* 19(6): 829–59.

Dermott, E. (2008) *Intimate Fatherhood: A Sociological Analysis*. London: Routledge.
Donaldson, M. (1993) 'What is hegemonic masculinity?', *Theory and Society*, 22(5): 643–57.
Doucet, A. (2006) *Do Men Mother? Fathering, Care, and Domestic Responsibility*. Toronto: University of Toronto Press.
dell'Agnese, E. and Ruspini, E. (eds.) (2005) *Turismo al Maschile, Turismo al Femminile. L'esperienza del Viaggio, il Mercato del Lavoro, il Turismo Sessuale [Male Tourism, Female Tourism]*. Padova: Cedam.
dell'Agnese, E. and Ruspini, E. (eds.) (2007) *Mascolinità all'Italiana. Costruzioni, Narrazioni, Mutamenti (Italian Masculinities)*. Torino: Utet.
Eisenstadt, S. N. (ed.) (2005) *Multiple Modernities*. New Brusnwick: Transaction Publishers.
Featherstone, B. (2009) *Contemporary Fathering: Theory, Policy and Practice*. Bristol: The Policy Press.
Gervasio, M. (2011) 'La crisi apre il risiko del turismo [Risks and Crisis in Tourism]'. *IlSole24Ore*, n. 97, 10.4.2011: 19, Available at: http://www.tandfonline.com/doi/abs/10.1080/13586840120085685 and http://www.uri.edu/artsci/wms/hughes/demand_sex_trafficking.pdf.
Ghigi, R. (2008) *Per Piacere. Storia culturale della Chirurgia Estetica [The Cultural History of Aesthetic Surgery]*. Bologna: Il Mulino.
Giddens, A. (1991) *Modernity and Self-identity. Self and Society in the Late Modern Age*. Cambridge: Polity Press.
Giddens, A. (1992) *The Transformation of Intimacy: Sexuality, Love, and Eroticism in Modern Societies*. Cambridge: Polity Press.
Hantrais, L. (2004) *Family Policy Matters: Responding to Family Change in Europe*. Bristol: The Policy Press.
Hearn, J. (1998) *The Violences of Men*. London: Sage.
Hearn, J. (1999) 'Educating men about violence against women', *Women's Studies Quarterly*, 27(1/2): 140–51.
Hughes, D. M. (2004) *Best Practices to Address the Demand Side of Sex Trafficking*. Women's Studies Program, Rhode Island: University of Rhode Island, Available at: http://g.virbcdn.com/_f/files/3d/FileItem-149914-BestPracticesAddressdemand.pdf.
Hughes, H. L. (2002) 'Gay men's holiday destination choice: A case of risk and avoidance', *International Journal of Travel Research*, 4(4): 299–312.
Hughes, H. L., Monterrubio, J. C. and Miller, A. (2010) 'Gay tourists and host community attitudes', *International Journal of Tourism Research*, 12(6): 774–86.
Janssens, A. (2002) *Family and Social Change: The Household as a Process in an Industrializing Community*. New York: Cambridge University Press.
Jõe-Cannon, I. (2006) *Primer on the Male Demand for Prostitution*, North Amherst, US: The Coalition Against Trafficking in Women.
Kimmel, M. S. (1995) *The Politics of Manhood*. Philadelphia: Temple University Press.
Kimmel, M. S. (1996) *Manhood in America. A Cultural History*. New York: Free Press.
Kimmel, M. S. and Levine, M. P. (1992) 'Men and AIDS', in M. S. Kimmel and M. Messner (eds) *Men's Lives*. New York: Macmillan, pp. 318–29.

Kimmel, M. S. and Aronson, A. (2010) *The Gendered Society Reader*, 4th ed. New York: Oxford University Press.
Kinnaird, V. and Hall, D. (eds.) (1994) *Tourism: A Gender Analysis*. Chichester: Wiley.
Lamanna, M. A. and Riedmann, A. (2009) *Marriages & Families: Making Choices in a Diverse Society*. Belmont: Thomson Wadsworth.
Lyotard, J.-F. (1979) *La Condition Postmoderne: Rapport Sur le Savoir,* Paris: Minuit (English translation: *The Postmodern Condition: A Report on Knowledge,* Manchester: Manchester University Press, 1984).
Månsson, S-A. (2004) 'Men's practices in prostitution and their implications for social work', in S.-A. Månsson and C. Proveyer Cervantes (eds) *Social Work in Cuba and Sweden: Achievements and Prospects*. Göteborg University: Department of Social Work, pp. 267–79.
Melián-González, A., Moreno-Gil, S. and Araña, J.E. (2011) 'Gay tourism in a sun and beach destination', *Tourism Management*, 32(5): 1027–37.
Miller, L. (2003) 'Male beauty work in Japan', in J. E. Robertson and N. Suzuki (eds) *Men and Masculinities in Contemporary Japan: Dislocating the Salaryman Doxa*. London, Routledge, pp. 37–58.
Monto, M. and McRee, J. (2005) 'A comparison of the male customers of female street prostitutes with national samples of men', *International Journal of Offender. Therapy and Comparative Criminology*, 49(5): 505–29.
O'Connell Davidson, J. (2001) 'The sex tourist, the expatriate, his ex-wife and her "other": The politics of loss, difference and desire', *Sexualities*, 4(1): 5–24.
Oinonen, E. (2008) *Families in Converging Europe. A Comparison of Forms, Structures and Ideas*. Basingstoke: Palgrave Macmillan.
Oppenheim Mason, J. and Jensen, A.-M. (eds) (1995). *Gender and Family Change in Industrialized Countries*. New York: Clarendon Press.
Perra, S. and Ruspini, E. (eds) (2013) 'Men who work in non-traditional occupations', *International Review of Sociology-Revue Internationale de Sociologie*, 23(2): 265–70.
Pleck, J. H. (1985) *Working Wives/Working Husbands*, Beverly Hills: Sage.
Pritchard, A. and Morgan, N. J. (2000) 'Privileging the male gaze. Gendered tourism landscapes', *Annals of Tourism Research*, 27(4): 884–905.
Pritchard, A., Morgan, N.J., Sedgely, D. and Jenkins, A. (1998) 'Reaching out to the gay tourist: Opportunities and threats in an emerging market segment', *Tourism Management*, 19(3): 273–82.
Pritchard, A., Morgan, N.J., Ateljevic, I. and Harris, C. (eds) (2007) *Tourism and Gender: Embodiment, Sensuality and Experience*. Wallingford: CAB International.
Rao, N. (1995) 'Commoditisation and commercialisation of women in tourism: Symbols of victimhood', *Contours*, 7(1): 30–2.
Ricardo, C. and Barker, G. (2008) *Men, Masculinities, Sexual Exploitation and Sexual Violence. A Literature Review and Call for Action*, Available at: http://www.hapinternational.org/pool/files/ricardobarkermenmasculinitiessexualexploitation-sexualviolence.pdf.
Richter, L. K. (1995) 'Gender and race: Neglected variables in tourism research', in R. Butler and D. Pearce (eds) *Change in Tourism. People, Places, Processes*. London: Routledge, pp. 71–91.
Richter, L. K. (2005) 'Perché studiare genere e turismo [Why Should We Study the Relationship Between Gender and Tourism]', in E. dell'Agnese and E. Ruspini

(eds) *Turismo al Maschile, Turismo al Femminile. L'esperienza del Viaggio, il Mercato del Lavoro, il Turismo Sessuale [Male Tourism, Female Tourism]*. Padova: Cedam, pp. 3–22.

Ruspini, E. (2005) 'Genere e relazioni di genere nel turismo [Gender and Gender Relations in Tourism]', in E. dell'Agnese and E. Ruspini (eds), *Turismo al Maschile, Turismo al Femminile. L'esperienza del Viaggio, il Mercato del Lavoro, il Turismo Sessuale [Male Tourism, Female Tourism]*. Padova: Cedam, pp. 23–44.

Ruspini, E. (2009) 'Italian forms of masculinity between familism and social change', *Culture, Society & Masculinities,* 1(2): 121–36.

Ruspini, E., Hearn, J., Pease, B. and Pringle, K. (eds) (2011). *Men and Masculinities Around the World. Transforming Men's Practices.* Basingstoke: Palgrave Macmillan.

Ruspini, E. (2013) *Diversity in Family Life. Gender, Relationships and Social Change.* Bristol: The Policy Press.

Rivers-Moore, M. (2012) 'Almighty gringos: Masculinity and value in sex tourism', *Sexualities,* 15(7): 850–70.

Russell, P. (2001) 'The world gay travel market', *Travel and Tourism Analyst,* 2: 37–58.

Sönmez, S., Apostolopoulos, Y., Yu, C.H., Yang, S., Mattila, A. and Yu, L.C. (2006) 'Binge drinking and casual sex on spring-break', *Annals of Tourism Research,* 33(4): 895–917.

Taylor-Gooby, P. (2005) 'Pervasive uncertainty in second modernity: An empirical test', *Sociological Research Online,* 10: 4.

Touraine, A. (2005) *Un Nouveau Paradigme, Pour Comprendre le Monde Aujourd'hui,* Paris: Fayard [English translation: *New Paradigm for Understanding Today's World.* New York: John Wiley and Sons, 2007].

Thurnell-Read, T. (2011) 'Off the leash and out of control: Masculinities and embodiment in Eastern European stag tourism', *Sociology,* 45(6): 977–91.

Wernick, A. (1987) 'From voyeur to narcissist: Imagining men in contemporary advertising', in M. Kaufman (ed.) *Beyond Patriarchy: Essays by Men on Pleasure, Power and Change*. Toronto: Oxford University Press, pp. 277–88.

Yokoto, F. (2006) 'Sex behavior of male Japanese tourists in Bangkok, Thailand', *Culture, Health and Sexuality*, 8(2): 115–31.

Zajczyk, F. and Ruspini, E. (2008) *Nuovi Padri? Mutamenti della Paternità in Italia e in Europa [New Fathers? Fatherhood Changes in Italy and in Europe].* Milano: Baldini Castoldi Dalai.

15
Afterword: Men's Touristic Practices: How Men Think They're Men and Know Their Place

Hazel Andrews

This book is a welcome contribution to the study of tourism, particularly in the way in which it brings together and foregrounds in one place issues relating travel and tourism to men, ideas of masculinity and male sexuality. At the same time, thinking through tourism (Scott and Selwyn, 2010) to examine these concepts and constructs provides a fruitful seam to mine in terms of illuminating the issues themselves. It is not a particularly profound point to observe that tourism requires people – although perspectives in the study of tourism that are driven by management/business concerns would at times do well to remember that – and that in this study of people the actions, dispositions and impacts of and on men will necessarily be considered. For example, Ulla Wagner (1977) explored how young, single Gambian men formed casual relationships with visiting Scandinavians, and that, on occasions, the 'romance' blossomed and led to the men migrating to the tourists' home country, where the relationships were subject to stresses and strains that threatened their viability. Phylis Passariello (1983) noted the hegemonic masculinity displayed in the form of loud, drunken behaviour by Mexican men when visiting the beach on the Pacific Coast of Chiapas at the weekends. More recently, Linda Malam (2004) examined the performance of masculinity on the beaches of Thailand. This is of course not an exhaustive list, and I only use it to illustrate the point that men have always been part of the study of tourism. However, their presence is less pronounced in relation to studies driven by feminist and gender concerns that reflect both developments within the study of tourism as well as the wider social arena in which tourism is set. The paper by Veijola and Jokinen (1994) challenged the apparent gender bias in the theorising of tourism, notably the idea that tourism was founded on ideas of difference and breaks from routine that did not take account of the role of women in the domestic sphere and as mothers, features of their lives that went on holiday with them. As Aitchison and Jordan point out 'both nationally and internationally, leisure and tourism practices continue to be used to reinforce gendered identities (1998: vi). We could point to many more examples of work (some already identified in the introduction to this volume) in which gender is a major consideration. What this volume does is to inform us much more about the ways in which tourism is gendered and the associated practices of being a man, being masculine in all its different ways and the role sexuality has within all of this.

The first section of the book tells us about how ideas of hegemonic masculinity are reinforced through tourism practices. Drawing on literary sources, Goodman reflects on historical changes to understandings of masculinity. Lozanski examines the role of risk-taking in tourism and how this can be seen to inform distinctions made between traveller and tourist and how in turn this is tied to the gendering of tourism. We learn from Thurnell-Read how ideas of hegemonic masculinity are performed and negotiated during the British premarital stag tour weekend. The Gay Games allows an arena for Jarvis to examine the ways in which sport can be used to reinforce social differences and not challenge dominant ideas, even when the activities are positioned as an alternative to the 'mainstream'.

In section two, we learn from the various studies about the interweaving of different aspects of identity and tourism with ideas of masculinity. Like all categories, masculinity is unsatisfactory when considered in isolation. As Casey and Thurnell-Read explore in the introduction, and is indeed apparent throughout the collection, there is no one way of being masculine, but equally what makes ways of being masculine is also related to class, race, sexuality and ethnicity. As Casey's chapter illustrates, understandings of masculinity need to be linked to class and to capabilities of reproduction. Both the chapter by Ward and the chapter by Waitt and Markell link masculinity to place and specific spaces, which is important in the world of tourism where ideas of changing spaces (home and away) are attendant with notions of freedom and indulgence. Costa looks at the way that gendered identities and associated family roles travel and remain visible as part of the holiday experience.

In the section entitled 'Sex, Sexuality, Tourism and Masculinity', Lin discusses 'wife finding-tours' that enable Taiwanese men to seek out suitable women for marriage from other countries in Southeast Asia.

Katsulis explores the hegemonic masculinity associated with sex tourism in Mexico. Both chapters demonstrate how the mobilities related to finding women serve to reinforce hegemonic masculinities and unequal gendered relationships. The questioning of the assumption of the economic power of gay tourism, or that such mobility is founded on sexual activity alone, is highlighted by Monterrubio and López-López.

The final section again focuses on dominant 'normative' forms of masculinity as Treadwell questions why football violence feeds off and into a particular kind of masculinity and how that violence is meaningful for the protagonists involved. Lastly, Gilli and Ruspini look to an important aspect of tourism – the appeal to the imaginations of potential tourists through the construction of tourism places found in the promotional material of travel brochures.

My purpose in this afterword is not to comment extensively on the chapters in the book as this has already been undertaken by Thurnell-Read and Casey in the introduction, and each separate chapter speaks more than adequately for itself. Having established some of the main themes to emerge from the collection and the importance of the book in contributing to understandings of social relations within touristic practices, I will now go on to reflect on the ideas of masculinity that presented themselves within my own research. I will begin by briefly setting the scene before moving on to practices of masculinity and finally drawing some concluding comments.

Setting the scene

My research has mainly been concerned with the social constructions of national identity through three key areas of social life: social space, the social body and practices of food and drink consumption. By corollary, such investigations also involve issues of class, gender, region and sexual identity. I have discussed in detail elsewhere (Andrews, 2011) the overall 'findings' of my participant observation in the resorts of Magaluf and Palmanova on the Mediterranean island of Mallorca and will return to these towards the end of my discussion. Both resorts are predominately 'British' in that the highest number of tourists come from the UK and much of the tourism offered caters to British tastes, with UK-based television programmes being aired, imported British foodstuffs – milk, bacon, bread – for sale, the dominance of English as the spoken and written language, as well as the ability to spend sterling, and numerous cafés and bars with distinctly British sounding names – The Willows, Eastenders, Scots Corner and so on. As well as being British, the tourists are predominately white, from areas outside of the southeast of England, working class and heterosexual. The masculinisation of the landscape of Magaluf and Palmanova – but particularly of Magaluf – is based on the ways in which the resorts are encoded with ideas relating to war and militarisation. The naming of bars that connect the visitor to past military victories by the British against European neighbours – Bar Trafalgar, Lord Nelson – and folk tales and sporting heroes – Robin Hood and Linnekars – and the complete absence of any acknowledgement of women's roles in real conflicts, sporting greatness or figures from folklore serve to send a message about the role of both men and women in constructing the nation. Further, aspects of the tourists' own actions including the entertainment they purchase – for example Pirates Adventure – serve to reinforce a sense of identity in which gendered differences are highly demarcated and an aggressive masculinity celebrated (see Andrews, 2009) and practiced. I want to now move on to explore how some of this aggressive masculinity is manifest in the resorts.

'Cause they think they're a man'

As noted, Magaluf and Palmanova are encoded by ideas that appeal to a certain kind of national male character: a heroic, strong and just person able to triumph in adversity. This toughness is played out in games that tourists are encouraged to play by hotel entertainers, bar workers, tour operator representatives (reps) and club and bar DJs. The role of the DJ in many of the nighttime entertainment venues was crucial. During the entire time of my fieldwork, I never encountered any women DJs; they were invariably men. As part of their shows, they would exchange banter with audiences, tell jokes and generally add to the entertainment. They often developed a 'following', and tourists would remain loyal to a particular bar or fun-pub for the duration of their holidays because of the DJ, and return to the same venues year after year because of a particular DJ. The magnetism that many of these young men enjoyed extended also to their abilities to 'pull the girls', and they shared, along with some tour operator reps, a credit card-like facility (as one ex-rep put it) to unlock the bedroom doors – so to speak – of numerous young women who clamoured for their attention.

Dylan (a pseudonym) was one such DJ in a fun-pub in Palmanova, introduced to me by a female bar owner of mutual acquaintance. A young man in his early twenties, he attracted a regular crowd of tourists. Dylan appeared easygoing and was likeable. Like many of the DJs in fun-pubs and bars, he held a commanding position through his jokey banter with his customers. This often involved talk of his sexual exploits, in which he openly bragged of the number of women he had slept with, playing games with the tourists, and encouraging them to dance. I was in the bar one evening when Dylan persuaded customers to form a conga line to dance out of the premises and along the street, eventually snake its way back to the pub.

The role of the DJs is to encourage tourists to drink and dance. Part of this is achieved through the banter they engage in and the games they play with the tourists. For example, Dylan would often encourage games such as drinking pints of cocktails in one go, downing a succession of shots, and games that did not involve alcohol, for example getting a group of people to cheer the loudest. The prize for the winner of these competitions was usually a bottle of champagne. Following one cheering contest, a group of tourists were unhappy with the result. The two men and two women felt that their group had made the loudest noise and the men – but one in particular – tried to remonstrate with Dylan. The latter stood his ground, and the encounter resulted in Dylan challenging the young man to come to the bar the next evening and enter a drinking competition with him that would involve drinking a pint of tequila and remaining standing. The challenge was initially accepted, but the partner of the male tourist came into the bar later and advised Dylan that it was not going to go ahead. I asked Dylan why he thought that anyone would agree to engage in an activity that would more than likely make him very ill. His response was simple: 'Cause they think they're a man'.

The link between proving manliness through drinking was found elsewhere in the resorts. As part of their need to supplement their wages and earn additional money for the tour operators, the reps would organise bar crawls around the resorts. These were promoted in the guise of showing the tourists the 'best bars' in the resorts. In reality, the bars frequented were those at which the reps had an arrangement with the owners or managers for a commission for each tourist introduced to the bar on crawls. The numbers on bar crawls could be as high as 250 tourists. The bars hoped that showing the tourists a good time and creating a link with them in the early stages of their holiday would encourage return visits for the duration of their stay. As in Dylan's bar, the way to encourage this loyalty was for both the reps and the DJ to play games with the tourists and enter into banter (often of a sexual nature) with the customers. The games played would vary but often relied on drawing attention to heterosexuality and differences between men and women.

This is highlighted and exaggerated by the assigning of names (such as, based on the mother and father of the cartoon family, The Flintstones, Wilma for women, Fred for men) and the requirement to use the toilets of the opposite sex. (That is women are expected to use the men's toilets in the clubs and the men are expected to use the women's toilets.) Failure to stick to the rules results in punishment: a public admonishment and a penalty, which usually involved alcohol. Such activities as these draw attention to differences between men and women and make the gendered relations that are being practiced the norm.

As Douglas attests '[i]t is only by exaggerating the difference between...male and female...that a semblance of order is created' (1966: 15). She also notes that transgression of the systems set up to establish order result in punishment: 'ideas about...punishing transgressions have as their main function to impose system on an inherently untidy experience' (Ibid.). During my fieldwork, the order was of conventional male-female roles and places in society. Similarly, in other games and conversations, the identification of women as 'heels' and 'skirts' and men as 'trousers' picks up on particular codes of dress (adornment of the body) and turns them into a way of ordering society. The use of the synecdoche 'wearing the trousers', for example, is indicative of power.

In one bar in Magaluf, the appeal to a hegemonic masculinity was based on proving manliness in terms of drinking. The bar was known for its 'deadly' cocktails: The Fireball, The Parrot and The Killer; the alcoholic combination of each was never divulged. According to one rep, drinking The Parrot was a marker of being a rep in Magaluf. As for the tourists, they were taken to the bar as part of bar crawls and encouraged to drink as much as possible. Drinking games were played and penalties given for breaking the bar crawl's rules as dictated by the reps. The cocktails were promoted as being very strong, and they had to be drunk in a certain way. The Fireball, for example, required having some of the liquid set fire to in the mouth. The Parrot involved inhaling fumes from the drink before drinking it down. One of the women reps demonstrated how to drink The Parrot. The tourists were then told that anyone who could drink all three of the cocktails in succession and not vomit 'will be a legend...will leave the bar and island a legend'. On one occasion, a young male tourist rose to the challenge. He drank the cocktails, and after the third, immediately rushed out of the bar to the street as if to be sick. Although he was not sick, he was, by now, very drunk. The next bar on the crawl was Heroes. The cocktails were made to appeal to a particular form of manliness, couched in terms of heroism alongside the idea that if a woman could hold the drinks (as the female rep had) than a man should be able to as well.

Expectations around and constructions of specific gendered identities is manifest elsewhere in the resorts based on embodied performances found in how people look, as well as how they act. For example, one morning I was advised by one of my tour operator rep informants, Paul (a pseudonym), 'You'll meet some real scum.' He described the problems he had encountered with a couple who had recently arrived for their holiday. They had been complaining, claiming to have had their child's buggy damaged during their flight and to have lost a bag on the journey. They had also raised with the rep that the brochure stated that their hotel room would have air conditioning, and it did not. According to Paul, the tourists refused to complete the necessary documentation in connection with the buggy, only reported the missing bag the day after their arrival at the destination, and air conditioning was not listed as a feature of the hotel in the tour operator's brochure. He went on to say that when they initially complained, they were abusive and allowed their children to run around screaming. Paul volunteered descriptions of the woman as 'fat, an inch taller than her husband and with just flat hair'. Although Paul's attitude toward this couple was based on his perceptions of their behaviour, he also used their appearance, and particularly that of the woman, to cast aspersions on their characters. She was fat and taller than her husband, which did not meet with the stereotypical picture of the wife being smaller. In this case, the woman does not physically conform to the ideal of the 'little woman' or verbally keep her place. Rather she is fat and out of control.

In a similar vein, 'deviance' from heterosexuality and associated expected behaviour and appearance is also considered unacceptable. The sorts of attitudes encountered can again be seen in some of the evening shows in hotels and café-bars. Two performers quip, 'We'd hate to be poof', because they would run the danger of being rejected by both men and women. Further, two of my male informants were in a homosexual relationship. Although the relationship was not a secret from fellow workers, one of the men worked in a hotel, and the guests were not allowed to know of his sexual orientation. This was in contrast to two openly male homosexual bar owners based in Palmanova, although it is important to note that this openness was the exception rather than the norm. Understanding some tourists' reactions to homosexuality and transexuality was evidenced during one late evening entertainment put on at a hotel, a drag act called 'Lady Man'. The group of tourists I sat with consisted of two married couples. The discussion that ensued as the performers came on stage was based on concerns that they would pick on audience members – particularly the men – in some way, should they catch someone's eye. One of the men decided to distance himself from the performance and left for his room to watch the more 'manly' pursuit of football. The remaining man also felt wary of the 'two poofs', and in a desire to avoid engagement with the performers, went to sit at the bar. The conversation then started to focus on the performers' lack of obvious male genitalia – where were they? How did they hide them? At the interval, the man who had gone to the bar returned to sit with the rest of the group. He was asked for his opinion on the 'missing' genitals. His response was, 'I don't think [about it]...as long as they keep away from me.' There was an implied fear of contamination in his comment, whether literally or by association, which was also exhibited in the withdrawal from the audience of both of the men in the group at the start of the act.

This was echoed in another hotel during the playing of an audience participation game involving women competing for the title of Miss Cala Blanca. They had to demonstrate how sexy they were, their ability to mimic a famous female entertainer, and their knowledge of the male body. This last aspect of the competition involved the women being blindfolded and then being asked to feel the body of a semi-naked male with a wig stuffed down his trousers. The justification for this part of the game was 'because these days you never know'. As with the other games already mentioned, and detailed elsewhere in my work (2009, 2011), this game served to reinforce ideas of a normative sexuality for both men and women as well as to heighten the awareness of what it means to be and to practice the gendered relationships upon which it is based.

Having set out some examples that illustrate the dominance and performance of hegemonic masculinity and heterosexuality, I want to go on to reflect on how this influenced my fieldwork practice and the epistemological foundations of my analysis.

Fieldwork reflections

For me, Magaluf and Palmanova, but especially the former, were hostile environments in which to conduct fieldwork. The overt sexism and dominant masculinity that I encountered often caused me to think, somewhat naively, that it would be a friendlier environment for a male (although not homosexual) researcher. At the heart of this were my feelings of vulnerability as a lone female researcher; I felt intimidated by the promotion of ideas related to the satisfaction of male heterosexual desires. It seemed (and in reality was) very much a man's world.

Ethnographic fieldwork is often undertaken in more than one go, and after my first major visit, I left the field indignant and angry at how I had seen women portrayed and used. The open exposure of women's naked bodies, the mocking of women's bodily functions, and the blatant sexism associated with that served to dehumanise women and make them objects for consumption by men. Upon my return to the field a few months later, I was more aware of what was perhaps a less one-sided story. There were naked men depicted on postcards and represented in the form of souvenirs (for example, a ring holder with the central stem shaped in the form of an erect penis), meaning that items symbolic of maleness, such as penises and condoms, entered my field of vision in a way that they had not in the previous year. Although men were still underrepresented in this way, I was more aware that men, too, were subject to being sexualised and commodified in an overt way. Nevertheless, the atmosphere of Magaluf and Palmanova and the touristic practices found there were about conforming to normative gendered roles, and it is this that has influenced much of my work.

In October 2013, I gave a guest lecture for the MA Anthropology of Travel, Tourism, and Pilgrimage at SOAS, University of London. The lecture had been based on the key elements that derived from my work in Mallorca relating to ideas of social space and place, the performance and embodiment of tourism and tourists' consumption practices, and went on to discuss notions of the misrecognition of and symbolic violence against women found in the resorts (Andrews, forthcoming). All of this is within the framework of ideas relating to understanding of constructions of national identity in particular but also the roles of gender and region within that. My overall conclusions remain those explored in my work of 2009 and 2011, which is exemplified in the closure to my monograph (2011: 241–2):

> In Magaluf and Palmanova, the tourists... interact with the signs of home, which they identify with, and continue to craft the resorts, via a dialogue with them, in their own image. They have 'escaped' to a place that offers them their understanding of what it is to be British. This understanding is based on the imaginings of a romanticised past in which Britain was Great, both on the world stage and in being able to fully nourish her children. In addition, men went to war and women knew their place.

At the end of the seminar my words were quoted back to me along with a question about whether, given the scenarios I had described, we might also think that the ways in which the gendering of the landscapes and touristic practices in Magaluf and Palmanova contributed to creating and reinforcing gendered identities, the environment being constructed was one in which men also know their place. In some respects, I think this a fair observation, which brings us to the tension between structure and agency. Who men are and what masculinity is both inform ideas of sexuality (and vice-versa) – as Rutherford argues, 'heterosexual men have inherited a language which can define the lives and sexualities of others' (1988: 22) – and is made in relation to the other, a binary opposition within a framework in which there are preexisting structural inequalities. Thus, for women to know their place, men must also know their place. We see from chapters in this volume that such structures are potent forces. For example, some men at the Gay Games (Jarvis, this volume) suggest that in order to mark out the differences of their masculinity and sexuality from that of heterosexual men, they should adopt 'traditionally' female attire in which to practice their sport. Questions about whether women can truly be travellers because of the association of risk (Lozanski, this volume) show how women are differently identified as in need of protection compared to men. In some cases, it is probably indeed the case that women do need protecting (for example, in my own feelings of vulnerability during my field work and see also Varney's (1996) powerful discussion about the consumability of women and issues of domestic violence) and, we might add, so do some men, such as the victims of gay hate crimes, cyberbullying and so on.

So women know their place, and men know their place, and within all the complexities of differences between men and women – and men and men – knowing one's place becomes part of everyday actions and dispositions. For men in Taiwan, knowing their place means also that they know what they want from a marriage partner (Lin, this volume) as they engage in their wife-finding tours to meet women who have undergone training to learn how to fulfil the men's expectations. Such activities reinforce gendered roles and strike me as not being that far removed from the alarming cases of grooming and trafficking of underage girls for sexual exploitation by gangs of men in the UK, which came to light in 2012–4.[1]

Men, Masculinities, Travel and Tourism undoubtedly adds to our understanding of gendered and sexual identities as they are manifest through different mobilities, over time, in different sociocultural contexts and of how identities such as class, for example, interweave with ideas of manhood to inform understandings of who we are and who we are not. What we often find is, on the one hand, men trying, whether consciously or not, to practice ideals of a form of dominant masculinity, or, on the other hand, in some cases finding ways to try to set themselves apart from the norm. Either way, in so doing, the men continue to contribute to constructing notions of gendered and sexual differences, and thus, through their travel and tourism, help to create a world in which everyone knows his/her place.

Note

1. Rotherham, http://www.bbc.co.uk/news/uk-england-south-yorkshire-23861466; Rochdale, http://www.bbc.co.uk/news/uk-england-manchester-25450512; Peterborough, http://www.bbc.co.uk/news/uk-england-cambridgeshire-25659042.

References

Aitchison, C. and Jordan, F. (1998) 'Introduction', in C. Aitchison and F. Jordan (eds) *Gender, Space and Identity: Leisure, Culture and Commerce*. Brighton: University of Brighton, pp. v–ix.

Andrews, H. (2009) 'Tits out for the boys and no back chat: Gendered space on holiday', *Space and Culture*, 12(2): 166–82.

Andrews, H. (2011) *The British on Holiday. Charter Tourism, Identity and Consumption*. Bristol: Channel View.

Andrews, H. (Forthcoming) 'The enchantment of violence: Tales from the Balearics' in H. Andrews (ed.) *Tourism and Violence*. Farnham: Ashgate.

Malam, L. (2004) 'Performing masculinity on the Thai beach scene', *Tourism Geographies*, 6(4): 455–71.

Passariello, P. (1983) 'Never on Sunday? Mexican tourists at the beach', *Annals of Tourism Research*, 10: 109–22.

Rutherford, J. (1988) 'Who's that man', in R. Chapman and J. Rutherford (eds) *Male Order. Unwrapping Masculinity*. London, Lawrence and Wishart, pp. 21–67.

Scott, J. and Selwyn, T. (2010) 'Introduction: Thinking through tourism – Framing the volume', in J. Scott and T. Selwyn (eds) *Thinking Through Tourism*. Oxford: Berg, pp. 1–26.

Varney, W. (1996) 'The briar around the strawberry patch', *Women's Studies International Forum*, 19(3): 267–76.

Veijola, S. and Jokinen, E. (1994) 'The body in tourism', *Theory, Culture and Society*, 11(3): 125–51.

Wagner, U. (1977) 'Out of time and place – Mass tourism and charter trips', *Ethnos*, 42: 38–52.

Index

Printed and bound by CPI Group (UK) Ltd, Croydon, CR0 4YY